WORKSHOPS IN COMPUTING
Series edited by C. J. van Rijsbergen

Springer
London
Berlin
Heidelberg
New York
Barcelona
Budapest
Hong Kong
Milan
Paris
Santa Clara
Singapore
Tokyo

Also in this series

Semantics of Specification Languages (SoSL)
Proceedings of the International Workshop on Semantics of Specification Languages, Utrecht, The Netherlands, 25–27 October 1993
D.J. Andrews, J.F. Groote and C.A. Middelburg (Eds)

Security for Object-Oriented Systems
Proceedings of the OOPSLA-93 Conference Workshop on Security for Object-Oriented Systems, Washington DC, USA, 26 September 1993
B. Thuraisingham, R. Sandhu and T.C. Ting (Eds)

Functional Programming, Glasgow 1993
Proceedings of the 1993 Glasgow Workshop on Functional Programming, Ayr, Scotland, 5–7 July 1993
John T. O'Donnell and Kevin Hammond (Eds)

Z User Workshop, Cambridge 1994
Proceedings of the Eighth Z User Meeting, Cambridge, 29–30 June 1994
J.P. Bowen and J.A. Hall (Eds)

6th Refinement Workshop
Proceedings of the 6th Refinement Workshop, organised by BCS-FACS, London, 5–7 January 1994
David Till (Ed.)

Incompleteness and Uncertainty in Information Systems
Proceedings of the SOFTEKS Workshop on Incompleteness and Uncertainty in Information Systems, Concordia University, Montreal, Canada, 8–9 October 1993
V.S. Alagar, S. Bergler and F.Q. Dong (Eds)

Rough Sets, Fuzzy Sets and Knowledge Discovery
Proceedings of the International Workshop on Rough Sets and Knowledge Discovery (RSKD'93), Banff, Alberta, Canada, 12–15 October 1993
Wojciech P. Ziarko (Ed.)

Algebra of Communicating Processes
Proceeedings of ACP94, the First Workshop on the Algebra of Communicating Processes, Utrecht, The Netherlands, 16–17 May 1994
A. Ponse, C. Verhoef and S.F.M. van Vlijmen (Eds)

Interfaces to Database Systems (IDS94)
Proceedings of the Second International Workshop on Interfaces to Database Systems, Lancaster University, 13–15 July 1994
Pete Sawyer (Ed.)

Persistent Object Systems
Proceedings of the Sixth International Workshop on Persistent Object Systems, Tarascon, Provence, France, 5–9 September 1994
Malcolm Atkinson, David Maier and Véronique Benzaken (Eds)

Functional Programming, Glasgow 1994
Proceedings of the 1994 Glasgow Workshop on Functional Programming, Ayr, Scotland, 12–14 September 1994
Kevin Hammond, David N. Turner and Patrick M. Sansom (Eds)

East/West Database Workshop
Proceedings of the Second International East/West Database Workshop, Klagenfurt, Austria, 25–28 September 1994
J. Eder and L.A. Kalinichenko (Eds)

Asynchronous Digital Circuit Design
G. Birtwistle and A. Davis (Eds)

Neural Computation and Psychology
Proceedings of the 3rd Neural Computation and Psychology Workshop (NCPW3), Stirling, Scotland, 31 August – 2 September 1994
Leslie S. Smith and Peter J.B. Hancock (Eds)

Image Processing for Broadcast and Video Production
Proceedings of the European Workshop on Combined Real and Synthetic Image Processing for Broadcast and Video Production, Hamburg, 23–24 November 1994
Yakup Paker and Sylvia Wilbur (Eds)

Recent Advances in Temporal Databases
Proceedings of the International Workshop on Temporal Databases, Zurich, Switzerland, 17–18 September 1995
James Clifford and Alexander Tuzhilin (Eds)

continued on back page...

Jörg Desel (Ed.)

Structures in Concurrency Theory

Proceedings of the International Workshop on Structures in Concurrency Theory (STRICT), Berlin, 11–13 May 1995

Published in collaboration with the
British Computer Society

BCS

Springer

Jörg Desel, Dr. rer. nat.
Institut für Informatik, Humboldt-Universität zu Berlin,
Unter den Linden 6, D-10099 Berlin, Germany

ISBN-13:978-3-540-19982-3 e-ISBN-13:978-1-4471-3078-9
DOI: 10.1007/978-1-4471-3078-9

British Library Cataloguing in Publication Data
Structures in Concurrency Theory : Proceedings of the International Workshop on
Structures in Concurrency Theory (STRICT), Berlin, 11-13 May 1995. -
(Workshops in Computing Series) I. Desel, Jörg II. Series
005.42

Library of Congress Cataloging-in-Publication Data
International Workshop on Structures in Concurrency Theory (1995 : Berlin, Germany)
Structures in concurrency theory : proceedings of the International Workshop on
Structures in Concurrency Theory (STRICT),
Berlin, 11-13 May 1995 / Jörg Desel, ed.
p. cm. - (Workshops in computing)
"Published in collaboration with the British Computer Society."
Includes bibliographical references.
ISBN-13:978-3-540-19982-3 (Berlin : pbk. : acid-free paper)
1. Parallel processing (Electronic computers)-Congresses. 2. Data structures
(Computer science)-Congresses. I. Desel, Jörg. II. British Computer
Society. III. Title. IV. Series.
QA769.58.I577 1995 95-32279
004'.35-dc20 CIP

Typesetting: Camera ready by contributors

34/3830-543210 Printed on acid-free paper

Preface

This book is the proceedings of the Structures in Concurrency Theory workshop (STRICT) that was held from 11th to 13th May 1995 in Berlin, Germany. It includes three invited contributions - by J. de Bakker, E. Best et al, and E.R. Olderog and M. Schenke - and all papers which were submitted and accepted for presentation.

Concurrency Theory deals with formal aspects of concurrent systems. It uses partly competing and partly complementary formalisms and structures. The aim of this workshop was to present and compare different formalisms and results in Concurrency Theory.

STRICT was organized by the Humboldt-University Berlin and the ESPRIT Basic Research Working Group CALIBAN. Original papers had been sought from all scientists in the field of Concurrency Theory. The Programme Committee selected twenty contributions with various different topics, including Petri Nets, Process Algebras, Distributed Algorithms, Formal Semantics, and others. I am grateful to the Programme Committee and to the other referees for the careful evaluation of the submitted papers. The Programme Committee had the following members:

Julian Bradfield, Edinburgh
Piotr Chrzastowski, Warsaw
Fiorella de Cindio, Milano
Jörg Desel, Berlin (chair)
Raymond Devillers, Brussels
Hartmut Ehrig, Berlin
Hans Fleischhack, Oldenburg
Paul Gastin, Paris
Rob van Glabbeek, Stanford
Jetty Kleijn, Leiden
Maciej Koutny, Newcastle
Madhavan Mukund, Madras
Arend Rensink, Hildesheim
Manuel Silva, Zaragoza
Antti Valmari, Tampere
Walter Vogler, Augsburg

I should like to express my gratitude to the invited speakers, Jaco de Bakker and Ernst-Rüdiger Olderog, for their interesting and stimulating lectures. I also acknowledge the survey on the results of CALIBAN which was collected and edited by Eike Best.

I am very happy that the Fraunhofer Gesellschaft / ISST gave us the opportunity to use their building which is located right in the historical centre of Berlin. The workshop would not have been possible without the help of the organizing team: Juliane Dehnert, Birgit Heene and Tobias Vesper. Katia Petruni helped in producing this book. The

Coordination Committee of STRICT consisted of Jörg Desel, Eike Best, who initiated and organized CALIBAN, Jetty Kleijn, who organized the Programme Committee meeting in Leiden and Maciej Koutny. The financial support from SIEMENS AG and from Deutsche Stiftung für Internationale Entwicklung is gratefully acknowledged.

May 1995, Berlin, Germany Jörg Desel

List of Referees

D. Barnard
E. Battiston
B. Berard
M. Bialasik
O. Botti
W. Brauer
G. Bruns
B. Caillaud
J. Campos
S. Christensen
J.M. Colom
J.-M. Couvreur
L. Czaja
P. Degano
G. De Michelis
R. De Nicola
C. Dietz
M. Droste
J. Elbro
J. Engelfriet
J. Esparza
H. Fauconnier
M. Fiore
G. Franceschinis
F. Garcia-Valles
R. Glas
H. Goeman
D. Gomm
S. Graf
A. Gronewold
M. Hesketh
H.J. Hoogeboom
R. Hopkins
M. Huhn
R. Janicki
T. Karvi
A. Kiehn
E. Kindler
S. Kleuker
P. Knijnenburg
D. Kuske
M. Kwiatkowska
K. Lautenbach
U. Lichtblau
J. Lilius
K. Lodaya
M. Löwe
A. Mader
A. Martini
A. Mazurkiewicz
D. Miller
P.R. Muro-Medrano
D. Murphy
K. Narayan Kumar
P. Niebert
A. Osterloh
J. Padberg
G. Pappalardo
E. Pelz
A. Petit
L. Petrucci
J. Power
M. Raczunas
M.-D. Radola
Y.S. Ramakrishna
R. Ramanujam
F. Regensburger
H. Reineke
W. Reisig
H. Ridder
L. Ribeiro
B. Rozoy
J.J.M.M. Rutten
N. Sabadini
M. Schenke
H. Schlinghoff
K. Schmidt
L. Sintonen
C. Simone
E. Smith
M. Srebrny
P. Stevens
C. Stirling
E. Teruel
P.S. Thiagarajan
M. Tienari
M. Tiusanen
D. Turi
M. Vardi
K. Varpaaniemi
C. Verhoef
R. Walter
H. Wehrheim
E. Wilkeit
J. Winkowski
H. Wimmel
U. Wolter
Z. Wu
A. Yakovlev
W. Zielonka

Contents

The Three Dimensions of Semantics

J. W. de Bakker[1]
CWI/VUA, Amsterdam

In Semantics, one studies functions $M : L \rightarrow P$, mapping the language L to a semantic domain P. We provide an overview of recent progress along the axes

- identifying, comparing and unifying the control flow principles of present day programming languages L
- designing the variety of semantic models P (input-output/trace-based, uniform/nonuniform, linear/branching, ...)
- the two ways of defining M (or is there only one way?).

We shall pay special attention to the mathematical structures exploited in the definitional framework, highlighting the use of

- (generalized) (ultra) complete metric spaces, and the hyperspaces of closed or compact sets
- labelled transition systems
- higher-order definitions of the semantic domains, functions and operators, and associated proof principles.

[1] Acknowledging joint work with/contributions by F.van Breugel, J.Rutten and E.de Vink

CALIBAN - Esprit Basic Research WG 6067

Eike Best[1], Raymond Devillers[2], Elisabeth Pelz[3], Arend Rensink[1], Manuel Silva[4] and Enrique Teruel[4]

Abstract

This paper describes the results of the Esprit Basic Research Working Group 6067 CALIBAN (Causal Calculi Based on Nets).

Introduction

This paper gives a survey of the results of the project CALIBAN (CAusal CalcuLI BAsed on Nets) which has been funded by the European Communities as an Esprit Basic Research Working Group (No.6067). According to its Technical Annex [1], the project CALIBAN

> ... aims to contribute towards the combination and the integration of causality based models of concurrency (such as Petri nets) and calculus oriented models of concurrency (such as process algebras). This involves:
>
> - On the foundational level: Investigations towards a unified behavioural semantics of causality, taking into account all major approaches (especially Petri net based ones), strengthening the existing and deriving new results based on the connections between them. The investigation of the connections between the structure and the behaviour of a concurrent system.
> - On the practically oriented level: The development of a causality based algebra (which has been defined in the precursor of this Action, the Esprit Basic Research Action 3148: DEMON – Design Methods Based on Nets) into a practicable method supporting the design of concurrent systems, emphasizing in particular the benefits of an explicit representation of causality. The integration of high-level nets into a calculus-oriented systematic specification method. The integration of logical specification techniques. Further, the demonstration of the advantages of this method by its application to substantial case studies.

To achieve these aims, the project has been structured into four main themes:

- Semantics of Causality.
- Structure Theory of Causal Models.
- Causal Calculi.
- Specification Based on High Level Nets.

The next sections are devoted to presenting the results of each of these themes.

[1] Institut für Informatik, Universität Hildesheim, Germany
[2] Département d'Informatique, Université Libre de Bruxelles, Belgium
[3] Université Paris Sud, L.R.I. bât. 490, F-91405 Orsay, also Université Paris Val de Marne, Equipe d'Informatique Fondamentale, F-94010 Créteil, France
[4] Dept. Informatica e Ingenieria de Sistemas, Universidad de Zaragoza, Spain

1 Semantics of Causality

Author/Editor: A.Rensink

The main aim of Strand 1 was a comparative study of causality, incorporating such subjects as traces, event structures, causal nets, elementary nets, causal transition systems and others. It was intended to derive inter-translatability and other results. A number of specific sub-themes were identified; see below.

These aims have for a large part been met, or significant progress has been made. Because of the fundamental nature of the research in this strand, a number of different approaches and basic formalisms were investigated. As a consequence, the results are less focussed than those of the other strands, and do not lend themselves easily to a uniform presentation. Rather than reviewing all research done in this area during the course of the project exhaustively, we pinpoint some typical results and describe them in as much detail as space allows. These descriptions are complemented with a brief summary of other directions of research.

1.1 Traces

This theme involves the consolidation of the work on (Mazurkiewicz) traces. It constitutes a major effort of Strand 1, which (in a joint project with the ESPRIT WG ASMICS) has resulted, according to the original aim, in the recent publication of the book [46] on the theory of traces, edited by Rozenberg and Diekert. It consists of tutorial-style contributions discussing all major trends in the theory of traces —many of them related to concurrent systems and in particular to Petri nets. Contributions from CALIBAN were especially made in the following themes covered by the book.

Basic Notions The basic notions of traces were introduced by Mazurkiewicz, in a series of papers starting in the beginning of the 1980s. A good overview can be found in [88]. The pivotal idea is to consider a set of abstract actions *together with* a fixed reflexive and symmetric dependency relation. Successive actions in a given sequential run of a system which are *not* dependent may commute, leading to the theory of *partially commutative monoids.* The chapter in [46] on basic notions is complemented by a discussion of Dependence Graphs by Hoogeboom and Rozenberg. Presented are basic properties of dependence graphs, the translation from words to graphs and back, some structural properties of dependence graphs, and the relation with graph grammars.

Concurrency and logic The relation between Mazurkiewicz traces and other models of concurrency, covered in [46] by Nielsen and Winskel, was studied in the context of CALIBAN by Rozoy in cooperation with Morin and Biermann. In [20], they investigate an extension of the trace model and its representation within prime event structures.

Generalisations The basic theory of traces has been generalised in several directions. Gastin and Petit report in [45] on complex trace languages and in [62] on infinite trace languages —which were also studied by Gastin and Rozoy in [63]. The latter subject has been included in the aforementioned book.

A recent approach, investigated by Bauget and Gastin in [7] and by Biermann and Rozoy in [21], is to relax the condition under which traces are considered equivalent, and thus allow more general partial orders to be modelled.

1.2 Equivalences and logics

This theme concerns especially the study of true concurrency models and corresponding logics.

One aspect of this theme is *action refinement*, the principle of replacing abstract actions by concrete, more involved behaviour. On the theoretical side, two formalisations of this principle exist. One can either think of refinement as generating a *morphism* over the algebra, in which case the effect is that of *syntactical substitution*, or as an *operator* of the algebra, in which case the effect is that of *semantic substitution* in some expressive enough model. These views are compared by Goltz, Gorrieri and Rensink in [64], where necessary and sufficient conditions are given for them to coincide. Furthermore, an SOS semantics for the refinement-as-operator view is given by Rensink in [101]. On the more practical side, in [100] Rensink describes a framework for action refinement that is more flexible than either refinement-as-morphism or refinement-as-operator.

Another fruitful line of research in this theme is built on work by Thiagarajan and others concerning *communicating sequential agents* (see for instance [87]) and partial order temporal logics based on these (see [116]. These models are basically event structures where a dependency relation over the labels is assumed very much like that underlying the trace model (see above). In a dual view, the maximal cliques of such a dependency relation can be identified with *sequential agents*. Based on this work, Niebert has defined a trace based μ-calculus a temporal logic in which modalities are *local* to such agents; a decidability result is presented in [91]. In a comparative work Niebert has shown together with Penczek [92] how such location based logics relate to subsets of standard interleaving temporal logics, for which *partial order reduction methods* like those of [95] can be applied. Such reductions are one successful way of coping with the state explosion problem in automatic verification. The results suggest that partial order temporal logics have the right level of abstraction for the formulation of efficiently testable properties.

Combining both subjects of action refinement and temporal logics in the setting of communicating sequential agents, Huhn and Wehrheim have studied a method for combining logical and process algebraic approaches to action refinement [73]. The refinement of an action on the abstract level is a product of local refinements, being a sequential process for each agent that participates in the abstract action. They then introduce a corresponding notion of action refinement on the logical side, and investigate under which conditions the two refinement operators are *compatible*, i.e., when the validity of a formula for the abstract system is equivalent to the validity of the refined formulae for the refined system.

1.3 The synthesis problem

This theme involves the problem of synthesising Elementary Net systems from (sequential) transition system specifications, by recognising *regions* in those specifications. Regions are subgraphs (more precisely, sets of states) with the property that equally labelled transitions either all enter it, or all stay within it, or all leave it, or all stay outside it. It was proved by Rozenberg, with Ehrenfeucht in [51, 52] and with Nielsen and Thiagarajan in [94], that such synthesis is possible; they also proposed an elegant categorical framework for the synthesis procedure.

Within CALIBAN, these results have been extended along various lines:

- The algebraic structure of the class of regions of a transition system is studied in [8], where it is shown that minimal regions (with respect to inclusion) are sufficient to synthesize an elementary net
- The algebraic structure of the class of *generalised regions* (whose definition is given in terms of multisets instead of sets) is studied in [9].
- It has been proved that generalised regions fully characterise the synchronic structure of a Net System [9].
- The synthesis procedure has been extended, through generalised regions, to classes of Place-Transition Net Systems (in a way that has similarities and differences with the one proposed by Mukund in [90]).
- A partial characterisation has been given in [10] of the synchronic and logic structure of a Transition System, through regions and through its arc-completion (adding to the Transition System all the arcs connecting two of its states).
- In [33], it has been shown how to generate all elementary Petri nets corresponding to a given transition system. If there is any such elementary Petri net, it is proven that there always exists a small one which has only polynomially many elements in the size of the transition system.

1.4 Basic Properties of Causality

This theme involves basic models of causality of which many other models are incarnations. Several different approaches have been and are being investigated, especially by Rozenberg in cooperation with Mazurkiewicz and by Koutny.

Foundational models In the spirit of unifying the major approaches to the semantics of causality, Rozenberg and Mazurkiewicz have investigated a new model of concurrency which would encompass only the very basic assumptions about concurrent systems and yet would serve as a common framework for many existing models (such as Petri nets, event structures, multitraces, and string synchronisation models). A basic formulation of the model and its behaviour have been obtained. Various known models of concurrency have been interpreted in this framework of enabling systems. Parts of these efforts are being reported in the first chapter of [46]; see 1.1 above.

Rozenberg has investigated the axiomatisation of state spaces of Petri nets, in particular state spaces of elementary net systems. It turns out that such state spaces have a rather clear algebraic structure related to some central laws of arithmetics. The systems formalising these ideas are called proportion systems.

In his research on dynamic labelled 2-structures (dl2s's), Rozenberg tries among other things to clarify the precise relationship between the computations in Petri nets and the computations in dl2s's. Dynamic labelled 2-structures have a natural interpretation as networks of processors working in a concurrent mode. The basic properties of dl2s's have been investigated ([48, 49, 47]). They concern the structure of configurations in such networks. This theory relies mathematically on the theory of groups. Dynamic labelled 2-structures with variable domains can be constructed from elementary ones by the operation of disjoint union (or the more general operation of amalgamated union), in combination with group-theoretic operations [50].

Discrete order structures In concurrency theory posets are used to model both specifications and observations of concurrent behaviour. Observations are usually modelled by total, stratified and interval orders with incomparability interpreted as *simultaneity*. Specifications often involve modelling some kind of *causality* structure. Causality is usually modelled as a partial order relation with incomparability interpreted as *concurrency*. The posets used to model causality relations are more general than those modelling observations. An additional important property which is usually required is that the posets be *discrete*, to exclude some pathological behaviours.

Although there is no doubt that in general the partial order approach to concurrency has been successful, there are situations, e.g. when priorities are involved, that pure partial orders may be insufficient to provide a full and adequate specification of all aspects of causality. In other words, causality can be more complex than the modelling power of partial order relations. In several papers relational structures of the type (X, R_1, R_2), where X is a set of event occurrences, $R_1 \subseteq R_2 \subseteq X \times X$ and R_1 is a poset, were defined and analysed to specify more adequately complex concurrent behaviours. R_1 was usually interpreted as 'precedence' or 'causality', while R_2 as 'weak precedence' or 'not later than' relation. Whereas R_1 was always a poset, the assumed properties of R_2 did vary. What also did vary was the degree of 'discreteness' assumed.

In modelling causality by posets allows one to take advantage of Szpilrajn's factorisation theorem that each poset is the intersection of its total extensions. The theorem has a straightforward interpretation: any behaviour specified by causality relation is completely defined by the set of its sequential observations. This provides a basis for such approaches as (Mazurkiewicz) traces and Shields' vector firing sequences. Janicki and Koutny have shown in [75, 78] that Szpilrajn's result can be extended to the discrete relational structures mentioned above; under specific conditions these are generated by sets of partial orders (stratified or interval ones) which represent observations of a concurrent behaviour.

The general results have been used by Janicki and Koutny to develop a causality-based semantics of Condition/Event-systems with inhibitor arcs [77] and nets with inhibitor arcs [76]. On the level of observations of concurrent behaviour, the study resulted in the development of a characterisation of discrete interval orders and semiorders in terms of convex intervals on a discrete time scale [74].

1.5 Event structure semantics for general nets

This theme involves the definition of behavioural semantics for Petri nets in terms of *event structures*. The subject was opened by the seminal paper of

Nielsen, Plotkin and Winskel [93], who established a very strong relation between the model of *prime event structures* and the subclass of *1-safe Petri nets*, through the intermediate level of *occurrence nets*. The aim of CALIBAN was to establish a generalisation of prime event structures that is related in a similar strong way to the general class of Place/Transition nets.

This aim has been reached in a joint research effort with the Dutch National Concurrency project REX. In [69, 70], Hoogers, Kleijn and Thiagarajan introduce a new class of event structures, the so-called *local event structures*.

A prime event structure is a structure $\langle E, \leq, \# \rangle$ where $\leq$ is a binary relation over E recording the *causal dependencies* between events and # is a binary relation over E recording the *conflicts* between events. Every safe Petri net can be unfolded into an *occurrence net*, roughly by recursively creating new instances of transitions plus their output places for every distinct way in which those transitions can be enabled. The relation between occurrence nets and event structures is then obtained by equating transition occurrences to events, such that if an output place of one transition is the input place of another then it is a causal predecessor, and if two transition share an input place then they are in conflict.

As soon as the Petri nets are no longer safe, there are different ways in which to generalise this principle. The key phrase in the above description is "every distinct way in which a transition can be enabled": the notion of distinctness has to be reevaluated. For if there are two tokens on a given place, does one distinguish between enablings using the one token and those using the other? The approach studied within CALIBAN is based on the assumption that such distinctions can or should not be made. The solution presented by Hoogers and others is to add to the prime event structures (or more accurately, to their families of configurations, which are $\leq$-left-closed and #-free subsets of E) a purely local concurrency axiom that states whether two enabled events are independent *in the current context*. Thus one has finer control over the notion of concurrency versus conflict, which turns out to be enough to cater for the added complexities of non-safe Petri nets. A subclass of local event structures possessing a certain unique occurrence property is identified. With each Petri net, a local event structure with the unique occurrence property can be associated leading to a proper generalisation of the result of Nielsen, Plotkin and Winskel. This event structure semantics is restricted in the sense that auto-concurrency is filtered out.

The results on local event structures and the earlier obtained results on a generalisation of the trace semantics of 1-safe nets to the more general Petri nets are part of Hoogers' Ph.D thesis [68].

In another line of research, the ideas underlying local traces and local event structure are being correlated to the work done by Gastin, Rozoy and others on traces, partial orders, and event structures ([103, 63, 2, 104]).

1.6 Petri net semantics for the π-calculus

The last decade has seen a growing interest in the application of Petri nets as models for process algebraic formalisms. A special challenge are the more recent higher-order and mobile algebras, such as in particular the π-calculus introduced in Milner, Parrow and Walker [89]. This theme has also been studied within CALIBAN.

In [53, 54] Engelfriet investigates a Petri net (multi set) semantics for the small π-calculus (no choice operator, and recursion is replaced by replication). One Petri net Mπ is defined and a compositional semantic mapping is given that associates a marking of Mπ with each process of the small π- calculus. The main results on this semantics are: the semantic mapping is a strong bisimulation between the interleaving transition system of the small π-calculus and the multiset transition system Mπ, and if two processes of the small π-calculus are structurally congruent, then they have the same semantics in Mπ. Recently, Engelfriet has shown together with Gelsema in [55] that the axioms of structural congruence can be extended in such a way that two processes are structurally congruent iff they have the same semantics in Mπ. Moreover, this structural congruence is decidable. Also an operational semantics for CCS which is more compact than the usual has been investigated.

2 Structure Theory of Causal Models

Authors/Editors: M.Silva, E.Teruel

For the purpose of this task in the Project, we mainly deal with the Petri net formalism of Place/Transition (P/T) net systems, that is, places are counters containing a number of unstructured tokens, and arcs may be weighted. However, several results are derived for diverse subclasses of these (e.g., Free Choice, 1-safe, ...).

Petri net systems consist of a few simple objects, relations, and rules, yet they can model very complex behaviours. Structure theory represents one of the crucial advantages of Petri nets compared to other causal models of concurrent systems, in the sense that it can provide a deep understanding on the system behaviour and it often leads to efficient algorithms for model verification because it does not require fully representing the behaviour. By model verification we mean either model checking, i.e., checking satisfaction of a formula from a given (temporal) logic, or deciding whether some property from a given kit holds. It comes out that the most powerful results are obtained when the scope is restricted to particular properties or/and net subclasses. By restricted properties we mean either a restricted logic for model checking or a selected set of properties from a kit. Regarding subclasses, they are often defined to limit the interplay between conflicts and synchronisations. [106] is an invited survey on the topic. [31] is an introductory book that focuses on the particular class of Free Choice Petri nets, which play a central role in structure theory. Main contributions to the field within CALIBAN concern:

- Deepening into state equation and invariant based descriptions of the behaviour. More precisely:
  - *Modulo-invariants* have been introduced as a generalisation of classical invariants, with the same decision power as the state equation over the integers for proving non-reachability of a marking.
  - The *state equation based algorithm for deadlock-freeness analysis* of general P/T net systems has been improved, by reducing the number of equation systems to check for absence of solutions. This leads to just one system in many non-trivial cases.

- Giving necessary or sufficient conditions for *well-formedness* (structural boundedness and liveness) of general P/T nets, based on the *rank of the incidence matrix.*

- Developing the structure theory of a number of *net subclasses*, where some of the previous results for general nets become specially powerful. The main ones are *Extended Free Choice* (EFC), *Choice-free* (CF), *Equal Conflict* (EC, includes the previous ones), and *Deterministic Systems of Sequential Processes* (DSSP). Many of them allow weighted arcs, hence the modelling of bulk services and arrivals.

- Developing *model checking* algorithms which avoid the state explosion problem by applying *partial order* techniques, and which are proven specially efficient for some net subclasses making use of results from the structure theory for these.

Next we describe these contributions in some detail.

2.1 State equation and invariants

The basic idea to analyse P/T systems by linear algebraic techniques is to describe, or approximate, the set of reachable markings by means of the set of solutions to the net state equation. Let $(\mathcal{N}, M_0)$ be a P/T system with set of places P and transitions T, and incidence matrix C. The state equation gives a necessary condition for a marking to be reachable: a vector $M \in \mathbb{N}^{|P|}$ such that $M = M_0 + C \cdot \vec{\sigma}$ has a solution $\vec{\sigma} \in \mathbb{N}^{|T|}$ is said to be potentially reachable. If, for efficiency sake, instead of requiring $\vec{\sigma} \in \mathbb{N}^{|T|}$ we ask for $\vec{\sigma} \in \mathbb{Q}^{|T|}$, then there are yet more solutions to the state equation not corresponding to actually reachable markings. Regarding the matter of considering $\mathbb{Z}$ or $\mathbb{N}$ (respectively, $\mathbb{Q}$ or $\mathbb{Q}^+$) it is easy to see that both descriptions coincide in presence of consistency, a most frequently required structural property. (See [27] for a summary of various linear descriptions of a net behaviour, together with their relations.)

Place-invariants (P-flows) are equivalent to the state equation in the rational domain. It is also known that in the presence of conservativeness we can equivalently concentrate on semi-positive place-invariants (P-semiflows). It is then clear that the state equation over the integers describes the reachability set more accurately than place-invariants. On the other hand, these are sometimes more intuitive and easy to interpret: if Y is a place-invariant, then every reachable marking must fulfill the following token conservation law: $Y \cdot M = Y \cdot M_0$. It is said that M agrees on the place-invariant. They are currently used for proving properties of concurrent systems (see [57] for the application to verification of mutual exclusion algorithms expressed in $B(PN)^2$).

Non-reachability and modulo-invariants: The picture of state equation and invariant based necessary conditions for reachability has been completed by the definition of modulo-invariants [32]. They operate in residue-classes modulo k instead of rational numbers. The set of markings which agree on modulo-invariants coincides with the set of solutions to the state equation over the integers, hence, provided consistency, with the potentially reachable markings. It is shown how to derive from each net a finite set of invariants — containing place-invariants and modulo-invariants — such that, if any invariant proves the non-reachability of a marking, then some invariant of this set proves that the marking is not reachable.

Linear algebraic deadlock-freeness analysis: Deadlock-freeness can be proven by verification of absence of potentially reachable deadlocks. The nature of the condition "a marking is a deadlock" leads to a possibly large number of linear equation systems to check for absence of solutions, in case the basic technique is applied naively. In [110, 108] several rules capable of greatly reducing this number of equation systems are given. In many non-trivial cases (1-safe, Simple, EC, DSSP,...) the rules lead to just one equation system to check. Moreover, the condition is not only sufficient but also necessary for some subclasses. This is the case of EC systems [112], for which the method can be used to decide on liveness, since it is equivalent to deadlock-freeness assuming strong connectedness and boundedness.

2.2 The rank theorems for general P/T nets

Well-formedness, that is, structural boundedness and liveness, is an important property of P/T nets, in the sense that a well-formed net can be boundedly and lively marked, as often required to reactive systems. There is no efficient characterisation of well-formedness for general P/T nets, but both a necessary and a sufficient condition exist. They have similar statements, and require polynomial time. Let $\mathcal{E}$ be the quotient set of the Equal Conflict relation between transitions, that is, "being the same transition or having the same non-null preincidence function", an let $\mathcal{C}$ be the quotient set of the Coupled Conflict relation between transitions, the transitive closure of the relation "being the same transition or having common input places". The net is asked to be conservative and consistent; rank$(C) \leq |\mathcal{E}| - 1$ is necessary for well-formedness, while rank$(C) = |\mathcal{C}| - 1$ is sufficient. Regarding terminology, in [31, 29, 28], clusters are used instead of conflict sets. (A cluster is a Coupled Conflict set together with its input places.)

The proof of the necessary condition for general (weighted) P/T nets was cleared up in [113]. The sufficient condition was introduced in [29] for ordinary nets, and generalised to the weighted case in [97]. In the ordinary case, a simple marking condition can be added to obtain a sufficient condition for boundedness and liveness, namely that every place-invariant is marked. Nets fulfilling this together with the rank condition are called Regular Marked nets in [29], and they share results of bounded and live EFC.

Unfortunately, there are many nets fulfilling the necessary but not the sufficient condition, so these are not enough to decide on well-formedness, and one must restrict to subclasses for stronger results. Actually, well-formedness is efficiently characterised in some way or another for all classes covered in next subsection.

2.3 Subclasses of P/T net systems

Many of the powerful results from structure theory have been derived for some ordinary subclasses. Nevertheless, weights may be very convenient to properly model systems with bulk services and arrivals, or batch consumption and production of resources. Although they can be implemented by means of ordinary nets, then the original structure of the net becomes too complex, even in the simplest cases, to be amenable of some interesting structural analysis techniques. Instead of spoiling clarity, the consideration of weights does often lead to simpler proofs, sheds new light over certain previous results, and suggests new ideas. This line of research was initiated by the end of the DEMON project with the study of Weighted T-systems [109].

Within the CALIBAN activity, more classes have been studied. We classify them in two groups. The first one contains subclasses defined by some local topological constraint, aiming at limiting the interplay between conflicts and synchronisations. The largest such subclass we deal with here is EC systems. The second group contains classes defined in a modular way. Simple modules cooperate or compete for resources in a restricted fashion.

Equal Conflict systems, and subclasses

Equal Conflict systems: EC systems [112, 113, 108, 114] can be defined as those where $\mathcal{E} = \mathcal{C}$, and they are a natural generalisation of EFC ones. They share many of their results, some of which are easier to prove at the EC level, although others are lost, paying for the gain in expressive power. It is proven that globally fair sequences of a well-behaved EC system are those that are locally fair. This leads to the equivalence of liveness and deadlock-freeness in bounded strongly connected EC systems. The potential reachability graph of a live EC system is directed, hence all the markings being potentially reachable from a live marking are also live. These results can be used for algebraic liveness analysis of bounded systems, as mentioned above. They also allow proving existence of home states, and liveness monotonicity in bounded EC systems. The generally only necessary condition for well-formedness is also sufficient for EC nets. Actual liveness can be expressed in terms of liveness of the P-components (a generalisation of the ordinary ones, with some linear-algebraic flavour), but liveness of a P-component is not as easy to verify as in the ordinary case where it suffices that it is marked. The decomposition and duality theorems — and some related results like T-semiflow realisability — appear as corollaries. Remarkably, Commoner's theorem — not generalised to the weighted case — and the decomposition theorem had been taken as starting point in all the ordinary theory [31]. Naturally, there are results from the Free Choice theory that do not hold in this more general setting, as for example the polynomiality of deciding boundedness and liveness based on liveness P-decomposition, as mentioned above. There are some others that we believe will shortly be generalised. We feel such is the case of the condition to reach the home space, leading to a reachability criterion for bounded, live, and reversible systems, or the elegant synthesis theory.

Extended Free Choice systems: The structure theory of EFC systems was greatly developed during the DEMON project. It has been completed during CALIBAN [28, 30], and collected in the book [31]. In [30], the following property of 1-safe Marked Graphs, Conflict-free and live EFC systems is proven (it has importance for the model-checking problem, see below): Given two reachable markings, M_1 and M_2, if some path leads from M_1 to M_2, then some path of polynomial length in the number of transitions of the net leads from M_1 to M_2.

Choice-free systems: CF systems [111, 108] have no forward branched places. They generalise the ordinary T-systems to a greater extent than weighted T-systems [109]. In addition to the results for EC systems, they enjoy the following properties. Well-formedness is equivalent to strong connectedness and consistency. Liveness of a well-formed system can be proven by firing a sequence grater than the unique minimal T-semiflow, what is also sufficient to ensure that the home space has been reached. Firability of the T-semiflow in the home space allows to decide that a potentially reachable marking is actually reachable from a reversible initial marking, thus leading to a reachability criterion which is polynomial on the length of the minimal T-semiflow. This can be expressed by linear equations, providing a unified framework for properties verification and even optimisation problems, like finding a "cheapest" bounded, live, and reversible initial marking.

Modular subclasses

Deterministic Systems of Sequential Processes: The DSSP considered in [115, 97] are an enlarged version of those in [98, 107]. A DSSP is formed by a collection of strongly connected 1-safe State Machines, the Sequential Processes (SP), communicating via message-passing through some places called buffers, with possibly weighted adjacent arcs. A buffer is restricted to be output-private (that is, its postset is included in one SP) but not input-private as in previous versions. It is also restricted to respect private conflicts in its sink SP. The main qualitative properties derived so far for the class are liveness equivalence to deadlock-freeness (when strongly connected and bounded), and the characterisation of well-formedness: the general necessary condition again happens to be sufficient. These and other qualitative results are important for the quantitative analysis of the subclasses, started in [115] and going on within other projects, as another example of the mutual, and beneficial, influence between functional and performance aspects [105]. The current/future work aims at extending the results to larger subclasses, mainly by allowing more general modules.

Systems of Simple Sequential Processes with Resources: These nets appear in the modelling of a wide set of flexible manufacturing systems. They are composed of a set of sequential processes that share a set of common resources, with some constraints in their use. In [58] deadlock problems in these systems are characterised in terms of empty siphons, and a deadlock prevention control policy is established. Within CALIBAN their structural properties are being studied further.

2.4 Model checking of 1-safe systems

The efficient 1-safe T-system model checker from [13] has been generalised to 1-safe systems [56]. The algorithm is based on net unfoldings, and does not build the state graph at all; this is in contrast to all comparable algorithms which, depending on the shape of the formula, still need to build a small or a large part of the state graph. The algorithm can be applied to verify properties of a temporal logic with a possibility operator, suitable for expressing reachability of a marking, mutual exclusion between places, concurrency of transitions, liveness of a transition, reversibility (cyclicity), etc. It is shown to be polynomial on the size of the net for the Conflict-free subclass. For this, the result on the length of shortest paths in the reachability graph [30] mentioned above is used.

Concerning also the analysis of 1-safe systems, a study of the complexity of several verification problems, like the reachability of a marking, liveness, or deadlock-freeness, has been carried out [25], both for general nets and diverse subclasses.

3 Causal Calculi

Author/Editor: R. Devillers

One of the main outcomes of the DEMON project was the design of a process algebra of Box expressions, together with a similar algebra of Petri Box nets, with the aim at serving as the semantic domain of a compositional semantics of high level concurrent programming languages and providing a compositional semantics in terms of Petri Nets and their associated concurrent behaviour.

The process algebra was very much inspired by Milner's CCS but presented distinguished features: the synchronisation operator is separated from the parallel composition one – the visible part of a transition is a multiaction, i.e., a finite multiset of elementary actions, instead of an unstructured (possibly silent) action – this allows to define multiway synchronisations implying more than two transitions – a true sequencing operator replaces the usual prefixing one – as a consequence, the recursions are not limited to tail-end ones – an iteration operator avoids heavy usage of recursions – a refinement operator avoids syntactic substitutions in many situations.

It assumes a countably infinite set of action names, A, to be given. The set of communication labels is the set $\mathcal{L} = \mathcal{M}_f(A)$ of finite multisets over A. Relabellings are functions $f : A \rightarrow A$. There are also variable names from a set $\mathcal{V}$. Box expressions are then constructed from basic building blocks, i.e., multiactions $\beta \in \mathcal{L}$ and variables $X \in \mathcal{V}$, and from various operators, i.e., sequences $(E_1; E_2)$, choices $(E_1 \,\Box\, E_2)$, concurrent compositions $(E_1 \| E_2)$, synchronisations $(E\ \mathbf{sy}\ a)$, restrictions $(E\ \mathbf{rs}\ a)$, scopings $([a : E])$, relabellings $(E[f])$, refinements $(E_1[X \leftarrow E_2])$, recursions $(\mu X.E)$ and iterations $([E_1 * E_2 * E_3])$.

The standard models for Box expressions are equivalence classes of labelled Petri nets, called Petri Boxes or Boxes for short, equipped with a place and transition labelling. The Petri Box domain is provided with the same operators as above, and there exists a semantic homomorphism from expressions to Boxes in the usual style. For our purpose, a labelled Petri net is a quadruple $\Sigma = (S, T, W, \lambda)$, where (S, T, W) is an arc-weighted net and λ is a labelling function such that $\lambda : S \rightarrow \{e, \emptyset, x\}$ and $\lambda : T \rightarrow \mathcal{L} \cup \{ f(X) \mid X \in \mathcal{V}$ and f a relabelling function$\}$. ${}^\bullet\Sigma = \{s \in S \mid \lambda(s) = \{e\}\}$ forms the entry interface. $\Sigma^\bullet = \{s \in S \mid \lambda(s) = \{x\}\}$ forms the exit interface. The transitions with $\lambda(t) \in \mathcal{L} \setminus \{\emptyset\}$ form the communication interface (which drives the synchronisation/restriction/scoping/relabelling mechanism), the transitions with a label of the $f(X)$ kind form the hierarchical interface (which drives the refinement and the recursion mechanisms), while expressing 'pending relabellings' (in general f will be the identity function, and will be omitted). A label $\emptyset$ always denotes an internal place or an internal (silent) transition.

It is assumed that each transition has input and output places, that entry places have no input transition, that exit places have no output transition, and that the entry/exit interfaces are nonempty. Examples of such nets are given in figure 1.

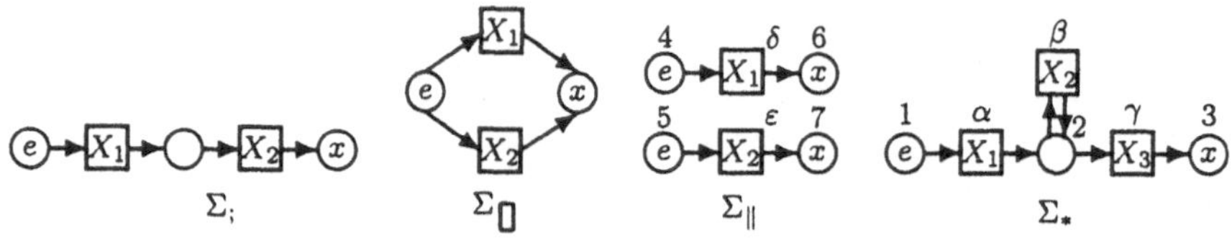

Figure 1: The operative A-nets

An equivalence relation was also defined on those labelled nets, for various purposes. First, we do not want to distinguished isomorphic labelled nets. But also it will be allowed to add/drop places or communication transitions which duplicate each other.

Petri Boxes are defined as unmarked, but they may be equipped with an initial marking which puts one token on each entry place, and no token elsewhere. The final marking is defined similarly on exit places. The standard Petri net transition rule applies to yield new markings from the initial one.

The Petri Box Calculus (PBC for short) was then in its infancy however, and among the CALIBAN objectives we find the development of this calculus, comprising (a) compositional behavioural semantics; (b) inference rules; (c) logics; and (d) an equational theory. And it may be said that important improvements have been obtained in these fields, both at the low level and at higher levels of the net theory. This presentation will concentrate on the low level achievements; the main sites which contributed to it are (in alphabetical order) Brussel, Hildesheim, Newcastle and Paris.

3.1 The labelled tree device and the general refinement/recursion operators

A first deficiency of the original theory was that it was only possible to define refinements of Petri Boxes, in an iterative manner, when there was finitely many transitions to be refined, no self loop and no weight around them, and it was not clear why the end result would be unique. In order to overcome these difficulties, it was necessary to generalise the matricial place interface which has been classically used. The solution was to introduce special labelled trees as the interface places, whose stucture recalls how and why they were obtained [11, 34, 71, 72]. This immediately allowed to drop all the restrictions, and to allow simultaneous refinements of various variables: $\Sigma[X_i \leftarrow \Sigma_i | i \in I]$. The interface places are then all the labelled trees of the following form:

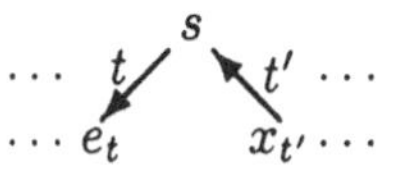

i.e., the root is labelled by a place $s \in S$, the arcs are labelled by a transition and a direction; for each $i \in I$ and for each (if any) $t \in s^\bullet$ with a label of the form $f(X_i)$, there is an arc labelled t going (down) to (a node labelled by) some (arbitrarily chosen) entry place e_t of Σ_i and for each (if any) $t' \in {}^\bullet s$ with a label of the form $f(X_i)$, there is an arc labelled t' coming (up) from (a node labelled by) some (arbitrarily chosen) exit place $x_{t'}$ of Σ_i. Such a tree may be represented by a set of sequences $\{s, t.e_t, \cdots, t'.x_{t'}, \cdots\}$, describing the root and all the children labels together with the corresponding arc labels.

There are also places and transitions of the form $t.s_i$ and $t.t_i$, for each transition $t \in T$ with a label $f(X_i)$ and each internal place s_i or transition t_i of Σ_i, copying the interior of Σ_i for each transition to be replaced by it. The connectivity and the label of all these elements is directly driven by their form, i.e. their origin.

This is illustrated by the net in Figure 2.

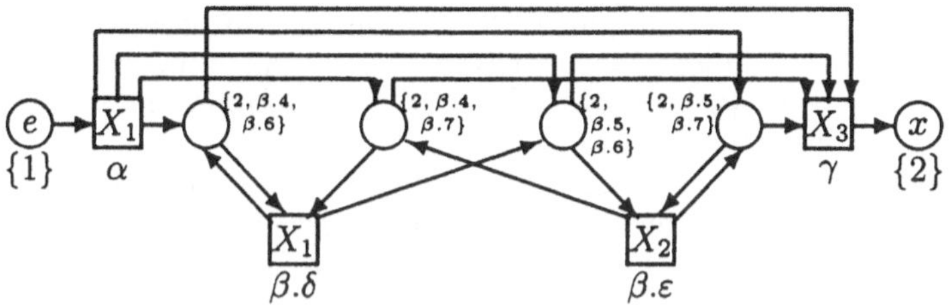

Figure 2: $\Sigma_*[X_2 \leftarrow \Sigma_{\|}]$

Many properties of the general refinement operator were derived, like the Expansion Law for Successive Refinements:

$$\Sigma[X_i \leftarrow \Sigma_i \mid i \in I][W_j \leftarrow \Sigma'_j \mid j \in J]$$
$$\equiv \Sigma[X_i \leftarrow \Sigma_i[W_j \leftarrow \Sigma'_j \mid j \in J], W_k \leftarrow \Sigma'_k \mid i \in I, k \in K]$$
if $K = \{k \in J | W_k \notin \mathcal{X}_I\}$.

But also, it occurred that many of the other operators can be synthesized from the simultaneous refinement, and that the properties of the formers may be directly derived from the general properties of the latter:

Let $\Sigma_;$, $\Sigma_{\Box}$, $\Sigma_{\|}$, Σ_* and Σ'_* be the Boxes shown in Figure 1

(i) $\Sigma_1; \Sigma_2 = \Sigma_;[X_1 \leftarrow \Sigma_1, X_2 \leftarrow \Sigma_2]$ (sequence)

(ii) $\Sigma_1 \Box\ \Sigma_2 = \Sigma_{\Box}\ [X_1 \leftarrow \Sigma_1, X_2 \leftarrow \Sigma_2]$ (choice)

(iii) $\Sigma_1 \| \Sigma_2 = \Sigma_{\|}[X_1 \leftarrow \Sigma_1, X_2 \leftarrow \Sigma_2]$ (concurrent composition)

(iv) $[\Sigma_1 * \Sigma_2 * \Sigma_3] = \Sigma_*[X_1 \leftarrow \Sigma_1, X_2 \leftarrow \Sigma_2, X_3 \leftarrow \Sigma_3]$ (iteration)

Extending the idea used to generalize the refinement operator, it was also possible to define a general simultaneous recursion operator $\mu\{X_i.\Sigma_i \mid i \in I\}\Sigma$; simply, in that case, the labelled trees giving the various kinds of places may have any (possibly infinite) depth, and the sequences defining the various transitions may be of any finite length. Their connectivity and labels are again directly driven by those structures. And again general properties were derived for this operator, like the Substitution Property

$$\mu\{X_i.\Sigma_i \mid i \in I\}\Sigma \equiv \Sigma[X_i \leftarrow \mu\{X_j.\Sigma_j \mid j \in I\}\Sigma_i \mid i \in I]$$

which generalizes the classical fixpoint equation:

$$\mu X.\Sigma \equiv \Sigma[X \leftarrow \mu X.\Sigma]$$

3.2 The S-invariant analysis

Rather early in the development of the theory, it was observed that most of the nets constructed to model Box expressions were 1-safe from the initial marking, but also that in some circumstances they could be 2-safe (see figure 2). The question then arose to know if the situation could still be worse, and to be able to characterize the origin of this unsafeness. In order to do so, the idea was to conduct an S-invariance analysis of the domain, i.e., to synthesize S-invariants and S-components of constructed nets, from the knowledge of similar characteristics for the composing nets.

The problem was first conducted for finite refinements [35], was then extended to infinite refinements [40, 42], and finally to any recursions and other operators [36, 43]; rather general techniques were devised to construct such invariants and the main derived results were the Coverability Property:
If S is a set of operative nets S-covered by S-components, as it is the case for the modelling of Box expressions (see figure 1), for any net $\tilde{\Sigma}$ constructed from them

- *$\tilde{\Sigma}$ is T-covered by S-components, so that from the initial marking the net does not have self-concurrency,*

- *${}^\bullet\tilde{\Sigma} \cup \tilde{\Sigma}^\bullet$ is covered by S-components, so that from the initial marking the net is clean: the exit places are 1-safe and if they are all marked there is no other token left in the net,*

- *$\forall \tilde{s} \in \tilde{\Sigma}$: there is a semi-positive 1-conservative S-invariant $\tilde{\nu}$ such that $\tilde{\nu}(\tilde{s}) \geq \frac{1}{2}$, so that from the initial marking the net is at most 2-safe.*

Moreover, if there is no side loop with a hierarchical label in S this may be reinforced as

- *Σ is S-covered by S-components, so that from the initial marking the net is at most 1-safe.*

As a consequence, it was clear that the unsafeness was due to the side loop in the operative net for iteration, and it was decided to replace it by another one, with the same global behaviour but without side loop, as shown in figure 3.

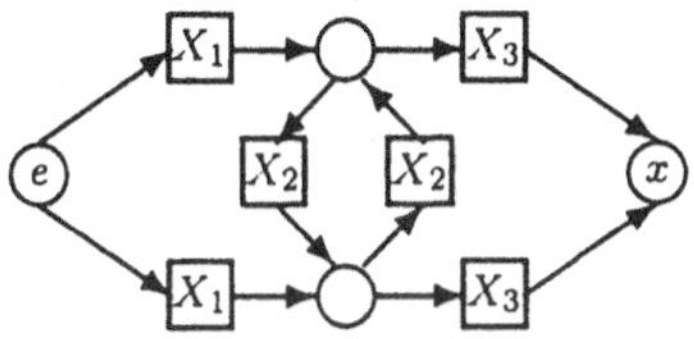

Σ'_*

Figure 3: The 1-safe iteration net

3.3 The process semantics

From the previous results, since the concurrent semantics of 1-safe nets is faithfully represented by its processes (in the Petri net theory sense), it was possible to also derive a compositional calculus for the processes of the Boxes modelling the Box expressions (from the initial marking) [19, 86]. A single case resisted however in order to get a full theory : for ungarded recursions it was not always possible to construct exactly all the infinite processes.

3.4 The synchronisation revisited

While the synchronisation operator defined for the Petri Box Calculus nicely generalizes the ideas of CCS to multiactions, it was soon recognized that other generalizations were possible, and that while the chosen definition was in some sense economical it was possible however to get infinite Boxes when synchronizing finite ones. First, it is possible to allow multiple synchronisation channels between two transitions instead of only one, to allow cycles of channels, but also channels internal to a single transition; while this seems to increase the cardinality problem, it was also noticed that if the equivalence classes are extended in order to allow to add/drop transitions equivalent to steps of other ones, then suddenly, with the more liberal channel policy, synchonizing a finite Box always respects finiteness [38].

Other generalisations were also explored [60, 61, 59] in order to allow other synchronisation principles than the usual conjugation mechanism, either from synchronisation algebras or by using more general functions (or relations) on multisets of multiactions. Moreover, while synchronisations do not commute nicely with refinements, due to the presence of hierarchical transitions, it was also noticed that since we are essentially interested in the end in the behaviour of Boxes without hierarchical transitions, modelling expressions without free variables, it was possible to modify the synchronisation principle in order to recover the desired commutativity in that case [41].

Finally, it was also observed that the restriction operator may be reduced to a kind of relabelling, using partial functions instead of total ones [37].

3.5 The denotational and the operational semantics

From the observations and results mentionned above, it finally occurred that all the desired Box operations could be generated from the combination of a generalised relabelling mechanism (coping with the transition interfaces) and a complete recursive system of simple refinement equations (complete in the sense that all used variables occur as left hand sides, simple in the sense that the right hand sides are single refinements of operative Boxes by basic Boxes, with a single variable or multiaction transition) [17, 18].

The solution of such a system may be obtain by an iterative procedure and it was shown that if the first step leads to a net including the initial one, then the inclusion property will be preserved for all the succeeding steps, leading to a well-defined limit Box net which is indeed a solution of the considered system. In many circumstances, this limit is unique and finite or enumerable, but for true unguarded recursions we may have a lattice of solutions with the same (finite) behaviours. The (unique) least solution is generally unacceptable however, because it does not fulfill all the conditions for a Box, but there are always acceptable (finite or) enumerable solutions; there is always a unique maximal solution, which corresponds to the one produced by the general simultaneous recursion operator, but it may be continuously infinite (in terms of places) [39].

It is also possible to provide an operational semantics for the ground Box expressions domain, in terms of derivation rules on marked expressions, with transformations labelled by multisets of multiactions, and it was possible to show the coherence of this semantics with the step semantics of the corresponding Box nets [83, 82, 81].

3.6 Practical achievements

Finally, in order to show the usefulness and practicality of the above theories, various concrete applications were realised [57].

First, in collaboration with the German PEP project, computer tools were implemented [102, 66, 117] in order to automatically construct nets corresponding to Box expressions or to net operations, to visualize them, to drive their evolutions and to perform model checking (for which theoretical developments in the logics field were also realised [56, 84]).

Moreover, a new (simple) programming notation (called $B(PN)^2$) was devised [16] together with a Box and Petri Box semantics, nicely exhibiting the usefulness of the multiway synchronisation. The resulting nets may be huge however, and this led to the introduction of higher level Petri Box formalisms, allowing more compact representations and analyses, which will be described in the next strand.

4 Specification Based on High Level Nets

Author/Editor: E.Pelz

In the DEMON project, two high level net specification or programming formalisms had been developed: OBJ-SA (by the Milano group) and CO-OPN (by the Paris group). The main task of CALIBAN Strand 4 was the integration of high-level nets into a calculus-oriented systematic specification method.

More precisely, the objectives were to enhance high level Petri nets from being purely a descriptive and analytical tool into being also a constructive and synthetical technique for the specification and stepwise implementation of systems. It was intended that activity should be centred around a single model of high level nets. We wished to (a) study the properties of special - compositionally defined - classes of such nets; (b) develop a causal temporal logic for reasoning about the properties of high level nets and for making their use as a specification language more flexible; and (c) study the causal semantics of high level nets depending on the possible semantics of the (algebraically specified) data part. Moreover, there was the intention to study object orientation in HL nets, cf. [1].

All these objectives are now largely achieved: As a common basic high level model we chose algebraic nets as proposed by Vautherin/Reisig [99, 118], where tokens of different colours are represented by elements of different sorts in a model of the specification of an abstract data type.

This model has been enriched with structuring capabilities based on hierarchy by Guelfi [67] and with object oriented features by Chizzoni, Buchs and Biberstein [26, 22], and exercised in substantial case studies [3, 24]. Also, its relation to low level Petri Nets in terms of foldings was studied from a categorical point of view, by Lilius [85], and as an application, a method for constructing deadlock preserving skeletons was obtained. Our recent work is strongly concentrated on object orientation by enriching the previous formalisms: OBJ-SA, becoming CLOWN (at Milano); and CO-OPN, becoming CO-OPN/2 (at Paris[5]). This work has been highly collaborative in order to develop similar notions and complementary formalisms, as will be detailed in section 4.1. Both formalisms exhibit (different) nice and strong modular features, but not in the sense of compositionality as required for a Calculus.

Thus a completely new approach was started at Paris [79], building on the PBC. After intensive discussions, several CALIBAN workshops devoted to Box-Calculi and mutual visits, we came to results unexpected in the beginning of CALIBAN: we suceeded in the integration of high-level nets into the PBC in a way totally coherent with the developments on the low level PBC (presented in section 3). The usefulness of algebraic nets for very compact data representation allowed us to take the same common net model, and to enrich it with a Box like flavor for the desired compositionality.

In fact, we propose two (coherent) versions of high level PBC:

- M-nets, a class with Box like interfaces for compositions but supporting only simple data representation. This is shown to be rich enough for a compositional semantics of $B(PN)^2$ which is consistent with that given by low level Boxes.
- A-nets, which generalize M-nets by data specification through arbitrary abstract data types, and which could be used to give semantics to more complex concurrent programming languages, e.g. for extensions of $B(PN)^2$.

The low and high level Box Calculi represent now a quite important part of CALIBAN outcomes, and result from a very strong collaboration between the sites of (in alphabetical order) Brussel, Hildesheim, Newcastle and Paris. To illustrate the mutual influence of work done in Strands 3 and 4 and the coherence of results, the following text will mainly present these HL-PBC's.

[5] together with research groups in Geneva and Lausanne, directed by D. Buchs (who was formerly a member in the DEMON-site Paris), where very important financial support from swiss FNRS allowed rapid development of the theory and of a powerful tool, called SANDS

4.1 HL-nets with object orientation

As mentioned above, the recent developments from CO-OPN and OBJ-SA stress object orientation and have been made in collaboration between Milano, Paris and the Swiss group. The differences between the formalisms are justified by the different abstraction levels chosen in each approach.

CLOWN [26] is a concurrent object oriented language based on OBJ-SA [5]. In contrast, CO-OPN/2 [23] gives an abstract specification formalism based on software engineering techniques and object oriented notions, extending the combination of Petri nets and algebraic specifications. Let us illustrate, e.g. different interpretations of the oo inheritance notion: in CLOWN inheritance is a subtype notion, characterized by a proper preorder relation based on a 'simulation' principle [4]; whereas in CO-OPN/2, inheritance is viewed as a duplication mechanism, while subtyping is considered at a semantic level as a relation between models.

There has been some progress in temporal logic techniques for the verification of concurrent systems specified by algebraic nets. The theoretical and practical results on the use of temporal logic for Petri net analysis have been extended to algebraic nets by Racloz [96] and implemented in the CO-OPN development environment, SANDS. That extension uses symbolic techniques for handling CTL formulas, although it is actually restricted to a reduced algebraic net class. Also, verification techniques based on a CTL-like logic have been developed for OBJ-SA nets 'by Gorni [65] and implemented in the OBJ-SA Net Environment, ONE (analogous to SANDS).

Work is planned to extend to Hierarchical Algebraic Nets [67] these results for the analysis of general CO-OPN/2 specifications .

Theoretical work on causal semantics of these structured algebraic net classes has been developed by Guelfi [67], based on a particular notion of distributed transition system. Operational and denotational semantics for OBJ-SA nets have been studied by De Cindio et al. [6].

4.2 The M-net model

The first common proposition for a high level version of PBC was the Calculus of M-nets by Best, Fleischhack, Frączak, Hopkins, Klaudel and Pelz [14], developed with the aim of providing a compositional high level semantics for $B(PN)^2$ programs, as is done in [15].

In traditional high-level net models there are place/transition *annotations* which determine the transition rule of the model and which also drive the 'vertical' unfolding of a high-level net into an elementary net. The M-net model is an extension of the traditional high-level model in that it allows the specification of 'horizontal' as well as vertical annotations. Horizontal annotations, which will be called *labels*, allow the modular construction of large high-level nets from smaller modules. Thus, every element (place or transition) of an M-net carries an *inscription* which is a pair of the following form:

$$\textit{inscription} \;=\; (\textit{label} \mid \textit{annotation}).$$

We first define the basic concepts which are important for the inscriptions of the transitions, the places and the arcs of an M-net.

Let VAL be a fixed set of *values* containing at least the distinguished element $\bullet$. This set contains all the 'colours' of the high level net model we define. A *type* is a subset of VAL. Let us point out that 'data types' in the M-net model are not structured: we have a single sort, VAL, containing without distinction all types we like.

Let VAR be the set of (symbols for) net variables and let **VT** (for value terms) be the set of well formed terms built from values, variables and a set of operators OP inductively in a standard way. Terms τ without occurrence of variables are called *ground terms.* Every value is also a ground term.

We assume the existence of a fixed but sufficiently large set **A** of *action symbols.* We assume that each action symbol A from **A** has an arity $ar(A)$, which gives the number of associated arguments. The set **A** is the carrier of a bijection $\bar{}: \mathbf{A} \to \mathbf{A}$ called *conjugation* which satisfies $\forall A \in \mathbf{A}$:
$\overline{A} \neq A$ and $\overline{\overline{A}} = A$, as well as $ar(A) = ar(\overline{A})$.

An *action term* is, by definition, a construct $A(\tau_1, \ldots, \tau_{ar(A)})$, where A is an action symbol and τ_j $(1 \leq j \leq ar(A))$ are value terms. If in an action term, all occuring τ's are constants in VAL then it will be called an *elementary action.*

An M-net is a triple (P, T, ι) such that P is a set of places, T is a set of transitions with $P \cap T = \emptyset$, and ι is an inscription function with domain $P \cup (P \times T) \cup (T \times P) \cup T$ such that:

- *For every place $p \in P$, $\iota(p)$ is a pair $(\lambda_p \mid \alpha_p)$, where λ_p is a (place-)label and α_p a (place-)annotation. λ_p is an element of the set $\{e, \emptyset, x\}$ (for 'entry', 'internal' and 'exit' place) and α_p, the type of p, is a subset of VAL.*

- *For every arc $f \in (P \times T) \cup (T \times P)$, $\iota(f)$ is a finite multiset of variables from VAR.*

- *For every transition $t \in T$, $\iota(t)$ is a pair $(\lambda_t \mid \alpha_t)$, where λ_t is its label and α_t its annotation. λ_t is a finite multiset of action terms, and α_t is a finite set of value terms.*

Note that an arc inscription may be empty (i.e., the empty multiset). This means that no tokens may ever flow along that arc and that no effective connection exists along it. Thus arcs with an annotation $\emptyset$ are considered as non existent.

We impose the following restrictions to yield Box-like M-nets. There must be at least one entry and one exit place; no entry place may have incoming arcs and no exit place may have outgoing arcs; all entry and exit places have $\{\bullet\}$ as annotation, to insure un-problematic compositions; and thus, in an M-net, the initial marking and also the final marking (if it can be reached) must consist of black tokens.

The class of M-nets had been defined is such a way that low level Boxes could be considered as a particular subclass, namely that of *elementary M-nets* for which: (i) the type of each place is a singleton; (ii) all action terms occurring in the net are elementary actions; (iii) all transition annotations in the net are empty.

The *unfolding* of an M-net (driven as usual by the annotations) yields an elementary M-net (action terms in the original label unfold to a set of elementary action terms). *Semantics* are as usual for algebraic nets.

4.3 M-net Calculus and properties

Using the labels of an M-net, a series of operations were defined. Most of them are counterparts of operations known from programming languages, such as sequential or parallel composition and iteration (illustrated in [15], this volume). The most novel operation is *transition synchronization* which allows transitions of a high-level net to be merged in a systematic way. This operation generalizes the synchronization operation found in process algebras such as CCS and serves a useful purpose for semantics of block structured languages [16].

We demonstrated that (label-driven) composition and (annotation-driven) unfolding were defined in harmony with each other. More precisely, we proved that they satisfy not just a set of algebraic laws but also a property of coherence with respect to each other. Informally, this property states that the unfolding of a composite net equals the composition of the unfoldings of the component nets; that is, if, say, U denotes unfolding and $\odot$ denotes some, e.g. binary, operation on M-nets N_1, N_2 then $U(N_1 \odot N_2) = U(N_1) \odot U(N_2)$. In order to satisfy this property, the synchronization operation needed to be defined judiciously, involving simple unification, logical conjunction and multiset addition of appropriate expressions, formulae and (multi)sets [14].

M-nets offer 'abbreviations' which reduce the size of nets. This being particularly important for the nets produced as semantics of $B(PN)^2$ programs, see [16], to which they were applied [15] (in this volume). M-net Calculus has been used to give both: a high level semantics (by unrestricted M-nets), and a low level net semantics (by elementary M-nets), of the specification and programming language $B(PN)^2$. The consistency of these two semantics [15] and thus the consistency of the high level net semantics with the original semantics of [16] has been shown.

All these results made it possible to integrate a coherent M-net component in the PEP-tool, see [15]. Work is in progress to develop (and integrate) verification techniques to be applied directly to the M-net semantics of $B(PN)^2$ programs.

The M-net model is the 'lowest' possible high-level approach, ('mid' level might be a more appropriate term), in that it is strongly dependent on the interpretation of symbols for the low level semantics. Although M-nets are a fairly powerful Petri net model, the lack of support for handling complex data structures would be a serious drawback if one wanted to use it as a semantic domain for a real programming language (which could be formulated as some extension of $B(PN)^2$). As sequel, the A-net model (a generalization of M-nets with a full abstract data type orientation) was developed.

4.4 A-nets, A-net-Calculus and properties

In contrast to M-nets whose design was bottom-up, Klaudel and Pelz proposed in [80] rather a top-down analysis, going from a (reasonably) abstract model to a concrete (low level) one. The aim was to allow abstractions in data specification, by means of abstract data types. This approach provides a formal way of describing properties of the values and operations, independently of any particular implementation. It also allows reasoning about programs using such data types and proofs of implementations.

Informally, an *abstract data type*, shortly adt, can be considered as a many-sorted algebraic theory which is given by a *presentation* (S, F, E, X), including: a set of *sorts* (type names) S; a *signature* F, i.e. a set of operation (function) names with their corresponding typed arities $ar : F \rightarrow S^+$; and a set of *axioms* E, i.e., equations between terms comprising operations and free variables from a set of typed variables X.

A *model* of an adt is a many-sorted algebra D which satisfies each axiom in E. In particular, such models include $T_{F,X}/_E$ and $T_F/_E$, the quotient algebras of $T_{F,X}$ (the set of terms with variables) and of T_F (the set of *ground* terms), modulo the smallest congruence generated by the set of axioms E. Moreover, $T_F/_E$ is initial in the class of all its models.

Let **A** be the set of action symbols (as in M-nets), provided with the conjugation bijection and an arity function. The difference from M-nets is that the arity is typed, i.e., $\forall A \in \mathbf{A}\ \ ar(A) \in S^*$, meaning that each action symbol can handle some number of parameters of corresponding sorts.

Let A_{HL} be the set of *parameterized actions* of the form $(A; \vec{u})$ where A is an action symbol, and $\vec{u}$ is a (possibly empty) vector of terms (with variables) of corresponding sorts, i.e., if $ar(A) = s_1 \cdots s_n$ then

$$\vec{u} = u_1, \ldots, u_n \in (T_{F,X})_{s_1} \times \ldots \times (T_{F,X})_{s_n}$$

where $(T_{F,X})_s$ denotes the set of terms of sort s. The M-net action terms can be viewed as a restricted case of parameterized actions where a parameter is only a variable or a constant (but not a more complex term).

An A-net looks like an M-net except the following small but significant differences: (i) the place annotations are sorts from S, (ii) the transition labels are finite multisets of parameterized actions from A_{HL}, (iii) the transition annotations are sets of boolean terms (predicates), and (iv) the arc inscriptions are finite multisets of terms from $T_{F,X}$.

Semantics of A-nets depends on a particular model D of the abstract data type used. A marking of an A-net $N = (P, T, \lambda)$ associates to each place $p \in P$ a multiset of ground terms of its corresponding sort. Its behavior is defined by giving the usual transition rule for algebraic nets, cf. for instance [44] in this volume.

In [80] a Calculus of A-nets was defined including the same set of operations as for M-nets. The definition of the synchronization operation, necessarily complex, became crucial. Between various ways of choosing synchronization on A-nets, a 'syntactic' one was taken, based on the syntactic unification of terms, which had the advantages that: it coincides with M-net synchronization; it allows also a reasonable degree of abstraction; it can be defined equally well symmetrically or asymmetrically, which offers real facilities for proofs. Some non-trivial algebraic properties were proved [80], e.g. commutativity and idempotence of synchronization, which were necessary in order to use A-nets in a compositional context.

However, in the theory of A-nets as described above, as well as in that of M-nets, the algebraic structure was not complete, since refinements and recursions were missing. This problem was overcome, by Devillers and Klaudel in [44], by defining these operators at the A-net level (thus also for the embedded case of M-nets). They use the same devices as for the low level Petri Boxes, in particular, labeled trees and sequences. The semantics and unfoldings of A-nets has been defined for any model of the specification, together with equivalence relations and various properties directly inherited from the low level theory. The precise results can be found in this volume [44].

5 CALIBAN Partners

CALIBAN has been a cooperative effort between the following partners:

- Universität Hildesheim (E. Best, J. Esparza (now at Technische Universität München), H.G. Linde-Göers, B. Graves, St. Römer, Th. Thielke, U. Goltz, A. Rensink, P. Niebert, H. Wehrheim).
- Technische Universität München (W. Brauer, J. Esparza, M.D. Radola, W. Vogler (now at the Universität of Augsburg)).
- Humboldt-Universität zu Berlin (J. Desel, W. Reisig, E. Kindler, R. Walter).
- Rijksuniversiteit te Leiden (G. Rozenberg, J. Engelfriet, H.J. Hoogeboom, H.C.M. Kleijn, P.W. Hoogers, Tj. Gelsema).
- Università degli Studi di Milano (F. De Cindio, E. Battiston, A. Chizzoni, G. De Michelis, K. Petruni (presently on leave in Berlin), L. Bernardinello (presently in Rennes), S. Vigna and C. Ferigato, L. Pomello, C. Simone, C. Diamantini).
- University of Newcastle upon Tyne (M. Koutny, R.P. Hopkins, M. Pietkiewicz-Koutny, M. Hesketh).
- Université Libre de Bruxelles (R. Devillers, A. Sinachopoulos).
- LRI, Centre National de la Recherche Scientifique (E. Pelz, I. Biermann, J. Fanchon, W. Fraczak, P. Gastin, N. Guelfi, H. Klaudel, A. Petit, B. Rozoy, D. Buchs (now in Lausanne)).
- Universidad de Zaragoza (M. Silva, J. Manuel Colom, E. Teruel).

It has been done in liaison with the following partners who are not directly supported by the European Communities:

- Helsinki University of Technology (L. Ojala, J. Lilius (presently in Hildesheim), K. Varpaaniemi, T. Pyssysalo).
- Bull France (D. Bolignano).
- University of Warszawa, Polish Academy of Sciences (A. Mazurkiewicz, P. Chrzastowski-Wachtel).
- Hewlett Packard Laboratories, USA (V. Kotov, L. Cherkasova).

Acknowledgements

All CALIBAN members have participated in producing this report. The support and encouragement by the European Union is gratefully acknowledged.

References

[1] Various authors: Technical Annex of the Esprit Basic Research Working Group No.6067 CALIBAN (1992).

[2] R. Assous, V. Bouchitte, C. Charretton, B. Rozoy: Finite Labelling Problem in Event Structures. Theoretical Computer Science 123:9-19 (1994).

[3] E. Battiston, O. Botti, E. Crivelli, F. De Cindio: An incremental specification of a Hydroelectric Power Plant Control System using a class of modular algebraic nets. Proc. of Petri Nets'95, Torino, (eds. G.De Michelis, M.Diaz), LNCS, Springer-Verlag (1995).

[4] E. Battiston, A. Chizzoni, F. De Cindio: Inheritance and Concurrency in CLOWN. To appear in Proc. of the Workshop on Object-Oriented Programming and Models of Concurrency, Torino (1995).

[5] E. Battiston, F. De Cindio, G. Mauri: Specifying concurrent systems with OBJSA Nets. CNR, Progetto Finalizzato Sistemi Informatici e Calcolo Parallelo, Technical Report i/4/72 (1992).

[6] E. Battiston, V. Crespi, F. De Cindio, G. Mauri: Semantics frameworks for a class of modular algebraic nets. Proc. of AMAST'93, (eds. M. Nivat, C.Rattay, T.Rus, G.Scollo), Workshops in Computing, Springer-Verlag (1993).

[7] S. Bauget, P. Gastin: On congruences and partial orders. To appear in MFCS'95, LNCS, Springer-Verlag (1995).

[8] L. Bernardinello: Synthesis of Net Systems. Proc. of Petri Net'93, Chicago, LNCS 691:89-105, Springer-Verlag (1993).

[9] L. Bernardinello, G. De Michelis, K. Petruni, S. Vigna: On the Synchronic Structure of Transition Systems. In this volume.

[10] L. Bernardinello, G. De Michelis, K. Petruni, S. Vigna: On Synchronic and Enlogic Structure of Transition Systems, DSI Internal Report, University of Milan (1994).

[11] E. Best, R. Devillers, J. Esparza: General Refinement and Recursion Operators for the Petri Box Calculus. Proc. of STACS'93, LNCS 665:130-140, Springer-Verlag (1993).

[12] E. Best, R. Devillers, J. Hall: The Box Calculus: a New Causal Algebra with Multi-label Communication. In Advances in Petri Nets 1992, LNCS 609:21-69, Springer-Verlag (1992).

[13] E. Best, J. Esparza: Model checking of persistent Petri nets. Proc. of the Vth Workshop Computer Science Logic-91, LNCS 626:35-52, Springer-Verlag (1992).

[14] E. Best, H. Fleischhack, W. Frączak, R.P. Hopkins, H. Klaudel, E. Pelz: A Class of Composable High Level Petri Nets. Proc. of Petri Nets'95, Torino, (eds. G.De Michelis, M.Diaz), LNCS, Springer-Verlag (1995).

[15] E. Best, H. Fleischhack, W. Frączak, R.P. Hopkins, H. Klaudel, E. Pelz: An M-net Semantics of $B(PN)^2$. In this volume.

[16] E. Best, R.P. Hopkins: $B(PN)^2$ - a Basic Petri Net Programming Notation. Hildesheimer Informatik-Bericht 27/93 (1993). Proc. of PARLE-93, (eds. A. Bode, M. Reeve, G. Wolf), LNCS 694:379-390, Springer-Verlag (1993).

[17] E. Best, M. Koutny: A Refined View of the Box Calculus. Proc. of Petri Nets'95, Torino, (eds. G.De Michelis, M.Diaz), LNCS, Springer-Verlag (1995).

[18] E. Best, M. Koutny: Solving Recursive Net Equations. Proc. of ICALP'95, Szeged, (ed. F.Gecseg), LNCS, Springer-Verlag (1995).

[19] E. Best, H.G. Linde-Göers: Compositional Process Semantics of Box Expressions. Presented at the IXth MFPS (Mathematical Foundations of Programming Semantics), New Orleans (1993), LNCS 802:250-270, Springer-Verlag (1994). Extended version: Hildesheimer Informatikbericht 19/93.

[20] I. Biermann, B. Rozoy: Context Traces and Transition Systems. ISCIS, Antalya (1994).

[21] I. Biermann, B. Rozoy: Graphs for Generalized Traces. In this volume.

[22] D. Buchs, O. Biberstein: An Object Oriented Specification Language based on Hierarchical Petri Nets. IVth International IS-CORE WORKSHOP, Amsterdam (1994).

[23] D. Buchs, O. Biberstein, N. Guelfi: COOPN2: an object oriented specification language for distributed system development. Submitted article.

[24] D. Buchs, N. Guelfi: Formal development of Actor programs using structured algebraic Petri nets. Proc. of PARLE'93, (eds. A. Bode, M. Reeve, G. Wolf), LNCS 694:379-390, Springer-Verlag (1993).

[25] A. Cheng, J. Esparza, J. Palsberg: Complexity Results for 1-safe Petri Nets. To appear in Theoretical Computer Science.

[26] A. Chizzoni: CLOWN: CLass Orientation With Nets. Master Degree Thesis, Dept. of Comp. Sci., Univ. of Milan (1994).

[27] J.M. Colom, M. Silva: Improving the linearly based characterization of P/T nets. In Advances in Petri Nets 1990, LNCS 483:113–145, Springer-Verlag (1991).

[28] J. Desel: A proof of the Rank Theorem for extended free choice nets. Proc. of Petri Nets'92, Sheffield, (ed. K.Jensen), LNCS 616:134-153, Springer-Verlag (1992).

[29] J. Desel: Regular Marked Petri Nets. Proc. of Graph-Theoretic Concepts in Computer Science (WG'93), Utrecht, (ed. Jan van Leeuwen), LNCS 790:276-287, Springer-Verlag (1994).

[30] J. Desel, J. Esparza: Shortest paths in reachability graphs. Proc. of Petri Nets'93, Chicago, (ed. M. Ajmone Marsan), LNCS 691:224-241, Springer-Verlag (1993). To appear in: Journal of Computer and System Science (1995).

[31] J. Desel, J. Esparza: Free Choice Petri Nets. Cambridge Tracts in Theoretical Computer Science 40, Cambridge University Press (1995).

[32] J. Desel, M.D. Radola: Proving non-reachability by modulo-place-invariants. Proc. of the XIVth Conference on the Foundations of Software Technology and Theoretical Computer Science, Madras, (ed. P.S.Thiagarajan), LNCS 880:366-377, Springer-Verlag (1994). Also to appear in Theoretical Computer Science (1995).

[33] J. Desel, W. Reisig: The Synthesis Problem of Petri Nets. Proc. of STACS'93, (eds. P.Enjalbert, A.Finkel, K.W.Wagne), LNCS 665:120-129, Springer-Verlag (1993).

[34] R. Devillers: Tree Interfaces in the Petri Box Calculus. TR-LIT-265, Université Libre de Bruxelles (1992).

[35] R. Devillers: Construction of S-invariants and S-components for Refined Petri Boxes. Proc. of Petri Nets'93, Chicago, LNCS 691, Springer-Verlag (1993).

[36] R. Devillers: S-invariant Analysis of Petri Boxes. TR-LIT-273, Université Libre de Bruxelles (1993).

[37] R. Devillers: Towards a General Relabelling Operator for the Petri Box Calculus. TR-LIT-274, Université Libre de Bruxelles (1993).

[38] R. Devillers: On a more Liberal Synchronisation Operator for the Petri Box Calculus. TR-LIT-281, Université Libre de Bruxelles (1993).

[39] R. Devillers: Modelling Petri Boxes are not too infinite. Submitted article (1993).

[40] R. Devillers: Analysis of General Refined Petri Boxes. Proc. of the XIIIth International Conference of the Chilean Society of Computer Science (La Serena, Chile), 419-434 (1993). Also in Computer Science 2: Research and Applications (Plenum Publishing), 411-428 (1994).

[41] R. Devillers: The Synchronisation Operator Revisited for the Petri Box Calculus TR-LIT-290, Université Libre de Bruxelles (1994).

[42] R. Devillers: S-invariant Analysis of General Refined Petri Boxes. TR-LIT-293, Université Libre de Bruxelles (1994).

[43] R. Devillers: S-invariant Analysis of General Recursive Petri Boxes. TR-LIT-294, Université Libre de Bruxelles (1994). To appear in Acta Informatica (1994 or 1995).

[44] R. Devillers and H. Klaudel: Refinement and Recursion in a High Level Petri Box Calculus. In this volume.

[45] V. Diekert, P. Gastin, A. Petit: Rational and Recognizable complex trace languages. To appear in Information and Computation.

[46] V. Diekert, G. Rozenberg (eds.): Book on Traces, World Academic Publishing, Singapore 1995.

[47] A. Ehrenfeucht, T. Harju, G. Rozenberg: Invariants of dynamic labelled 2-structures. Submitted article.

[48] A. Ehrenfeucht, G. Rozenberg: An introduction to dynamic labelled 2-structures. Proc. of MFCS'93, LNCS 711:156-173, Springer-Verlag (1993).

[49] A. Ehrenfeucht, G. Rozenberg: Dynamic labeled 2-structures. Mathematical Structures of Computer Science 4:433-455 (1995).

[50] A. Ehrenfeucht, G. Rozenberg: Dynamic labeled 2-structures with variable domains. LNCS 812:97-123, Springer-Verlag (1993).

[51] A. Ehrenfeucht, G. Rozenberg: Partial (Set) 2-Structures - Part 1: Basic Notions and the Representation Problem. Acta Informatica 26:315-342 (1990).

[52] A. Ehrenfeucht, G. Rozenberg: Partial (Set) 2-Structures - Part 2: State Spaces of Concurrent Systems. Acta Informatica 26:343-368 (1990).

[53] J. Engelfriet: A multiset semantics for the π-calculus with replication. Proc. of CONCUR'93, LNCS 715:7-21, Springer-Verlag (1993).

[54] J. Engelfriet: A multisets semantics for the π-calculus with replication. RUL Report I94-26.

[55] J. Engelfriet, Tj. Gelsema: Multisets and structural congruence of the π-calculus with replication. RUL Report I95-02.

[56] J. Esparza: Model Checking based on Branching Processes. Habilitation, Hildesheim (1993). Published as: J. Esparza: Model Checking Using Net Unfoldings. Proc. of TAPSOFT'93 (1993), (eds. M.C. Gaudel, J.P. Jouannaud), LNCS 668:613-628, Springer-Verlag (1993). Also in Science of Computer Programming 23:151-195 (1994).

[57] J. Esparza, G. Bruns: Trapping Mutual Exclusion in the Box Calculus. To appear in Theoretical Computer Science (1995).

[58] J. Ezpeleta, J.M. Colom, J. Martínez: A Petri Net Based Deadlock Prevention Policy for Flexible Manufacturing Systems. IEEE Trans. Robotics and Automation 11,2:173-184 (1995).

[59] W. Frączak: Synchronization Algebra and Multi-actions. LRI-TR 896, Université Paris Sud, Orsay (1994).

[60] W. Frączak, H. Klaudel: General Synchronization Operator in the Petri-Box Calculus. LRI-TR 821, Université Paris Sud, Orsay (1993).

[61] W. Frączak, H. Klaudel: A Multi-Action Synchronization Schema and its Application to the PBC. ESDA'94, ASME, Methodologies, techniques, and tools for design development, London (1994).

[62] P. Gastin, A. Petit, W. Zielonka: An extension of Kleene's and Ochmański's theorems to infinite traces. Theoretical Computer Science 125:167-204 (1994).

[63] P. Gastin, B. Rozoy: The PoSet of Infinite Traces. Theoretical Computer Science 120:101-121 (1993).

[64] U. Goltz, R. Gorrieri, A. Rensink: On syntactic and semantic action refinement. In Theoretical Aspects of Computer Software, (eds. M. Hagiya, J. C. Mitchell), LNCS 789:385-404. Springer-Verlag (1994). Full report version: Hildesheimer Informatik-Berichte 17/92, Institut für Informatik, Universität Hildesheim.

[65] U. Gorni: Sviluppo ed integrazione di strumenti di supporto alla verifica di proprieta' su reti modulari di alto livello. Master Degree Thesis. Dept. of Comp. Sci., Univ. of Milan (1994).

[66] B. Graves: Implementation of a Model Checking Algorithm Based on Partial Order Semantics. In Algorithmen und Werkzeuge für Petrinetze, (eds. J. Desel, A. Oberwies, W. Reisig), Workshop der GI-Fachgruppe 0.01 "Petrinetze und verwandte Systemmodelle", Bericht 309 Universität Karlsruhe, Berlin (1994).

[67] N. Guelfi: Les réseaux Algèbriques Hiérarchiques: un formalisme de spécifications structurées pour le développement de systèmes concurrents. Ph.D. Thesis, Laboratoire de recherche en Informatique (LRI), Université Paris Sud, Orsay (1994).

[68] P.W. Hoogers: Behavioural aspects of Petri nets. Ph.D. Thesis, Leiden University (1994).

[69] P.W. Hoogers, H.C.M. Kleijn, P.S. Thiagarajan: Local event structures and Petri nets. Proc. of CONCUR'93, LNCS 715:462- 476, Springer-Verlag (1993).

[70] P.W. Hoogers, H.C.M. Kleijn, P.S. Thiagarajan: An event structure semantics for general Petri nets. RUL Report I93-13 (revised version of I92-22) (1993). Accepted by Theoretical Computer Science.

[71] R.P. Hopkins: Recursion and Refinement for a Generalisation of the Petri Box Calculus. Technical Report TR-440, Dept. of Comp. Sci., University of Newcastle upon Tyne (1993).

[72] R.P. Hopkins: Formalising Refinement and Composition of Concurrent Software. Draft paper.

[73] M. Huhn, H. Wehrheim: On Refining Logical Specifications. Hildesheimer Informatik-Bericht, University of Hildesheim 1995. In preparation.

[74] R. Janicki, M. Koutny: Representations of Discrete Interval Orders and Semi-Orders. TR-9302, Department of Comp. Sci. and Systems, McMaster University (1993).

[75] R. Janicki, M. Koutny: Order Structures and Generalisations of Szpilrajn's Theorem. Proc. of FSTTCS'93, LNCS 761, Springer-Verlag (1993). Also: TR-425, Dept. of Comp. Sci., University of Newcastle upon Tyne (1993).

[76] R. Janicki, M. Koutny: Deriving Histories of Nets with Priority Relation. Proc. of PARLE'94, LNCS, Springer-Verlag (1994).

[77] R. Janicki. M. Koutny: Semantics of Inhibitor Nets. TR-9401, Dept. of Comp. Sci. and Systems, McMaster University (1994).

[78] R. Janicki, M. Koutny: Fundamentals of Modelling Concurrency Using Discrete Relational Structures. TR-9402, Dept. of Comp. Sci. and Systems, McMaster University (1994).

[79] H. Klaudel: Traitement de Donnés dans l'Algèbre des Petri-Boxes. Rapport de stage de DEA, LRI, Université Paris Sud, Orsay (1992).

[80] H. Klaudel, E. Pelz: Communication as Unification in the Petri Box Calculus, To appear in FCT'95, Dresden, August 1995, LNCS, Springer-Verlag. Full version in LRI-TR 967, Université Paris Sud, Orsay (1995).

[81] M. Koutny: Syntactic Derivation of Petri Boxes. Memo, Newcastle upon Tyne (1994).

[82] M. Koutny: Partial Order Semantics of Box Expressions. Proc. of Petri Nets'94, Zaragoza, LNCS, Springer-Verlag (1994).

[83] M. Koutny, J. Esparza, E. Best: Operational Semantics for the Petri Box Calculus. Proc. of CONCUR'94, LNCS, Springer-Verlag (1994). Long version: Hildesheimer Informatik-Berichte 13/93 (1993).

[84] M. Koutny, M. Pietkiewicz-Koutny: On the Sleep Sets Method for Partial Order Verification of Concurrent Systems. Technical Report, Dept. of Comp. Sci., Newcastle upon Tyne (1994).

[85] J. Lilius: On the Structure of High-level Nets. Doctoral Dissertation, Helsinki University of Technology, Dept. of Comp. Sci. (1995). Also as Technical Report A30, Helsinki University of Technology, Digital Systems Laboratory, Espoo (1995).

[86] H.G. Linde-Göers: Process and Branching Process Semantics of Box Expressions. Ph.D. Thesis (1993).

[87] K. Lodaya, R. Ramanujam, P. S. Thiagarajan: Temporal logics for communicating sequential agents: I. International Journal of Foundations of Computer Science 3(2):117-159 (1992).

[88] A. Mazurkiewicz: Basic notions of trace theory. In Linear Time, Branching Time and Partial Order in Logics and Models for Concurrency, (eds. J. W. de Bakker, W.-P. de Roever, G. Rozenberg), LNCS 354:285-363, Springer-Verlag (1989).

[89] R. Milner, J. Parrow, D. Walker: A calculus of mobile processes, I. Information and Computation 100:1-40 (1992).

[90] M. Mukund: Petri Nets and Step Transition Systems. Int. Journal of Foundations of Computer Science 3.4 (1992).

[91] P. Niebert: A μ-calculus with local views for systems of sequential agents. To be presented at MFCS'95, LNCS, Springer-Verlag (1995).

[92] P. Niebert, W. Penczek: On the Connection of Partial Order Logics and Partial Order Reduction Methods. Technical University Eindhoven. Technical Report (1994).

[93] M. Nielsen, G.D. Plotkin, G. Winskel: Petri nets, event structures and domains, part 1. Theoretical Computer Science 13:85-108 (1981).

[94] M. Nielsen, G. Rozenberg, P. S. Thiagarajan: Elementary Transition Systems. Theoretical Computer Science 96/1 (1992).

[95] D. Peled: All from one and one from all: on model checking using representatives. Proc. of the International Workshop on Computer Aided Verification CAV, LNCS 409-423, Springer-Verlag (1993).

[96] P. Racloz: Analyse des Reseaux Algebriques a l'aide de la logique temporelle. Ph.D. Thesis, CUI, Geneve, Suisse, 1994. Also in P. Racloz, D.Buchs, Symbolic Proof of CTL Formulae over Petri Nets, Proc. of the VIIIth ISCIS (eds. Gun Levent, Onvural Raif, Gelembe Erol) 189-196 (1993).

[97] L. Recalde, E. Teruel, M. Silva: On Well-formedness Analysis: The Case of Deterministic Systems of Sequential Process. In this volume.

[98] W. Reisig: Deterministic buffer synchronization of sequential processes. Acta Informatica 18:117-134 (1982).

[99] W. Reisig: Petri Nets and Algebraic Specifications. Theoretical Computer Science 80:1-34 (1991).

[100] A. Rensink: Methodological aspects of action refinement. In Programming Concepts, Methods and Calculi, IFIP, North-Holland Publishing Company (1994).

[101] A. Rensink: An event-based SOS for a language with refinement. In this volume.

[102] St. Römer: Implementierung der Branching-Box-Algebra. Master Degree Thesis, Hildesheim (1993).

[103] B. Rozoy: The Distributed monoid, a Model for Parallelism, Elsevier. Book to appear.

[104] B. Rozoy : On the Star Problem in trace monoids, Automata and Models for Concurrency. Dagstuhl (1993).

[105] M. Silva: Interleaving Functional and Performance Structural Analysis of Net Models. Proc. of Petri Nets'93, Chicago, LNCS 691:17-23, Springer-Verlag (1993).

[106] M. Silva, E. Teruel: Analysis of Autonomous Petri Nets with Bulk Services and Arrivals. In Procs. of the XIth Int. Conf. on Analysis and Optimization of Systems. Discrete Event Systems, Lecture Notes in Control and Information Sciences 199:131-143, Springer-Verlag (1994).

[107] Y. Souissi, N. Beldiceanu: Deterministic systems of sequential processes: Theory and tools. In Concurrency 88, LNCS 335:380-400, Springer-Verlag (1988).

[108] E. Teruel: Structure Theory of Weighted Place/Transition Net Systems: The Equal Conflict Hiatus. Dissertation, DIEI. Univ. Zaragoza (1994).

[109] E. Teruel, P. Chrzastowski-Wachtel, J.M. Colom, M. Silva: On weighted T-systems. Proc. of Petri Nets'92, Sheffield, LNCS 616:348-367, Springer-Verlag (1992).

[110] E. Teruel, J.M. Colom, M. Silva: Linear Analysis of Deadlock-Freeness of Petri Net Models. Proc. of the IIth European Control Conference '93, Groningen, 2:513-518, North Holland (1993).

[111] E. Teruel, J.M. Colom, M. Silva: Modelling and Analysis of Deterministic Concurrent Systems with Bulk Services and Arrivals. In Decentralized and Distributed Systems, IFIP Transactions A-39:213-224, Elsevier (1994).

[112] E. Teruel, M. Silva: Liveness and Home States in Equal Conflict Systems. Proc. of Petri Nets'93, Chicago, LNCS 691:415-432, Springer-Verlag (1993).

[113] E. Teruel, M. Silva: Well-formedness of Equal Conflict Systems. Proc. of Petri Nets'94, Zaragoza, LNCS 815:491-510, Springer-Verlag (1994).

[114] E. Teruel, M. Silva: Structure Theory of Equal Conflict Systems. To appear in Theoretical Computer Science (1995).

[115] E. Teruel, M. Silva, J.M. Colom, J. Campos: Functional and Performance Analysis of Cooperating Sequential Processes. Procs. of the XIth Int. Conf. on Analysis and Optimization of Systems. Discrete Event Systems. Lecture Notes in Control and Information Sciences 199:169-175, Springer-Verlag (1994).

[116] P.S. Thiagarajan: A trace based extension of Linear Time Temporal Logic. Proc. of the IXth annual IEEE symposium on Logic in Computer Science (LICS) (1994).

[117] Th. Thielke: Modelchecking als Komponente der petrinetzbasierten Entwicklungs- und Programmierumgebung PEP. In Algorithmen und Werkzeuge für Petrinetze, (eds. J. Desel, A. Oberweis, W. Reisig), Workshop der GI-Fachgruppe 0.01 "Petrinetze und verwandte Systemmodelle", Bericht 309 Universität Karlsruhe, Berlin (1994).

[118] J. Vautherin: Parallel systems specification with colored Petri nets and algebraic specification. LNCS 266, Springer-Verlag (1987).

Design of Real-Time Systems: Interface between Duration Calculus and Program Specifications*

E.-R. Olderog and M. Schenke
FB Informatik, Universität Oldenburg
D-26111 Oldenburg, Germany

Abstract

We present a transformational approach to the design of real-time systems. The starting point are requirements formulated in a subset of Duration Calculus called *implementables* and the target are program specifications in a language SL that combines regular expressions with action systems and time conditions. While Duration Calculus is state-based, SL is event-based and can be seen as a stepping stone towards a timed occam-like programming language. The approach is illustrated by the example of a computer controlled gas burner.

1 Introduction

For systems with real-time restrictions a variety of specification formalisms have been developed, among them process algebraic approaches [2, 11], the temporal agent model [18], duration calculus [20], generalised Hoare triples and metrical temporal logic [7].

However, the design of realistic systems typically requires a suitable combination of several of such specification techniques each of which is best suited for a certain level of abstraction. This raises the risk of introducing errors when crossing the interfaces between different levels of abstraction. Eliminating these risks has been the major motivation of the ESPRIT project ProCoS II "Provably Correct Systems" [6]. In this project the links between representatives for several levels of abstraction covering requirement, specification, programming, machine code and hardware have been studied. For the description of time dependent requirements the Duration Calculus (DC) [20], a real-time interval temporal logic, has been developed. At the programming level a timed occam-like language [8, 4] has been chosen.

Since the formalisms of DC and occam are on very distant abstract levels, intermediate levels have been developed in the ProCoS project. On the one hand, a subset of Duration Calculus called *implementables* together with a methodology how such implementables can be obtained form general DC formulas has been introduced by [15]. On the other hand, a program specification language SL that is more abstract and flexible than occam itself has

*This work is partially funded by the Commission of the European Communities (CEC) under the ESPRIT Basic Research Action No. 7071: "ProCoS II: Provably Correct Systems".

been developed by [13, 16]. The specification language SL combines regular expressions with ideas from action systems [1] and with time conditions; it allows to describe the distributed architecture of the intended implementation of the real-time system and the timed communication between the different components.

In this paper we consider the interface between DC implementables and SL. While DC and hence DC implementables are state-based, SL and occam are event-based. We present a transformational approach how to gradually transform a requirements' description given in DC implementables into a program specification given in SL. How to transform SL specifications into occam-like programs is reported elsewhere [17].

2 DC and Gas Burner

We illustrate our approach by the example of a computer controlled gas burner due to [14]. In this paper we consider a version of that example described in [6]. The gas burner is controlled through a thermostat, and can directly control a gas valve and monitor the flame. This physical system is illustrated by the following diagram taken from [6]:

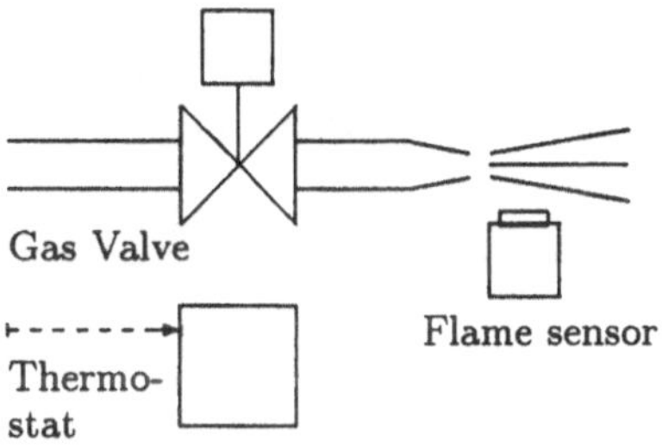

Besides several functional requirements the gas burner should satisfy the following safety requirement:

Safe: Gas must not leak for more than 4 seconds in any 30 second period.

A formal description of a real-time system using Duration Calculus starts by choosing a number of time dependent states variables or observables **obs** of a certain type. An interpretation $\mathcal{I}$ assigns to each state variable a function

$$\mathbf{obs}_{\mathcal{I}} : \mathbf{Time} \rightarrow \mathbf{D}$$

where **Time** is the time domain, here the non-negative reals, and **D** is the type of **obs**. For the gas burner we use the following Boolean observables:

$$\mathbf{hr}, \mathbf{gas}, \mathbf{fl} : \mathbf{Time} \rightarrow \mathbf{Bool}$$

which express the state of the thermostat (heat request), the gas valve and the flame.

State assertions $\mathbf{P}$ are obtained by applying propositional connectives to elementary assertions of the form $\mathbf{obs} = \mathbf{v}$. For a given interpretation $\mathcal{I}$ state assertions denote functions

$$\mathbf{P}_{\mathcal{I}} : \mathbf{Time} \rightarrow \mathbf{Bool}$$

Leakage is modelled by the state assertion $\mathbf{Leak} \stackrel{\text{def}}{=} \mathbf{gas} \wedge \neg \mathbf{fl}$. For a given interpretation of **fl** and **gas** the semantics of **Leak** can be illustrated by the following timing diagram:

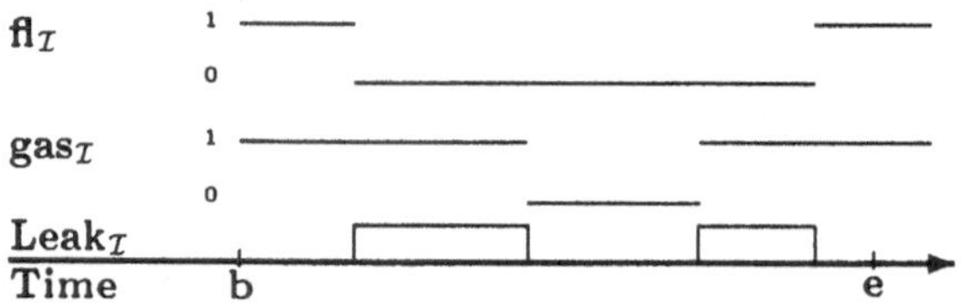

Duration formulas $\mathbf{F}$ are evaluated in a given interpretation $\mathcal{I}$ and a given time interval $[\mathbf{b}, \mathbf{e}] \subseteq \mathbf{Time}$. The basic syntax of duration formulas is as follows:

- **Duration:** $\int \mathbf{P} = \mathbf{k}$ expresses that the duration of the state assertion $\mathbf{P}$ in $[\mathbf{b}, \mathbf{e}]$ is $\mathbf{k}$. Semantically, duration is the measurement $\int_{\mathbf{b}}^{\mathbf{e}} \mathbf{P}_{\mathcal{I}}(\mathbf{t})\mathbf{dt}$.

- **Chop:** The composite duration formula $\mathbf{F}_1 \; ; \; \mathbf{F}_2$ (read $\mathbf{F}_1$ "chop" $\mathbf{F}_2$) holds in $[\mathbf{b}, \mathbf{e}]$ if this interval can be divided into an initial subinterval $[\mathbf{b}, \mathbf{m}]$ where $\mathbf{F}_1$ holds and a final subinterval $[\mathbf{m}, \mathbf{e}]$ where $\mathbf{F}_2$ holds.

- **Connectives:** Duration formulas are closed under propositional connectives.

Besides this basic syntax various abbreviations are used:

length:	ℓ	$\stackrel{\text{def}}{=}$	$\int \mathbf{true}$
point interval:	$\lceil\rceil$	$\stackrel{\text{def}}{=}$	$\ell = 0$
everywhere:	$\lceil \mathbf{P} \rceil$	$\stackrel{\text{def}}{=}$	$\int \mathbf{P} = \ell \wedge \ell > 0$
somewhere:	$\Diamond \mathbf{F}$	$\stackrel{\text{def}}{=}$	$\mathbf{true} \; ; \; \mathbf{F} \; ; \; \mathbf{true}$
always:	$\Box \mathbf{F}$	$\stackrel{\text{def}}{=}$	$\neg \Diamond \neg \mathbf{F}$

A duration formula $\mathbf{F}$ *holds* in an interpretation $\mathcal{I}$ if $\mathbf{F}$ evaluates to true in $\mathcal{I}$ and every interval of the form $[0, \mathbf{t}]$ with $\mathbf{t} \in \mathbf{Time}$.

To formalize the safety requirement we have to express that for every interval $[\mathbf{b}, \mathbf{e}]$ of time of length $\mathbf{e} - \mathbf{b} \leq 30$ the duration of **Leak** within that interval is at most 4. This is done as follows:

$$\mathbf{Safe} \stackrel{\text{def}}{=} \Box (\ell \leq 30 \Rightarrow \int \mathbf{Leak} \leq 4)$$

3 DC Implementables

In general it is quite difficult to find implementations that satisfy DC requirements involving durations such as $\int \mathbf{Leak}$. Therefore in the ProCoS project a subset of DC has been developed which is much closer to an implementation. The formulas of this subset are called *implementables* [15].

Implementables make use of the following abbreviations:

- *followed-by*: $\mathbf{F} \longrightarrow \lceil \mathbf{P} \rceil \stackrel{\text{def}}{=} \Box\neg(\mathbf{F}\,;\ \lceil \neg\mathbf{P} \rceil)$
- *timed leads-to*: $\mathbf{F} \stackrel{\mathbf{t}}{\longrightarrow} \lceil \mathbf{P} \rceil \stackrel{\text{def}}{=} (\mathbf{F} \wedge \ell = \mathbf{t}) \longrightarrow \lceil \mathbf{P} \rceil$
- *timed up-to*: $\mathbf{F} \stackrel{\leq \mathbf{t}}{\longrightarrow} \lceil \mathbf{P} \rceil \stackrel{\text{def}}{=} (\mathbf{F} \wedge \ell \leq \mathbf{t}) \longrightarrow \lceil \mathbf{P} \rceil$

As before we have $\mathbf{t} \in \mathbf{Time}$. Intuitively, $\mathbf{F} \longrightarrow \lceil \mathbf{P} \rceil$ expresses that whenever a pattern given by a formula **F** is observed, then it will be "followed by" an interval where **P** holds. In the "leads-to" form this pattern has a fixed length **t**, and in the "up-to" form the pattern is bounded by a length "up to" **t**.

Implementables now stipulate that *phases* of an abstract control automaton denoted by π or $\pi_{\mathbf{i}}$ have been identified so that other state assertions denoted by φ can be related to these phases. Implementables suppose that each requirement is given in one of the following formats:

- *Initialisation*; $\lceil\rceil \vee \lceil \varphi \rceil$; **true**
  saying that the system must start in a state where φ holds.
- *Sequencing*: $\lceil \pi \rceil \longrightarrow \lceil \pi \vee \pi_1 \vee \ldots \vee \pi_{\mathbf{n}} \rceil$
  saying that a system in a phase π can either remain in this phase or at most evolve into a phase $\pi_1 \vee \ldots \vee \pi_{\mathbf{n}}$.
- *Progress*: $\lceil \pi \wedge \varphi \rceil \stackrel{\mathbf{t}}{\longrightarrow} \lceil \neg\pi \rceil$
  saying that a system in a phase π must leave this phase within **t** time units as soon as other observables fulfill the formula φ.
- *Stability*: $\lceil \neg\pi \rceil$; $\lceil \pi \wedge \varphi \rceil \stackrel{\leq \mathbf{t}}{\longrightarrow} \lceil \pi \vee \pi_1 \vee \ldots \vee \pi_{\mathbf{n}} \rceil$
  saying that if the system enters into a phase π while φ holds it is guaranteed for **t** time units to evolve only to phases where $\pi \vee \pi_1 \vee \ldots \vee \pi_{\mathbf{n}}$ (if there are no $\pi_{\mathbf{i}}$ this a condition for the stability of phase π). The **t** can be dropped. Then there is no upper bound for the guarantee.
- *Synchronisation*: $\lceil \pi_1 \vee \ldots \vee \pi_{\mathbf{n}} \rceil \stackrel{\mathbf{t}}{\longrightarrow} \lceil \varphi \rceil$
  saying that if the observable under consideration fulfills $\pi_1 \vee \ldots \vee \pi_{\mathbf{n}}$ then within **t** time units other observables must fulfill φ.

The individual requirements of a system are then joined together by a conjunction. Syntactically we shall adopt a Z-style notation for the implementables.

For implementables there is a fully developed theory how they can be obtained from general DC formulas [15]. Also implementables are carefully designed to fulfill a variety of algebraic laws so that they can be used to calculate further properties of the intended systems.

We shall now describe our running example by means of implementables. The phases are modelled as the values of a distinguished observable, here called **phase** and ranging over values **Idle, Purge, Ignite, Burn**. Additionally we observe **hr, fl, gas** and need a time constant ε. Using a Z-like schema notation, we record this as follows:

GB_Intern

$\mathbf{hr}, \mathbf{fl}, \mathbf{gas} : \mathbf{Time} \to \mathbf{Bool}$
$\mathbf{phase} : \mathbf{Time} \to \{\mathbf{Idle}, \mathbf{Purge}, \mathbf{Ignite}, \mathbf{Burn}\}$
$\varepsilon : \mathbf{Time}$

Formally, a phase is a state assertions of the form **phase = Value** which will be abbreviated to **value**. With this convention the required behaviour of the gas burner is specified by

GB_0

GB_Intern

$\mathbf{Init\_0} \wedge \mathbf{Seq\_0} \wedge \mathbf{Prog\_0} \wedge \mathbf{Stab\_0} \wedge \mathbf{Syn\_0}$

where the initial, sequencing, progress, stability and synchronisation requirements **Init_0**, **Seq_0**, **Prog_0**, **Stab_0**, **Syn_0** are given as follows. Initially the burner is idle, there is no heat request, no gas is leaking and no flame is burning.

Init_0

$\lceil\rceil \vee \lceil \mathbf{idle} \rceil; \mathbf{true}$
$\lceil\rceil \vee \lceil \neg \mathbf{hr} \rceil; \mathbf{true}$
$\lceil\rceil \vee \lceil \neg \mathbf{fl} \rceil; \mathbf{true}$
$\lceil\rceil \vee \lceil \neg \mathbf{gas} \rceil; \mathbf{true}$

The burner cycles (in this order) through the four phases **idle** (nothing is being done), **purge** (a short phase during which a dangerously high concentration of gas can disperse, before the flame is switched on), **ignite** (the gas burner is trying to ignite the gas), **burn** (the gas is burning).

Seq_0

$\lceil \mathbf{idle} \rceil \longrightarrow \lceil \mathbf{idle} \vee \mathbf{purge} \rceil$
$\lceil \mathbf{purge} \rceil \longrightarrow \lceil \mathbf{purge} \vee \mathbf{ignite} \rceil$
$\lceil \mathbf{ignite} \rceil \longrightarrow \lceil \mathbf{ignite} \vee \mathbf{burn} \rceil$
$\lceil \mathbf{burn} \rceil \longrightarrow \lceil \mathbf{burn} \vee \mathbf{idle} \rceil$

The **purge** phase must not last longer than 30 s and the **ignite** phase must not last much longer than 1 s.

Prog_0

$$\lceil \mathbf{purge} \rceil \xrightarrow{30} \lceil \neg \mathbf{purge} \rceil$$
$$\lceil \mathbf{ignite} \rceil \xrightarrow{1} \lceil \neg \mathbf{ignite} \rceil$$

The according lower bounds for the **purge** and **ignite** phases are guaranteed. The burner remains idle as long as there is no heat request, it remains in the **burn** phase as long as there is a flame burning (no flame failure) and the heat request is not turned off.

Stab_0

$$\lceil \neg \mathbf{idle} \rceil ; \lceil \mathbf{idle} \wedge \neg \mathbf{hr} \rceil \longrightarrow \lceil \mathbf{idle} \rceil$$
$$\lceil \neg \mathbf{purge} \rceil ; \lceil \mathbf{purge} \rceil \xrightarrow{\leq 30} \lceil \mathbf{purge} \rceil$$
$$\lceil \neg \mathbf{ignite} \rceil ; \lceil \mathbf{ignite} \rceil \xrightarrow{\leq 1} \lceil \mathbf{ignite} \rceil$$
$$\lceil \neg \mathbf{burn} \rceil ; \lceil \mathbf{burn} \wedge \mathbf{fl} \wedge \mathbf{hr} \rceil \longrightarrow \lceil \mathbf{burn} \rceil$$

The **idle** phase is left if a heat request occurs. In the **idle** or **purge** phases the gas is switched off almost immediately. In the **ignite** or **burn** phases the gas is switched on almost immediately. The **burn** phase is left if a flame failure occurs or heat request is turned off.

Syn_0

$$\lceil \mathbf{idle} \wedge \mathbf{hr} \rceil \xrightarrow{\varepsilon} \lceil \neg \mathbf{idle} \rceil$$
$$\lceil \mathbf{idle} \vee \mathbf{purge} \rceil \xrightarrow{\varepsilon} \lceil \neg \mathbf{gas} \rceil$$
$$\lceil \mathbf{ignite} \vee \mathbf{burn} \rceil \xrightarrow{\varepsilon} \lceil \mathbf{gas} \rceil$$
$$\lceil \mathbf{burn} \wedge \neg \mathbf{hr} \rceil \xrightarrow{\varepsilon} \lceil \neg \mathbf{burn} \rceil$$
$$\lceil \mathbf{burn} \wedge \neg \mathbf{fl} \rceil \xrightarrow{\varepsilon} \lceil \neg \mathbf{burn} \rceil$$

How these requirements of the gas burner can be derived from more abstract ones can be seen in [6]. For example **Safe** will be guaranteed by the following informal argument:

A leak is possible only during the **ignite** phase and in case of a flame failure during the **burn** phase. The time for both situations is bounded by the implementables with the bounds 1s and εs. Each consecutive occurrence of two ignites or flame failures is separated by a **purge** phase which lasts 30s. So the altogether leak time in any 30s interval does not exceed $1 + \varepsilon$s. The formal proof relies on the properties of the standard forms.

Architectural idea. At this stage we have to take a decision about the distributed architecture of the implementation, i.e. which of the above mentioned observables are to be controlled by one component of a distributed system. We could have aimed at a highly parallel implementation in which each observable corresponds to one parallel component or at a sequential implementation. For didactic reasons we make a compromise, so that our overall plan is now to achieve an architecture with three components:

- a *heat controller* monitoring the observable *hr*,
- a *flame controller* monitoring the observable *fl*,
- a *main controller* monitoring the observables *phase* and *gas*.

4 Interleaving

Since the main controller now has to monitor the two observables *phase* and *gas* their a priori independent changes have to be sequentialised or *interleaved*. Formally, this amounts to building the Cartesian product space of all single observables and construct the subautomaton of all reachable states. This automaton is controlled by a new observable *newphase*. Its new states are encodings of the old ones, e.g.

$$\mathbf{newphase} = \mathbf{Idle} \Leftrightarrow \mathbf{phase} = \mathbf{Idle} \wedge \neg\mathbf{gas},$$
$$\mathbf{newphase} = \mathbf{Waitidle} \Leftrightarrow \mathbf{phase} = \mathbf{Idle} \wedge \mathbf{gas}.$$

In general we shall call such relationships *linking invariants*. The above invariant we call $\mathbf{Link}_1$.

As with **phase** we abbreviate **newphase** = **Value** to **value**. Furthermore some simple estimations for the quantitative time restrictions can be obtained easily. The gas burner thereafter looks as follows:

GB_1

GB_Intern

$\mathbf{Init\_0} \wedge \mathbf{Seq\_1} \wedge \mathbf{Prog\_1} \wedge \mathbf{Stab\_0} \wedge \mathbf{Syn\_0}$

with the following new sequencing and progress synchronisation requirements:

Seq_1

$\lceil \mathbf{waitidle} \rceil \longrightarrow \lceil \mathbf{waitidle} \vee \mathbf{idle} \rceil$
$\lceil \mathbf{idle} \rceil \longrightarrow \lceil \mathbf{idle} \vee \mathbf{purge} \rceil$
$\lceil \mathbf{purge} \rceil \longrightarrow \lceil \mathbf{purge} \vee \mathbf{waitignite} \rceil$
$\lceil \mathbf{waitignite} \rceil \longrightarrow \lceil \mathbf{waitignite} \vee \mathbf{ignite} \rceil$
$\lceil \mathbf{ignite} \rceil \longrightarrow \lceil \mathbf{ignite} \vee \mathbf{burn} \rceil$
$\lceil \mathbf{burn} \rceil \longrightarrow \lceil \mathbf{burn} \vee \mathbf{waitidle} \rceil$

Prog_1

$\lceil \mathbf{waitidle} \rceil \xrightarrow{\varepsilon} \lceil \mathbf{idle} \rceil$
$\lceil \mathbf{purge} \rceil \xrightarrow{30} \lceil \neg\mathbf{purge} \rceil$
$\lceil \mathbf{waitignite} \rceil \xrightarrow{\varepsilon} \lceil \mathbf{ignite} \rceil$
$\lceil \mathbf{ignite} \rceil \xrightarrow{1} \lceil \neg\mathbf{ignite} \rceil$

The old initial and stability requirements have not changed, except that $\lceil\rceil \vee \lceil\neg\mathbf{gas}\rceil$; **true** has become superfluous because it is implied by the initial value of newphase. The old synchronisation requirements have become redundant now, but in general there will remain synchronisation requirements, and indeed we shall introduce some in the next section.

In this step we have defined a linking invariant $\mathbf{Link_1}$ with the property

$$\mathbf{GB_1} \wedge \mathbf{Link_1} \;\Rightarrow\; \mathbf{GB_0}.$$

Also in future steps we shall use linking invariants $\mathbf{Link_i}$ for simulation purposes with

$$\mathbf{GB_i} \wedge \mathbf{Link_i} \;\Rightarrow\; \mathbf{GB_{i-1}}.$$

5 Phase Splitting

In an occam-like distributed system each of the components heat controller, flame controller and main controller should work concurrently and exchange information only when necessary. For example, look at the second progress requirement:

(P) $\lceil \mathbf{idle} \wedge \mathbf{hr} \rceil \xrightarrow{\varepsilon} \lceil \neg \mathbf{idle} \rceil$.

To change the phase in the main controller information about **hr** of the heat controller is needed. In an occam-like distributed system this information exchange has to be achieved by a synchronous communication. To prepare for this communication, we would like to split the phase **idle** into one part where this exchange of information has not yet taken place and a subsequent part where the change of the heat request has been noticed.

In general we have the following situation: As a preparation for a parallel decomposition let the observables be partitioned into several blocks. Suppose a progress or stability constraint of a phase π from a block **B** contains a condition φ involving observables not local to **B**. Then it is necessary that the component described by **B** knows when φ becomes true. To this end, the phase π is split into two parts π_1 and π_2 where π_2 indicates that the component under consideration has been informed about the occurence of φ and π_1 indicates the absence of such an information.

Formally, a phase π is *split* or sequentially decomposed into two parts π_1 and π_2, abbreviated

$$\pi = \mathbf{Seq}[\pi_1, \pi_2],$$

if the following holds:

- (S1) $\Box(\lceil\pi\rceil \;\Leftrightarrow\; \lceil\pi_1 \vee \pi_2\rceil)$
- (S2) $\pi_1 \wedge \pi_2 \Rightarrow \mathbf{false}$
- (S3) $\lceil\pi_1\rceil \longrightarrow \lceil\pi_1 \vee \pi_2\rceil$.

- (S4) $\lceil \pi_2 \rceil \longrightarrow \lceil \pi_2 \vee \neg\pi \rceil$.

- (S5) $\lceil \neg\pi \rceil \longrightarrow \lceil \neg\pi \vee \pi_1 \rceil$.

Here S1 and S2 define a simulation mapping and S3)–S5 are sequencing constraints. We define

$$\mathbf{Link}_2 = \mathbf{S1} \wedge \mathbf{S2}.$$

Definition 1 Suppose that $\pi = \mathbf{Seq}[\pi_1, \pi_2]$.
1.) A progress constraint $\lceil \pi \wedge \varphi \rceil \xrightarrow{t} \lceil \neg\pi \rceil$ is *implemented by phase splitting*, if the synchronisation constraint $\lceil \pi_1 \wedge \varphi \rceil \xrightarrow{t/2} \lceil \neg\pi_1 \rceil$ and the progress constraint $\lceil \pi_2 \rceil \xrightarrow{t/2} \lceil \neg\pi_2 \rceil$ hold.
2.) A stability constraint $\lceil \neg\pi \rceil; \lceil \pi \wedge \neg\varphi \rceil \longrightarrow \lceil \pi \rceil$ is *implemented by phase splitting*, if the stability constraint $\lceil \neg\pi_1 \rceil; \lceil \pi_1 \wedge \neg\varphi \rceil \longrightarrow \lceil \pi_1 \rceil$ holds.
3.) An initialisation constraint $\lceil\rceil \vee \lceil \pi \rceil;$ **true** is *implemented by phase splitting*, if the initialisation constraint $\lceil\rceil \vee \lceil \pi_1 \rceil;$ **true** holds. □

In our example we split the phase **idle** as follows:

(∗) **idle = Seq[stayidle, leaveidle]**

The progress requirement

(**P**) $\lceil \mathbf{idle} \wedge \mathbf{hr} \rceil \xrightarrow{\varepsilon} \lceil \neg\mathbf{idle} \rceil$

of the gas burner is now implemented by means of the synchronisation requirement

(**P1**) $\lceil \mathbf{stayidle} \wedge \mathbf{hr} \rceil \xrightarrow{\varepsilon/2} \lceil \neg\mathbf{stayidle} \rceil$

and the progress requirement

(**P2**) $\lceil \mathbf{leaveidle} \rceil \xrightarrow{\varepsilon/2} \lceil \neg\mathbf{leaveidle} \rceil$

The stability requirement

(**S**) $\lceil \neg\mathbf{idle} \rceil; \lceil \mathbf{idle} \wedge \neg\mathbf{hr} \rceil \longrightarrow \lceil \mathbf{idle} \rceil$

is now implemented by means of the stability requirement

(**S1**) $\lceil \neg\mathbf{stayidle} \rceil; \lceil \mathbf{stayidle} \wedge \neg\mathbf{hr} \rceil \longrightarrow \lceil \mathbf{stayidle} \rceil$

It can be checked that indeed

$$(*) \wedge \mathbf{P1} \wedge \mathbf{P2} \Rightarrow \mathbf{P}$$
$$(*) \wedge \mathbf{S1} \Rightarrow \mathbf{S}$$

holds.

Phase Splitting
$\mathbf{Req}_1 = \mathbf{Init} \wedge \mathbf{Seq} \wedge \mathbf{Prog} \wedge \mathbf{Stab} \wedge \mathbf{Syn}$ $\Uparrow$ $\mathbf{Req}_2 = \mathbf{Init}' \wedge \mathbf{Seq}' \wedge \mathbf{Prog}' \wedge \mathbf{Stab}' \wedge \mathbf{Syn}' \wedge \mathbf{Link}$
if the constraints in **Prog** are identical to constraints in **Prog′** or implemented by phase splitting, accordingly **Init** and **Init′**, **Stab** and **Stab′**. **Seq′** equals **Seq** plus the constraints of type (**S3**) – (**S5**). **Link** is a linking constraint of type (**S1**) ∧ (**S2**). (We also say that $\mathbf{Req}_1$ is *split into* $\mathbf{Req}_2$.)

Here and elsewhere in this paper $\Leftarrow$ reads "implements" and is logical implication between DC formulas under the declared types of the observables, in our example given by **GB_Intern**.

In the main controller component we deal with **¬hr** and **¬fl** simultaneously by splitting **burn** into the new phases **stayburn** and **leaveburn** with **leaveburn** representing **¬hr** ∨ **¬fl**.

Then the complete picture we get for the gas burner is

GB_2

GB_Intern

Init_2 ∧ **Seq_2** ∧ **Prog_2** ∧ **Stab_2** ∧ **Syn_2**

Init_2

$\lceil\rceil \vee \lceil \mathbf{stayidle} \rceil ; \mathbf{true}$

$\lceil\rceil \vee \lceil \neg\mathbf{hr} \rceil ; \mathbf{true}$

$\lceil\rceil \vee \lceil \neg\mathbf{fl} \rceil ; \mathbf{true}$

Seq_2

$\lceil \mathbf{waitidle} \rceil \longrightarrow \lceil \mathbf{waitidle} \vee \mathbf{stayidle} \rceil$

$\lceil \mathbf{stayidle} \rceil \longrightarrow \lceil \mathbf{stayidle} \vee \mathbf{leaveidle} \rceil$

$\lceil \mathbf{leaveidle} \rceil \longrightarrow \lceil \mathbf{leaveidle} \vee \mathbf{purge} \rceil$

$\lceil \mathbf{purge} \rceil \longrightarrow \lceil \mathbf{purge} \vee \mathbf{waitignite} \rceil$

$\lceil \mathbf{waitignite} \rceil \longrightarrow \lceil \mathbf{waitignite} \vee \mathbf{ignite} \rceil$

$\lceil \mathbf{ignite} \rceil \longrightarrow \lceil \mathbf{ignite} \vee \mathbf{stayburn} \rceil$

$\lceil \mathbf{stayburn} \rceil \longrightarrow \lceil \mathbf{stayburn} \vee \mathbf{leaveburn} \rceil$

$\lceil \mathbf{leaveburn} \rceil \longrightarrow \lceil \mathbf{leaveburn} \vee \mathbf{waitidle} \rceil$

Prog_2

$\lceil \mathbf{waitidle} \rceil \xrightarrow{\varepsilon} \lceil \mathbf{stayidle} \rceil$

$\lceil \mathbf{leaveidle} \rceil \xrightarrow{\varepsilon/2} \lceil \neg \mathbf{leaveidle} \rceil$

$\lceil \mathbf{purge} \rceil \xrightarrow{30} \lceil \neg \mathbf{purge} \rceil$

$\lceil \mathbf{waitignite} \rceil \xrightarrow{\varepsilon} \lceil \mathbf{ignite} \rceil.$

$\lceil \mathbf{ignite} \rceil \xrightarrow{1} \lceil \neg \mathbf{ignite} \rceil$

$\lceil \mathbf{leaveburn} \rceil \xrightarrow{\varepsilon/2} \lceil \neg \mathbf{leaveburn} \rceil$

Stab_2

$\lceil \neg \mathbf{stayidle} \rceil; \lceil \mathbf{stayidle} \wedge \neg \mathbf{hr} \rceil \longrightarrow \lceil \mathbf{stayidle} \rceil$

$\lceil \neg \mathbf{purge} \rceil; \lceil \mathbf{purge} \rceil \xrightarrow{\leq 30} \lceil \mathbf{purge} \rceil$

$\lceil \neg \mathbf{ignite} \rceil; \lceil \mathbf{ignite} \rceil \xrightarrow{\leq 1} \lceil \mathbf{ignite} \rceil$

$\lceil \neg \mathbf{stayburn} \rceil; \lceil \mathbf{stayburn} \wedge \mathbf{hr} \wedge \mathbf{fl} \rceil \longrightarrow \lceil \mathbf{stayburn} \rceil$

Syn_2

$\lceil \mathbf{stayidle} \wedge \mathbf{hr} \rceil \xrightarrow{\varepsilon/2} \lceil \neg \mathbf{stayidle} \rceil$

$\lceil \mathbf{stayburn} \wedge \neg \mathbf{hr} \rceil \xrightarrow{\varepsilon/2} \lceil \neg \mathbf{stayburn} \rceil$

$\lceil \mathbf{stayburn} \wedge \neg \mathbf{fl} \rceil \xrightarrow{\varepsilon/2} \lceil \neg \mathbf{stayburn} \rceil$

6 Introduction of Events

Up to now we have a purely state based description of the system, but the target occam-like programming language is event based, i.e. only communications are observable. Events are introduced here as certain state changes. An *event* will be described by a labelled transition

$$\mathbf{pre} \xrightarrow{\mathbf{ev}} \mathbf{post}$$

where **ev** is an arbitrarily chosen event name or *communication* and **pre** and **post** are state assertions describing the enable condition and the effect of the event. By minor changes of the syntax these labelled transitions can be transformed into a construct of the specification language SL, called *communication assertions.*

Example 2 The sequencing constraint $\lceil \mathbf{waitidle} \rceil \longrightarrow \lceil \mathbf{waitidle} \vee \mathbf{stayidle} \rceil$ of the main controller can be modelled by the event

$$\mathbf{waitidle} \xrightarrow{\mathbf{GasOff}} \mathbf{stayidle}.$$

Similarly all other sequencing events of the main controller can be modelled by events.

In the heat controller only the Boolean observable **hr** may change. These changes can be modelled by two events:

$$\neg\mathbf{hr} \stackrel{\mathbf{HeatOn}}{\longrightarrow} \mathbf{hr} \quad \text{and} \quad \mathbf{hr} \stackrel{\mathbf{HeatOff}}{\longrightarrow} \neg\mathbf{hr}.$$

Similarly the changes of the Boolean observable **fl** of the flame controller are dealt with. □

In Duration Calculus the semantics of events can be described by introducing a distinguished observable **tr** that ranges over *traces*, i.e. finite sequences of events names or communications of a given set **Comm**:

$$\mathbf{tr} : \mathbf{Time} \to \mathbf{Comm}^*$$

Initially, **tr** should be empty:

$$\lceil\rceil \vee \lceil \mathbf{tr} = \epsilon \rceil;\ \mathbf{true}.$$

For a given set of labelled transitions

$$\mathbf{CA} = \{\mathbf{pre}_1 \stackrel{\mathbf{ev}_1}{\longrightarrow} \mathbf{post}_1, ..., \mathbf{pre}_n \stackrel{\mathbf{ev}_n}{\longrightarrow} \mathbf{post}_n\}$$

and an event name $\mathbf{ev} \in \{\mathbf{ev}_1, ..., \mathbf{ev}_n\}$ we define:

$$\mathbf{pre}(\mathbf{ev}) \stackrel{\text{def}}{=} \bigwedge_{\mathbf{ev}=\mathbf{ev}_i} \mathbf{pre}_i \quad \text{and} \quad \mathbf{post}(\mathbf{ev}) \stackrel{\text{def}}{=} \bigwedge_{\mathbf{ev}=\mathbf{ev}_i} \mathbf{post}_i.$$

Semantically **CA** is then identified with:

$$\begin{array}{lll} \mathbf{CA} & \stackrel{\text{def}}{=} & \bigwedge_{\mathbf{ev}} \quad (\Diamond \exists \mathbf{h} \bullet \lceil \mathbf{tr} = \mathbf{h} \rceil;\ \lceil \mathbf{tr} = \mathbf{h}.\mathbf{ev} \rceil \\ & & \qquad \Leftrightarrow \\ & & \qquad \Diamond \lceil \mathbf{pre}(\mathbf{ev}) \rceil;\ \lceil \mathbf{post}(\mathbf{ev}) \rceil) \end{array}$$

This duration formula formalises our intention that each event **ev** denotes a change from **pre(ev)** to **post(ev)**.

The old phases and the new events are linked by the following linking invariant:

$$\mathbf{Link}_3 \quad \equiv \quad (\mathbf{idle} \Leftrightarrow \mathbf{tr} \in \mathcal{L}[\![\mathbf{idle}]\!]) \wedge (\mathbf{purge} \Leftrightarrow ...)$$

where a phase π is described by the set of all traces of events that drive the system into π. This set is a language $\mathcal{L}[\![\pi]\!]$ over an alphabet consisting of events.

Lemma 3 (Decomposition) For any event **ev** and predicates $\mathbf{pre}_1$, $\mathbf{pre}_2$, $\mathbf{post}_1$, $\mathbf{post}_2$ labelled transitions can be decomposed as follows:

$$\{\mathbf{pre}_1 \wedge \mathbf{pre}_2 \stackrel{\mathbf{ev}}{\longrightarrow} \mathbf{post}_1 \wedge \mathbf{post}_2\} \quad \equiv \quad \{\mathbf{pre}_1 \stackrel{\mathbf{ev}}{\longrightarrow} \mathbf{post}_1, \mathbf{pre}_2 \stackrel{\mathbf{ev}}{\longrightarrow} \mathbf{post}_2\}.$$

Definition 4 A phase sequencing constraint $\lceil \pi \rceil \longrightarrow \lceil \pi \vee \pi_1 \vee ... \vee \pi_m \rceil$ and a set of labelled transitions $\mathbf{CA} = \{\mathbf{pre}_1 \stackrel{\mathbf{ev}_1}{\longrightarrow} \mathbf{post}_1, ..., \mathbf{pre}_n \stackrel{\mathbf{ev}_n}{\longrightarrow} \mathbf{post}_n\}$ are called *compatible* if for all $\mathbf{i} \in \{1, ..., \mathbf{n}\}$ whenever

$$\pi \Rightarrow \mathbf{pre}(\mathbf{ev}_i)$$

holds then also

$$\mathbf{post}(\mathbf{ev}_i) \Rightarrow \bigvee_{\mathbf{k}} \pi_{\mathbf{k}}.$$

A set of phase sequencing constraints and set **CA** of labelled transitions are *compatible* if each phase sequencing constraint is compatible with **CA**. □

Then we can state the following transformation rule:

Introduction of Events
$\mathbf{Req_1 = Init \wedge Seq \wedge Prog \wedge Stab \wedge Syn}$ $\Uparrow$ $\mathbf{Req_2 = Init \wedge CA \wedge Prog \wedge Stab \wedge Syn \wedge Link}$
where **Seq** is compatible with the set **CA** of labelled transitions and **Link** is a linking invariant like $\mathbf{Link_3}$.

Applying this transformation to the gas burner yields:

GB_3

GB_Intern

$\mathbf{Init\_2 \wedge CA\_3 \wedge Prog\_2 \wedge Stab\_2 \wedge Syn\_2}$

CA_3 for the main controller

$\mathbf{waitidle} \stackrel{GasOff}{\longrightarrow} \mathbf{stayidle}$

$\mathbf{leaveidle} \stackrel{\mathtt{startpurge}}{\longrightarrow} \mathbf{purge}$

$\mathbf{purge} \stackrel{\mathtt{startign}}{\longrightarrow} \mathbf{waitignite}$

$\mathbf{waitignite} \stackrel{GasOn}{\longrightarrow} \mathbf{ignite}$

$\mathbf{ignite} \stackrel{\mathtt{startburn}}{\longrightarrow} \mathbf{stayburn}$

$\mathbf{leaveburn} \stackrel{\mathtt{startidle}}{\longrightarrow} \mathbf{waitidle}$

CA_3 for the heat controller

$\neg\mathbf{hr} \stackrel{\mathbf{HeatOn}}{\longrightarrow} \mathbf{hr}$

$\mathbf{hr} \stackrel{\mathbf{HeatOff}}{\longrightarrow} \neg\mathbf{hr}$

CA_3 for the flame controller

$\neg\mathbf{fl} \stackrel{\mathbf{FlOn}}{\longrightarrow} \mathbf{fl}$

$\mathbf{fl} \stackrel{\mathbf{FlOff}}{\longrightarrow} \neg\mathbf{fl}$

7 Removal of Synchronisation

In our target occam-like programming language, concurrent components can synchronise only via internal communications. In the synchronisation requirements of GB_2 and hence GB_3, however, synchronisation is still by common inspection of observables from different controller components. For example, the synchronisation requirement

$$\lceil \mathbf{stayidle} \wedge \mathbf{hr} \rceil \xrightarrow{\varepsilon/2} \lceil \mathbf{leaveidle} \rceil$$

involves the observable **newphase** of the main controller and the observable **hr** of the heat controller. The idea is to replace this by two events with the same name, say **yesheat**:

- $\mathbf{stayidle} \xrightarrow{\mathbf{yesheat}} \mathbf{leaveidle}$ in the main controller,
- $\mathbf{hr} \xrightarrow{\mathbf{yesheat}} \mathbf{hr}$ in the heat controller,

the first one changing the phase in the main controller, the second one checking the condition in the heat controller. Following the occam paradigm, communications with the same name happen simultaneously and thus realise the desired communication. The timing restriction $\varepsilon/2$ is guaranteed by some assupmtion on the speed of the communication which is not elaborated upon here. An event like yesheat is called a *replacing event.*

We can visualise the architectural aspect described by the three controller components and the connecting communications as follows:

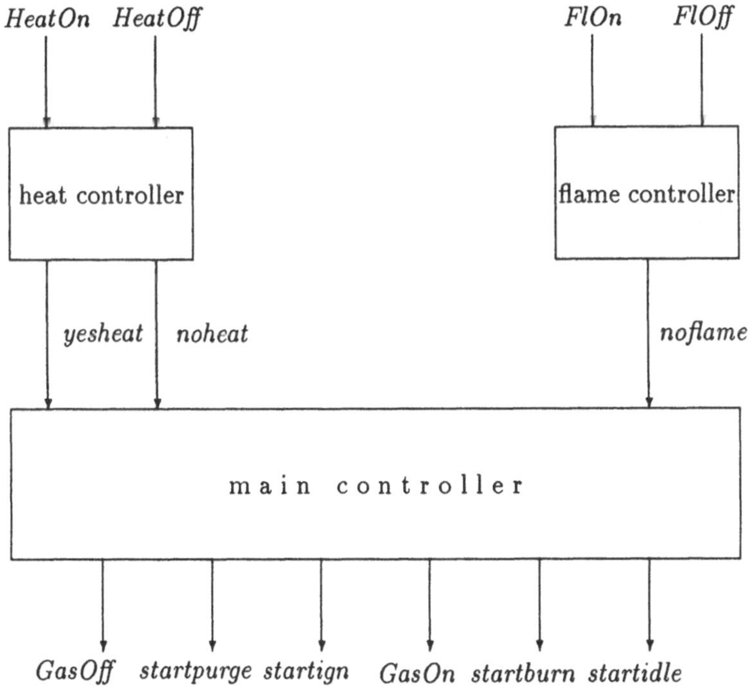

We use the following transformation rule where $\mathbf{Link}_4$ is a linking invariant defined analogously to the section on the introduction of events.

Removal of Synchronisation
$\mathbf{Req_1 = Init \wedge CA \wedge Prog \wedge Stab \wedge Syn}$ $\Uparrow$ $\mathbf{Req_2 = Init \wedge CA' \wedge Prog \wedge Stab \wedge Link}$
where **CA′** is the join of **CA** and a complete set of replacing events for **Syn** and **Link** is a linking invariant like $\mathbf{Link_4}$.

The complete view of the gas burner is now

GB_4

GB_Intern

Init_1 ∧ CA_4 ∧ Prog_2 ∧ Stab_2

CA_4 for the main controller

$\mathbf{waitidle} \xrightarrow{GasOff} \mathbf{stayidle}$

$\mathbf{stayidle} \xrightarrow{\mathbf{yesheat}} \mathbf{leaveidle}$

$\mathbf{leaveidle} \xrightarrow{\mathbf{startpurge}} \mathbf{purge}$

$\mathbf{purge} \xrightarrow{\mathbf{startign}} \mathbf{waitignite}$

$\mathbf{waitignite} \xrightarrow{GasOn} \mathbf{ignite}$

$\mathbf{ignite} \xrightarrow{\mathbf{startburn}} \mathbf{stayburn}$

$\mathbf{stayburn} \xrightarrow{\mathbf{noheat}} \mathbf{leaveburn}$

$\mathbf{stayburn} \xrightarrow{\mathbf{noflame}} \mathbf{leaveburn}$

$\mathbf{leaveburn} \xrightarrow{\mathbf{startidle}} \mathbf{waitidle}$

CA_4 for the heat controller

$\neg\mathbf{hr} \xrightarrow{\mathbf{HeatOn}} \mathbf{hr}$

$\mathbf{hr} \xrightarrow{\mathbf{yesheat}} \mathbf{hr}$

$\mathbf{hr} \xrightarrow{\mathbf{HeatOff}} \neg\mathbf{hr}$

$\neg\mathbf{hr} \xrightarrow{\mathbf{noheat}} \neg\mathbf{hr}$

CA_4 for the flame controller

$\neg\mathbf{fl} \xrightarrow{\mathbf{FlOn}} \mathbf{fl}$

$\mathbf{fl} \xrightarrow{\mathbf{FlOff}} \neg\mathbf{fl}$

$\neg\mathbf{fl} \xrightarrow{\mathbf{noflame}} \neg\mathbf{fl}$

The changes of the observable **newphase** under the corresponding events is summarised in the following diagram:

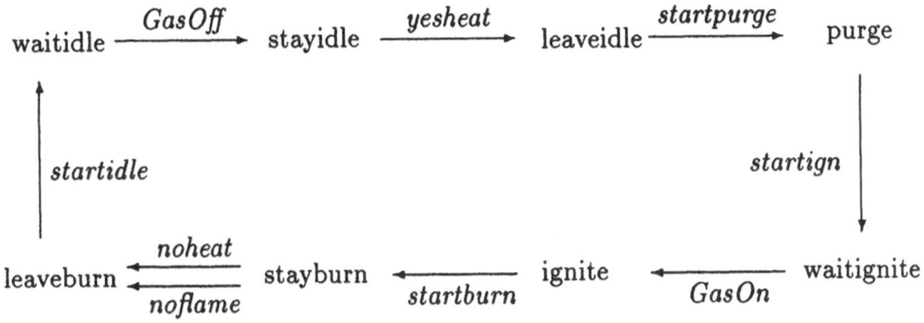

8 Introduction of Trace Assertions

Here we introduce a new part of a specification, the *trace assertions*. A trace assertion has the form

TRACE re

where **re** is a regular expression over the alphabet of event names. It states that the system should engage in the events mentioned in **re** only in the order as allowed by one of the words in the language described by **re**.

In this section we aim at eliminating communication assertions in favour of trace assertions. The decision which trace assertions should be introduced is arbitrary. In principle the whole information of a requirement in the format that we have reached so far can remain in the communication assertions. It is, however, clearer to separate a trace part describing the channel sequencing from a part describing the communicated values. Such a separation enhances the readability of the specification and eases the transformation from SL to occam, because trace assertions help very much to find a good structure for an occam program that implements a given specification. The rule that will be given in this section is a good heuristic which takes a set of appropriately chosen communication assertions and makes a trace assertion from them.

Definition 5 For a set $\mathbf{CA} = \{\pi_1 \xrightarrow{\mathbf{ev_1}} \pi'_1, \ldots, \pi_\mathbf{n} \xrightarrow{\mathbf{ev_n}} \pi'_\mathbf{n}\}$ of communication assertions obeying to

$$(*) \qquad \forall \mathbf{i}, \mathbf{j} : (\pi_\mathbf{i} \wedge \pi'_\mathbf{j} = \mathbf{false}) \vee (\pi_\mathbf{i} = \pi'_\mathbf{j})$$

we define the following automaton:

$$\mathbf{Aut(CA)} \stackrel{\text{def}}{=} (\mathbf{Alph(CA)}, \mathbf{States(CA)}, \rightarrow, \mathbf{Init})$$
$$\mathbf{Alph(CA)} \stackrel{\text{def}}{=} \{\mathbf{ev_1}, \ldots, \mathbf{ev_n}\}, \mathbf{States(CA)} \stackrel{\text{def}}{=} \{\pi_1, \ldots, \pi_\mathbf{n}\}, \mathbf{Init} \in \{\pi_1, \ldots, \pi_\mathbf{n}\},$$
$$\rightarrow \stackrel{\text{def}}{=} \{(\pi_\mathbf{i}, \mathbf{ev_i}, \pi_\mathbf{j}) \mid \mathbf{i} \in \{1, \ldots, \mathbf{n}\} \wedge \pi'_\mathbf{i} = \pi_\mathbf{j}\}$$

The condition $(*)$ is not a severe one. It can be replaced by weaker ones which in turn can be achieved by other rules. □

Introduction of Trace Assertions
$$\mathbf{Req_1 = Init \wedge TA \wedge CA \cup CA' \wedge Prog \wedge Stab}$$ $$\Uparrow$$ $$\mathbf{Req_2 = Init \wedge TA \cup \{ta\} \wedge CA \wedge Prog \wedge Stab}$$
where **ta** is of the form **TRACE pref re** and **re** is a regular expression describing the language of **Aut(CA′)**.

At this step we drop the initialisation information **Init**. In the gas burner it is already contained in the trace assertions (regular expressions are equivalent to automata which already contain initialisation information). In general we would need a step by which the initial values of certain local variables are defined. The rules lead to the following trace assertions and the complete view of the gas burner is now:

GB_5

GB_Intern

TA_5 ∧ Prog_2 ∧ Stab_2

where the set **TA_5** consists of three trace assertions, one for the main controller, one for the heat controller and one for the flame controller:

TA_5

**TRACE pref(yesheat.startpurge.startign.GasOn.startburn.
(noheat + noflame).startidle.GasOff)***

TRACE pref(HeatOn.yesheat*.HeatOff.noheat*)*

TRACE pref(FlOn.FlOff.noflame*)*

In general some communication assertions will remain. Only due to the simplicity of the gas burner specification all communication assertions could be removed in favour of trace assertions.

9 Introduction of Time Restrictions

As defined with $\mathbf{Link_3}$ a phase π can be described by the set of all traces of events that drive the system into π,the language $\mathcal{L}[\![\pi]\!]$. In our case $\mathcal{L}[\![\pi]\!]$ can be identified with a regular expression because we only refer to states that can be reached by prefixes of the regular language described by the trace assertions.

To express waiting time we use *time restrictions* of the form

$$\texttt{AFTER}\ \mathbf{re}\ \texttt{WAIT}\ (\mathbf{ev}, \mathbf{t})$$

which means that after engaging in a trace belonging to the language described by the regular expression **re**, the system will not engage in event **ev** before time **t** has elapsed.

Definition 6 The time restrictions $\texttt{AFTER}\ \mathcal{L}[\![\pi]\!]\ \texttt{WAIT}\ (\mathbf{ev_i}, \mathbf{t})$ with $\mathbf{i} = 1, ..., \mathbf{n}$ are called the *transformed* of the timed stability requirement

$$\lceil \neg\pi \rceil;\ \lceil \pi \rceil \xrightarrow{\leq \mathbf{t}} \lceil \pi \rceil$$

if the $\mathbf{ev_i}$ with $\mathbf{i} = 1, ..., \mathbf{n}$ are the only events by which π can be left. □

Introduction of Time Restrictions
$\mathbf{Req_1} = \mathbf{TA} \wedge \mathbf{CA} \wedge \mathbf{TR} \wedge \mathbf{Prog} \wedge \mathbf{Stab}$ $\Uparrow$ $\mathbf{Req_2} = \mathbf{TA} \wedge \mathbf{CA} \wedge \mathbf{TR} \cup \{\mathbf{tir_1}, ..., \mathbf{tir_n}\} \wedge \mathbf{Prog} \wedge \mathbf{Stab'}$
where $\{\mathbf{tir_1}, ..., \mathbf{tir_n}\}$ is the set of all transformed of some stability information, and $\mathbf{Stab'}$ is $\mathbf{Stab}$ without this stability information.

Example 7 In the gas burner the following stability requirements are unchanged from the beginning:

$$\lceil \neg\mathbf{purge} \rceil;\ \lceil \mathbf{purge} \rceil \xrightarrow{\leq 30} \lceil \mathbf{purge} \rceil$$
$$\lceil \neg\mathbf{ignite} \rceil;\ \lceil \mathbf{ignite} \rceil \xrightarrow{\leq 1} \lceil \mathbf{ignite} \rceil$$

The languages that characterise the phases are

$$\mathcal{L}[\![\mathbf{purge}]\!] = \mathbf{cycle}^*.\mathit{yesheat}.\mathit{startpurge}$$
$$\mathcal{L}[\![\mathbf{ignite}]\!] = \mathbf{cycle}^*.\mathit{yesheat}.\mathit{startpurge}.\mathit{startign}.\mathit{GasOn}$$

with

$$\mathbf{cycle} = \mathit{yesheat}.\mathit{startpurge}.\mathit{startign}.\mathit{GasOn}.\mathit{startburn}.$$
$$(\mathit{noheat}+\mathit{noflame}).\mathit{startidle}.\mathit{GasOff}\ .$$

In each case there is only one event leaving the phase. So the corresponding time restrictions are

$$\texttt{AFTER}\ \mathcal{L}[\![\mathbf{purge}]\!]\ \texttt{WAIT}\ (\mathit{startign}, 30)$$
$$\texttt{AFTER}\ \mathcal{L}[\![\mathbf{ignite}]\!]\ \texttt{WAIT}\ (\mathit{startburn}, 1).$$

□

Finally we consider the progress constraints. Here we require readiness for communication explicitly. This is done by time restrictions of the form

AFTER **re** READY (**ev**, **t**)

meaning that after engaging in a trace in **re**, the system becomes ready for **ev** within time **t**.

Example 8 In the gas burner the following progress requirements are unchanged from the beginning:

$$\lceil \mathbf{purge} \rceil \xrightarrow{30} \lceil \neg \mathbf{purge} \rceil$$
$$\lceil \mathbf{ignite} \rceil \xrightarrow{1} \lceil \neg \mathbf{ignite} \rceil$$

The **purge** phase has a stability of 30. After that a communication on **startign** is not prevented by the lower bound. We must force a communication on **startign** 30 seconds after the beginning of the **purge** phase. The requirements are then guaranteed by

AFTER $\mathcal{L}[\![\mathbf{purge}]\!]$ READY (*startign*, 30)
AFTER $\mathcal{L}[\![\mathbf{ignite}]\!]$ READY (*startburn*, 1)

The four other timing requirements from **Prog_2** are guaranteed, if we make some assumption on the speed of the communication. □

Summarising the gas burner specification looks now as follows:

GB_6

GB_Intern

TA_5 ∧ **TR_6**

where the set **TR_6** of timing restrictions is as follows:

TR_6

AFTER $\mathcal{L}[\![\mathbf{purge}]\!]$ WAIT (**startign**, 30)
AFTER $\mathcal{L}[\![\mathbf{ignite}]\!]$ WAIT (*startburn*, 1)

AFTER $\mathcal{L}[\![\mathbf{purge}]\!]$ READY (**startign**, 30)
AFTER $\mathcal{L}[\![\mathbf{ignite}]\!]$ READY (**startburn**, 1)

10 Specification Language SL

In the previous sections we introduced more and more parts of the specification language SL which serves as an intermediate stage in our transformation from DC implementables to an occam-like programming language. In this section we give an outline of SL and summarize the gas burner specification obtained so far in SL.

A simple SL specification is of the form **S** = `SPEC` Δ *TA CA TR* `END` where Δ is the interface, *TA* the trace part, *CA* the state part and *TR* the timing part of **S**.

Interface. An *interface* declares the communication channels **ch** of the component, for example for the main controller the interface is given by

`INPUT OF Signal` *yesheat, noheat, noflame*

`OUTPUT OF Signal` *startidle, startpurge, startburn, startign, GasOn, GasOff*

The input signal *startpurge* from the heat controller starts one heating cycle, the output communications *GasOn, GasOff* make the gas valve open or close.

The desired behaviour of the system components is described using a *trace part*, a *state part* and a *timing part.*

Trace Part. The trace part specifies the sequencing constraints on the channels whereas the communicated values are ignored. This is done by stating one or more *trace assertions* of the form

`TRACE` **re**

By stating several such trace assertions, we can specify different aspects of the intended system behaviour in a modular fashion.

State Part. This part describes what the exact values are that can be exchanged over the interface channels. To this end local state variables may be introduced, for example

`VAR` *type x* `INIT` e.

The expression **e** represents the initial value of **x**. These variables constitute the state space of the specification but need not appear in an implementation of the specified system.

For the last specification of the gas burner the state part is empty, but we can illustrate this part of SL by a communication assertion which appeared during the development process. In SL syntax that communication assertion would read

`COM` *GasOff* `WHEN` **newphase = waitidle** `THEN` **newphase′ = stayidle**

saying that for a communication *GasOff* to be enabled the `WHEN` predicate **newphase = waitidle** must be fulfilled. The effect will be that at termination of the communication the variable **newphase** is set to **idle**. Note that we use the prime notation of Z [19] to refer to the value of a variable at the moment of termination. In this article we dropped the prime notation and wrote such a communication assertion simply as $\mathbf{waitidle} \stackrel{GasOff}{\longrightarrow} \mathbf{idle}$.

Timing Part. In the timing part it is specified when channels are ready for communication. Lower bounds are expressed by

AFTER re **WAIT** (ch, t)

which means that after the communication of a trace belonging to the language described by the regular expression **re**, the system will not communicate on **ch** before time **t** has elapsed. Upper bounds are expressed by

AFTER re **READY** (ch, t)

which mean that after communication of a trace in **re**, the system becomes ready to communicate on **ch** within time **t**. Since in general we assume immediate readiness this restriction is only sensible in connection with an according **WAIT** restriction.

We have already seen the timing restrictions for the gas burner in the specification **GB_6**.

11 Conclusion

In general we pursue a transformational approach where a given specification is transformed stepwise into a program. Our work is in the tradition of Dijkstra's approach to refinement, and the work originated by Burstall and Darlington and pursued further to practical application in projects like CIP (standing for Computer-aided Intuition-guided Programming) [3] and PROSPECTRA (standing for PROgram development by SPECification and TRAnsformation) [9] but our emphasis is on concurrency, communication and real time.

In this paper we have presented some key techniques of a general approach to transforming DC implementables into program specifications. Not all of these techniques need to be applied to the gas burner. For example, phase splitting is an important design idea but needed only if the begin and the end of the phase which is going to be split should be made noticeable to the environment. This is not the case for the gas burner where the values of the observable **newphase** are of interest only for the internal control. However, in this paper we have not clearly distinguished between internally and externally observable values or events.

The correctness of all transformations and hence of the resulting occam-like programs is ultimately based on a combined *state-trace-readiness model* [13] for reactive systems extended to a timed semantics. However, a user of the transformations will not be concerned with such semantic details but will deal only with the syntactical rules.

References

[1] R.J.R. Back, Refinement Calculus, Part II: Parallel and Reactive Programs, in: J.W. de Bakker, W.-P. de Roever, G. Rozenberg, Eds., Stepwise Refinement of Distributed Systems: Models, Formalisms, Correctness, LNCS 430 (Springer-Verlag, 1990) pp. 67-93.

[2] J.C.M. Baeten, J.A. Bergstra. Real time process algebra. Formal Aspects of Computing 3(2), 1991, 142-188.

[3] F.L. Bauer et al., The Munich Project CIP, Vol. II: The Transformation System CIP-S, LNCS 292 (Springer-Verlag, 1987).

[4] M. Fränzle and B. v. Karger. Proposal for a Programming Language Core for ProCoS II. ProCoS Project Document [MF 11/3], Universität Kiel, 1993.

[5] M.R. Hansen, E.-R. Olderog, M. Schenke. et al. A Duration Calculus Semantics for Real-Time Reactive Systems. ProCoS Project Document [OLD MS 16/1], Universität Oldenburg , 1994.

[6] Jifeng He, C.A.R. Hoare, M. Fränzle, M. Müller-Olm, E.-R. Olderog, M. Schenke, M.R. Hansen, A.P. Ravn, and H. Rischel. Provably correct systems. In: H. Langmaack, W.-P. de Roever, and J. Vytopil, Eds., Formal Techniques in Real-Time and Fault-Tolerant Systems, LNCS 863 (Springer-Verlag, 1994) 288–335.

[7] J.J.M. Hooman. Specification and Compositional Verification of Real-Time Systems. LNCS 558 (Springer-Verlag, 1991).

[8] INMOS Ltd., occam 2 Reference Manual (Prentice Hall, 1988).

[9] B. Krieg-Brückner. Algebraic specification and functionals for transformational program and meta program development. In: J. Diaz, F. Orejas, Eds., Proc. TAPSOFT '89, LNCS 352 (Springer-Verlag, 1989).

[10] C. Morgan. Programming from Specifications. Prentice Hall, 1990.

[11] X. Nicollin, J. Sifakis, S. Yovine. From ATP to Timed Graphs and Hybrid Systems. REX 1991, LNCS 600 (Springer-Verlag, 1991), 549-572.

[12] E.-R. Olderog. Interfaces between Languages for Communicating Systems. ICALP 1992, LNCS 623 (Springer-Verlag, 1992), 641-655, invited paper.

[13] E.-R. Olderog, S. Rössig, J. Sander, M. Schenke. ProCoS at Oldenburg: The Interface between Specification Language and occam-like Programming Language. Technical Report Bericht 3/92, Univ. Oldenburg 1992.

[14] A.P. Ravn, H. Rischel, K.M. Hansen. Specifying and Verifying Requirements of Real-Time Systems. IEEE Transactions on Software Engineering, vol. 19,1 (1993) 41-55.

[15] A.P. Ravn. Design of Embedded Real-Time Computing Systems. Manuscript, DTU Lyngby, 1994.

[16] M. Schenke. A Timed Specification Language for Concurrent Reactive Systems. In: D.J. Andrews, J.F. Groote, C.A. Middelburg, Eds., Proc. Semantics of Specification Languages, Workshops in Computer Science (Springer-Verlag, 1994) 152–167.

[17] M. Schenke. Specification and Transformation of Reactive Systems with Time Restrictions and Concurrency. In: H. Langmaack, W.-P. de Roever, and J. Vytopil, Eds., Formal Techniques in Real-Time and Fault-Tolerant Systems, LNCS 863 (Springer-Verlag, 1994) 605–621.

[18] D. Scholefield, H. Zehan, He Jifeng. A Specification Oriented Semantics for the Refinement of Real-Time Systems, to appear in Theoretical Computer Science 1994.

[19] J.M.Spivey. The Z Notation: A Reference Manual. Prentice Hall, 1989.

[20] Zhou Chaochen, C.A.R.Hoare, A.P.Ravn. A Calculus of Durations. IPL 40/5 1991, 269-276.

Conformance: A Precongruence close to Bisimilarity

S. Arun-Kumar*
Department of Computer Science and Engineering
Indian Institute of Technology
Hauz Khas, New Delhi 110 016, India.

V. Natarajan
Department of Computer Science
North Carolina State University
Raleigh, NC 27606, U.S.A.

Abstract

In a previous paper we had defined the notion of an efficiency preorder for concurrent systems. In this paper, we present a coarser relation, called the elaboration preorder, which is finer than observational equivalence. Further, this preorder is incomparable with the almost-weak bisimulation preorder of Sangiorgi and Milner. In particular, the elaboration preorder is preserved under all contexts except summation. The largest precongruence contained in it, which we call conformance, is obtained by the usual means and a complete axiomatization for conformance of finite processes is given. The paper ends with an example to show the use of this relation.

1 Introduction

In [1] the efficiency preorder was defined on processes and a proof system was given and shown to be complete for finite processes. It was shown that it was possible to compare efficiencies of different implementations of the same specification with little extra effort than that required to prove their correctness. However, the efficiency preorder (as it was called in the paper) is open to certain criticisms. For one, it is too fine to be really useful for comparing many different implementations. Secondly, unlike the case of strong bisimulation and bisimilarity, its proof system even for finite processes necessitates the introduction of a rule of inference (called R0 in [1]).

[1] has shown that there are issues which go beyond merely equivalences and nondeterminism which are worthy of investigation. One such is a crude notion of efficiency of implementations. Intuitively two correct implementations of the same specification may differ from each other and the specification, only in terms of the so-called "silent actions". Since silent actions represent the amount of internal communication a system needs to perform in order to

*Supported by project grant SR/OY/M-08/92 of the Department of Science and Technology, Govt. of India.

achieve a certain desired external behaviour, the relative complexity of the implementation may be measured in terms of the amount of internal computations (communications in the case of concurrent systems) the systems perform.

Sangiorgi and Milner [7] have raised another issue – an upto technique to reduce the sizes of relations obtained when reasoning with weak bisimulations. They reject the efficiency preorder because it is too fine. They have defined a new notion, almost-weak bisimulation, as a supporting relation for bisimilarity. Their objective is to obtain a new relation which is contained within observational equivalence and is as close to it as possible. However their preorder has the disadvantage of not being preserved under composition contexts.

In this paper we introduce a new preorder on processes, called elaboration (and denoted $\lessapprox$), to overcome the disadvantages of the efficiency preorder (denoted $\lesssim$). The elaboration preorder is also an efficiency-based one – efficiency is measured in terms of the amount of internal communication a system must perform in order to achieve a certain desired functionality.

Loosely speaking, two processes p and q are related in the fashion $p \lessapprox q$ if they are observationally equivalent and for every run of q there exists a run of p which is at least as "slow" (i.e. it requires a possibly larger amount of internal computation). This is a relaxation from the conditions imposed on the efficiency preorder in which, $p \lesssim q$ only if for every run of p there exists a run of q that is no "slower", and for every run of q there exists a run of p that is no "faster".

The elaboration preorder is closely allied to observational equivalence in two ways. Firstly, the techniques used in the theory of the latter relation can be readily adapted to obtain a theory of the elaboration preorder. Secondly, considered as a relation on processes, it is incomparable with the almost-weak bisimulation preorder and is strictly coarser than any of the following preorders – efficiency, expansion ([4], [7]) and contraction [5].

The largest precongruence contained in the elaboration preorder is called conformance. Conformance is very close to observational congruence and for finite processes, a complete proof system for conformance is obtained by making a very small change to Milner's proof system for observational congruence [3]. Moreover, unlike efficiency precongruence, conformance requires no inference rules.

The paper is organised as follows. In section 2 we define the language of CCS and its operational semantics. We also fix our notation in this section. In section 3 we define the new preorder and compare it with some of the existing relations on processes. In section 4 we define and characterise the largest precongruence contained in the elaboration preorder. In particular we give a proof that recursion preserves this precongruence. This is followed by a complete axiomatization for finite processes. Throughout the paper we state without proof several results that are already known from [1] and [3]. The penultimate section treats a small example which also illustrates the inadequacy of the efficiency preorder. The last section is the conclusion.

$P.$	$a.e \xrightarrow{a} e$
$S1.$	$e \xrightarrow{a} e' \Rightarrow e+f \xrightarrow{a} e'$
$S2.$	$f \xrightarrow{a} f' \Rightarrow e+f \xrightarrow{a} f'$
$C1.$	$e \xrightarrow{a} e' \Rightarrow e \mid f \xrightarrow{a} e' \mid f$
$C2.$	$f \xrightarrow{a} f' \Rightarrow e \mid f \xrightarrow{a} e \mid f'$
$C3.$	$e \xrightarrow{\alpha} e',\ f \xrightarrow{\overline{\alpha}} f' \ \Rightarrow\ e \mid f \xrightarrow{1} e' \mid f'$
$H.$	$e \xrightarrow{a} e',\ a \notin L \cup \overline{L} \ \Rightarrow\ e\backslash L \xrightarrow{a} e'\backslash L$
$R.$	$e \xrightarrow{a} e' \Rightarrow e[h] \xrightarrow{h(a)} e'[h]$
$\mu.$	$e_i\{\mu\vec{x} : \vec{e}/\vec{x}\} \xrightarrow{a} e' \Rightarrow \mu_i\vec{x} : \vec{e} \xrightarrow{a} e'$

Figure 1: The operational semantics of CCS expressions

2 Syntax and Semantics of CCS

This section describes the syntax and operational semantics of CCS exactly as in [3] except for some notational differences. Let Λ and $\overline{\Lambda}$ (the complement of Λ) be infinite disjoint sets in bijection under the complementation operation "$\overline{\ \ }$". Then $V = \Lambda \cup \overline{\Lambda}$ is the set of *visible actions* and $A = V \cup \{1\}$ is the set of all *actions*, where 1 is a distinguished action called the *internal* (*silent* or *invisible*) action. Complementation is extended to the whole of A so that $\overline{1} = 1$ and $\overline{\overline{a}} = a$ for all $a \in A$.

Unless otherwise mentioned, we use the following notational conventions. Typically lower case greek letters (possibly decorated) denote visible actions, initial lower case roman letters ($a, b, c, \ldots$, suitably decorated) range over (visible or invisible) actions and $x, y, z, \ldots$ (possibly decorated) represent process variables. The language of CCS expressions is then given by the following BNF.

$$e ::= x \mid \mathbf{0} \mid a.e \mid e+e \mid e \mid e \mid e\backslash L \mid e[h] \mid \mu_i\vec{x} : \vec{e}$$

where $\mathbf{0}$ is a constant and $L \subseteq V$. $\vec{x}$ and $\vec{e}$ denote vectors of process variables and process expressions respectively, and $\mu\vec{x} : \vec{e}$ denotes the solution of a system of recursive process equations, whose i-th component is $\mu_i\vec{x} : \vec{e}$. In the relabelling operation e[h], $h : A \to A$ is the relabelling function such that $h(1) = 1$ and for all $\alpha \in V$, $\overline{h(\alpha)} = h(\overline{\alpha})$ and $h(\alpha) \neq 1$.

Any term generated by the above BNF is called a *process expression* and $\mathcal{E}$, ranged over by $e, f, g, \ldots$ (possibly decorated), denotes the set of process expressions. For any process expression e, $FPV(e)$ is the set of free process variables in e. *Processes* are closed process expressions (i.e. expressions with no free process variables) and $p, q, r, \ldots$ (possibly decorated) range over the set $\mathcal{P}$ of all processes.

We have defined the language of pure CCS since it is sufficient for our purposes. It is possible to extend our results to value-passing without any difficulty. The operational semantics of the language is defined in terms of labelled transition systems in the usual fashion.

Let $\langle \mathcal{E}, A, \{\xrightarrow{a} | a \in A\}\rangle$ be a labelled transition system (LTS) where the transition relation $\longrightarrow \subseteq \mathcal{E} \times A \times \mathcal{E}$ is the smallest relation satisfying the axioms and rules of inference in Figure 1. In the rule for recursion we use $f\{\vec{r}/\vec{x}\}$ to denote the syntactic substitution of all free occurrences of the variable x_j ($x_j \in \vec{x}$) in the expression f by r_j ($r_j \in \vec{r}$) for each j in the indexing set of the vectors $\vec{x}$ and $\vec{r}$.

Let A^* denote the set of all finite sequences of actions (including the empty sequence ε) and $A^+ = A^* - \{\varepsilon\}$ the set of nonempty sequences of actions. $|s|$ denotes the length of the sequence $s \in A^*$ and $\hat{s}$ is the sequence obtained from s by deleting all occurrences of 1. If s contains no visible action then $\hat{s}$ yields ε. For $a \in A$, $\hat{a} = \varepsilon$ if $a = 1$ and a otherwise.

Definition 2.1 *If $s, t \in A^*$, $a \in A$ then the transitions $p \xrightarrow{s} p'$ and $p \overset{s}{\Longrightarrow} p'$ are defined by induction on the length of s as follows.*

- $p \xrightarrow{\varepsilon} p$ *for all* p.
- $p \xrightarrow{s} p'$ *for* $s = ta$ *iff* $\exists p'' : p \xrightarrow{t} p'' \xrightarrow{a} p'$.
- $p \overset{\varepsilon}{\Longrightarrow} p'$ *iff* $\exists m \geq 0 : p \xrightarrow{1^m} p'$.
- $p \overset{a}{\Longrightarrow} p'$ *iff* $\exists m, n \geq 0 : p \xrightarrow{1^m a 1^n} p'$.
- $p \overset{s}{\Longrightarrow} p'$ *for* $s = ta$ *iff* $\exists p'' : p \overset{t}{\Longrightarrow} p'' \overset{a}{\Longrightarrow} p'$.

The following definition enables us to extend every relation we may define on closed terms to open terms as well.

Definition 2.2 *Let $\preceq$ be any binary relation on processes. Then for any two process expressions e, f such that $FPV(e) \cup FPV(f) \subseteq \vec{x}$, $e \preceq f$ if and only if for every vector of processes $\vec{p}$, $e\{\vec{p}/\vec{x}\} \preceq f\{\vec{p}/\vec{x}\}$.*

3 Elaborations

Definition 3.1 *$R \subseteq \mathcal{P} \times \mathcal{P}$ is an* **elaboration** *(or* $\lesssim$**-bisimulation***) iff for every $\langle p, q\rangle \in R$ the following conditions hold for every action $a \in A$.*

$$p \xrightarrow{a} p' \Rightarrow \exists q' : q \overset{\hat{a}}{\Longrightarrow} q' \wedge p' R q' \tag{1}$$

$$q \xrightarrow{a} q' \Rightarrow \exists p' : p \overset{a}{\Longrightarrow} p' \wedge p' R q' \tag{2}$$

Clause (1) is the usual first clause in the definition of weak bisimulations [3] whereas clause (2) is the second clause in the characterization of efficiency prebisimulations (proposition 3.6 in [1]). The following proposition gives an alternate characterization of elaborations.

Proposition 3.1 *$R \subseteq \mathcal{P} \times \mathcal{P}$ is an elaboration iff for every $\langle p, q\rangle \in R$ the following conditions hold for every action sequence $s, t \in A^*$.*

$$p \xrightarrow{s} p' \Rightarrow \exists q' : \exists t : \hat{s} = \hat{t} \wedge q \xrightarrow{t} q' \wedge p' R q'$$
$$q \xrightarrow{t} q' \Rightarrow \exists p' : \exists s : \hat{s} = \hat{t} \wedge |t| \leq |s| : p \xrightarrow{s} p' \wedge p' R q'$$

Proof. The proof is standard. □

Recall from [1] that an efficiency prebisimulation is a binary relation satisfying the following conditions for every $s, t \in A^*$.

$$p \xrightarrow{s} p' \Rightarrow \exists q' : \exists t : \hat{s} = \hat{t} \wedge |s| \geq |t| \wedge q \xrightarrow{t} q' \wedge p'Rq'$$
$$q \xrightarrow{t} q' \Rightarrow \exists p' : \exists s : \hat{s} = \hat{t} \wedge |t| \leq |s| : p \xrightarrow{s} p' \wedge p'Rq'$$

The following corollary and the next proposition are easily proven.

Corollary 3.2 (Granularity)

1. *Every strong bisimulation is an efficiency prebisimulation.*

2. *Every efficiency prebisimulation is an elaboration.*

3. *Every elaboration is a weak bisimulation.*

Proposition 3.3 (The largest elaboration)

1. *The union of a family of* $\lessapprox$*-bisimulations is a* $\lessapprox$*-bisimulation.*

2. *The composition of any two* $\lessapprox$*-bisimulations is a* $\lessapprox$*-bisimulation.*

3. *There exists a maximum* $\lessapprox$*-bisimulation, denoted by* $\lessapprox$*, which is the union of all* $\lessapprox$*-bisimulations and is a preorder.*

4. $p \lessapprox q$ *iff* $\langle p, q \rangle$ *belongs to some* $\lessapprox$*-bisimulation.*

5. $\lessapprox$ *is compatible with all operators of CCS except summation.*

We refer to $\lessapprox$, the largest elaboration as the **elaboration preorder**. It is easy to see from the usual example given in [3] that $1.\alpha.\mathbf{0} \lessapprox \alpha.\mathbf{0}$, but $1.\alpha.\mathbf{0} + \beta.\mathbf{0} \not\lessapprox \alpha.\mathbf{0} + \beta.\mathbf{0}$. The preemptive power of the silent action makes the relation incompatible under summation contexts.

The notation $\sim$ for strong congruence, $\lesssim$ for the efficiency preorder and $\approx$ for observational equivalence is used throughout the rest of the paper. Further, $\simeq$ (efficiency equivalence) and $\approxeq$ (elaboration equivalence) denote the kernels of the preorders $\lesssim$ and $\lessapprox$ respectively.

Proposition 3.4 (Inclusion)

1. $\sim \subset \lesssim \subset \lessapprox \subset \approx$

2. $\sim \subset \simeq \subset \approxeq \subset \approx$

Proof. The inclusions follow from corollary 3.2 and the strictness of the inclusions follow from the following pairs of processes.

- $\alpha.0 \approx \alpha.1.0$ but $\alpha.0 \not\lessapprox \alpha.1.0$
- $\alpha.0 + \alpha.1.1.0 \lessapprox \alpha.1.0$ but $\alpha.0 + \alpha.1.1.0 \not\lesssim \alpha.1.0$
- $\alpha.1.0 \lesssim \alpha.0$ but $\alpha.1.0 \not\simeq \alpha.0$
- $\alpha.0 + \alpha.1.1.0 \simeq \alpha.0 + \alpha.1.0 + \alpha.1.1.0$ but
  $\alpha.0 + \alpha.1.1.0 \not\simeq \alpha.0 + \alpha.1.0 + \alpha.1.1.0$
- $\alpha.0 + \alpha.1.0 \underline{\approx} \alpha.1.0$ but $\alpha.0 + \alpha.1.0 \not\underline{\approx} \alpha.1.0$
- $\alpha.1.0 \approx \alpha.0 + \alpha.1.1.0$ but $\alpha.1.0 \not\underline{\approx} \alpha.0 + \alpha.1.1.0$

□

Proposition 3.5 *The following inequations hold for all p and q.*

$$\begin{array}{ll} 1.p \lessapprox p & 1.p \lessapprox p + 1.p \lessapprox p \\ p + 1.p \lessapprox 1.p & a.1.p \lessapprox a.p \\ a.(p + 1.q) \lessapprox a.(p + 1.q) + a.q & a.(p + 1.q) + a.q \lessapprox a.(p + 1.q) \end{array}$$

As is clear from corollary 3.2, it suffices to produce efficiency prebisimulations in each case to prove proposition 3.5.

Definition 3.2 $R \subseteq \mathcal{P} \times \mathcal{P}$ *is an* **elaboration upto** $\lessapprox$ *if* $\lessapprox \circ R \circ \lessapprox$ *is an elaboration (where* $\circ$ *denotes relational composition).*

It follows from the definition that if R is an elaboration upto $\lessapprox$, then R is also contained in $\lessapprox$. It is then easy to prove the following proposition.

Proposition 3.6 $R \subseteq \mathcal{P} \times \mathcal{P}$ *is an elaboration upto* $\lessapprox$ *if the following conditions are satisfied for each* $\langle p, q \rangle \in R$ *and* $a \in A$.

$$\begin{array}{l} p \xrightarrow{a} p' \Rightarrow \exists q' : q \stackrel{\hat{a}}{\Longrightarrow} q' \wedge p'(R \circ \lessapprox)q' \\ q \xrightarrow{a} q' \Rightarrow \exists p' : p \stackrel{a}{\Longrightarrow} p' \wedge p'(\lessapprox \circ R)q' \end{array}$$

Proposition 3.6 shows that $\lessapprox$ supports an "upto-technique" (in the terminology of [7]), i.e. to prove $p \lessapprox q$ it suffices to find an elaboration upto $\lessapprox$ containing $\langle p, q \rangle$ (rather than an elaboration containing $\langle p, q \rangle$).

4 Conformance

It is clear that $\lessapprox$ is not compatible with summation because of the preemptive power of the silent action. As was done with $\approx$ in [3] and $\lesssim$ in [1], we define conformance as the largest precongruence contained in $\lessapprox$.

Definition 4.1 *For $\preceq \in \{ \lesssim, \lessapprox, \approx \}$, $p \preceq^+ q$ if and only if for some visible action α not occurring in p or q, $p + \alpha.0 \preceq q + \alpha.0$. The relation $\lessapprox^+$ is called* **conformance**. *The relations $\lesssim^+$ and $\approx^+$ are known as efficiency precongruence [1] and observation congruence [3] respectively.*

In the following proposition we give a behavioural characterization of conformance. The proof is similar to that for the efficiency precongruence (proposition 3.11.3 in [1]).

Proposition 4.1 *$p \lessapprox^+ q$ iff for every $a \in A$, the following conditions are satisfied.*

$$p \xrightarrow{a} p' \Rightarrow \exists q' : q \stackrel{a}{\Longrightarrow} q' \wedge p' \lessapprox q' \tag{3}$$

$$q \xrightarrow{a} q' \Rightarrow \exists p' : p \stackrel{a}{\Longrightarrow} p' \wedge p' \lessapprox q' \tag{4}$$

The kernels of the precongruences $\lesssim^+$ and $\lessapprox^+$ are denoted respectively by $\simeq^+$ and $\approxeq^+$ (it is obvious that they are congruence relations). It follows from proposition 3.4 and the examples given in its proof that the following strict containment relations hold.

Proposition 4.2 (Containment)

1. $\sim \subset \lesssim^+ \subset \lessapprox^+ \subset \approx^+$
2. $\sim \subset \simeq^+ \subset \approxeq^+ \subset \approx^+$

The following example gives some idea of the distinctions between the various compatible relations that we are considering.

Example 4.1 *Let a be an action. Then*

1. $a.0 + a.1.0 \lesssim^+ a.1.0$ *but not the converse.*
2. $a.1.0 \lessapprox^+ a.0$ *but the converse does not hold.*
3. $a.0 + a.1.1.0 \lessapprox^+ a.1.0$ *but* $a.0 + a.1.1.0 \not\lesssim^+ a.1.0$
4. $a.1.0 \approx^+ a.0 + a.1.1.0$ *but* $a.1.0 \not\lessapprox^+ a.0 + a.1.1.0$
5. $a.0 + a.1.1.0 \simeq^+ a.0 + a.1.0 + a.1.1.0$ *but*
   $a.0 + a.1.1.0 \not\sim a.0 + a.1.0 + a.1.1.0$
6. $a.0 + a.1.0 \approxeq^+ a.1.0$ *but* $a.0 + a.1.0 \not\simeq^+ a.1.0$
7. $a.1.0 \approx^+ a.0 + a.1.1.0$ *but* $a.1.0 \not\approxeq^+ a.0 + a.1.1.0$

We end this section with the proof that $\lessapprox^+$ is indeed a precongruence. The technique is quite similar to that adopted in [3] and [1].

Lemma 4.3 *Let e and f be process expressions such that $e \lesssim^+ f$ and $FPV(e) \cup FPV(f) \subseteq \vec{x}$. Then $\mu_i \vec{x} : e \lesssim^+ \mu_i \vec{x} : f$.*

Proof outline. We give an outline of the proof only for the case where e and f have at most one free variable, say x. That is, we prove that $e \lesssim^+ f$ implies $p \lesssim^+ q$, where $p \equiv \mu x : e$ and $q \equiv \mu x : f$. For any process expression g with $FPV(g) \subseteq \{x\}$ and any process r, let $g(r)$ denote $g\{r/x\}$. Now consider the relation

$$R = \{\langle g(p), g(q)\rangle | g \in \mathcal{E}, FPV(g) \subseteq \{x\}\}$$

By induction on the depth of the inferences $g(p) \xrightarrow{a} p'$ and $g(q) \xrightarrow{a} q'$ respectively, it is possible to show that R satisfies the following conditions for all $a \in A$.

$$g(p) \xrightarrow{a} p' \Rightarrow \exists q' : g(q) \stackrel{a}{\Longrightarrow} q' \wedge p'(R \circ \lesssim)q' \tag{5}$$

$$g(q) \xrightarrow{a} q' \Rightarrow \exists p' : g(p) \stackrel{a}{\Longrightarrow} p' \wedge p'(\lesssim \circ R)q' \tag{6}$$

It follows that R is an elaboration upto $\lesssim$. Letting $g \equiv x$, we have $\langle p, q\rangle \in R$ and hence $p \lesssim q$. Further from (3), (4), (5) and (6) it follows that $p \lesssim^+ q$. □

Theorem 4.4 $\lesssim^+$ *is a precongruence over CCS expressions.*

5 Proof System for Finite Processes

In this section we give a sound and complete inequational proof system for the conformance relation on finite processes (those in which there is no occurrence of recursion). Since such axiomatisations already exist for the other three relations we also reproduce them for comparison.

Milner [3] shows that finite or recursion-free processes (i.e. processes in which there is no occurrence of the recursion operator "μ") can be expressed upto strong congruence as finite serial processes. By a finite serial process, we mean one which is made up of only the operators **0**, prefixing and summation. That is, every finite process p has a normal form and $p \sim \sum_{i \in I} a_i.p_i$, for I a finite index set, and for each $i \in I$, $a_i \in A$ and p_i is in normal form. The reader may verify from the following two propositions (stated without proof) that this is indeed true. Since composition and summation are associative and commutative, we use the unary operators Γ and Σ over sets of processes to denote their composition and summation respectively.

Proposition 5.1 (The Static Laws)

1. *Let $p \equiv \Gamma P$, where $P = \{p_i | 1 \leq i \leq n\}$. Then*

$$\begin{aligned} p \;\sim\; & \Sigma\{a_i.p' | p_i \xrightarrow{a_i} p_i', 1 \leq i \leq n, p' \equiv \Gamma P'\} + \\ & \Sigma\{1.p'' | p_i \xrightarrow{\alpha} p_i', p_j \xrightarrow{\overline{\alpha}} p_j', i < j, p'' \equiv \Gamma P''\} \end{aligned}$$

where $P' = ((P - \{p_i\}) \cup \{p_i'\})$ and $P'' = ((P - \{p_i, p_j\}) \cup \{p_i', p_j'\})$

$$
\begin{array}{llll}
A1. & x+y=y+x & A3. & x+x=x \\
A2. & x+(y+z)=(x+y)+z & A4. & x+0=x
\end{array}
$$

Figure 2: The proof system $\mathcal{A}_{\sim}$.

$$
\begin{array}{llll}
A5\lessapprox^{+}. & a.1.x \sqsubseteq a.x & A6. & 1.x = x + 1.x \\
A5\approx^{+}. & a.1.x = a.x & A7. & a.(x+1.y) = a.(x+1.y)+a.y
\end{array}
$$

Figure 3: Other axioms for conformance and observational congruence.

2. $(a.p)\backslash L \sim \begin{cases} 0 & \textit{if } a \in L \cup \overline{L} \\ a.(p\backslash L) & \textit{otherwise} \end{cases}$

3. $(a.p)[h] \sim h(a).p[h]$

4. $(p+q)\backslash L \sim p\backslash L + q\backslash L$

5. $(p+q)[h] \sim p[h] + q[h]$

As with any equational or inequational axiom system, provability is assumed to be closed under reflexivity, transitivity, substitutivity and instantiation. Further for $\sqsubseteq$ -inequational systems we also have that $x = y$ if and only if $x \sqsubseteq y$ and $y \sqsubseteq x$.

We denote the proof systems for the various compatible relations ($\preceq$) of interest by $\mathcal{A}_{\preceq}$. The axioms for $\sim$ are given in Figure 2. Other axioms for $\mathcal{A}_{\lessapprox^+}$ and $\mathcal{A}_{\approx^+}$ are given in Figure 3 and those for $\mathcal{A}_{\lesssim^+}$ along with the rule R0 are in Figure 4. Axioms that are specifically applicable only to a particular relation have been labelled appropriately. Note that $\mathcal{A}_{\lessapprox^+}$ is identical to $\mathcal{A}_{\approx^+}$ except that axiom A5$\approx^+$ is replaced by A5$\lessapprox^{+}$.

As we see from the axiomatization, the conformance relation is very close to observational congruence. It is also complete for finite processes as the following theorem shows.

Lemma 5.2 *$p \lessapprox q$ implies $p \lessapprox^{+} q$ or $p \lessapprox^{+} 1.q$ for all processes p, q.*

$$
\begin{array}{llll}
A5 \lesssim^{+}. & a.(x+1.x) \sqsubseteq a.x & A7 \lesssim^{+}. & a.(x+1.y) \sqsubseteq a.(x+1.y)+a.y \\
A6 \lesssim^{+}. & 1.x \sqsubseteq x+1.x & R0 \lesssim^{+}. & x \sqsubseteq x+y+z \Rightarrow x \sqsubseteq x+z
\end{array}
$$

Figure 4: Other axioms and rule R0 for efficiency precongruence.

Proof outline. An easy variation of lemma 5.3 in [1]. □

Lemma 5.3 (Saturation). $p \stackrel{a}{\Longrightarrow} p'$ *implies* $\mathcal{A}_{\underset{\approx}{<}^+} \vdash p = p + a.p'$

Proof. Identical to proof of lemma 16 (page 163) in [3]. □

Theorem 5.4 (Completeness) *The proof system* $\mathcal{A}_{\underset{\approx}{<}^+}$ *is sound and complete for the conformance relation on finite processes, i.e.* $p \underset{\approx}{<}^{+} q$ *iff* $\mathcal{A}_{\underset{\approx}{<}^+} \vdash p \underline{\sqsubseteq} q$.

Proof. The proof of soundness, viz. that $\mathcal{A}_{\underset{\approx}{<}^+} \vdash p \underline{\sqsubseteq} q$ implies $p \underset{\approx}{<}^{+} q$, is easy. The proof of completeness proceeds by induction on the sum of the depths of p and q, where the depth of a finite process in normal form is the maximum number of nested prefixes in it. Assume that p and q have the standard forms

$$p \equiv \sum\{a_i.p_i | 1 \leq i \leq m\}, \quad q \equiv \sum\{b_j.q_j | 1 \leq j \leq n\}$$

where all summands that are strongly congruent to **0** have been removed. We then have the following two claims.

Claim. $\mathcal{A}_{\underset{\approx}{<}^+} \vdash a_i.p_i + q \underline{\sqsubseteq} q$ for each i, $1 \leq i \leq m$.

$\vdash$ Since $a_i.p_i$ is a summand of p and $p \underset{\approx}{<}^{+} q$ we have that $p \stackrel{a_i}{\longrightarrow} p_i$ implies for some $j, 1 \leq j \leq n, b_j = a_i$ and $q \stackrel{a_i}{\Longrightarrow} q'_j$ and $p_i \underset{\approx}{<} q'_j$. By lemma 5.3 it follows that $\mathcal{A}_{\underset{\approx}{<}^+} \vdash q = q + a_i.q'_j$. By lemma 5.2 we have either $p_i \underset{\approx}{<}^{+} q'_j$ or $p_i \underset{\approx}{<}^{+} 1.q'_j$. In the former case the result follows trivially, and in the latter case we have

$$\begin{aligned} \mathcal{A}_{\underset{\approx}{<}^+} & \vdash p_i \underline{\sqsubseteq} 1.q'_j \\ & \vdash a_i.p_i \underline{\sqsubseteq} a_i.1.q'_j \\ & \vdash a_i.p_i \underline{\sqsubseteq} a_i.q'_j \\ & \vdash a_i.p_i \underline{\sqsubseteq} a_i.q'_j \\ & \vdash a_i.p_i + q \underline{\sqsubseteq} a_i.q'_j + q \\ & \vdash a_i.p_i + q \underline{\sqsubseteq} q \end{aligned}$$

$\dashv$

Claim. $\mathcal{A}_{\underset{\approx}{<}^+} \vdash p \underline{\sqsubseteq} p + b_j.q_j$ for each j, $1 \leq j \leq n$.

$\vdash$ Similar to the above. $\dashv$

By summing up the terms for each i, reordering them and repeatedly applying axiom A3 we get $\mathcal{A}_{\underset{\approx}{<}^+} \vdash p + q \underline{\sqsubseteq} q$. Similarly by summing up the terms for each j and reordering them we get $\mathcal{A}_{\underset{\approx}{<}^+} \vdash p \underline{\sqsubseteq} p + q$. The result then follows by transitivity. □

6 An Example: The 3-element Buffer

We consider the following specification (called $FIFO3$) of a simple 3-element buffer which receives messages from an unspecified process SOURCE via a port α and delivers them to another unspecified process DEST via the port δ. For convenience and simplicity, we simply use signals to denote the fact of a message being conveyed from one process to another. The state of the process $FIFO3$ is represented by a number varying from 0 to 3 indicating the number of messages received but not delivered. The initial state of this process is $FIFO3(0)$.

$$FIFO3(n) = \begin{cases} \alpha.FIFO3(1) & \text{if } n = 0 \\ \alpha.FIFO3(n+1) + \bar{\delta}.FIFO3(n-1) & \text{if } 1 \leq n \leq 2 \\ \bar{\delta}.FIFO(2) & \text{if } n = 3 \end{cases}$$

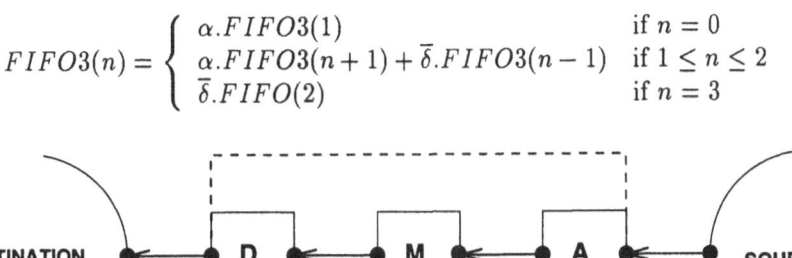

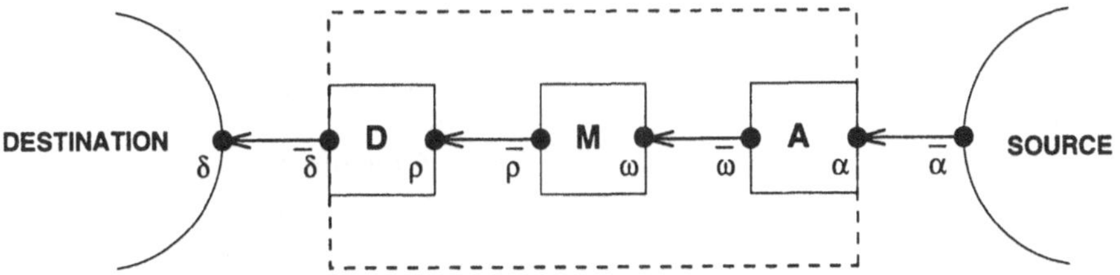

Figure 5: Implementation of PIPE3

One possible implementation of this specification is obtained by interconnecting three cells end to end (see Figure 5). We call this implementation $PIPE3$ and the state here is represented by a 3-bit binary sequence indicating whether a given cell is occupied or empty. We use the symbols d, m, a to denote respectively the states of D, M, and A in the implementation. The initial state is the string 000.

$$\begin{aligned} C(0) &\triangleq \iota.C(1) \\ C(1) &\triangleq \bar{o}.C(0) \\ D(d) &\triangleq C(d)[\rho/\iota, \delta/o] \\ M(m) &\triangleq C(m)[\omega/\iota, \rho/o] \\ A(a) &\triangleq C(a)[\alpha/\iota, \omega/o] \\ PIPE3(dma) &\triangleq (D(d) \mid M(m) \mid A(a)) \backslash \{\rho, \omega\} \end{aligned}$$

It is easy to see that this implementation starting in the initial state 000 is efficiency precongruent to FIFO3(0) by the following efficiency prebisimulation.

$$R = \{\langle PIPE3(dma), FIFO3(f(dma))\rangle | d, m, a \in \{0, 1\}\}$$

where $f(dma) = d + m + a$.

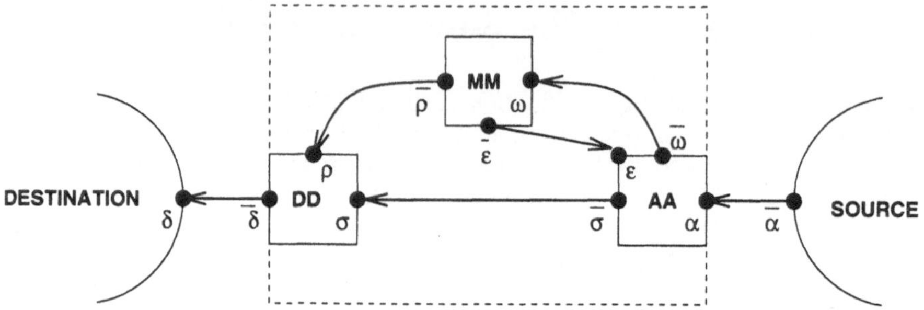

Figure 6: Implementation of RJF3

We now consider a small variation of PIPE3. If M is empty (and A 'knows' that it is empty) then A could send an incoming message directly to D if D were waiting for one. However, this requires that M convey its emptiness to A via some message. Then we have the following design in which the signal ϵ is used by M to indicate that it is empty. An additional port σ is used to send the message directly from A to D. We call this implementation $RJF3$ (short for "Receive and Jump Forward-3"; see Figure 6).

The state of $RJF3$ can be represented by a 4-bit binary string, where the two left-most bits represent the occupancy or emptiness of DD and MM respectively. The two rightmost bits represent the state of AA – the right one denoting the occupancy of AA and the left one indicating whether MM is empty (the bit is '0' if AA has received an 'emptiness' signal from MM since the last time a message was sent to MM from AA, otherwise it is '1'). The initial state of $RJF3$ is 0000.

$$
\begin{array}{lcl}
DD(0) & \triangleq & \rho.DD(1) + \sigma.DD(1) \\
DD(1) & \triangleq & \overline{\delta}.DD(0) \\
MM(0) & \triangleq & \omega.MM(1) \\
MM(1) & \triangleq & \overline{\rho}.\overline{\epsilon}.MM(0) \\
AA(00) & \triangleq & \alpha.AA(01) \\
AA(01) & \triangleq & \overline{\omega}.AA(10) + \overline{\sigma}.AA(00) \\
AA(10) & \triangleq & \alpha.AA(11) + \epsilon.AA(00) \\
AA(11) & \triangleq & \epsilon.AA(01) \\
RJF3(dmua) & \triangleq & (DD(d) \mid MM(m) \mid AA(ua))\backslash\{\rho,\omega,\sigma,\epsilon\}
\end{array}
$$

Note that states in which $m = 1$ and $u = 0$ are impossible in RJF3. Also note that the behaviour of $AA(01)$ is nondeterministic. The fact that $RJF3$ is a correct implementation of $FIFO3$ is clear from the following efficiency prebisimulation.

$$S = \{\langle RJF3(dmua), FIFO3(g(dmua))\rangle | mu \neq 10, d, m, u, a \in \{0,1\}\}$$

where $g(dmua) = d + m + a$.

The following relation T between the states of $PIPE3$ and $RJF3$ is an elaboration and demonstrates that $RJF3 \lesssim PIPE3$.

$$T = \{\langle RJF3(dmua), PIPE3(g(dmua))\rangle | mu \neq 10, d, m, u, a \in \{0,1\}\}$$

In their respective initial states, the only action they are both capable of is α. Hence $RJF3(0000) \lessapprox^{+} PIPE3(000)$.

However there exists no efficiency prebisimulation relating the two processes. Neither is it possible to prove $PIPE3(000) \lessapprox RJF3(0000)$. The reason is simply that for every visible trace of $PIPE3$, there exists a longer trace (with the same visible action sequence but packed with more silent actions) that $RJF3$ can perform. At the same time there also exist traces in the execution of $RJF3$ with the visible content $\alpha\overline{\delta}$ which are shorter than any trace in $PIPE3$ with the same visible content. Hence $RJF3$ and $PIPE3$ are incomparable under the efficiency preorder, but since in the worst-case $RJF3$ could be "slower" than $PIPE3$, they are related by the elaboration preorder.

Intuitively, there are extra overheads on $RJF3$ (because of the extra connections) which make $RJF3$ "slower" in general than $PIPE3$. But in situations where messages are delivered as fast as they are received, $RJF3$ could execute "faster" than $PIPE3$, because of these short-circuit connections.

7 Conclusion

Our motivation in defining elaborations has been to obtain an efficiency-based preorder to compare observationally equivalent processes. That is, we regard the silent action as signifying the amount of communication that the components of a process need to perform amongst themselves in order to give the process a desired external behaviour. Hence loosely speaking, if $p \lesssim q$ or $p \lessapprox q$ then we could take it to mean that both p and q are bisimilar to the same specification and that in a certain sense p could be considered to require a greater degree of communication (perhaps because of the complexity of its internal architecture) than q. This is a viewpoint that has been illustrated adequately in [1], by taking different implementations of the same specification.

Under the elaboration preorder, $p \lessapprox q$ merely implies that for every sequence of visible actions, the least efficient run of q could be at worst as inefficient as the least efficient run of p. Unlike in the case of the efficiency preorder, this does not rule out the possibility of p having at the same time several more efficient runs than any that q possesses. This yields a coarser relation than $\lesssim$.

Contrary to the usual methodology in vogue, viz. that correctness and efficiency are separate issues and have to be evaluated one after the other (with correctness preceding efficiency), our preorders enable one to combine the two issues into one for which the effort required is no more than that required for finding an appropriate bisimulation. In fact, the two preorders and precongruences have been implemented on the Edinburgh Concurrency Workbench [6].

However, while implementing the precongruences, we have departed from the method used to implement observational congruence [2], by actually defining a new special action (which is internally generated) and taking recourse to definition 4.1. This yields a simpler and more general method that could also be used for observational congruence.

A few other preorders lying between strong congruence and observational equivalence have been defined in the recent past by Milner and his followers. We mention only expansions, contractions and almost-weak bisimulations and compare them with the elaboration preorder. Expansion is merely the converse of the efficiency preorder and from the definition of contraction [5] it is clear that the contraction preorder is a slight weakening of expansion [4]. The elaboration preorder is coarser than any of them. The following example shows that $\lesssim$ is incomparable with the almost-weak bisimulation relation $\leqq$ of [7].

Example 7.1 *Let* $p \equiv 1.1.1.\alpha.\mathbf{0} + 1.\alpha.\mathbf{0}$ *and* $q \equiv 1.1.\alpha.\mathbf{0} + \alpha.\mathbf{0}$. *In this case an almost-weak bisimulation cannot be constructed because* $q \xrightarrow{\alpha} \mathbf{0}$ *whereas* p *has no* α*-derivative. However an elaboration does exist, which is the following relation*
$\{\langle p,q\rangle, \langle \mathbf{0},\mathbf{0}\rangle, \langle 1.1.\alpha.\mathbf{0}, 1.\alpha.\mathbf{0}\rangle, \langle 1.1.\alpha.\mathbf{0}, q\rangle, \langle 1.\alpha.\mathbf{0}, q\rangle, \langle 1.\alpha.\mathbf{0}, 1.\alpha.\mathbf{0}\rangle, \langle \alpha.\mathbf{0}, \alpha.\mathbf{0}\rangle\}$.
The example given in [7] suffices to show that $p \leqq q$ *does not imply* $p \lesssim q$. *Take* $p \equiv 1.\alpha.\mathbf{0} + 1.1.\beta.\mathbf{0}$ *and* $q \equiv 1.1.\alpha.\mathbf{0} + 1.\beta.\mathbf{0}$. *It is clear that there exists no elaboration containing either* $\langle p,q\rangle$ *or* $\langle q,p\rangle$.

References

[1] S. Arun-Kumar, M. Hennessy, An efficiency preorder for processes, **Acta Informatica**, 29:737-760, 1992.

[2] R. Cleaveland, J. Parrow, B. Steffen, The concurrency workbench: A semantics-based verification tool for finite-state systems, **Proceedings of the Workshop on Automated Verification Methods for Finite-state Systems**, Lecture Notes in Computer Science 407, Springer-Verlag, 1989.

[3] R. Milner, **Communication and Concurrency**, Prentice-Hall International, 1989.

[4] R. Milner, Expansions, handwritten notes, 1990.

[5] R. Milner, Contractions, handwritten notes, 1990.

[6] F. Moller, The Edinburgh Concurrency Workbench (Version 6.1), Department of Computer Science, University of Edinburgh, 1992.

[7] D. Sangiorgi, R. Milner, The problem of "weak bisimulation up to", **CONCUR '92**, Lecture Notes in Computer Science 630, Springer-Verlag, 1992.

On The Synchronic Structure of Transition Systems

Luca Bernardinello
Campus de Beaulieu, IRISA, Rennes, France
Luca.Bernardinello@irisa.fr

Giorgio De Michelis Katia Petruni Sebastiano Vigna
Dipartimento di Scienze dell'Informazione, Università di Milano, Italia
gdemich@hermes.dsi.unimi.it {petruni,vigna}@dsi.unimi.it

Abstract

Net Theory was introduced in the early sixties by Carl Adam Petri [1] as a form of general system theory based on the notion of concurrency. Net Theory has been widely developed during these years, becoming very popular as a framework for the analysis and specification of concurrent systems. Among the basic notions of the theory, stands the synchronic structure of a system. It characterizes dependencies between sets of its events in terms of a distance measuring their degree of synchronization. In this paper we show that a natural generalization of regions introduced by Ehrenfeucht and Rozenberg exactly corresponds to synchronic distances and that this notion of region can be used to axiomatise a class of transition systems corresponding to bounded place/transition nets without loops.

1 Introduction

Net theory has been introduced in the early sixties by Carl Adam Petri [1] with the purpose of providing a conceptual tool to derive "global system properties from given local properties, and detailed system requirements from global specifications" [2]. The application of Net Theory, in Petri's words, "consists in the description, analysis and synthesis of systems and processes in the real world." [2]. Net Theory has been widely developed during these years, becoming very popular as a framework for the analysis and specification of concurrent systems. Not all the basic concepts of Net Theory have received the same attention in these developments: among the ones to which very little attention was paid, are the notions of synchronic and enlogic structure. The synchronic structure of a net system characterizes all the dependencies between sets of its events in terms of a distance measuring their degree of synchronization, while its enlogic structure allows to distinguish its events and their admitted combinations from the transformations of local conditions characterizing violations of their behaviour and from the impossible transformations, characterizing the factually valid statements that can be made in terms of its local conditions. In other terms the synchronic and enlogic structures allow to characterize in structural terms the behavioural properties of a system.

The fundamental work of G. Rozenberg, with A. Ehrenfeucht [3] and with M. Nielsen and P.S. Thiagarajan [4], showing that an elementary net system (or a condition/event system) can be synthesized through the basic notion of *region* from a (sequential) transition system satisfying some separation conditions (namely from an elementary transition system) suggests that it is possible to characterize both the synchronic and the enlogic structure of transition systems, opening interesting possibilities to the study of the structural properties of the concurrent system having a given (sequential) behaviour. Transitions systems are, in fact, models of sequential observations of system behaviours and, therefore, their synchronic and enlogic structures can be considered as extensions of the synthesis procedure defined by the above authors.

In this paper we are concerned with the synchronic structure of a net system (and of the corresponding transition system); we will show that a natural generalization of regions exactly corresponds to synchronic distances. This is not surprising, of course, since it was known that the synchronic structure of a net system is strongly related to the so-called S-completion of a net, that is the operation consisting in adding to a net the maximal set of places compatible with the behaviour of the net; here we consider the marking graph of a net, up to isomorphism, as its behaviour. So, by means of generalized regions, we are able to build the "place saturated" version of a condition/event system. The set of generalized regions has particularly good algebraic properties: it is in fact a finitely generated abelian group. This was the key observation leading Badouel, Bernardinello and Darondeau [5] to develop polynomial algorithms solving the synthesis problem for bounded place/transition nets without loops.

The same notion of region can be used to axiomatise a class of transition systems corresponding to bounded place transition nets without loops. This class of nets has been already used by Desel and Merceron [6], in connection with synchronic distances, in order to define an abstraction operation on nets. On the other hand, a notion of region very similar to the one we have defined was proposed by Mukund [7], with different aims.

In [8, 9] various synchronic relations between events are considered, which give rise to linear spaces similar to the ones discussed in the present paper. However, those notions are not related to the S-completion of a net, because only directed paths are used in their computation. Due to this fact, the definition of sychronic distance we use is not reducible to any of the synchronic relations described there.

The paper is structured as follows: in Section 2 we recall basic definitions concerning transition systems and net systems, the notion of region introduced by Ehrenfeucht and Rozenberg, and the synchronic structure of a condition/event net system. Section 3 introduces a new notion of region and shows its tight relation with synchronic distances. The possibility of using this type of region to build a subclass of place/transition nets is then presented.

2 Preliminaries

2.1 Transition Systems

Sequential transition systems can be considered as models of sequential observation of system behaviours. Different classes of transition systems characterize therefore different classes of systems. In this section we focus our attention on condition/event transition systems (shortly, CETS). Condition/event transition systems are elementary transition systems without initial state and therefore with a relaxed reachability condition and an enforced firing condition [3, 4].

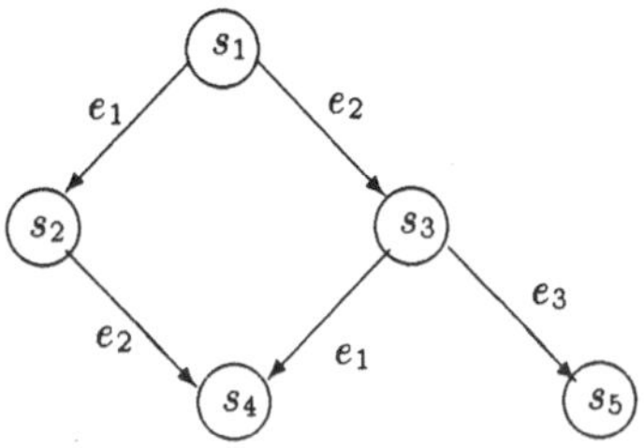

Figure 1: A transition system

The properties of elementary transition systems can be easily extended to condition/event systems.

Definition 2.1 (*transition system*) A transition system is a structure $A = (S, E, T)$, where S is a set of *states*, E is a set of *events*, $T \subseteq S \times E \times S$ is a set of *transitions*.

A transition system is *finite* if S and E are finite. In the rest of the paper we will only consider finite transition systems satisfying the following properties:

1. $\forall\ (s, e, s') \in T \quad s \neq s'$
2. $\forall\ (s, e_1, s_1), (s, e_2, s_2) \in T \quad s_1 = s_2 \Rightarrow e_1 = e_2$
3. $\forall\ e \in E\ \exists\ (s, e, s') \in T$

The notion of region is basic in studying properties of transition systems. It was introduced by Ehrenfeucht and Rozenberg in [3]. A region is a set of states such that every occurrence of a given event must have the same crossing relation (entering, leaving or non crossing) with respect to the region itself, and this property holds for all events.

Definition 2.2 (*region*) Let $A = (S, E, T)$ be a transition system. A set of states $r \subseteq S$ is said to be a *region* iff $\forall\ (s_1, e, s_1'), (s_2, e, s_2') \in T$ we have $(s_1 \in r \wedge s_1' \notin r) \Leftrightarrow (s_2 \in r \wedge s_2' \notin r)$ and $(s_1 \notin r \wedge s_1' \in r) \Leftrightarrow (s_2 \notin r \wedge s_2' \in r)$.

Remark. It is easy to verify that, for each transition system, both the empty set and the set of all states are regions. They are called the trivial regions. We are interested in the set of non-trivial regions, which will be denoted by R_A. For each $s \in S$, R_s will denote the set of non-trivial regions containing s.

Example 2.3 Consider the transition system of Figure 1. The set $\{s_3, s_4\}$ is a region. On the other hand, $r = \{s_1\}$ is not a region, since an occurrence of e_1 "leaves" r, while another occurrence of the same event neither "leaves" nor "enters" r.

The condition defining regions allows us to define a relation between events and regions formalizing the crossing relation. This is captured by the notions of pre- and post-sets of regions and of pre- and post-sets of events.

Definition 2.4 (*pre- and post-sets of events, pre- and post-set of regions*) Let $A = (S, E, T)$ be a transition system. Let r be a region of A. Then the pre-set of r, denoted by ${}^\bullet r$, and the post-set of r, denoted by $r^\bullet$, are defined by:

$$
\begin{aligned}
{}^\bullet r &= \{e \in E \mid \exists (s, e, s') \in T : s \notin r\ \wedge\ s' \in r\} \\
r^\bullet &= \{e \in E \mid \exists (s, e, s') \in T : s \in r\ \wedge\ s' \notin r\}.
\end{aligned}
$$

Let $e \in E$. Then the pre-set and the post-set of e, denoted by, respectively, ${}^\bullet e$ and $e^\bullet$, are defined by:

$${}^\bullet e = \{r \mid r \in R_A \wedge e \in r^\bullet\} \quad \text{and} \quad e^\bullet = \{r \mid r \in R_A \wedge e \in {}^\bullet r\}.$$

Example 2.5 Consider again the transition system of Figure 1. Its non-trivial regions are:

$$\begin{array}{llll} r_1 = \{s_1, s_2\} & r_2 = \{s_1, s_3\} & r_3 = \{s_2, s_4\} & r_4 = \{s_3, s_4\} \\ r_5 = \{s_1, s_3, s_5\} & r_6 = \{s_3, s_4, s_5\} & r_7 = \{s_2, s_4, s_5\} & r_8 = \{s_5\} \\ r_9 = \{s_1, s_2, s_5\} & r_{10} = \{s_1, s_2, s_3, s_4\}. & & \end{array}$$

Some examples of pre- and post-sets of regions:

$${}^\bullet r_4 = \{e_2\} \quad {}^\bullet r_7 = \{e_1, e_3\} \quad r_7{}^\bullet = \varnothing.$$

Definition 2.6 (*isomorphism*) Let $A_1 = (S_1, E_1, T_1)$ and $A_2 = (S_2, E_2, T_2)$ be two transition systems. They are said to be *isomorphic* iff there exist two bijections, $\beta : S_1 \to S_2$ and $\eta : E_1 \to E_2$ such that:

1. $\forall\, (s, e, s') \in T_1 \quad (\beta(s), \eta(e), \beta(s')) \in T_2$
2. $\forall\, (s, e, s') \in T_2 \quad (\beta^{-1}(s), \eta^{-1}(e), \beta^{-1}(s')) \in T_1$

In order to distinguish different classes of transition systems, we introduce some reachability relations.

Definition 2.7 (*reachability relations*) $(s_1, s_2) \in K_1 \Leftrightarrow \exists\, (s_1, e, s_2) \in T$; $K := (K_1 \cup K_1^{-1})^*$.

K is an equivalence relation. The equivalence class of K containing s will be denoted by $[s]_K$.

The following definition characterizes condition/event transition systems, which correspond, in a sense to be explained later, to condition/event net systems.

Definition 2.8 (*condition/event transition system*) A transition system $A = (S, E, T)$ is a condition/event transition system, denoted CETS, iff it satisfies the following axioms:

A1 $\forall\, s \in S \quad [s]_K = S$

A2 $\forall\, s, s' \in S \quad R_s = R_{s'} \Rightarrow s = s'$

A3 $\forall\, s \in S\ \forall e \in E \quad {}^\bullet e \subseteq R_s \Rightarrow \exists\, s' \in S \quad (s, e, s') \in T$

A4 $\forall s \in S\ \forall e \in E \quad e^\bullet \subseteq R_s \Rightarrow \exists\, s' \in S \quad (s', e, s) \in T$.

2.2 Nets

In this section we briefly recall the basic definitions about nets [10].

Definition 2.9 (*net*) A net is a triple $N = (B, E, F)$ where B and E are finite sets such that $B \cap E = \varnothing$, $B \cup E \neq \varnothing$, $F \subseteq (B \times E) \cup (E \times B)$ is the flow relation, $\mathrm{dom}(F) \cup \mathrm{ran}(F) = B \cup E$.

B is the set of *conditions*, E is the set of *events*, F is the *flow relation*. We will use the standard graphical notation for nets.

Definition 2.10 (*pre- and post-sets*) Given a net $N = (B, E, F)$ and an element $x \in B \cup E$ we define the pre- and post-set of x, denoted by, respectively, ${}^\bullet x$ and $x^\bullet$:

$$^\bullet x = \{y \mid (y, x) \in F\}, \qquad x^\bullet = \{y \mid (x, y) \in F\}.$$

This notation is extended to subsets X of $B \cup E$ by setting ${}^\bullet X = \bigcup_{x \in X} {}^\bullet x$ ($X^\bullet$ is defined analogously).

Definition 2.11 (*simple, pure net*) A net $N = (B, E, F)$ is:

- *simple* iff $\forall x, y \in B \cup E \quad {}^\bullet x = {}^\bullet y$ and $x^\bullet = y^\bullet \Rightarrow x = y$
- *pure* iff $\forall x \in B \cup E \; {}^\bullet x \cap x^\bullet = \varnothing$

The occurrence of an event changes the state according to the transition rule defined as follows.

Definition 2.12 (*transition rule, transition relation*) Let $N = (B, E, F)$ be a net, $e \in E$, $c \subseteq B$.

1. c is called a *case*.
2. e is said to be enabled at c, denoted $c[e\rangle$, iff ${}^\bullet e \subseteq c$ and $e^\bullet \cap c = \varnothing$.
3. If e is enabled at c, then the occurrence of e leads from c to c', denoted $c[e\rangle c'$, iff $c' = (c \setminus {}^\bullet e) \cup e^\bullet$.
4. The transition relation of $N, \rightarrow_N \subseteq 2^B \times E \times 2^B$ is given by: $\rightarrow_N = \{(c, e, c') \mid c[e\rangle c'\}$.
5. $(c_1, c_2) \in Q_1 \Leftrightarrow c_1[e\rangle c_2$ for some $e \in E$. $Q := (Q_1 \cup Q_1^{-1})^*$ is called the *full reachability* relation.

2.3 Condition/Event Net Systems

We can define the reachable cases of a net system as an equivalence class of the full reachability relation, so that we will get a net model of a concurrent system, namely a condition/event net system (briefly CENS).

Definition 2.13 (*CENS*) A CENS is a tuple $\Sigma = (B, E, F, C)$, where (B, E, F) is a simple, pure net, C is an equivalence class of the full reachability relation given in Definition 2.12 and for each $e \in E$ there is a case of C at which e is enabled.

The sequential behaviour of a CENS Σ can be operationally described by a transition system whose nodes are the reachable cases of Σ and whose arcs are labelled by events.

Definition 2.14 (*sequential case graph*) Let $\Sigma = (B, E, F, C)$ be a CENS. The sequential case graph of Σ is the transition system $SCG(\Sigma) = (C, E, \rightarrow_\Sigma)$, where $\rightarrow_\Sigma$ is $\rightarrow_N$ restricted to $C \times E \times C$.

Figure 2 shows a CENS and its sequential case graph.

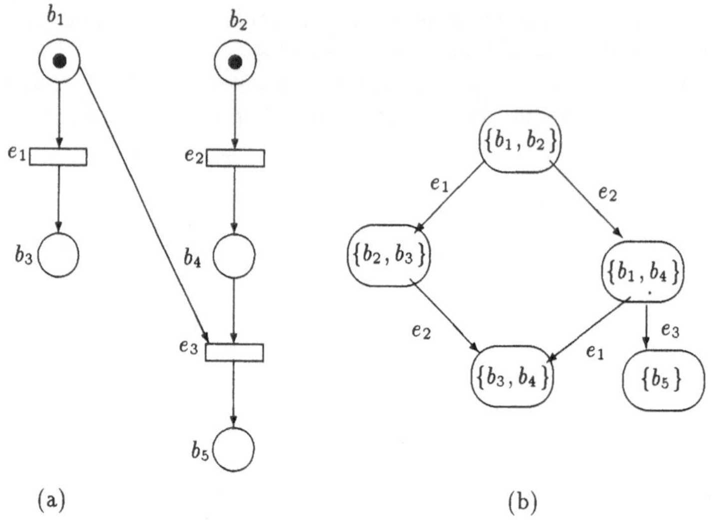

Figure 2: A CENS and its sequential case graph

2.4 Synthesis

In this section we rephrase for CETS the solution given to the synthesis problem (given a transition system A, find a net system Σ such that the sequential case graph of Σ is isomorphic to A) by Ehrenfeucht, Rozenberg, Nielsen and Thiagarajan [3, 4]. The key idea behind it is that regions are actually local states. The net system constructed according to this procedure has a condition for each region in the transition system. The flow relation is determined by pre- and post-sets of regions. We will call a net system obtained in this way from a transition system A the *saturated net version* of A. It has a very peculiar property: it is maximal with respect to the number of conditions among all the net systems whose sequential case graph is isomorphic to A.

Definition 2.15 (*saturated net version*) Let $A = (S, E, T)$ be a CETS. The *saturated net version* of A is the structure $\Sigma_A = (R_A, E, F, C)$ where $F = \{(r, e) \mid r \in {}^\bullet e\} \cup \{(e, r) \mid r \in e^\bullet\}$ and $C = \{R_s \mid s \in S\}$.

Proposition 2.16 Let $A = (S, E, T)$ be a CETS. Then Σ_A is a CENS and its sequential case graph is isomorphic to A.

Proof. See [4]. ∎

Example 2.17 The saturated net version of the transition system of Figure 1 is given in Figure 3. It is easy to see that its sequential case graph is isomorphic to the transition system of Figure 1.

2.5 The Synchronic Structure of CENS

The notion of *synchronic distance* was introduced by C. A. Petri in order to measure dependencies among sets of events in a system. Its value is computed by taking differences in the counting of occurrences of the two sets of events in

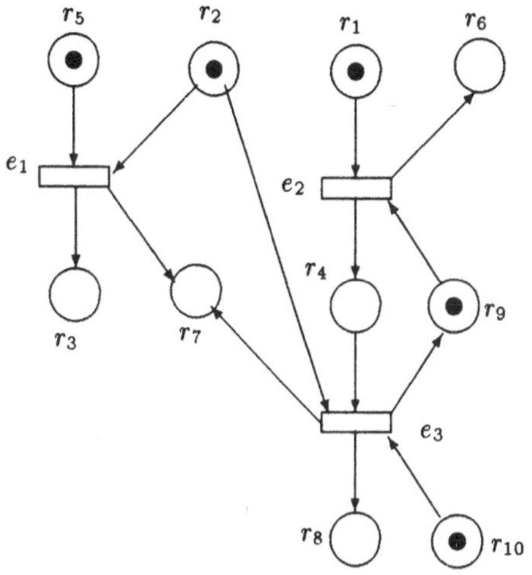

Figure 3: A saturated CENS

all possible executions of the system. In cyclic systems it can be interpreted as a relative frequency, hence it can be used, for instance, to explicitly model timing constraints without introducing an *ad hoc* formalism, just using the fundamental tools of net theory. In this section we recall the definition of synchronic distance and some of its basic properties.

There exist various definitions of synchronic distance; though slightly different, they can all be traced back to the original idea of S-completion of a net system [11]. Here we stick to the most recent one, given by Petri [12], but we extend it to weighted synchronic distances. Before giving the definition we need some preliminary notions.

Definition 2.18 Let $A = (S, E, T)$ be a CETS; a (non-directed) path in A is a non-empty sequence

$$\pi = (s_0, e_0, s_1)(s_1, e_1, s_2)\dots(s_k, e_k, s_{k+1})$$

such that

$$\forall i \in \{0, \dots, k\} \quad (s_i, e_i, s_{i+1}) \in T \vee (s_{i+1}, e_i, s_i) \in T;$$

s_0 is the initial state of π and s_{k+1} is its final state. $\Pi(A)$ is the set of all paths of A. $-\pi$ is the path $(s_{k+1}, e_k, s_k)\dots(s_1, e_0, s_0)$.

Definition 2.19 Given a CETS $A = (S, E, T)$, a total function $g : E \to \mathbf{N}$, a subset E_1 of E and a path $\pi \in \Pi(A)$, we define a function *count* as follows:

1. if $\pi = (s_1, e, s_2)$, then

$$count(g, \pi, E_1) = \begin{cases} g(e) & \text{if } e \in E_1 \text{ and } (s_1, e, s_2) \in T \\ -g(e) & \text{if } e \in E_1 \text{ and } (s_2, e, s_1) \in T \\ 0 & \text{if } e \notin E_1 \end{cases}$$

2. if $\pi = \pi_1\pi_2$ then $count(g, \pi, E_1) = count(g, \pi_1, E_1) + count(g, \pi_2, E_1)$.

Notice that in a CETS, if $(s_1, e_1, s_2), (s_2, e_2, s_1) \in T$, then $e_1 \neq e_2$; hence *count* is well-defined.

Definition 2.20 (*weighted synchronic distance*) Let $\Sigma = (B, E, F, C)$ be a CENS and A its sequential case graph. Let E_1, E_2 be subsets of E and let $g : E \rightarrow \mathbf{N}$ be a weight function. The g-weighted synchronic distance of E_1 and E_2 is defined as follows:

$$\sigma_g(E_1, E_2) = \sup_{\pi \in \Pi(A)} \mid count(g, \pi, E_1) - count(g, \pi, E_2) \mid,$$

if the supremum exists, and ∞ otherwise.

This definition is closely related to the notion of S-completion of a net [11]. We will not discuss in detail this notion but just mention that, if $\sigma_g(E_1, E_2) = h$, then we can add a place p to Σ, with an arc from each event $e \in E_1$ to p with weight $g(e)$, and an arc from p to each event $e' \in E_2$ with weight $g(e')$. By a suitable choice of the initial marking, we obtain a place/transition system whose marking graph is isomorphic to the case graph of Σ.

It is easy to show that σ_g is indeed a symmetric distance (i.e., it satisfies the symmetric and triangular properties, and $\sigma_g(E, E) = 0$). Moreover, synchronic distances are actually computable [13].

Example 2.21 In the net system of Figure 2(a), $\{e_1\}$ and $\{e_2\}$ have synchronic distance 2, while $\{e_2\}$ and $\{e_3\}$ have synchronic distance 1 (assuming as weight function $g = 1$).

3 Aspects of The Synchronic Structure of Transition Systems

3.1 Generalized Regions

In this section we generalize the notion of region recalled in section 2.1, switching from sets to multisets of states of a transition system. The "crossing" property of regions translates into a "gradient" property.

Definition 3.1 (*generalized region*) Let $A = (S, E, T)$ be a CETS, let $\mathbf{N}$ be the set of integers. A multiset $r : S \longrightarrow \mathbf{N}$ is a generalized region iff

$$\forall\ (s_1, e, s_1'), (s_2, e, s_2') \in T \quad r(s_1) - r(s_1') = r(s_2) - r(s_2').$$

GR_A is the set of generalized regions of A.

This means that a generalized region is a multiset of states such that each occurrence of a label uniformly translates its multiplicity. Notice that the zero multiset, here denoted simply by 0, is always a generalized region. "Standard" regions represent a special case in which the multiplicities have been reduced to 0 and 1. In this sense the definition generalizes the one given by Ehrenfeucht and Rozenberg [3]. In the rest of the paper we will call the generalized regions simply regions.

In [7] the reader can find a definition of region similar to ours. There, in order to characterize transition systems corresponding to the class of general place/transition nets, the author has to add information in the structure of the transition systems, under the form of explicit concurrent steps. Symmetrically, the regions he uses are specified not only by the map on the set of states, but also by a map on the set of events, which is not uniquely determined by the former. In our approach we have chosen to generalize elementary transition systems while keeping the extensionality of events and the finiteness of the structures under consideration. We argue that the class of bounded systems is interesting to the system designer for two reasons; the boundedness naturally corresponds to the limitation of resources, which is often a key feature in real systems, and the availability of polynomial algorithms solving the synthesis problem [5] makes them viable as an effective specification tool when the aim is to obtain a distributed model in terms of a Petri net.

We use the following notation for multisets: if $h : \{s_1, s_2, s_3\} \to \mathbf{N}$ is such that $h(s_1) = 3$, $h(s_2) = 1$ and $h(s_3) = 0$, we will write $h = 3s_1 \cup s_2$.

Example 3.2 Consider the CETS of Figure 1 and the following multisets:

$$r_1 = s_1 \cup 2s_2 \cup s_4 \quad r_2 = 2s_1 \cup s_2 \cup s_3 \cup 2s_5 \quad r_3 = s_1 \cup 2s_3 \cup 3s_5.$$

r_1 and r_2 are regions, while r_3 is not a region since $(s_1, e_1, s_2), (s_3, e_1, s_4) \in T$, but $r_3(s_1) - r_3(s_2) \neq r_3(s_3) - r_3(s_4)$. Note that in this example r_1 and r_2 are sums of standard regions, but this is not always true in the general case.

Definition 3.3 (*pre- and post-multiset*) Let $A = (S, E, T)$ be a CETS and let r be a region of A; the pre- and post-multiset of r are, respectively, the mappings ${}^\bullet r, r^\bullet : E \to \mathbf{N}$ defined as follows: if $(s, e, s') \in T$

$${}^\bullet r(e) = \max\{0, r(s') - r(s)\} \quad \text{and} \quad r^\bullet(e) = \max\{0, r(s) - r(s')\}.$$

Due to the properties of regions (from Definition 3.1), this definition is independent from the choice of (s, e, s').

Example 3.4 Consider the CETS of Figure 1 and the region r_1 of the example 3.2; then ${}^\bullet r_1 = e_1$ and $r_1{}^\bullet = e_2$.

We remark here that it follows immediately from Definition 3.3 that for any $r \in GR_A$ and $e \in E$ one has ${}^\bullet r(e) r^\bullet(e) = 0$.

Definition 3.5 (*normal form*) Let $A = (S, E, T)$ be a CETS. A region r of A is in normal form iff $\exists\, s \in S \quad r(s) = 0$.

Let $\equiv$ be the equivalence relation defined as follows: $r_1 \equiv r_2$ iff ${}^\bullet r_1 = {}^\bullet r_2$ and $r_1{}^\bullet = r_2{}^\bullet$; then, for each $r \in GR_A$, the equivalence class of r under $\equiv$ contains one and only one region in normal form. This happens because if r, r' are regions lying in the same equivalence class, then $r - r'$ is a constant function.

Definition 3.6 (*canonical region*) Let $A = (S, E, T)$ be a CETS. We will call the equivalence classes of GR_A under $\equiv$ canonical regions. CGR_A is the set of canonical regions: $CGR_A = GR_{A/\equiv}$. We will denote each element of CGR_A through its normal form representative.

Definition 3.7 (*sum and complement*) Let $A = (S, E, T)$ be a CETS and let $r, r' \in CGR_A$. Then the sum of r and r', $r \oplus r'$, is the canonical region such that $(r \oplus r')(s) = r(s) + r'(s) - k$, where $k = \min\{|r(s) + r'(s)| \mid s \in S\}$; the complement of r, $\bar{r}$, is the canonical region such that $\bar{r}(s) = k - r(s)$, where $k = \max\{r(s) \mid s \in S\}$

We can easily see that generalized regions form an abelian group.

Proposition 3.8 The sum of canonical regions defines an abelian group on the set of canonical regions, in which the unit is the null region and the inverse of a region r is its complement $\bar{r}$.

Definition 3.9 (*subregion*) Let $A = (S, E, T)$ be a CETS. Let r, r' be canonical regions of A.

1. r' is a *subregion* of r, denoted by $r' \subseteq r$, iff $\forall s \in S \quad r'(s) \leq r(s)$;

2. r' is a *proper subregion* of r, denoted by $r' \subset r$, iff r' is a subregion of r and $\exists s \in S \quad r'(s) < r(s)$.

The notion of subregion induces a partial order on the set of canonical regions. We define the set of minimal canonical regions and give some of its properties.

Definition 3.10 (*minimal canonical region*) Let $A = (S, E, T)$ be a CETS. A canonical region r is minimal iff $r \neq 0$ and $\forall r' \in CGR_A \backslash \{0\} \quad r' \not\subset r$. $MCGR_A$ is the set of minimal canonical regions of A.

Example 3.11 Consider the CETS of Figure 1: following the notation of Example 2.5, its minimal regions are r_1, r_2, r_3, r_4 and r_8. Note that in general a minimal region might not be a standard region.

Proposition 3.12 Every canonical region can be represented as a linear combination of minimal canonical regions.

Proof. Let r be a canonical region; if r is not minimal then there exists $r_1 \in MCGR_A$ such that $r_1 \subset r$. Let k_1 be the greatest natural number such that $k_1 r_1 \subseteq r$. Then $r = r_2 \oplus k_1 r_1 = r_2 \cup k_1 r_1$, where r_2 is again a region. If r_2 is not minimal then we can apply the same rule to r_2. Hence after a finite number of steps we obtain $r = k_1 r_1 \oplus k_2 r_2 \oplus \ldots \oplus k_n r_n$ where $r_1, r_2, \ldots, r_n$ are minimal canonical regions. ▮

This proposition shows that minimal canonical regions are sufficient to construct all canonical regions. Moreover this set is finite because all the antichains of $\mathbf{N}^k$ are finite (Dickson's Lemma). Using well known abelian group properties we can reinforce the latter finiteness result, showing that canonical regions have a finite basis and giving some bounds for it.

Let us recall some basic properties of abelian groups. The notation $\langle X \mid Y \rangle$, where X is a set of letters (generators) and Y is a set of formal sums (relations) $\sum_{x \in X} n_x x$, with $n_x \in \mathbf{Z}$, denotes the abelian group generated by X and satisfying the equations $\{y = 0\}_{y \in Y}$. For instance, $\langle \{x\} \mid \{px\} \rangle$ is the cyclic group of order p. Such a group is best described as having as elements functions $g : X \rightarrow \mathbf{Z}$, which are identified up to pointwise addition or subtraction of the functions $x \mapsto n_x$ defined by the relations.

Proposition 3.13 Let $A = (S, E, T)$ be a CETS. The abelian group of the canonical regions of A is isomorphic to a subgroup of $G_A = \langle S \mid \sum_{s \in S} s \rangle$, i.e., the abelian group having as generators the set of states, and as unique relation the sum of all states.

Proof. We can extend the characteristic map of a region via the monoid injection $\mathbf{N} \rightarrow \mathbf{Z}$. This construction yields a well defined map $\phi : CGR_A \rightarrow G_A$. It is easy to see that ϕ is a group morphism mapping $\oplus$ to $+$. ▮

We now have a clean algebraic representation of generalized regions. Moreover, we know that CGR_A is a subgroup of a free abelian group. We recall a basic definition [14]:

Definition 3.14 A *basis* of an abelian group G is a subset X of G such that every element of G is a finite linear combination of elements of X (i.e., for all $g \in G$ we have $g = n_1x_1 + n_2x_2 + \cdots + n_kx_k$ with $n_i \in \mathbf{Z}$, $x_i \in X$), and if for distinct $x_1, x_2, \ldots, x_k \in X$ and $n_i \in \mathbf{Z}$ $n_1x_1 + n_2x_2 + \cdots + n_kx_k = 0$, then $n_i = 0$ for every i.

It can be shown that any two bases of an abelian group have the same cardinality. This cardinality is then called the *rank* of the group. For instance, $\langle S \mid \rangle$ has clearly rank $|S|$. A classic theorem says that subgroups of free abelian groups are finitely generated abelian groups, with smaller or equal rank. Since a standard result shows that $\langle S | \sum_{s \in S} s \rangle$ is isomorphic to $\langle S \setminus \{\bar{s}\} \mid \rangle$ for any $\bar{s} \in S$ (i.e., the equation $\sum_{s \in S} s = 0$ cancels exactly one generator) we have that

Theorem 3.15 Let $A = (S, E, T)$ be a CETS. The group CGR_A of all canonical regions has rank $\leq |S| - 1$.

Thus, we have a first upper bound on the number of generators for CGR_A. Now we turn our attention to another abelian group, the group of *differentials*: it is defined as the free abelian group $\mathcal{D}_A = \langle E \mid \rangle$

The "gradient" property of regions allows us to define a differential function Δ_r which associates to each event $e \in E$ the variation in multiplicity between target and source. This remains true also for the free group representation, so for $r = \sum_{s \in S} n_s s$, and for $(s, e, s') \in T$ we can define $\Delta_r(e) = n_{s'} - n_s$. Therefore Δ_r can be seen as an element of the group of differentials, and $\Delta_{(-)}$ is a function from CGR_A to $\mathcal{D}_A$. And it is really something more:

Proposition 3.16 $\Delta_{(-)}$ is a group morphism $CGR_A \to \mathcal{D}_A$.

Proof. It is clear that $\Delta_{(0)} = 0$. Moreover, if $r = \sum_{s \in S} n_s s$, $r' = \sum_{s \in S} n'_s s$ and $(s, e, s') \in T$

$$\Delta_{(r \oplus r')}(e) = (n'_{s'} + n_{s'}) - (n'_s + n_s) =$$
$$= (n'_{s'} - n'_s) + (n_{s'} - n_s) = \Delta_r(e) + \Delta_{r'}(e). \blacksquare$$

The kernel of a group morphism is the subgroup of elements which is mapped to the zero of the target group. A morphism is injective iff its kernel is just composed by the zero of the source group.

Proposition 3.17 $\mathrm{Ker}(\Delta_{(-)}) = 0$.

Proof. If $\Delta_r = 0$, then for all transitions $(s, e, s') \in T$ we have $n_s = n_{s'}$. But r is a canonical region, and therefore $\exists\ s \in S\ r(s) = 0$. Axiom **A1** of CETS (see Definition 2.8) guarantees that all states have 0 as coefficient. This implies $r = 0$. $\blacksquare$

Theorem 3.18 Let $A = (S, E, T)$ be a CETS. The group CGR_A of all canonical regions has rank $\leq |E|$.

Thus, we have the following limitation on the number of generators:

$$\mathrm{rank}(CGR_A) \leq \min(|E|, |S| - 1).$$

3.2 Structural Synchronic Distance

In this section we discuss the strong relation between the set of canonical regions of a condition/event transition system A and the synchronic structure [11] of any condition/event net system whose sequential case graph is isomorphic to A. The main result of this section consists in showing that canonical regions can be used to characterize the synchronic distances in CETS.

Definition 3.19 Let $A = (S, E, T)$ be a CETS. Let $g : E \to \mathbf{N}$ be a weight function and $E_1 \subseteq E$; we define a mapping $g_{E_1} : E \to \mathbf{N}$ as follows:

$$g_{E_1}(e) = \begin{cases} g(e) & \text{if } e \in E_1 \\ 0 & \text{otherwise.} \end{cases}$$

Definition 3.20 (*structural synchronic distance*) Let $A = (S, E, T)$ be a CETS. Let E_1, E_2 be subsets of E and let $g : E \to \mathbf{N}$ be a weight function. The g-weighted structural synchronic distance of E_1 and E_2 is defined as follows:

$$\begin{array}{ll} \delta_g(E_1, E_1) = 0 & \forall E_1 \in 2^E \\ \delta_g(E_1, E_2) = \max\{r(s) \mid s \in S\} & \text{if } \exists r \in CGR_A \text{ s.t. } {}^\bullet r = g_{E_1 \setminus E_2} \\ & \text{and } \; r^\bullet = g_{E_2 \setminus E_1} \\ \delta_g(E_1, E_2) = \infty & \text{otherwise.} \end{array}$$

In order to show that the previous definition coincides with the one given in Definition 2.20 we introduce some interesting properties.

Definition 3.21 Let $A = (S, E, T)$ be a CETS. Let r be a region of A, ${}^\bullet r$ and $r^\bullet$ as in Definition 3.3; then $[{}^\bullet r] = \{e \in E \mid {}^\bullet r(e) > 0\}$ and $[r^\bullet] = \{e \in E \mid r^\bullet(e) > 0\}$.

Lemma 3.22 Let $A = (S, E, T)$ be a CETS. Let r be a region of A and let s_0, s_1 be, respectively, the initial and final state of a path $\pi \in \Pi(A)$. Then

$$r(s_0) - r(s_1) = count(r^\bullet, \pi, [r^\bullet]) - count({}^\bullet r, \pi, [{}^\bullet r]).$$

Proof. We proceed by induction on the length of the path π.
Induction basis: Let $\mid \pi \mid = 1$ and suppose $\pi = (s_0, e, s_1)$; then

$$count(r^\bullet, \pi, [r^\bullet]) = \begin{cases} r^\bullet(e) & \text{if } r^\bullet(e) > 0 \text{ and } (s_0, e, s_1) \in T \\ -r^\bullet(e) & \text{if } r^\bullet(e) > 0 \text{ and } (s_1, e, s_0) \in T \\ 0 & \text{otherwise.} \end{cases}$$

But suppose $r^\bullet(e) > 0$; then,

$$r^\bullet(e) = \begin{cases} r(s_0) - r(s_1) & \text{if } (s_0, e, s_1) \in T \\ r(s_1) - r(s_0) & \text{if } (s_1, e, s_0) \in T \end{cases}$$

and

$$count(r^\bullet, \pi, [r^\bullet]) = \begin{cases} r(s_0) - r(s_1) & \text{if } r^\bullet(e) > 0 \\ 0 & \text{otherwise.} \end{cases}$$

In the same way we obtain:

$$count({}^\bullet r, \pi, [{}^\bullet r]) = \begin{cases} r(s_1) - r(s_0) & \text{if } {}^\bullet r(e) > 0 \\ 0 & \text{otherwise.} \end{cases}$$

Since $r^\bullet(e) > 0 \Rightarrow {}^\bullet r(e) = 0$ (and viceversa), then

$$count(r^\bullet, \pi, [r^\bullet]) - count({}^\bullet r, \pi, [{}^\bullet r]) = r(s_0) - r(s_1).$$

Induction hypothesis: $\forall \pi_n = (s_0, e_0, s_1)(s_1, e_1, s_2)\ldots(s_{n-1}, e_{n-1}, s_n)$ with $n \geq 1$ we have $count(r^\bullet, \pi_n, [r^\bullet]) - count({}^\bullet r, \pi_n, [{}^\bullet r]) = r(s_0) - r(s_n)$.

Induction step: Let $|\ \pi_{n+1}\ | = n+1$; then $\pi_{n+1} = \pi_n(s_n, e_n, s_{n+1})$ and

$$\begin{aligned} & count(r^\bullet, \pi_{n+1}, [r^\bullet]) - count({}^\bullet r, \pi_{n+1}, [{}^\bullet r]) = \\ = \quad & count(r^\bullet, \pi_n, [r^\bullet]) - count({}^\bullet r, \pi_n, [{}^\bullet r]) + \\ & + count(r^\bullet, (s_n, e_n, s_{n+1}), [r^\bullet]) - count({}^\bullet r, (s_n, e_n, s_{n+1}), [{}^\bullet r]) \\ = \quad & r(s_0) - r(s_n) + r(s_n) - r(s_{n+1}) = r(s_0) - r(s_{n+1}) \end{aligned}$$

as required. ∎

Lemma 3.23 Let $A = (S, E, T)$ be a CETS, g a weight function, $E_1, E_2 \subseteq E$, and r a canonical region of A such that ${}^\bullet r = g_{E_1 \setminus E_2}$ and $r^\bullet = g_{E_2 \setminus E_1}$; then

$$\sigma_g(E_1, E_2) = \max\{r(s) \mid s \in S\}.$$

Proof. By a property of synchronic distance [13]:

$$\begin{aligned} & \sigma_g(E_1, E_2) = \sigma_g(E_1 \setminus E_2, E_2 \setminus E_1) = \\ = \quad & \sup_{\pi \in \Pi(A)} \mid count(g_{E_1 \setminus E_2}, \pi, E_1 \setminus E_2) - count(g_{E_2 \setminus E_1}, \pi, E_2 \setminus E_1) \mid \\ = \quad & \sup_{\pi \in \Pi(A)} \mid count({}^\bullet r, \pi, [{}^\bullet r]) - count(r^\bullet, \pi, [r^\bullet]) \mid \quad \text{(by hypothesis)} \\ = \quad & \sup_{\pi \in \Pi(A)} \mid r(s_0) - r(s_n) \mid \qquad \text{(by Lemma 3.22)} \\ = \quad & \max\{r(s) \mid s \in S\}, \end{aligned}$$

where s_0, s_n are respectively the initial and the final state of π. ∎

We are now ready to prove the main theorem of this section, that allows us to characterize the synchronic distance in terms of generalized regions.

Theorem 3.24 Let $A = (S, E, T)$ be a CETS. Then $\forall E_1, E_2 \subseteq E$ and $\forall g : E \to \mathbf{N}$

$$\delta_g(E_1, E_2) = \sigma_g(E_1, E_2).$$

Proof. Obviously $\sigma_g(E_1, E_1) = \delta_g(E_1, E_1) = 0$. We prove that $\sigma_g(E_1, E_2) = n \Leftrightarrow \delta_g(E_1, E_2) = n$. Assume first $\sigma_g(E_1, E_2) = n$; then we can construct a region r that satisfies the conditions of Definition 3.20. We proceed by steps. 1) Given E_1, E_2, g, we define a function $p : \Pi_A \to \mathbf{Z}$ as follows: $p(\pi) = count(g_{E_1}, \pi, E_1) - count(g_{E_2}, \pi, E_2)$. We show that if π_1 and π_2 are paths of A from s to s' and $\sigma_g(E_1, E_2) = n$ then $p(\pi_1) = p(\pi_2)$. The path $\pi = \pi_1 \cdot (-\pi_2)$ is a cycle such that $p(\pi) = p(\pi_1) - p(\pi_2)$. If $p(\pi) \neq 0$ then $\sigma_g(E_1, E_2) = \infty$. So we have $p(\pi_1) = p(\pi_2)$. 2) We define a function $f : S \to \mathbf{Z}$ as follows: let $\hat{s}$ be any state of S; then $f(\hat{s}) = 0$ and for all $s \in S$ $f(s) = p(\pi)$, where π is a path from $\hat{s}$ to s. By point 1), and as A is connected, f is well-defined. 3) We associate to f a function $r_f : S \to \mathbf{N}$ such that $r_f(s) = f(s) - m$ where $m = \min\{f(s) \mid s \in S\}$. It is easy to show that r_f is a region such that $\max\{r_f(s) \mid s \in S\} = n$, ${}^\bullet r = g_{E_1 \setminus E_2}$ and $r^\bullet = g_{E_2 \setminus E_1}$.

Now suppose that $\delta_g(E_1, E_2) = n$. In this case there is an $r \in CGR_A$ such that ${}^\bullet r = g_{E_1 \setminus E_2}$, $r^\bullet = g_{E_2 \setminus E_1}$ and $\delta_g(E_1, E_2) = \max\{r(s) \mid s \in S\}$, but by Lemma 3.23 $\sigma_g(E_1, E_2) = \max\{r(s) \mid s \in S\}$. This completes the proof. ∎

4 Further Developments: The Synthesis of P/T Systems

In a place/transition system (shortly, P/T system) S-elements represent counters which may hold a variable number of "tokens"; the occurrence of transitions, represented by T-elements, changes the distribution of tokens. In accordance with the previous section we will consider non initialized P/T systems, having as global states the elements of an equivalence class of the full reachability relation.

Definition 4.1 (*P/T net*) A P/T net is a quintuple $N = (S, T, F, W, K)$ where (S, T, F) is a net, $W : F \to \mathbf{N} \setminus \{0\}$ is a *weight* function and $K : S \to \mathbf{N} \cup \{\infty\}$ is a *capacity* function. The S-elements are called *places* and the T-elements are called *transitions.*

For the rest of this section we will consider a subclass of P/T nets in which there are no loops and all capacities are finite.

Definition 4.2 (*firing rule*) Let $N = (S, T, F, W, K)$ be a P/T net. Then, a function $M : S \to \mathbf{N}$ is called a *marking* of N iff $\forall s \in S\ M(s) \leq K(s)$. A transition $t \in T$ is enabled at M iff $\forall s \in S\quad W'(s,t) \leq M(s) \leq K(s) - W'(t,s)$, where

$$W'(x,y) = \begin{cases} W(x,y) & \text{if } (x,y) \in F \\ 0 & \text{otherwise.} \end{cases}$$

If t is a transition enabled at a marking M then t may occur yielding a new marking M' such that $\forall s \in S\quad M'(s) = M(s) - W'(s,t) + W'(t,s)$. The occurrence of t is denoted $M[t\rangle M'$.

Definition 4.3 (*P/T net system*) A sextuple $\Sigma = (S, T, F, W, K, \mathcal{M})$ is a (non initialized) P/T net system iff (S, T, F, W, K) is a P/T net, and $\mathcal{M}$ is an equivalence class of the full reachability relation Q defined as follows:

$$(M_1, M_2) \in Q_1 \Leftrightarrow M_1[t\rangle M_2 \text{ for some } t \in T; \quad Q := (Q_1 \cup Q_1^{-1})^*.$$

The behaviour of a place/transition net system can be described by a graph whose nodes are reachable markings and arcs are labelled by transitions. The details of the construction can be found in [15]. In the following we will call this graph the *sequential marking graph* of Σ, $SMG(\Sigma)$.

Definition 4.4 (*elementary P/T net system*) A (non initialized) P/T net system $\Sigma = (S, T, F, W, K, \mathcal{M})$ is *elementary* iff (S, T, F) is a pure and simple net, and, for all $s \in S$, $K(s)$ is finite.

Elementary P/T net systems can be generated from a (non initialized) elementary P/T transition systems by means of canonical regions

Definition 4.5 (*elementary P/T transition system*) $A = (S, E, T)$ is a (non initialized) elementary P/T transition system (shortly, EPTTS) iff axiom **A1** of Definition 2.8 is satisfied together with the following three ones:

R1 $\forall s, s' \in S\ \forall r \in CGR_A \quad r(s) = r(s') \Rightarrow s = s'$

R2 $\forall s \in S\ \forall e \in E\ \forall r \in CGR_A \quad r(s) \geq r^{\bullet}(e) \Rightarrow \exists s' \in S\ : (s, e, s') \in T$

R3 $\forall s \in S\ \forall e \in E\ \forall r \in CGR_A \quad r(s) \geq {}^{\bullet}r(e) \Rightarrow \exists s' \in S\ : (s', e, s) \in T.$

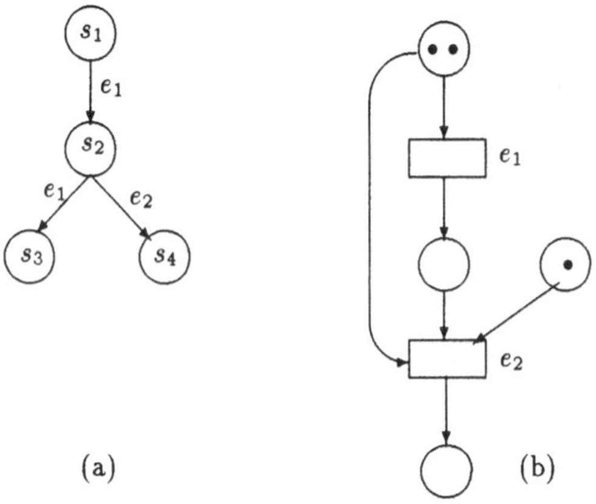

Figure 4: An EPTTS and its minimal saturated net version

Definition 4.6 (*pre- and post-regions*) Let $A = (S, E, T)$ be a EPTTS and $e \in E$; the sets of pre- and post-regions are defined as follows: ${}^\bullet e = \{r \in CGR_A \mid r^\bullet(e) > 0\}$ and $e^\bullet = \{r \in CGR_A \mid {}^\bullet r(e) > 0\}$.

The minimal saturated net version of a transition system generalizes the construction made in [16]; the key idea is to use minimal canonical regions as places of the net. The crossing relation between regions and events determines the flow relation of the net and the arc weights. Notice that the set of canonical regions is infinite; we consider minimal canonical region to obtain a finite net structure.

Definition 4.7 (*minimal saturated EPTTS net version*) Let $A = (S, E, T)$ be a EPTTS; the minimal saturated net version of A is the structure $\Sigma_A = (MCGR_A, E, F, W, K, \mathcal{M})$ where:

1. $F = \{(r, e) \mid r \in {}^\bullet e\} \cup \{(e, r) \mid r \in e^\bullet\}$
2. $W(r, e) = r^\bullet(e)$ and $W(e, r) = {}^\bullet r(e)$
3. $K(r) = \max\{r(s) \mid s \in S\}$
4. $\mathcal{M} = \{M_s \mid s \in S\}$, where M_s is defined by $M_s(r) = r(s)$ for all $r \in MCGR_A$.

Theorem 4.8 Let $A = (S, E, T)$ be a EPTTS and let Σ_A be its minimal saturated net version. Then $SMG(\Sigma_A)$ is isomorphic to A.

The proof can be found in [17].

Example 4.9 The transition system of Figure 4(a) is an EPTTS. The net system of Figure 4(b) is its minimal saturated net version.

Acknowledgments

This research has been conducted under the financial support of the Italian Ministero dell'Università e della Ricerca Scientifica e Tecnologica, and of the ESPRIT-BRA WG-6067 CALIBAN. Luca Bernardinello, on leave from the University of Milano, has been partially supported by an H.C.M. fellowship. Katia

Petruni, on leave at the Humboldt University of Berlin, has been supported by a "Borsa di perfezionamento all'estero per laureati" of the University of Milan.

References

[1] C.A. Petri. Kommunikation mit automaten. Schriften des Institutes für Instrumentelle Mathematik, Bonn, 1962.

[2] C.A. Petri. Concepts of net theory. In *Mathematical Foundations of Computer Science. Proceedings of Symposium and Summer School*, pages 137–146, Bratislava, 1973. Mathematical Institute of Slovak Academy of Sciences.

[3] A. Ehrenfeucht and G. Rozenberg. Partial (set) 2-structures. I and II. *Acta Informatica*, 27(4):315–368, 1990.

[4] M. Nielsen, G. Rozenberg, and P.S. Thiagarajan. Elementary transition systems. *Theoretical Computer Science*, 96:3–33, 1992.

[5] E. Badouel, L. Bernardinello, and P. Darondeau. Polynomial algorithms for the synthesis of bounded nets. Technical Report PI-847, IRISA, July 1994. To appear in Proceedings of CAAP '95.

[6] J.Desel and A.Merceron. P/T-systems as abstractions of C/E-systems. In G. Rozenberg, editor, *Advances in Petri Nets 1989*, number 424 in LNCS. Springer-Verlag, 1990.

[7] M. Mukund. Petri nets and step transition systems. *International Journal of Foundations of Computer Science*, 96(1), 1992.

[8] M. Silva and T. Murata. B-fairness and structural B-fairness in Petri net models of concurrent systems. *Journal of Computer and System Sciences*, 44:447–477, 1992.

[9] I. Suzuki and T. Kasami. Three measures for synchronic dependence in Petri nets. *Acta Inform.*, 19:325–338, 1983.

[10] W. Reisig. *Petri Nets. An Introduction*. EATCS Monographs on Theoretical Computer Science, Springer–Verlag, 1985.

[11] P.S. Thiagarajan H.J. Genrich, K. Lautenbach. Elements of general net theory. In W. Brauer, editor, *Net Theory and Applications*, number 84 in LNCS. Springer-Verlag, 1980.

[12] C.A. Petri. An outline of general net theory. Unpublished manuscript, 1992.

[13] U. Goltz. Synchronic distance. In W. Brauer, W. Reisig, and G. Rozenberg, editors, *Petri Nets: Central Models and Their Properties*, number 254 in LNCS. Springer-Verlag, 1987.

[14] T.W. Hungeford. *Algebra*. Springer-Verlag, 1974.

[15] W. Reisig. Place/transition systems. In W. Brauer, W. Reisig, and G. Rozenberg, editors, *Petri Nets: Central Models and Their Properties*, number 254 in LNCS. Springer-Verlag, 1987.

[16] L. Bernardinello. Synthesis of net systems. In *Proceedings of 14th International Conference on Application and Theory of Petri Nets*, Chicago (USA), 1993.

[17] K. Petruni. Verso una caratterizzazione algebrica della distanza sincronica. Università degli studi di Milano, Tesi di Laurea, 1993.

An M-net Semantics of $B(PN)^2$

Eike Best[1], Hans Fleischhack[2], Wojciech Frączak[3], Richard P. Hopkins[4], Hanna Klaudel[3] and Elisabeth Pelz[5]

This work is partly supported by ESPRIT WG CALIBAN, no 6067.

Abstract

Using a class of high level Petri nets, M-nets, endowed with composition operators resembling those of CCS, we give the compositional semantics of $B(PN)^2$ - a syntactically simple but semantically powerful concurrent programming language. We also give an associated low level net semantics and show the consistency of these high and low level semantics, as well as consistency with a previously defined low level semantics of $B(PN)^2$.

1 Introduction

$B(PN)^2$ (a basic Petri net programming notation, [2]) is a notation for the specification and programming of concurrent algorithms which incorporates within a simple syntax many of the constructs used in concurrent programming languages - (nested) parallel composition, iteration, guarded commands, and communication via both handshake and buffered communication channels, as well as shared variables. $B(PN)^2$ has an existing compositional Petri net semantics, given via the Petri Box Calculus (PBC, [3]) which is a formalism developed in order to support the application of net theoretical methods to the verification of concurrent algorithms and also to achieve the compositional semantics of languages needed to express them. This formalism supports operators, similar to the operators of CCS [15], on a class of labelled 1-safe place/transition nets. Furthermore, the PBC has an action structure allowing partial synchronisation between multisets of actions, similar to that in SCCS [15], but within an asynchronous framework. This plays a crucial role in obtaining a parsimonious compositional semantics for the powerful shared variable accessing provided in $B(PN)^2$.

The PBC semantics have been applied to program verification, in the *PEP* system [4] which supports: generation of a low level net from a $B(PN)^2$ program; interactive simulation of the net; verification of a temporal logic formula using a fast model checker. For any practical use of the semantics such as this, the size of the nets produced by the semantics becomes an issue in view of their machine representability and amenability to automated techniques: the elementary net may need to be infinite, namely, when the program contains a declaration of a variable with infinite value domain.

[1] Institut für Informatik, Universität Hildesheim, Germany
[2] Fachbereich Informatik, Universität Oldenburg, Germany
[3] LRI, Université Paris Sud, Orsay, France
[4] Department of Computing Science, University of Newcastle upon Tyne, UK
[5] LRI, Université Paris Sud, Orsay, and UPVM, Créteil, France

An obvious way to deal with this problem was to give the language a high level net semantics [10, 11, 17, 18]. Because compositionality remained of primary importance to us, it was necessary to view high level nets not just as abbreviations of low level nets, but to define suitable operations (corresponding to those of the PBC) on the domain of these nets. In the literature there are some high level net models providing some forms of compositionality, for instance those involving the composition of net modules by place and/or transition fusion [8, 16] or by synchronisation expressions as in [7]. However, there is no model providing the systematic action synchronisation required for the $B(PN)^2$ semantics, i.e., the partial synchronisation on action multisets described later in this paper. For this reason, we have developed and studied two classes of composable high level nets: M-nets (multilabelled high level nets [5]), a class of nets useful for this paper; and A-nets (algebraic multilabelled nets [9, 13]), a generalisation of M-nets supporting abstract data types. M-nets are a class of high level nets providing both vertical unfolding as usual, and also suitable horizontal composition operations. In [5], the coherence of these two kinds of operations has been shown there; that is, all horizontal operations commute with unfolding. Following this initial development, it is now the purpose of the present paper to demonstrate the use of M-nets in giving the high-level net semantics of programming languages, by giving an M-net semantics for $B(PN)^2$. Whilst there are benefits of a high level net semantics, there are also benefits of a low level net semantics (for a program yielding a net of tractable size) due to the greater availability of analysis techniques. Thus we also give here a compositional low level net semantics in terms of a restricted class of M-nets which correspond to labelled place/transition nets. We present a theorem which shows the consistency of the two semantics, and thus also the consistency between the new high level net semantics and the existing PBC semantics [2]. This consistency theorem relies on the coherence results proved in [5].

The structure of this paper is as follows. In Section 2 we give an extended informal introduction to both the $B(PN)^2$ programming language and to M-nets[6]. Sections 3 to 6 describe the syntax and M-net semantics of $B(PN)^2$. Section 7 describes the main result, while some concluding remarks can be found in Section 8.

2 Key Concepts

In this section we consider the following example program fragment:

$\mathcal{F}$ **begin var** $x : \{0,1,2\}$; **begin var** $y : \{0,1\}$; $\langle 'x = 0 \wedge x' = 'y + 1 \wedge y' = 'y \rangle$
end
end.

The construct **var** x:$\{0,1,2\}$ is the declaration of a program variable x with type $\{0,1,2\}$ as the set of its possible values. The **var** y:$\{0,1\}$ is similarly the inner declaration of y. The $\langle 'x = 0 \wedge x' = 'y + 1 \wedge y' = 'y \rangle$ denotes an atomic action comprising a logical expression. This accesses the variables x and y with $'x$ and $'y$ denoting their values prior to execution of the action ('pre-values')

[6]Formal definitions can be found in [1, 5].

and x' and y' denoting their values afterwards ('post-values'). The action can be executed, and establishes particular post-values, if the existing pre-values and the established post-values are such that the logical expression evaluates to **true**. In this case execution depends on x being initially 0, by $'x = 0$, and the effect is $x := y + 1$ (where + is addition modulo 3), by $x' = 'y + 1 \wedge 'y = y'$ (the latter condition forces the value of y to be unchanged). We will also allow this to be written $\langle 'x = 0 \wedge x' = y + 1\rangle$, where the use of plain y means $'y$ with the constraint $y' = 'y$.

2.1 The low level net semantics

In Figure 1(i)-(iii) we show the low level net semantics for each of the constructs of the example individually. The net (i) for the declaration of x has three places, x_0, x_1 and x_2 for its three possible values, and nine transitions for each of the possible accesses to x. For example, the transition **tx** labelled $\{\overline{x^A}(0,1)\}$ represents an access in which the value of x changes from 0 to 1; that labelled $\{\overline{x^A}(0,0)\}$ represents an access in which the value is and remains 0; and so on. The net (iii) for the declaration of y is analogous.

The net (ii) for the atomic action of the program fragment has: places a and b for representing program control being respectively immediately before and immediately after its execution; and a transition for each potential execution, i.e., for each combination of the potential pre- and post-values of the accessed variables for which the expression yields **true**. For example, the transition **t** labelled $\{x^A(0,1), y^A(0,0)\}$ corresponds to an execution with $'x = 0$, $x' = 1$, $'y = 0$, $y' = 0$.

A transition label, such as $\{\overline{x^A}(0,1)\}$ or $\{x^A(0,1), y^A(0,0)\}$, is a set of actions (generally, a multi-set, but here we may and will restrict ourselves to sets). Each such action can be thought of as a synchronisation capability, as actions in CCS. The action $x^A(0,1)$ occurring in the label of transition **t** signifies that **t** needs to be synchronised with a transition having the conjugate action $\overline{x^A}(0,1)$ in its label - for instance, transition **tx** from the net for the declaration of x. We can think of action $x^A(0,1)$ as representing the need for an access to x in which it changes from 0 to 1, and $\overline{x^A}(0,1)$ as representing that access. The $y^A(0,0)$ in the label of **t** means that **t** also needs to be synchronised with a transition having the conjugate action $\overline{y^A}(0,0)$ in its label - e.g., transition **ty** from the net for the declaration of y.

The complete semantics for the declaration of x and y with this atomic action as their scope would give the net shown in (viii) which can be obtained in terms of the operators on low level nets by parallel composition (i.e., juxtaposition) of the three nets (i)-(iii) and then applying synchronisation and then restriction, both over the appropriate action set - namely all actions of form $x^A(\ldots)$, $\overline{x^A}(\ldots)$, $y^A(\ldots)$ or $\overline{y^A}(\ldots)$.

The synchronisation would add the new transitions shown in (viii) as **t1** and **t2**. The transition **t1** brings together **tx**, **t** and **ty** which mutually satisfy each others' synchronisation needs, leaving no residual needs, and thus the empty set as label. The **t2** is likewise derived from **ux**, **u** and **uy**. These are the only possible complete synchronisations. Restriction over the action set then deletes any transitions having an action from that action set in its label, i.e., all the original transitions from (i)-(iii), leaving just the net shown in (viii).

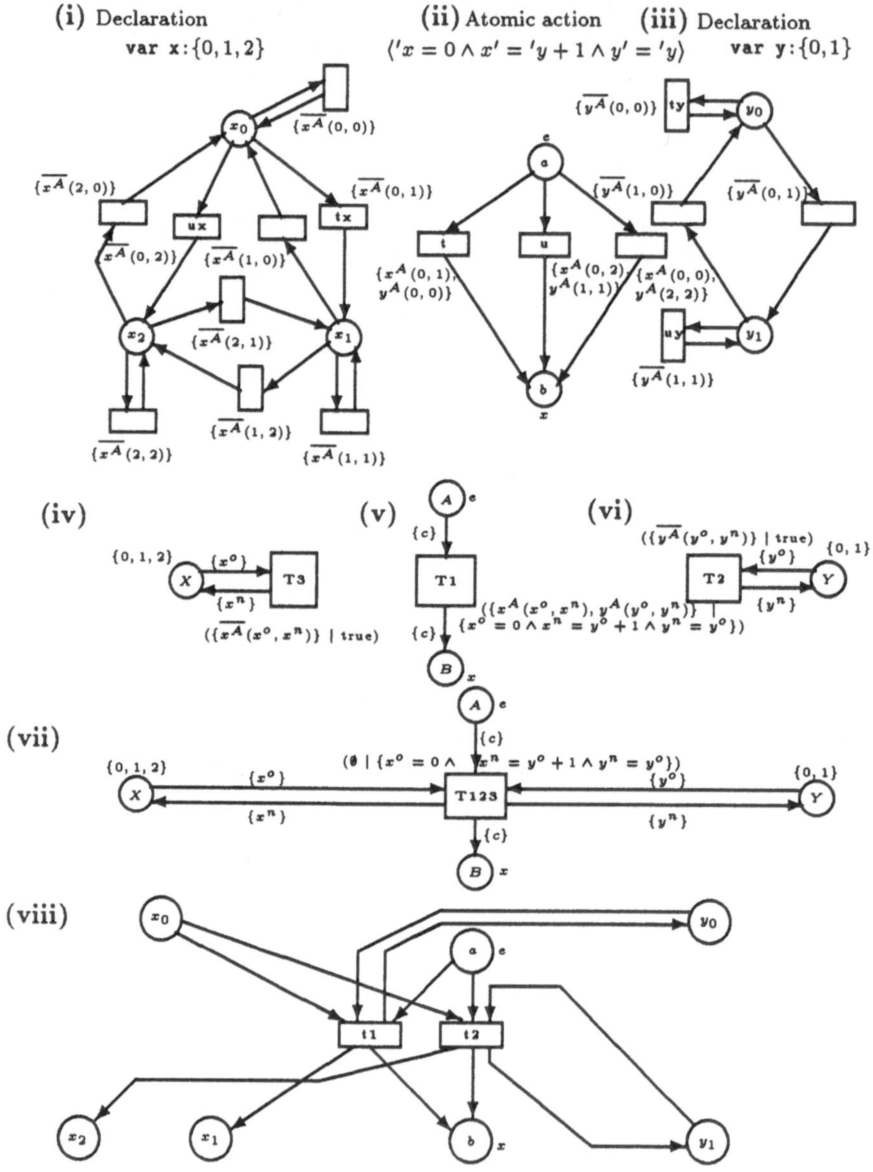

Figure 1: Low ((i)-(iii), (viii)) and high ((iv)-(vii)) level net semantics of $\mathcal{F}$

The transition **t1** represents the execution of the atomic action in the case of initially $x = 0$ and $y = 0$, requiring tokens on the corresponding places x_0 and y_0, and moving the former to the place x_1 for the updating of x to the value 1, and restoring the latter token to preserve the value of y. The transition **t2** is similarly the only other possible execution, in the case of initially $x = 0$ and $y = 1$.

There are two additional points to be made about the example so far.

The nets used have place labels as well as transition labels. In the net of (ii): Place a has label e, an 'entry' place[7]; place b has label x, an 'exit' place[8]. The third possibility for a place is to be 'internal' - for example, x_0, x_1, x_2, y_0, y_1 in (ii)/(iii) are internal. These have the empty label, which is omitted on figures. These labels are used in control structure compositions of nets. E.g., sequential composition of two nets is merging the exit places of the first net with the entry places of the second net.

The nets shown in (i)/(iii) for the variable declarations are not quite complete in that there would be two additional places, an entry place and an exit place, with initialisation and termination transitions. These additional elements have been omitted in order to avoid cluttering the figure, but they will be described later.

2.2 The high level net semantics

Figure 1(iv)-(vi) and (vii) show the high level net semantics analogously to the low level net semantics of (i)-(iii) and (viii), respectively. In (iv), X is the place holding the value of program variable x, and x^o and x^n (old x and new x) are net variables for its pre- and post- value; similarly in (vi), Y, y^o and y^n for variable y; A or B in (v) represent the program control points before and after the atomic action, respectively; they are able hold a black token for which net variable c is used. Let us consider (vii) and its unfolding to (viii) to cover those aspects of our high level nets which are similar to other high level net models, namely:

1. A place has a type, e.g. $\{0, 1, 2\}$ for X, which is the possible values for any token on that place. Omission of a type, as for A, means the singleton type $\{\bullet\}$, i.e. the place only sustains 'black' tokens.

2. An arc has an inscription, e.g. $\{x^o\}$ for the arc $X \rightarrow$ **T123** from X to **T123**, which in general is a multi-set of net variables. However, only singleton multisets are relevant here.

3. A transition has an annotation, e.g. the singleton set

$$\{x^o = 0 \wedge x^n = y^o + 1 \wedge y^n = y^o\}$$

for **T123**, which is a set of conjuncts defining an occurrence condition. The empty set will be shown as `true`.

4. A place unfolds to low level places for each value in its type. E.g., X unfolds to x_0, x_1 and x_2, denoting $x = 0$, $x = 1$ and $x = 2$, respectively.

[7] The initial marking of a net is a token on every entry place and no token elsewhere.
[8] By definition, a marking is final if all exit places (and no others) are marked.

5. A transition unfolds to low level transitions for each binding of its variables such that: all its conjuncts evaluate to **true**; each variable of an incident arc has a value consistent with the type of the connected place. E.g., **T123** unfolds to **t1** and **t2**, for bindings $\{x^o = 0, x^n = 1, y^o = 0, y^n = 0\}$ and $\{x^o = 0, x^n = 2, y^o = 1, y^n = 1\}$ respectively.

6. An arc $P \rightarrow \mathbf{T}$ with variable x unfolds to produce an arc $p \rightarrow \mathbf{t}$ for every instance of p, **t** and value v, such that p is the unfolding of P for v and **t** is an unfolding of **T** for a binding in which $x = v$. E.g., the arc $X \rightarrow \mathbf{T123}$ unfolds (for $x^o = 0$) to $x_0 \rightarrow \mathbf{t1}$ and $x_0 \rightarrow \mathbf{t2}$. The unfolding of an arc $\mathbf{T} \rightarrow P$ is analogous.

7. The behaviour of a high level net is consistent with that of the low level net to which it unfolds. E.g., if X, Y and A have tokens 0, 1 and $\bullet$, then there is an appropriate binding for the variables of **T123**, giving all its conjuncts as **true**, which allows its occurrence (corresponding to the occurrence of **t2**) by: removing the 0 token from X ($x^o = 0$); removing the 1 token from Y ($y^o = 1$); removing the $\bullet$ from A, and placing a $\bullet$ on B ($c = \bullet$); placing a 2 token on X ($x^n = 2$); placing a 1 token on Y ($y^n = 1$).

The aspects in which the high level net model here is less usual, are the place or transition inscriptions including not only type or occurrence condition, but also a label. The place labels identify each place as entry, exit or internal, and these have exactly the same roles as in the low level nets, that is (as already mentioned) to serve for the control structure compositions of nets. To illustrate high level transition labels we turn to Figure 1(iv)/(v) which shows the high level nets corresponding to the low level nets in (i)/(ii). A label, e.g. $\{x^A(x^o, x^n), y^A(y^o, y^n)\}$ for **T1**, is a multi-set of action terms, each being a 'parameterised' action which, for a particular binding of the variables in its parameter list, evaluates to an elementary action. E.g., for the binding $x^o = 0, x^n = 2$, the label $\overline{x^A}(x^o, x^n)$ of **T3** evaluates to $\overline{x^A}(0, 2)$, which is the label of the low level transition **ux** to which **T3** unfolds for that binding.

Our low level net domain, more precisely, is the class of elementary M-nets, which corresponds to labelled place/transition nets. An M-net is *elementary* if every place has singleton type and for every transition, its label consists of just elementary actions, and its annotation is the empty set. Types, transition annotations and arc inscriptions can thus be omitted from the representation.

2.3 Incremental partial synchronisation

In the program example, the variables x and y are declared by two distinct declarations in the program, and thus there will be two distinct semantic steps in the construction of (viii) from (i)-(iii), or (vii) from (iv)-(vi), i.e., first a partial synchronisation and restriction over y^A actions (for the declaration of y) and then a further partial synchronisation and restriction over x^A actions (for the declaration of x). In Figure 2 we show intermediate steps for the high level semantics, describing the high level net synchronisation mechanism. (The synchronisation mechanism for low level nets is a simplified version of that for high level nets.)

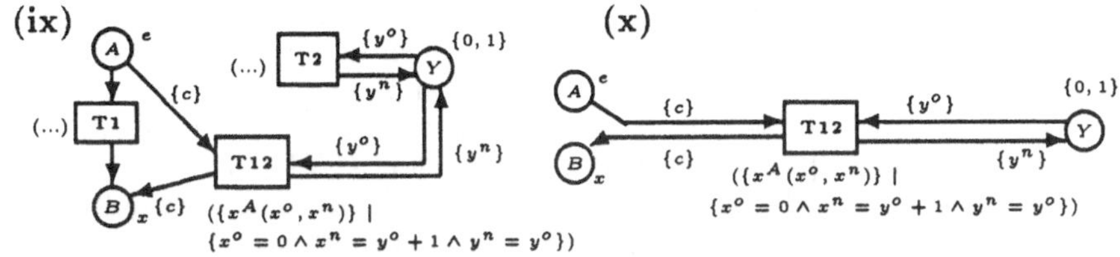

Figure 2: Synchronisation ((ix)) of (v) and (vi), followed by restriction ((x))

The parallel composition of the high level nets for the atomic action and the declaration of y would give the juxtaposition of nets (v) and (vi); and (ix) is the result of applying synchronisation over y^A actions to that. The conjugate actions $\overline{y^A}(y^o, y^n)$ in **T2** and $y^A(y^o, y^n)$ in **T1** allow these two to come together as **T12** which has: as its label, the multi-set sum of the labels of its two constituent transitions, but with the elimination of the pair of actions, y^A and $\overline{y^A}$, which allowed the synchronisation, leaving just $(\{x^A(x^o, x^n)\})$ as its annotation, the union of the annotations of the two constituent transitions, i.e., giving the occurrence condition as the conjunction of the constituents' occurrence conditions.

In (x) we see the result of applying restriction over y^A actions to (ix), which deletes all transitions with labels involving y^A actions. This completes the semantics of combining the declaration of y with the atomic action. The use of variable y has been internalised - the only external requirements are for access to a variable x, shown by the $x^A(x^o, x^n)$ action label. In particular, an outer declaration of a variable y would not affect this net.

In (vii) we see the result of further adding the declaration of x - parallel composition of (x) with (iv), then the application of synchronisation over x^A actions, and then the application of restriction over x^A actions – leaving just the transition **T123** which is the synchronisation of **T12** with **T3**.

This example is sufficient for understanding the use of the general synchronisation mechanism in the semantics of $B(PN)^2$, although the full definition of synchronisation is substantially more sophisticated, see [5].

3 Preliminaries to Syntax and Semantics

In the rest of the paper we present the syntax and both high and low level semantics of $B(PN)^2$ programs. We proceed top-down through the syntax, starting with program blocks and declarations (Section 4); then command connectives for parallel, sequence, choice and iteration (Section 5); and finally atomic actions (Section 6). The semantic functions are $Net_H(P)$ and $Net_L(P)$, yielding, respectively, the high level net and the low level net associated with some construct P.

The semantic definitions employ the unary net operations already described, namely for an M-net N and a set of action names A: unfolding, $U(N)$; synchronisation N **sy** A; restriction, N **rs** A. Unfolding is with respect to a set of values over which all net variables range, which here is taken as the set of all values supported by the particular realisation of $B(PN)^2$.

The semantics will also use some non-unary net composition operators, namely: parallel, $N1 \| N2$; sequential, $N1; N2$; choice, $N1 \,\Box\, N2$; and iteration, $[NI * NR * NF]$. Parallel composition is simply the disjoint union of the two nets (with renaming of the places and transitions to obtain the disjointness). Sequential composition, shown in Figure 3(i)-(iii), is parallel composition followed by a standard place multiplication construct (as in e.g. [12, 14]), to merge the exit places of the first net with the entry places of the second. Choice is a standard construction, similar to sequential composition, but with two place multiplications: of the entry places from one net with the entry places from the other; and of the exit places from one with the exit places from the other. Iteration, $[NI * NR * NF]$, is illustrated in Figure 3(iv)-(v). It involves three nets: initial, NI; repeated, NR; final, NF. The effect is one execution of NI; followed by zero or more executions of NR; followed by one execution of NF. The construction is to take two copies of each of the original nets and put them together, with the appropriate place multiplications, into the structure shown in (v) where **u1** and **u2** are copies of **u** etc.

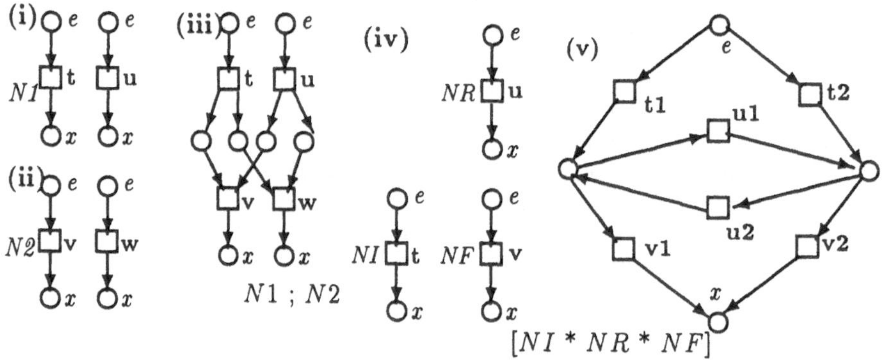

Figure 3: Illustrations of control structure compositions

4 Programs, Blocks and Declarations

The following is the syntax pertaining to programs *Prog*, blocks *Block*, (block) bodies *B* and declarations *Decl*:

Prog ::=	*Block*	::=	`begin` *B* `end`
	B	::=	*Decl* ; *B* | *Com*
	Decl	::=	`var` *name* : *set* | `var` *name* : `chan` *b* `of` *set*
	b	::=	$0 \mid 1 \mid \ldots \mid \omega$

A declaration involves a variable identifier *name* and the specification *set* of a type, i.e., a set of possible values (e.g. `int` or $\{0,1,2\}$). Either ordinary variables, or channel variables, may be declared. The declarations may be followed by a command *Com*. For the syntax and semantics of *Com*, the reader is referred to the next section.

4.1 Blocks

The main idea in describing a block is to juxtapose the nets for its declarations and the net for its command, and then to synchronise all matching transitions and to restrict in order to make them invisible outside the block (see Section 2). Just as a variable declared inside a block may gain an initial value on block entry, it must lose its value on block exit, because otherwise value inconsistencies might arise (for instance if the block in question is nested inside a loop). For this reason, we also associate a termination part $\tau(D)$ with every declaration D. More precisely, for both high and low level semantics, the following are the steps in obtaining the net semantics of a block body $D; B$, where D is the declaration of a program or channel variable w.

1. The sequential composition of the net for B and the net $\tau(D)$ for the termination of w, which comprises a single transition labelled with a special termination action for w called w^{A_T}.

2. The parallel composition of that with the net for the declaration.

3. The application to that of synchronisation and restriction over $\delta(D)$, the set of all actions pertaining to w.

Definition 4.1 Using the above notation and the parameterised class of nets $Net_H(D)$ defined in sections 4.2 and 4.3 below, and for I $\in$ {H,L}:

$$Net_I(\texttt{begin}\ B\ \texttt{end}) = Net_I(B)$$
$$Net_I(D;B) = ((Net_I(D) \parallel (Net_I(B);\tau(D)))\ \texttt{sy}\ \delta(D))\ \texttt{rs}\ \delta(D)$$
$$\text{and}\quad Net_L(D) = U(Net_H(D)),$$

where $\tau(D) = \Big($ 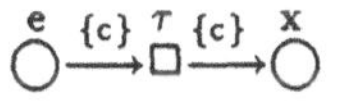

$\Big)$

with $\tau = (\{w^{A_T}\} \mid \texttt{true})$ and $\delta(D) = \{w^{A}, w^{A_?}, w^{A_!}, w^{A_T}\}$. ◇

Note the low level net for a declaration is defined via unfolding[9]. Next we complete the definition by treating the two types of variable declarations separately.

4.2 Ordinary variables

Definition 4.2 $Net_H(\texttt{var w : set}) = M_{data}(w, \texttt{set})$, that is, the net shown in Figure 4. ◇

The high level net semantics of an ordinary variable thus consists of a central place for the value of the variable together with a transition, **T1**, for the possible value changes. In addition, there are initialising and terminating transitions **T0**, **T2** and **T3**. The transitions **T2** and **T3** have to synchronise with the termination action τ associated to D via the net $\tau(D)$ according to Definition 4.1. The initialisation transition **T0** represents the first access to the variable,

[9]For a more explicit representation of this net, compare [2].

which may assume any pre-value of the variable. E.g. in any of the following the first atomic action could successfully be executed -

$\texttt{var}\ w : \{0,1,2\}; \langle 'w = 0 \wedge w' = 2\rangle \ldots$
$\texttt{var}\ w : \{0,1,2\}; \langle 'w = 1 \wedge w' = 2\rangle \ldots$
$\texttt{var}\ w : \{0,1,2\}; \langle w' = 2\rangle \ldots$

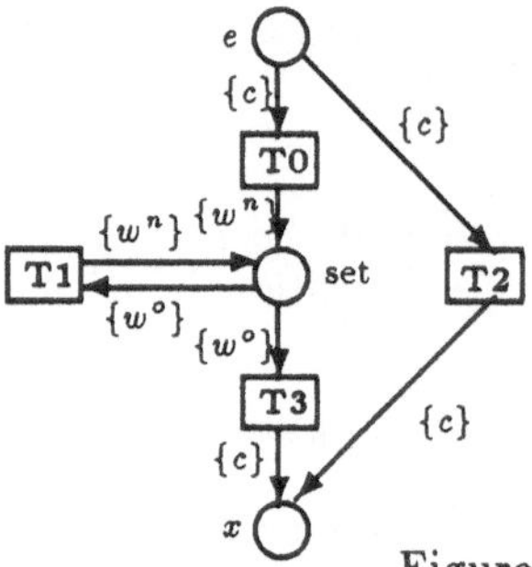

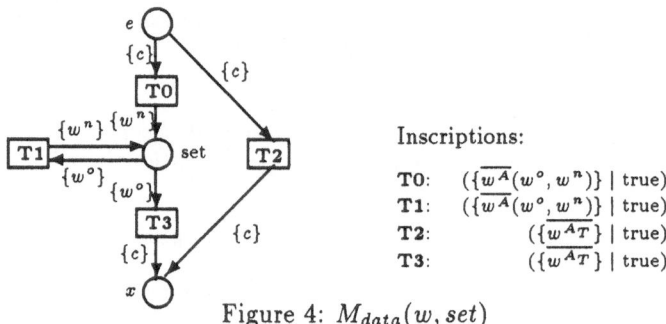

Inscriptions:

T0:	$(\{\overline{w^A}(w^\circ, w^n)\} \mid \text{true})$
T1:	$(\{\overline{w^A}(w^\circ, w^n)\} \mid \text{true})$
T2:	$(\{\overline{w^{AT}}\} \mid \text{true})$
T3:	$(\{\overline{w^{AT}}\} \mid \text{true})$

Figure 4: $M_{data}(w, set)$

4.3 Channel variables

The declaration, **var w : chan b of set**, with $b \neq 0$, declares a channel of capacity b which acts as a FIFO buffer of values. A $w!$ (send) which appears in an atomic action within the scope of the declaration causes a value to be appended to the end of the buffer and a $w?$ (receive) causes the first value to be removed from the buffer. The special case of $b = \omega$ is for a buffer of unbounded capacity. For the case of $b = 0$, the buffer has capacity 0 which enforces handshake sychronisation between a send and a receive.

Definition 4.3 Net_H(**var w : chan b of set**) $= M_{chan}(w, b, set)$, the nets shown in Figures 5 and 6. ◇

In Figure 5, to model the queue of values in a channel we consider a (complex) type $set*$ of tokens. $set*$ is the set of all sequences with entries from set. In general, sequences are denoted by s; a value $v \in set$ is identified with a sequence of length 1 (singleton sequence); the sequential composition of sequences is denoted by a dot .; ϵ denotes the empty sequence; and, finally, the length of sequence s, is denoted by $lh(s)$.

To enable concurrent send and receive actions, the contents of a channel is modelled by two sequences: that held on place $P2$, either empty or the first element in the channel; that held on place $P1$, the rest of the channel (which is the whole channel if $P2$ holds empty). The channel is initialised by the first sending of a value v to the channel. The initialisation transition **T0** places this value on $P1$, and the empty sequence on $P2$. Then there is one transition, **T1**, for the general send action, appending the communicated value to the content of place $P1$, provided this could not result in the channel capacity being exceeded - condition $lh(s_2) < b$ which will always yield **true** if $b = \omega$ (unbounded capacity), and otherwise ensures at most $b-1$ values on $P1$.

There is one transition, **T3**, for the receive action, removing the communicated value from place $P2$ and replacing it by ϵ. An internal transition, **T2**, transfers the first entry of the sequence on place $P1$ to place $P2$, if this contains ϵ. In addition, we have two terminating transitions, **T4** and **T5** which are analogous to those of an ordinary variable declaration (and which allow termination even when the channel is non-empty).

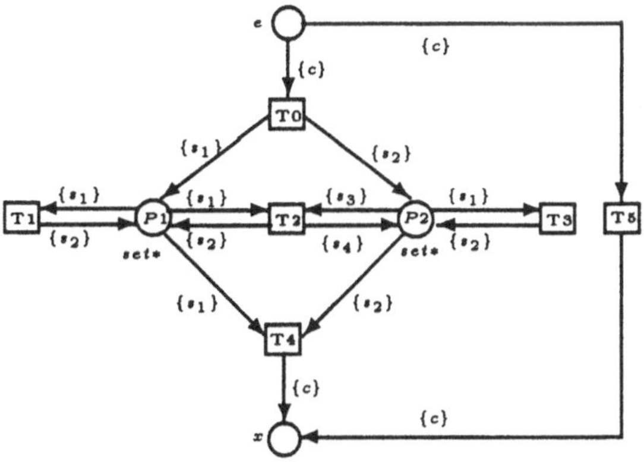

T0: $\{\overline{w^{A!}(w^!)}\} \mid \{s_1 = w^!, s_2 = \epsilon\}$
T1: $(\{\overline{w^{A!}(w^!)}\} \mid \{lh(s_2) < b,$
$s_2 = s_1.w^!\})$
T2: $(\emptyset \mid \{s_1 = v.s, lh(v) = 1,$
$s_2 = s, s_3 = \epsilon, s_4 = v\})$
T3: $(\{\overline{w^{A?}(w^?)}\} \mid \{s_1 = w^?, s_2 = \epsilon\})$
T4: $(\{\overline{w^{A}T}\} \mid \text{true})$
T5: $(\{\overline{w^{A}T}\} \mid \text{true})$

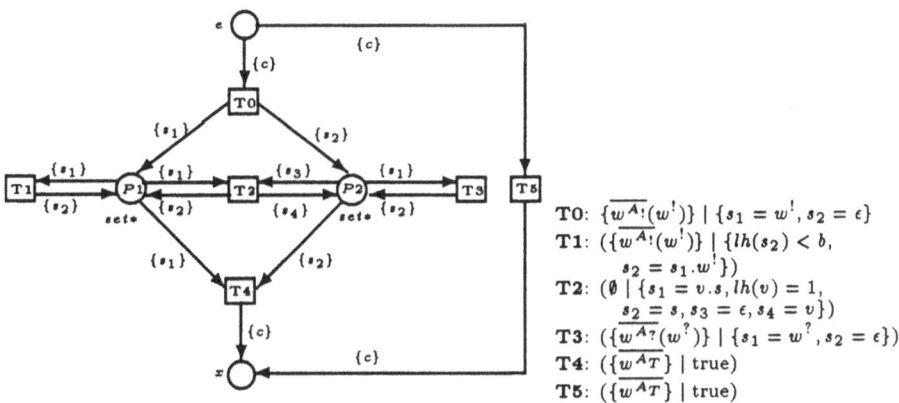

Figure 5: $M_{chan}(w, b, set), b \neq 0, b \neq 1$

For the case, $b = 1$, the channel may either be empty, or contain a singleton sequence of values. The net for this case is shown in Figure 6 (i). It can be seen as being obtained from $M_{chan}(w, b, set)$ by merging the two places $P1$ and $P2$.

For the case, $b = 0$ (i. e., in the case of synchronous communication), the net of Figure 6(i) is modified in the following way to give Figure 6(ii). The value set of the internal place becomes $\{\bullet\}$. The send transition and the receive transition are folded into a single transition, which has to synchronise with a send and a receive action and whose occurrence condition assures that both the communicated values coincide.

5 Command Connectives

The following is the syntax for a command, which can be: a block (see previous section); an atomic action, *AtAct*, which is the basic unit of commands and is described in the next section; or a combination of commands by one of the three connectives dealt with in this section, namely sequential composition, parallel composition and choice/iteration (which are both described by the do–od construct):

$$Com \;::=\; Block \mid AtAct \mid Com; Com \mid Com \| Com \mid \mathtt{do}\ alt\text{-}set\ \mathtt{od}$$

$$alt\text{-}set \;::=\; alt\text{-}set \ \square\ alt\text{-}set \mid Com; \mathtt{exit} \mid Com; \mathtt{repeat}.$$

Within a do–od construct, we may have **repeat** clauses, the 'repeat-set', and **exit** clauses, the 'exit-set'. The effect is repeated executions of arbitrarily chosen executable clauses of the repeat-set, possibly followed by a single

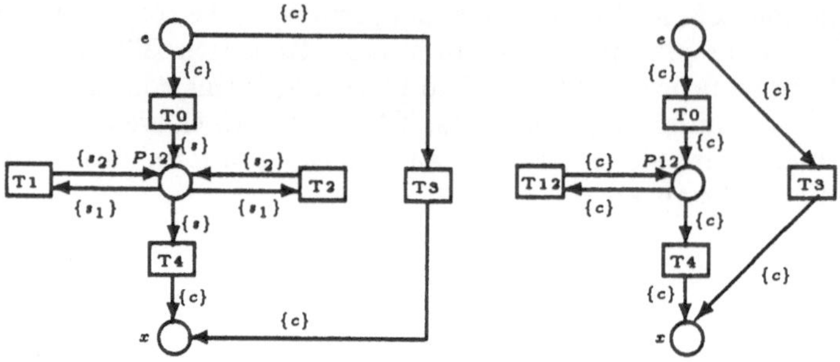

(i) (ii)

$M_{chan}(w, 1, set)$:

$P12$:	$\{s \in set* \mid lh(s) \leq 1\}$
T0:	$(\{\overline{w^{A!}}(w^!)\} \mid \{s = w^!\})$
T1:	$(\{\overline{w^{A!}}(w^!)\} \mid s_1 = \epsilon, s_2 = w^!\})$
T2:	$(\{\overline{w^{A?}}(w^?)\} \mid \{s_1 = w^?, s_2 = \epsilon\})$
T3:	$(\{\overline{w^{AT}}\} \mid \text{true})$
T4:	$(\{\overline{w^{AT}}\} \mid \text{true})$

$M_{chan}(w, 0, set)$:

$P12$:	$\{\bullet\}$
T0:	$(\{\overline{w^{A!}}(w^!), \overline{w^{A?}}(w^?)\} \mid \{w^! = w^?, w^! \in set\})$
T12:	$(\{\overline{w^{A!}}(w^!), \overline{w^{A?}}(w^?)\} \mid \{w^! = w^?, w^! \in set\})$
T3:	$(\{\overline{w^{AT}}\} \mid \text{true})$
T4:	$(\{\overline{w^{AT}}\} \mid \text{true})$

Figure 6: (i) $M_{chan}(w, 1, set)$ (ii) $M_{chan}(w, 0, set)$

execution of one clause of the exit-set. If a clause of the exit-set is chosen, termination of the whole construct ensues (unless, of course, this clause has a deadlock or an infinite loop). Two examples are -

1. do $\langle n > 0\rangle; \langle y' = {}'y + n\rangle; \langle n' = {}'n - 1\rangle;$ `repeat` $\Box$ $\langle n \leq 0\rangle;$ `exit` od,

   which is equivalent to `while` $n > 0$ do `begin` $y := y + n; n := n - 1$ end.

2. do $\langle c? = x'\rangle; \langle n' = {}'n + 1\rangle;$ `exit` $\Box$ $\langle d? = x'\rangle; \langle m' = {}'m + 1\rangle;$ `exit` od,

   which is equivalent to do $c?x \rightarrow n := n + 1 \;\Box\; d?x \rightarrow m := m + 1$ od.

The semantics of the $B(PN)^2$ do–od construct involves two supplementary semantic functions relating to the net iteration operator $[NI * NR * NF]$ used, namely R to give, for use in NR part, the net which is choice over the nets from all the repeat-set clauses, and E similarly for the exit-set clauses and the NF part. The NI part is the net for a single silent transition. Thus, for $I \in \{H, L\}$:

$$\begin{aligned}
R_I(\textit{alt-set}_1 \Box \textit{alt-set}_2) &= R_I(\textit{alt-set}_1) \Box R_I(\textit{alt-set}_2)\\
E_I(\textit{alt-set}_1 \Box \textit{alt-set}_2) &= E_I(\textit{alt-set}_1) \Box E_I(\textit{alt-set}_2)\\
R_I(Com; \texttt{repeat}) = E_I(Com; \texttt{exit}) &= Net_I(Com)\\
R_I(Com; \texttt{exit}) = E_I(Com; \texttt{repeat}) &= N_{stop} = \left(\overset{e}{\bigcirc} \;\; \overset{x}{\bigcirc} \right).
\end{aligned}$$

The high and low level semantics of program connectives are the same, involving applications of the appropriate net operators:

Definition 5.1 For $I \in \{H, L\}$:

$$\begin{aligned}
Net_I(P_1 \| P_2) &= Net_I(P_1) \| Net_I(P_2)\\
Net_I(P_1; P_2) &= Net_I(P_1); Net_I(P_2)\\
Net_I(\text{do } \textit{alt-set} \text{ od}) &= [N_{silent} * R_I(\textit{alt-set}) * E_I(\textit{alt-set})]
\end{aligned}$$

where $N_{silent} =$ (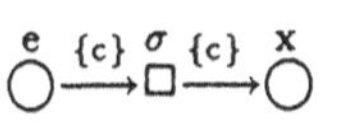) with $\sigma = (\emptyset \mid \mathtt{true})$. ◇

6 Atomic Actions

The following is the syntax for atomic actions, *AtAct*, and constituent expressions, *expr*:

$$\begin{aligned} AtAct &::= \langle\, expr \,\rangle \\ expr &::= {}'w \mid w' \mid w? \mid w! \mid const \mid expr\ op\ expr \mid op\ expr. \end{aligned}$$

Operators (*op*, e.g. +) and constants (*const*, e.g. 1) are standard programming constructs, details of which depend on a particular realisation of $B(PN)^2$. The identifier w is a program variable, which may be declared as an ordinary variable, in which case the forms $'w$ and w' are meaningful, as discussed in Section 2. A program variable may also be declared as a channel variable in which case the forms $w?$ and $w!$ are meaningful. The construct $w?$ requires an input on the channel w and denotes the value received. The construct $w!$ requires an output on the channel w and denotes the value sent. E.g., $\langle w? = x' \rangle$ and $\langle w! = x + 2 \rangle$ correspond to, respectively, **occam** notation $w?x$ and $w!(x+2)$.

First we define a supplementary semantic function, *Ins*, to yield the semantics of an expression as an action-expression pair $(AS|E)$, which is similar to a transitions inscription. AS is a set of action terms, the action names used being: w^A for access to an ordinary variable w; $w^{A_?}$ for input from, and $w^{A_!}$ for output to, a channel variable w. E is a term, which is e with program variables appropriately replaced by net variables: w^o for $'w$; w^n for w'; $w^?$ for $w?$; $w^!$ for $w!$.

$$\begin{aligned} Ins('w) &= (\{w^A(w^o, w^n)\} \mid w^o) \\ Ins(w') &= (\{w^A(w^o, w^n)\} \mid w^n) \\ Ins(w?) &= (\{w^{A_?}(w^?)\} \mid w^?) \\ Ins(w!) &= (\{w^{A_!}(w^!)\} \mid w^!) \\ Ins(const) &= (\emptyset \mid const) \\ Ins(e_1\, op\, e_2) &= (AS_1 \cup AS_2 \mid E_1\, op\, E_2) \text{ for } Ins(e_i) = (AS_i \mid E_i) \\ Ins(op\, e) &= (AS \mid op\, E) \text{ for } Ins(e) = (AS \mid E). \end{aligned}$$

E.g., for $\langle 'x = 0 \wedge x' = c? + 1 \rangle$, the semantics is obtained via:

$$\begin{aligned} Ins('x = 0) &= (\{x^A(x^o, x^n)\} \mid x^o = 0) \\ Ins(x' = c^? + 1) &= (\{x^A(x^o, x^n), c^{A_?}(c^?)\} \mid x^n = c^? + 1) \\ Ins('x = 0 \wedge x' = c^? + 1) &= (\{x^A(x^o, x^n), c^{A_?}(c^?)\} \mid x^o = 0 \wedge x^n = c^? + 1) \end{aligned}$$

By the following definition, this would give a net like Figure 1(v), except that the transition inscription is $(\{x^A(x^o, x^n), c^{A_?}(c^?)\} \mid \{x^o = 0 \wedge x^n = c^? + 1)\})$.

Definition 6.1 For $Ins(expr) = (AS|E)$,

$Net_H(\langle expr \rangle)$ = (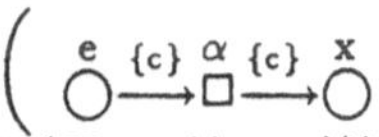), where $\alpha = (AS \mid \{E\})$

$Net_L(\langle expr \rangle) = U(Net_H(\langle expr \rangle)).$

◇

Note that, as for declarations, the low level semantics is defined via unfolding.

7 Consistency Theorem

The following theorem shows the consistency of the high and low level semantics.

Theorem 7.1 *(Commutativity of semantics and unfolding)*
For any syntactic construct P for which is defined $Net_H(P)$ (and so too $Net_L(P)$), we have $U(Net_H(P)) = Net_L(P)$.

Proof: By induction on the structure of P. The theorem is trivial for the base cases, viz. for declarations and for atomic actions, for which it is established by definition. The induction steps rely on some non-trivial results for M-nets [5]. The results used are: first, for elementary M-nets N (such as N_{silent} or N_{stop}), unfolding is identity, i.e., $U(N) = N$, and second, all operators commute with unfolding, i.e.:

$$\begin{array}{lcll} U(N_1\ op\ N_2) & = & U(N_1)\ op\ U(N_2) & \text{for } op \in \{ \parallel , ; , \Box \} \\ U(N\ op\ A) & = & U(N)\ op\ A & \text{for } op \in \{ \textbf{sy} , \textbf{rs} \} \\ U([NI * NR * NF]) & = & [U(NI) * U(NR) * U(NF)]. & \end{array}$$

For the declaration, D, of an (ordinary or channel) variable w together with its scope B, i.e. $D; B$, the induction step requires $Net_L(D; B) = U(Net_H(D; B))$, which is proved as follows:

$$\begin{array}{lll} & Net_L(D;B) & \\ = & (Net_L(D)\|(Net_L(B);\tau(D)))\ \textbf{sy}\ \delta(D)\ \textbf{rs}\ \delta(D) & \text{- definition of } Net_L(D;B) \\ = & (U(Net_H(D))\|(U(Net_H(B));U(\tau(D))))\ \textbf{sy}\ \delta(D)\ \textbf{rs}\ \delta(D) & \text{- induction hypoth.} \\ = & U(Net_H(D)\|(Net_H(B);\tau(D))\ \textbf{sy}\ \delta(D)\ \textbf{rs}\ \delta(D)) & \text{- commutativity of unfolding,} \\ & & \text{and } \tau(D) \text{ being elementary} \\ = & U(Net_H(D;B)) & \text{- definition of } Net_H(D;B) \end{array}$$

For the induction steps pertaining to the parallel and sequential constructs, viz.

$$Net_L(P_1\|P_2) = U(Net_H(P_1\|P_2)) \quad \text{and} \quad Net_L(P_1;P_2) = U(Net_H(P_1;P_2)),$$

the proofs are similar to the above. For iteration and choice, there need to be subsidiary inductions to show

$$U(R_H(alt\text{-}set)) = R_L(alt\text{-}set) \quad \text{and} \quad U(R_H(alt\text{-}set)) = R_L(alt\text{-}set).$$

8 Conclusion

In this paper we have used the domain of M-nets and the operations defined on this domain to give both a high level and low level net semantics of the specification and programming language $B(PN)^2$; and have shown the consistency of these two semantics and thus the consistency of the high level net semantics with the original semantics of [2]. These semantics (both high and low level) have four important properties -

P1 They are compositional in that every program construct has a semantics which is independent of its context, and (in the case of a compound construct) is defined only in terms of the semantics of its constituents.

P2 They yield a single transition occurrence for an atomic action execution, despite the generality of the atomic action construct in supporting write access to multiple shared variables together with simultaneous communication on multiple channels.

P3 They yield a concurrent semantics for the parallel execution construct of the language.

P4 They are systematic, and thus expressing an algorithm as a $B(PN)^2$ program fixes a standard Petri net representation of the algorithm (avoiding the need to invent *ad hoc* representations whenever some analysis is to be performed).

It is the synchronisation structure on multi-sets of actions which allows the combination of P1 and P2, as was illustrated in the discussion of Figure 2. In contrast, if we used the simpler CCS style, then property P2 could not be achieved (as indeed it is not in the CCS approach to programming language semantics illustrated in [15]). Moreover, if we used the SCCS style, which has a somewhat similar synchronisation on multi-sets of actions, then property P3 could not be achieved.

The semantics was developed not just for theoretical reasons, but also for practical application in program verification, to which all of the above properties are relevant (P1 – P4 all contributing to the comprehensibility of the net produced by the semantics and P3 enabling various efficiency techniques in state space searching). The *PEP* system, originally supporting simulation and verification on the low level net semantics of $B(PN)^2$, has now been enhanced to a system (exhibited at the Hannover Computer Fair CeBIT'95, March 1995) which incorporates: the high level net semantics defined here, yielding smaller and more readable net models for programs than does the low level net translation; a fast interpreter for a $B(PN)^2$ program via the simulation of its high level net semantics; the unfolding of those semantics for the purpose of verification on the resulting low level nets. Work is in progress on developing verification techniques to be applied directly to the M-net semantics.

Petri net research in general has so far yielded fewer (and less powerful) analysis techniques for high level nets than for low level nets. Thus at the present time there is a trade-off: using low-level nets gives more analysis power but less easy machine representability; while using high level nets allows nice visualisation of every program in a small net but less analysis capabilities. However, as there is an ever-growing body of results pertaining to the analysis of high level nets, investing in the implementation of a high level net semantics can be done in anticipation of future results in that area.

References

[1] Eike Best, Raymond Devillers, Elisabeth Pelz, Arend Rensink, Manuel Silva and Enrique Teruel: Causal Calculi Based on Nets (CALIBAN). In this volume.

[2] Eike Best and Richard P. Hopkins: B(PN)^2 - a Basic Petri Net Programming Notation. In *Proceedings of PARLE'93*, Volume 694 of LNCS, pages 379–390. Springer Verlag, 1993.

[3] Eike Best, Raymond Devillers and Jon Hall: The Box Calculus: a New Causal Algebra with Multi-label Communication. In *Advances in Petri Nets 1992*, Volume 609 of *LNCS*, pages 21-69. Springer Verlag, 1992.

[4] Eike Best and Hans Fleischhack: *PEP - Programming Environment Based on Petri Nets* (project supported by the Deutsche Forschungsgemeinschaft, 1993-1994).

[5] Eike Best, Hans Fleischhack, Wojciech Frączak, Richard P. Hopkins, Hanna Klaudel and Elisabeth Pelz: A Class of Composable High Level Petri Nets, to appear in *Proc. of 16th ICPN*, Torino, LNCS, Springer Verlag, June 1995.

[6] Eike Best, Hans Fleischhack, Wojciech Frączak, Richard P. Hopkins, Hanna Klaudel and Elisabeth Pelz: An M-net Semantics of $B(PN)^2$, Hildesheimer Informatikbericht 10/95, Hildesheim, 1995.

[7] Didier Buchs and Nicolas Guelfi: CO-OPN: A Concurrent Object-Oriented Petri Nets Approach for System Specification. In *Proc. of 12th ICPN*, Århus, 1991, also LRI Technical Report 616, Orsay, 1990.

[8] Soren Christensen and Laure Petrucci: Towards a Modular Analysis of Coloured Petri Nets. In *Advances in Petri Nets 1992*, Volume 609 of LNCS, pages 113-133. Springer Verlag, 1992.

[9] Raymond Devillers and Hanna Klaudel: Refinement and Recursion in a High Level Petri Box Calculus. In this volume.

[10] Hartmann Genrich: Predicate-transition Nets. In *High Level Petri Nets: Theory and Application*, pages 3-43. Springer Verlag, 1991.

[11] Kurt Jensen: Coloured Petri Nets. In *High Level Petri Nets: Theory and Application*, pages 44-122. Springer Verlag, 1991.

[12] Ursula Goltz: On Representing CCS Programs by Finite Petri Nets. In *Proc. MFCS'88*, Volume 324 of LNCS, pages 339-350. Springer Verlag, 1988.

[13] Hanna Klaudel and Elisabeth Pelz: *Communication as Unification in the Petri Box Calculus*, to appear in *Proc. FCT'95*, Dresden, LNCS, Springer Verlag, August 1995.

[14] Vadim Kotov: An Algebra for Parallelism based on Petri Nets. In *Proc. MFCS'78*, Volume 64 of LNCS, pages 39-55. Springer Verlag, 1978.

[15] Robin Milner: *Communication and Concurrency.* Prentice Hall (1989).

[16] Giuseppe Pinna: *Petri Nets and Their Composition Problem.* PhD Thesis: TD-2/90, Dipartimento di Informatica, Università degli Studi di Pisa, March 1990.

[17] Wolfgang Reisig: Petri Nets and Algebraic Specifications. In *Theoretical Computer Science* Vol.80, 1-34 (1991).

[18] Jacques Vautherin: Parallel systems specification with colored Petri nets and algebraic specification. In *Advances in Petri Nets 87*, Volume 266 of LNCS. Springer Verlag, 1987.

Graphs for Generalized Traces*

I. Biermann, B. Rozoy
L.R.I. URA 410 CNRS, Université Paris Sud
91405 - Orsay Cedex, France

Abstract

Mazurkiewicz traces form a model suitable for the investigation of some aspects of concurrency. We want to investigate one of its generalization based on the weakening of the condition that the equivalence relation is a congruence. The (quasi-)prefix ordering is then compared with the configuration graph induced by the partial ordering of symbol occurrences. We point out differences and exhibit necessary and/or sufficient conditions for them to be isomorphic: this is settled with the help of diamond properties.

1 Introduction

Various tools have been proposed to represent the computation of concurrent systems, such as transition systems, event structures and finally equivalence relations on words. When dealing with transition systems, the state space models the dynamic behavior of the system and concurrency is identified with non deterministic interleaving of actions: the investigation leads to consider several paths as expressions of the same run, the order of two non dependent actions being regarded as irrelevant. In order to consider models where concurrency is explicitly modelled, one may look at event structures or traces. With objects generically called event structures, different occurrences of transitions -the events- are considered for themselves with the relationship between them: concurrency, conflict, enabling relation, ... With the theory of traces (introduced in [Maz86], see also [AR88], [Die90]) a successful mathematical framework uses an alphabet of actions and a symmetric, irreflexive relation of concurrence on it: two strings are said to be equivalent if they differ only by the order of concurrent letters. Then, the consideration of the prefix relation between traces brings to light some strong properties: the associated PoSet is a finitary coherent and prime algebraic CPO, thus admits a good representation within Scott domain theory [RT91]. However, if restricted to a finite alphabet, classical Mazurkiewicz traces succeed to model concurrency for elementary one-safe Petri nets and cannot express more general nets and problems, such as the Producer/Consumer paradigm for example. Thus extensions of original traces have been looked at ([Arn91], [AH93], [BBR94], [BG95], [Dro89], [HKT92], [Sta89], [Vog91]) in relation with their possible adequacy to domains, event structures or Petri Nets. Unfortunately, from an algebraic point of view, on one hand, some of them (such as local traces in [HKT93]) suffer from a major drawback: events are not necessarily ordered, on the other hand others (such

*This work has been partially supported by the ESPRIT working group "Caliban" n° 6067 and the inter PRC project "Modèles du Parallélisme".

as infinite traces in [GP92], generalized traces in [BDK95], P-traces in [Arn91] and distributed traces in [Roz92]) more or less force the underlying structure to be restricted to prime or stable event structures.

In that spirit, we are interested in generalized traces that may be defined by an equivalence relation on a monoid of actions. The proposed generalization is not, till now, tied to a specific computational model; but its properties are generally satisfied by other attempts connected to Petri nets or event structures: it is the case for traces deriving from Winskel general event structures, from Droste and Stark automata and of Hoogers-Kleijn-Thiagarajan local traces. Thus the traces we introduce here exhibit necessary properties to be satisfied by any generalization. We hope they will help to understand what kind of requirements lead necessarily to such or such restriction. As within trace theory the well known occurrence graph gives a good account of the idea of process, we have studied carefully two graphs associated with a generalized trace: its graph of quasi-prefixes and the lattice of configurations associated with the intersection of orders of actions. The purpose of this paper is to point out differences and similarities between the two graphs, to prove necessary and/or sufficient conditions for them to be isomorphic and to exhibit some examples of generalized equivalences satisfying these conditions.

2 $\mathcal{G}$-Traces and Graphs of Quasi-Prefixes

The idea is to describe concurrency that may be neither a global property of the system nor a binary property on actions. Thus we are interested in generalized traces defined by equivalence relations that are not necessarily congruences. However, and as our purpose is to model executions of distributed systems, we will require first that an execution depends only on its past and not on its future: we get right semi-congruences, second that different executions differ only by the order in which the actions are executed: we get Parikh equivalences. If $u \in A^*$ and $a \in A$, $|u|_a$ will denote the number of occurrences of a in u.

Definition 1 Let A be an alphabet and $\sim$ be an equivalence relation over A^*. Then $\sim$ is a *$\mathcal{G}$-relation* if:

i) $\sim$ is a right semi-congruence: $\forall u, v, w \in A^*, u{\sim}v \Rightarrow uw{\sim}vw$

ii) $\sim$ is a Parikh equivalence : $\forall u, v \in A^*, u{\sim}v \Rightarrow \forall a \in A, |u|_a = |v|_a$

If $\sim$ is a $\mathcal{G}$-relation, a *$\mathcal{G}$-trace* is an equivalence class $[w]_\sim$. As the right semi-congruence $\sim$ preserves the length and the occurrences of letters, the notions of number of occurrences, alphabet and length can be naturally extended to $\mathcal{G}$-traces by applying them to any word of the equivalence class.

A $\mathcal{G}$-relation is not a congruence and there is no internal product in $A^*/_\sim$, but an external product $\diamond$ might be defined in the following way: $\forall [w] \in A^*/_\sim, \forall w' \in A^*, [w]\diamond w' = [w.w']$. Using this external product, we define the notion of quasi-prefix as follows: for t in $A^*/_\sim$, we say that a $\mathcal{G}$-trace t' is a quasi-prefix of t if there exists a word w in A^* such that $t = t'\diamond w$. Then $Q\_Pref(t) = \{t' \in A^*/_\sim \mid \exists w \in A^*, t = t'\diamond w\}$ is the set of quasi-prefixes of t. Finally, we say that $\sim$ is right half-cancellative if: $\forall t, t' \in A^*/_\sim, \forall a \in A,\ t\diamond a = t'\diamond a \ \Rightarrow\ t = t'$.

Since a $\mathcal{G}$-trace stands for an execution of a system, a natural idea is to associate a transition graph with any t in $A^*/_\sim$.

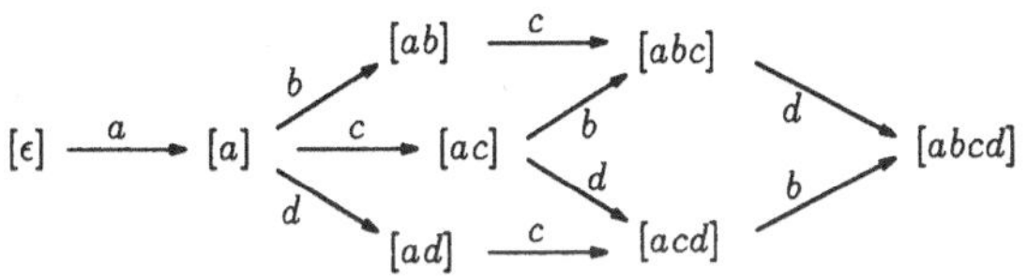

Figure 1: Graph of quasi-prefixes of $t = [abcd]_{\sim_1}$ (Example 1).

Definition 2 Let A be an alphabet, $\sim$ be a $\mathcal{G}$-relation over A^* and t be a $\mathcal{G}$-trace. The *graph associated with* t is $\mathcal{G}_p(t) = (Q\_Pref(t), \longrightarrow_p, A, [\epsilon], t)$ where $Q\_Pref(t)$ is the set of vertices, A the labels of the arcs, $\longrightarrow_p = \{(t, a, t\diamond a) \mid t\diamond a \in Q\_Pref(t)\}$ and ϵ is the empty word.

We write $s' \longrightarrow s''$ or $s' \xrightarrow{a} s''$ whenever there exists a in A such that (s', a, s'') is in $\longrightarrow$, and we write $s' \xrightarrow{*} s''$ or $s' \xrightarrow{a_1 a_2 \cdots a_n} s''$ whenever $s' = s''$ or there exist s_1, s_2, $\cdots$, s_{n-1} and a_1, a_2, $\cdots$, a_n in A such that $s' \xrightarrow{a_1} s_1 \xrightarrow{a_2} s_2 \cdots s_{n-1} \xrightarrow{a_n} s''$.

Similarly to classical traces where the investigation of occurrence graphs constitutes an important research line, we will see here that the graph associated with any $\mathcal{G}$-trace is quite powerful. We have the following properties:

- $\mathcal{G}_p(t)$ is acyclic, thus $\xrightarrow{*}_p$ is a partial order with a minimal element, $[\epsilon]$, and a maximal element, t.
- $\mathcal{G}_p(t)$ is deterministic: $\forall s, s', s'', \forall a, (s \xrightarrow{a}_p s' \wedge s \xrightarrow{a}_p s'') \Rightarrow (s' = s'')$,
- $\forall s, \forall w, w', ([\epsilon] \xrightarrow{w}_p s \wedge [\epsilon] \xrightarrow{w'}_p s) \Rightarrow (w \sim w' \wedge s = [w] = [w'])$,
- $\forall s, s', \forall w, w', (s \xrightarrow{w}_p s' \wedge s \xrightarrow{w'}_p s') \Rightarrow (\forall a \in A, |w|_a = |w'|_a)$,
- Quasi-prefixing behaves well towards graph inclusion: $(t' \in Q\_Pref(t)) \Rightarrow \mathcal{G}_p(t')$ is the restriction of $\mathcal{G}_p(t)$ to $Q\_Pref(t')$.

Let us remark that a $\mathcal{G}$-relation is right half-cancellative iff for any $\mathcal{G}$-trace t, $\mathcal{G}_p(t)$ is co-deterministic, i.e.: $\forall s', s'', s \in Q\_Pref(t)$, $\forall a \in A$, $(s' \xrightarrow{a}_p s \wedge s'' \xrightarrow{a}_p s) \Rightarrow (s' = s'')$. Moreover, one of the basic properties of the graph associated with a $\mathcal{G}$-trace is that the paths of the graph describe exactly the $\mathcal{G}$-trace and its prefixes. Set $Path(G) = \{w \in A^* \mid \exists s \in V, [\epsilon] \xrightarrow{w} s\}$ and $Path_{max}(G) = \{w \in A^* \mid [\epsilon] \xrightarrow{w} t\}$.

Lemma 3 *Let t be a $\mathcal{G}$-trace, then $Path_{max}(\mathcal{G}_p(t)) = t$.*

Example 1 Consider:

$A \quad = \{a, b, c, d\}$

$\sim_1$ congruence generated by $abc \equiv acb$, $acd \equiv adc$ and $cbd \equiv cdb$

$t \quad = [abcd]_{\sim_1} = \{abcd, acbd, acdb, adcb\}$

The graph of quasi-prefixes of t, $\mathcal{G}_p(t)$, is depicted in Figure 1.

3 Partial Orders and Lattices of Configurations

As we want to deal with partial ordered sets, a second and quite common idea is to look at the relative order of occurrences of letters, viewed as actions: two

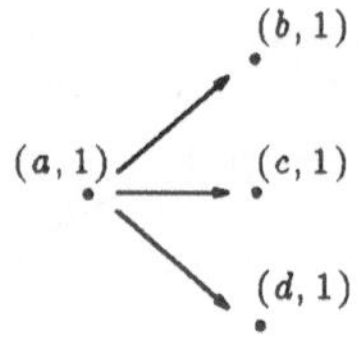

Figure 2: Partial order associated with $t = [abcd]_i$ (Example i), i=1, 2, 3.

actions are ordered iff they are ordered identically in any equivalent execution. Even if not settled down here in clear terms, this will appear later as the way of introducing concurrency as a binary relation between events.

Definition 4 Let A be an alphabet, $\sim$ be a $\mathcal{G}$-relation over A^* and t be a $\mathcal{G}$-trace. Then define:

- the *set of events associated with* t: $E_t = \{(a,i) \in A \times I\!N \mid a \in alph(t) \text{ and } 1 \leq i \leq |t|_a\}$.
- the *labelling function*: $\lambda : E_t \longrightarrow A$ defined by $\forall (a,i) \in E_t,\ \lambda((a,i)) = a$.
- the *labelled partial order associated with* t, $\theta(t) = (E_t, \prec_t, \lambda, A)$, where $\prec_t$ is defined by: $\forall (a,i),(b,j) \in E_t, (a,i) \prec_t (b,j)$ iff the i-th occurrence of a appears before the j-th occurrence of b in every word of t.

For a labelled partial order $\theta = (E, \prec, \lambda, A)$, the set of linearizations of θ, denoted by $Lin(\theta)$, is defined by: $Lin(\theta) = \{w \in A^* \mid \exists m \in E^*, w = \lambda(m),\ \forall e \in E, |m|_e = 1 \text{ and } \forall e, e' \in E, e \prec e' \Rightarrow e \text{ occurs before } e' \text{ in } m\}$. The question that immediately arises is to determine whether and how the order $\theta(t)$ describes a $\mathcal{G}$-trace t. It is clear that any $\mathcal{G}$-trace is included in its set of linearizations, but the converse may be not true.

Lemma 5 *Let t be a $\mathcal{G}$-trace, then $t \subseteq Lin(\theta(t))$.*

Proof of Lemma 5 is quite straightforward. For a counter-example to the converse, see Example 1.

Example 1 (continued) The set of events associated with t is $E_t = \{(a,1), (b,1), (c,1), (d,1)\}$, the partial order $\theta(t)$ is depicted in Figure 2. The set of linearizations of $\theta(t)$ is $Lin(\theta(t)) = \{abcd, abdc, acbd, acdb, adbc, adcb\}$ and $t = \{abcd, acbd, acdb, adcb\}$ is strictly included in $Lin(\theta(t))$.

As soon as an execution is considered as a partial ordered set of events, it is very well known that the notion of configurations[1] gives an account of global states of the system during the execution. Let us recall these definitions.

Definition 6 Let $\theta = (E, \prec, \lambda, A)$ be a labelled partial order.

- $\mathrm{C} \subseteq E$ is a *configuration* of θ if C is *left-closed* for $\prec$, i.e.: $\forall e \in \mathrm{C}, \forall e' \in E,\ (e' \prec e) \Rightarrow (e' \in \mathrm{C})$.
- $\mathcal{C}(\theta) = \{\mathrm{C} \subseteq E \mid \mathrm{C} \text{ configuration of } \theta\}$ is the *set of configurations* of θ.
- $\mathcal{G}_\mathcal{C}(\theta)$ *the graph associated with* θ is the lattice of configurations of θ, i.e.: $\mathcal{G}_\mathcal{C}(\theta) = (\mathcal{C}(\theta), \longrightarrow_\mathcal{C}, A, \emptyset, E)$ where $\longrightarrow_\mathcal{C} = \{(\mathrm{C}, \lambda(e), \mathrm{C} \cup \{e\}) \in \mathcal{C}(\theta) \times A \times \mathcal{C}(\theta) \mid e \notin \mathrm{C}\}$.

[1] or ideals, or consistent cuts

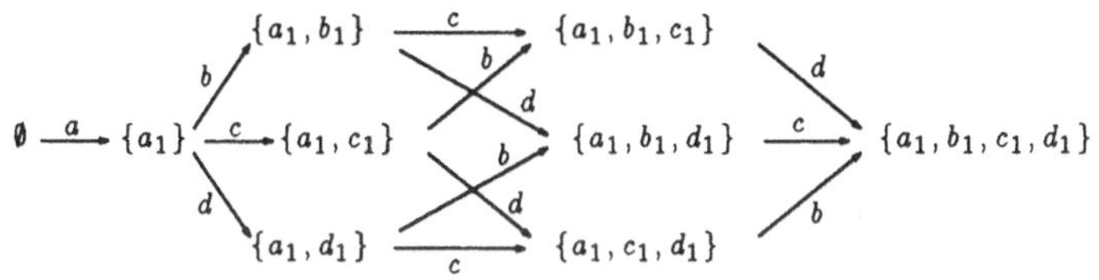

Figure 3: Lattice of configurations of $\theta([abcd]_i)$ (Example i), i=1, 2, 3.

Example 1 (continued) For sake of readability we will write a_i for the event (a, i); the set of configurations of $\theta(t)$ is $\mathcal{C}(\theta(t)) = \{\emptyset, \{a_1\}, \{a_1, b_1\}, \{a_1, c_1\}, \{a_1, d_1\}, \{a_1, b_1, c_1\}, \{a_1, b_1, d_1\}, \{a_1, c_1, d_1\}, \{a_1, b_1, c_1, d_1\}\}$ and the lattice of configurations of $\theta(t)$ is depicted in Figure 3.

It is well known that the lattice of configurations is a distributive lattice and that it is possible to rebuild the partial order from its prime elements [NPW81]. In the same way, for partially commutative monoids, occurrence graphs and classical traces are the two sides of the same object. This fact has been very often depicted for finite traces and is true for infinite ones (as soon as they are defined as the ideal completion of the finite case). Here and obviously, for any labelled partial order θ, $\mathcal{G}_\mathcal{C}(\theta)$ is an acyclic graph with root $\emptyset$ and leaf E. Moreover, the paths of this lattice give exactly the linearizations of the partial order set.

Lemma 7 *Let θ be a partial order, then $Lin(\theta) = Path_{max}(\mathcal{G}_\mathcal{C}(\theta))$.*

Whereas in classical trace theory, graph of prefixes and lattice of configurations are equal, for $\mathcal{G}$-traces, in general, the graph of quasi-prefixes is not isomorphic to the lattice of configurations. (see Example 1, $\mathcal{G}_p(t)$ (Figure 1) is not isomorphic to $\mathcal{G}_\mathcal{C}(\theta(t))$ (Figure 3))

4 A mapping between the two graphs

Thus, given a $\mathcal{G}$-trace t, we may want to consider two graphs: $\mathcal{G}_p(t)$, that describes t rather completely but has few nice properties with respect to an ordering relation, and $\mathcal{G}_\mathcal{C}(\theta(t))$, a distributive lattice that describes totally the order associated with t. As in general $t \neq Lin(\theta(t))$, our purpose is to look accurately at similarities between the two graphs.

In order to compare them, we will use a rather natural notion of mapping: a *mapping* between the graphs $G = (V, \longrightarrow, A, r, l)$ and $G' = (V', \longrightarrow', A, r', l')$[2] is an application $\mathcal{H} : V \longrightarrow V'$ such that $\mathcal{H}(r) = r'$, $\mathcal{H}(l) = l'$ and $\forall s_1, s_2 \in V, \forall a \in A$, $s_1 \xrightarrow{a} s_2 \Rightarrow \mathcal{H}(s_1) \xrightarrow{a}{}' \mathcal{H}(s_2)$. We then say that G is mapped by $\mathcal{H}$ into G'. If the application $\mathcal{H}$ is *injective* (resp. *surjective, bijective*), we will say that G is *injectively* (resp. *surjectively, bijectively*) *mapped* by $\mathcal{H}$ into G'. In case $\mathcal{H}$ is injective we will say that G is embedded into G' and write $G \subseteq G'$. Finally, a mapping is *full* if every edge in the target set is the image of at least one edge of the source set: $\forall s'_1, s'_2 \in \mathcal{H}(V), \forall a \in A, \quad (s'_1 \xrightarrow{a}{}' s'_2) \Rightarrow (\exists s_1 \in$

[2]Note that the mapping is defined only for graphs having the same labelling set A.

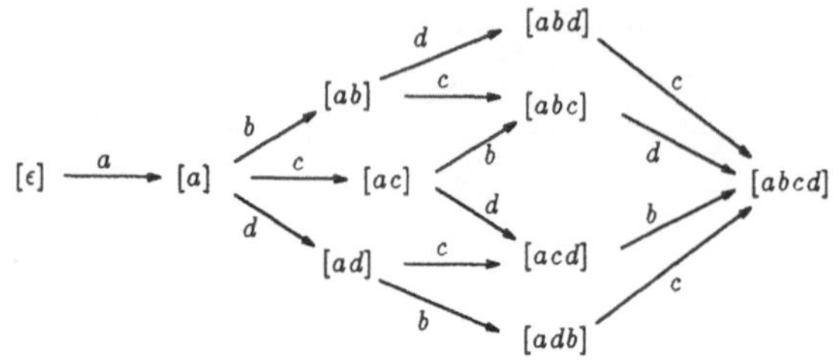

Figure 4: Graph of prefixes for $t = [abcd]_{\sim_2}$ (Example 2).

$\mathcal{H}^{-1}(s'_1)$, $\exists s_2 \in \mathcal{H}^{-1}(s'_2)$, $s_1 \xrightarrow{a} s_2$). In the same way, if the mapping from G to a G' is *full*, we will say that G is *fully mapped* by $\mathcal{H}$ into G'. If $\mathcal{H}$ is bijective and full, we say that G and G' are isomorphic: $G \approx G'$.

The point for us is to compare the two graphs $\mathcal{G}_p(t)$ and $\mathcal{G}_C(\theta(t))$. As $Path_{max}(\mathcal{G}_p(t)) = t \subseteq Lin(\theta(t)) = Path_{max}(\mathcal{G}_C(\theta(t)))$, a first but deceptive intuition is to think that $\mathcal{G}_p(t)$ is embedded into $\mathcal{G}_C(\theta(t))$: in fact, none of the two graphs is embedded into the other as illustrated by Examples 1 and 2. In these examples $\mathcal{G}$-traces $[abcd]_{\sim_1}$ and $[abcd]_{\sim_2}$ have the same associated partial order and thus the same lattice of configurations (depicted in Figure 3), but the graph of quasi-prefixes for $[abcd]_{\sim_1}$ (Figure 1) is strictly embedded into the lattice of configurations whereas the graph of quasi-prefixes for $[abcd]_{\sim_2}$ (Figure 4) is not emdedded into it.

Example 2 Consider:

$A \ = \{a, b, c, d\}$

$\sim_2$ the congruence generated by $bdc \equiv bcd$, $abc \equiv acb$, $cbd \equiv cdb$, $acd \equiv adc$, $dcb \equiv dbc$

$t \ = [abcd]_{\sim_2} = \{abdc, abcd, acbd, acdb, adcb, adbc\}$

$\mathcal{G}_p(t)$is depicted in Figure 4. The partial order associated with $[abcd]_{\sim_2}$ is the same than the one associated with $[abcd]_{\sim_1}$ (Example 1, Figure 2), thus they have the same lattice of configurations (Figure 3). Note that $\mathcal{G}_p(t)$ is not embedded into $\mathcal{G}_C(\theta(t))$.

However, $\mathcal{G}_C(\theta(t))$ cannot be really "smaller" than $\mathcal{G}_p(t)$ as settled in the following proposition.

Proposition 8 *Let A be an alphabet, $\sim$ be a $\mathcal{G}$-relation and t be a $\mathcal{G}$-trace. Then:*

$$\mathcal{G}_C(\theta(t)) \subseteq \mathcal{G}_p(t) \implies \mathcal{G}_C(\theta(t)) \approx \mathcal{G}_p(t).$$

Sketch of proof: Let t be a $\mathcal{G}$-trace such that $\mathcal{G}_C(\theta(t))$ is embedded into $\mathcal{G}_p(t)$: there exists $\mathcal{H}$ injective from $\mathcal{C}(\theta(t))$ to $Q\_Pref(t)$. We have to prove that $\mathcal{H}$ is surjective and full. Let $[w]$ be a quasi-prefix of t. By induction on the length of $[w]$ and with the help of lemmas 3, 5 and 7, we prove that it exists a path in $\mathcal{G}_C(\theta(t))$ from $\emptyset$ to the configuration $C = \{(x, i) \mid x \in alph([w])$ and $1 \leq i \leq |[w]|_x\}$. As $\mathcal{H}(\emptyset) = [\epsilon]$ and $\mathcal{G}_p(t)$ is deterministic, we have $\mathcal{H}(C) = [w]$ and $\mathcal{H}$ is surjective. Let now $[w] \xrightarrow{a}_p [w] \diamond a$ be in $\mathcal{G}_p(t)$, then $\mathcal{H}^{-1}([w] \diamond a) - \mathcal{H}^{-1}([w]) = \{(a, |[w.a]|_a)\}$, thus $\mathcal{H}^{-1}([w]) \xrightarrow{a}_C \mathcal{H}^{-1}([w] \diamond a)$ is in $\mathcal{G}_C(\theta(t))$ and $\mathcal{H}$ is full. □

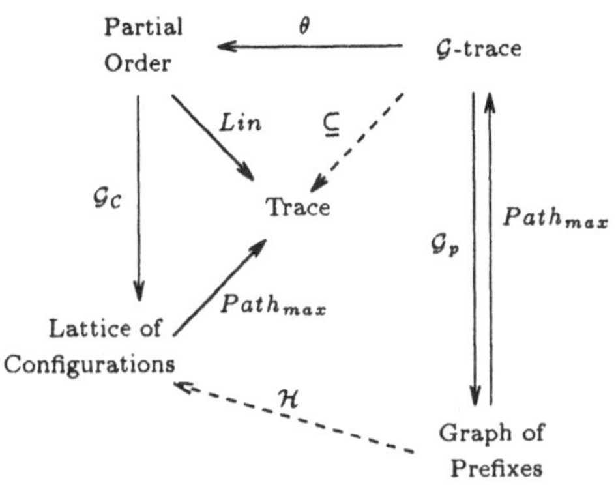

Figure 5: Links between different representations of a $\mathcal{G}$-trace.

Moreover, there exists a single way of mapping $\mathcal{G}_p(t)$ into $\mathcal{G}_C(\theta(t))$.

Proposition 9 mapping from $\mathcal{G}_p(t)$ to $\mathcal{G}_C(\theta(t))$. *For any $\mathcal{G}$-trace t, define $h : Q\_Pref(t) \longrightarrow \mathcal{C}(\theta(t))$ by $h([w]) = \{(a,i) \mid a \in alph(w) \text{ and } 1 \leq i \leq |w|_a\}$. Then h is the unique mapping from $\mathcal{G}_p(t)$ to $\mathcal{G}_C(\theta(t))$.*

Sketch of proof: First prove that, for any t, the application h in Proposition 9 is well defined. Second, let $\mathcal{H}$ be any mapping from $\mathcal{G}_p(t)$ to $\mathcal{G}_C(\theta(t))$, then $\mathcal{H}([\epsilon]) = \emptyset$ and, by induction on the length of $[w]$, $\mathcal{H}([w]\diamond a) = h([w]) \cup \{(a, |[w]\diamond a|_a)\}$; thus $\mathcal{H} = h$. ⊓

However, the mapping of the graph of prefixes of a $\mathcal{G}$-trace into the lattice of configurations of its partial order does not behave well. The corresponding mapping h can be injective and/or surjective and/or full and many possibilities exist. Examples 1, 2 and 3 exhibit some of them: h is injective, full and not surjective in Example 1, or h is surjective, full and not injective in Example 2, or h is injective and surjective but not full in Example 3.

Example 3 Consider:

$A \quad = \{a, b, c, d\}$

$\sim_3$ congruence generated by $abc \equiv acb$, $cbd \equiv cdb$, $acd \equiv adc$, $dcb \equiv dbc$

$t \quad = [abcd]_{\sim_3} = \{abcd, acbd, acdb, adcb, adbc\}$

$\mathcal{G}_p(t)$is depicted in Figure 6. The partial order associated with $[abcd]_{\sim_3}$ is the same than the one associated with $[abcd]_{\sim_1}$ (Example 1, Figure 2), thus the lattice of configurations are the same (Figure 3). We can see that the mapping is bijective and is not full, as arrow $\{a_1, b_1\} \xrightarrow{d}_C \{a_1, b_1, d_1\}$ is not reached.

As $t = Path_{max}(\mathcal{G}_p(t)) \subseteq Lin(\theta(t)) = Path_{max}(\mathcal{G}_C(\theta(t)))$, it is clear that if the mapping is bijective and full, then the $\mathcal{G}$-trace is equal to the linearizations of its partial order. The next section looks at some necessary and/or sufficient properties for the mapping to be injective, surjective or full.

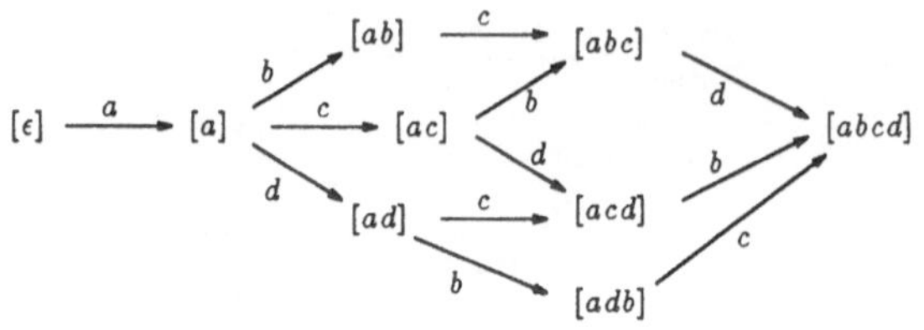

Figure 6: Graph of prefixes for $t = [abcd]_{\sim_3}$ (Example 3).

5 Properties of the mapping

As in general $t \subseteq Lin(\theta(t))$, a natural idea is to try to relate the possible equality $t = Lin(\theta(t))$ with the fact for the mapping to be surjective and full. We have the following result:

Proposition 10 *Let h be as defined in Proposition 9.*
Then: $\quad t = Lin(\theta(t)) \Longrightarrow$ *h is surjective and full.*

Sketch of proof: Let t be in $A^*/_\sim$ such that $t = Lin(\theta(t))$. By definition of h, $h([\epsilon]) = \emptyset$. We will prove by induction on the size of the configurations that any configuration and any arrow is in the image of $\mathcal{H}$. It is true for any $\emptyset \xrightarrow{\lambda(e)}_{\mathcal{C}} \{e\}$. If it is true until n, suppose now $C \xrightarrow{\lambda(e)}_{\mathcal{C}} C \cup \{e\}$ and $|C| = n$. Let $w = \lambda(e_1 \dots e_n)$ be a path of $\mathcal{G}_{\mathcal{C}}(\theta(t))$ leading to C, by induction hypothesis, it is a path of $\mathcal{G}_p(t)$ and $C = h([w])$. By lemmas 3, 5 and 7, we get that $w.\lambda(e)$ is a path of $\mathcal{G}_p(t)$ and thus that $h([w] \diamond \lambda(e)) = C \cup \{e\}$. Thus $C \xrightarrow{\lambda(e)}_{\mathcal{C}} C \cup \{e\}$ is $h([w]) \xrightarrow{\lambda(e)}_p h([w] \diamond \lambda(e))$ and h is surjective and full. □

This result is illustrated by Example 2 where $[abcd]_{\sim_2} = Lin(\theta([abcd]_{\sim_2}))$, note that in this example the associated mapping is not injective, thus the associated graphs are not isomorphic. The converse of Proposition 10 may be not true: in Example 4 and Figure 7, the mapping is surjective and full but the $\mathcal{G}$-trace is not equal to the linearizations of its associated partial order.

Example 4 Consider:
$A \quad = \{a, b\}$
$\sim_4 \quad$ the congruence over A^* generated by $abab \equiv baba$
$t \quad = [abab]_{\sim_4} = \{abab, baba\}$
Figure 7 depicts the graph of quasi-prefixes, the partial order and the lattice of configurations associated with t. The mapping is surjective and full, however t is strictly included into $Lin(\theta(t)) = \{abab, abba, baab, baba\}$.

The second idea is then to look at the possibility for h to be injective. We get the following:

Proposition 11 *Let A be an alphabet and $\sim$ be a $\mathcal{G}$-relation. Then:*

$$\forall t \in A^*/_\sim,\ h \text{ is injective} \Longrightarrow \forall t \in A^*/_\sim,\ \mathcal{G}_p(t) \text{ is co-deterministic}$$

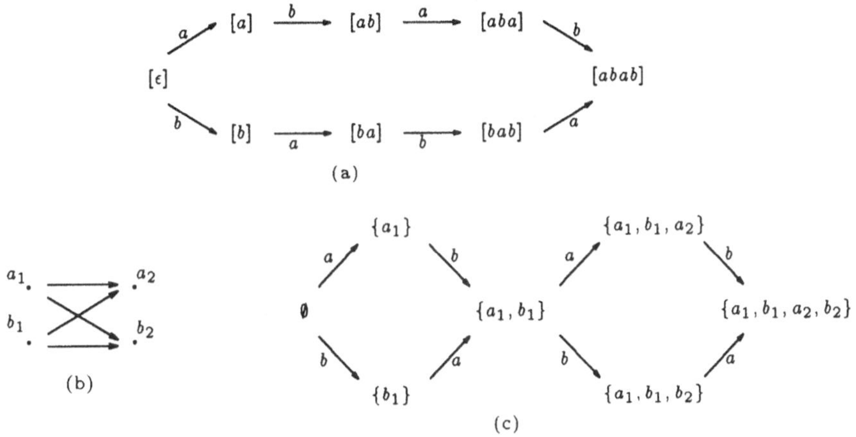

Figure 7: (a) $\mathcal{G}_p(t)$, (b) $\theta(t)$, (c) $\mathcal{G}_C(\theta(t))$ for $t = [abab]_{\sim_4}$ (Example 4).

Sketch of proof: Suppose that for any t in $A^*/_\sim$, h is injective. Let $[w_1], [w_2] \in A^*/_\sim$ and $a \in A$ such that $[w_1]\diamond a = [w_2]\diamond a$. Let $t' = [w_1]\diamond a = [w_2]\diamond a$, then $[w_1]$ and $[w_2]$ are quasi-prefixes of t' and $h([w_1]) = h([w_2]) = h(t') - \{(a, |t'_a|)\}$. Since h is injective, $[w_1] = [w_2]$. Therefore $\mathcal{G}_p(t)$ is co-deterministic. □

When dealing with $\mathcal{G}$-relations, half-cancellative and co-determinism for any graph of quasi-prefixes are equivalent notions, therefore Proposition 11 is equivalent to: $(\forall t \in A^*/_\sim,\ h$ is injective$) \Rightarrow \sim$ is half-cancellative.

Consider the example of classical traces, for any independence relation $||$ and any trace of $A^*/_{\sim_{||}}$ the mapping is injective and we know that the congruence $\sim_{||}$ is cancellative. Consider this time Example 2 and Figure 4, the $\mathcal{G}$- relation $\sim_2$ is not right half-cancellative ($[abd]\diamond c = [adb]\diamond c$, but $[abd] \neq [adb]$) and the mapping is not injective ($h([abd]) = h([adb])$ and $[abd] \neq [adb]$). But in example 4, $\sim_4$ is right half-cancellative but the mapping from $\mathcal{G}_p([abab]_{\sim_4})$ to $\mathcal{G}_C(\theta([abab]_{\sim_4}))$ is not injective, thus the converse of Proposition 11 is not true.

Previous examples show that the graph of prefixes and the lattice of configurations of a $\mathcal{G}$-trace may have a lot of differences. Our purpose now is to find conditions on their internal structure that will relate them closely. We will do that trough diamond properties which are, in various ways, very often required for transitions systems and automata.

Definition 12 Let $G = (V, \longrightarrow, A, r, l)$ be a graph,

- G has the *forward diamond property* if: $\forall s_1, s_2, s_3 \in V$, $\forall a, b \in A$, $((a \neq b)$, $(s_1 \xrightarrow{a} s_2)$ and $(s_1 \xrightarrow{b} s_3)) \Rightarrow (\exists s_4 \in V$, $(s_2 \xrightarrow{b} s_4)$ and $(s_3 \xrightarrow{a} s_4))$.

- G has the *backward diamond property* if: $\forall s_2, s_3, s_4 \in V$, $\forall a, b \in A$, $((a \neq b)$, $(s_2 \xrightarrow{b} s_4)$ *and* $(s_3 \xrightarrow{a} s_4)) \Rightarrow (\exists s_1 \in V$, $(s_1 \xrightarrow{a} s_2)$ and $(s_1 \xrightarrow{b} s_3))$.

If a graph G has the forward (resp. backward) diamond property, we will write that $FD(G)$ (resp. $BD(G)$) is true. It is easy to see that the two diamond conditions given above are satisfied by the lattice $\mathcal{G}_C(\theta(t))$. When looking at the graph of prefixes of a $\mathcal{G}$-trace t, the two diamond properties have equivalent

(a) (b)

Figure 8: (a) Forward Diamond (b) Backward Diamond

definitions on the $\mathcal{G}$-relation. If $\sim$ is a $\mathcal{G}$-relation, then:
$(\forall t \in A^*/_\sim,\ FD(\mathcal{G}_p(t))) \iff (\forall w \in A^*, \forall a,b \in A,\ (\exists u,v, wau{\sim}wbv) \Rightarrow (wab{\sim}wba))$ and $(\forall t \in A^*/_\sim,\ BD(\mathcal{G}_p(t))) \iff (\forall w',w'' \in A^*, \forall a,b \in A,\ (w'a{\sim}w''b) \Rightarrow \exists w,\ (wb{\sim}w')\ and\ (wa{\sim}w''))$.

Note that closely related diamond properties have very often been studied, the so called diamond property being even a paradigm used for the definition of binary concurrency ([Dro89], [Sta89], ...). Note also that the definition of any concurrency equivalence by $uabv \sim ubav$ may be viewed as another kind of "diamond property". Finally, several alternative notions have been studied such as cube and inverse cube in [AH93] and in [BDK95].

We will see now that the two diamond conditions have a strong influence on the mapping.

Theorem 13 *Let t be a $\mathcal{G}$-trace such that its graph of prefixes $\mathcal{G}_p(t)$ either satisfies the forward diamond property or satisfies the backward diamond property and is co-deterministic. Then the graph of prefixes $\mathcal{G}_p(t)$ is embedded into the graph of configurations $\mathcal{G}_\mathcal{C}(\theta(t))$ and the mapping h is full.*

Sketch of proof: We first have to prove that $FD(\mathcal{G}_p(t)) \Longrightarrow$ (h is injective and full). First prove that (1) $\forall s,s',s'' \in Q\_Pref(t), \forall w \in A^*, \forall a \in A : (s \xrightarrow{w}_p s'$, $s \xrightarrow{a}_p s''$ and $|w|_a \geq 1) \Longrightarrow (\exists w' \in A^* : s'' \xrightarrow{w'}_p s'$ and $a.w'$ is a permutation of w). The proof is done by induction on the length of the least prefix u of w such that $|u|_a = 1$. Then show that (2) $\forall s,s',s'' \in Q\_Pref(t), \forall w',w'' \in A^*$: $(s \xrightarrow{w'}_p s'$, $s \xrightarrow{w''}_p s''$ and $\forall a \in A, |w'|_a = |w''|_a) \Longrightarrow (s' = s'')$. The proof is done by induction on $|w'|$: if $w' = a.w'_1$ and $w'' = b.w''_1$ $(w'_1, w''_1 \in A^*)$, then close the forward diamond $s \xrightarrow{a}_p s'_1$ and $s \xrightarrow{b}_p s''_1$ $(s'_1, s''_1 \in Q\_Pref(t))$ by $s'_1 \xrightarrow{b}_p u$ and $s''_1 \xrightarrow{a}_p u$ $(u \in Q\_Pref(t))$, apply (1) to $s'_1 \xrightarrow{w'_1}_p s'$ and $s'_1 \xrightarrow{b}_p u$, what allows to apply again the induction hypothesis to the paths between u and s' and s''. Finally, $\forall w',w'' \in A^*$, if $h([w']) = h([w''])$ then $\forall a \in A, |w'|_a = |w''|_a$ and thus, by (2), $[w'] = [w'']$ and h is injective.
To prove that h is full, use a similar construction. Suppose that $[w'],[w''] \in Q\_Pref(t)$, $C',C'' \in \mathcal{C}(\theta(t))$ and $a \in A$ such that $C' = h([w']$, $C'' = h([w'']$ and $C' \xrightarrow{a}_\mathcal{C} C''$. Then show by induction on $|w'|$ and applying (1), that $\exists u \in A^*$ such that $[w'] \xrightarrow{u}_p [w'']$ and $w'.u$ is a permutation of w''. Thus $w = a$ and h is full.
Second, we have to prove that ($BD(\mathcal{G}_p(t))$ and $\mathcal{G}_p(t)$ is co-deterministic) $\Longrightarrow$ (h is injective and full). This is done in an equivalent way but departing from the leaf (and closing the backward diamond) instead of starting from the root (and closing the forward diamond). □

Figure 9: Lattices $M3$ (left) and $N5$ (right).

Thus for any graph of quasi-prefixes, if it satisfies the backward diamond property and if it is co-deterministic, then the associated mapping is injective (Theorem 13), and if the associated mapping is injective, then the graph is co-deterministic (Proposition 11)[3]. However, injectivity (and thus co-determinism) does not imply backward diamond property, as shown by Example 3. In this example, the corresponding mapping h is injective and the graph of quasi-prefixes (Figure 6) does not satisfy the backward diamond property. Moreover the same example shows that injectivity does not imply to be full either.

Theorem 14 gives now a strong relationship between all these conditions. This result may be considered in a sense as stemmed from prime event structure representation theorem in [Roz92], following the [NPW81] tradition. More or less, works that establish a connection between traces and prime event structures start by identifying "events" in the graph as equivalence classes of prime intervals; then the given graph is shown to be the distributive lattice of configurations built with these events. As pointed out by an anonymous referee, this result may also be viewed as a consequence of the fact that $\mathcal{G}_C(\theta(t))$ is a distributive lattice, as the ideal completion of a partial order. We give here a proof built on that idea. A third and combinatorial proof is also available.

Theorem 14 *Let t be a $\mathcal{G}$-trace. Then its graph of prefixes $\mathcal{G}_p(t)$ and its graph of configurations $\mathcal{G}_C(\theta(t))$ are isomorphic iff $\mathcal{G}_p(t)$ satisfies the forward and the backward diamonds properties.*

Sketch of proof: The fact that $\mathcal{G}_p(t) \approx \mathcal{G}_C(\theta(t))$ implies $FD(\mathcal{G}_p(t))$ and $BD(\mathcal{G}_p(t))$ can be easily derived from the properties of the lattice of configurations. Let now t be a $\mathcal{G}$-trace such that $\mathcal{G}_p(t)$ satisfies the forward and the backward diamond properties.

First show that if u closes the forward diamond for $s \xrightarrow{a}_p s'$ and $s \xrightarrow{b}_p s''$, then $u = s' \sqcup$ [4] s'' and if u closes the backward diamond for $s' \xrightarrow{b}_p s$ and $s'' \xrightarrow{a}_p s$, then $u = s' \sqcap s''$. This implies that any two vertices of $\mathcal{G}_p(t)$ admit a least upper bound and a greatest lower bound and thus $\mathcal{G}_p(t)$ is a lattice. Second prove that this lattice does not have the lattices M_3 or N_5 (Figure 9) as sublattices, which is a characteristic of distributive lattices ([MMT87]) and easily derived from forward and backward diamond properties.

[3]Note that, as a consequence of Proposition 11 and Theorem 13, we obtain that a $\mathcal{G}$-relation is right half-cancellative as soon as its graph of prefixes satisfies the forward diamond property.

[4]As the quasi-prefix relation is a partial order, usual notations for least upper bound, $\sqcup$, and greatest lower bound, $\sqcap$, are available

As a consequence $\mathcal{G}_p(t)$ is a distributive lattice and thus, through h, is isomorphic to $\mathcal{G}_C(\theta(t))$. □

This result is a characterization of the structure needed for a graph of quasi-prefixes of a $\mathcal{G}$-trace to be isomorphic to the lattice of configuration of its partial order. Of course it is also a sufficient condition for the $\mathcal{G}$-trace to be equal to the linearizations of its partial order (recall that this condition is not necessary (see Example 2)).

In the next section, we look at some examples of particular $\mathcal{G}$-relations illustrating these results.

6 Examples

6.1 The free partially commutative monoid

Let I be a symmetric irreflexive relation in $A \times A$ and $\sim_I$ be the reflexive and transitive closure of the relation $\{(ab, ba) \mid (a,b) \in I\}$. The relation $\sim_I$ is a congruence and the quotient monoid $A^*/_{\sim_I}$ is the classical partially commutative monoid or trace monoid. An element of $A^*/_{\sim_I}$ is called a trace. It is well known that for any trace t, $t = Lin(\theta(t))$, $\mathcal{G}_p(t) \approx \mathcal{G}_C(\theta(t))$ [RT91] and the partial order $\theta(t)$ is known as occurrence graph of the trace [Per89].

6.2 Context Traces

Let us now consider the case of context-traces introduced in [BR94]. Let $R \subseteq \{(abc, acb) \in A^3 \times A^3\}$ be a relation. A context relation $\sim_R$ is a congruence relation generated by R, an element in $A^*/_{\sim_R}$ is called a context trace. Clearly, context relations are $\mathcal{G}$-relations. The graph of prefixes of a context trace $[w]_{\sim_R}$ can be obtained from the graph $r \xrightarrow{w} l$ by iterating the following process: starting from r when finding $s_1 \xrightarrow{a} s_2 \xrightarrow{b} s_3 \xrightarrow{c} s_4$ in the graph with $(abc, acb) \in R$, then, if $s_2 \xrightarrow{c} s_3'$ exists but $s_3' \xrightarrow{b} s_4$ does not exist then add $s_3' \xrightarrow{b} s_4$ to the graph, if $s_2 \xrightarrow{c} s_3'$ does not exist add $s_2 \xrightarrow{c} s_3' \xrightarrow{b} s_4$ to the graph (the construction is performed such that in that case $s_3' \xrightarrow{b} s_4$ does not exists). This process ends and the resulting graph is the graph of quasi-prefixes of $[w]_{\sim_R}$ (up to a renaming of the vertices).

In general, for a context relation R, the graph of quasi-prefixes of a context trace t is not isomorphic to its lattice of configurations, thus $\mathcal{G}_p(t) \not\approx \mathcal{G}_C(\theta(t))$. Examples 1, 2 and 3 exhibit situations where context relations give rise to graphs of quasi-prefixes that do not satisfy forward diamond and/or backward diamond property.

If a context does not commute, that is if R is of the form $\{(a_ib_ic_i, a_ic_ib_i) \mid \forall i,j : \{a_i\} \cap \{b_j, c_j\} = \emptyset\}$, then any resulting graph of prefixes satisfies FD and BD, thus $\mathcal{G}_p(t) \approx \mathcal{G}_C(\theta(t))$.

Example 5

Consider: $A = \{a,b,c,d,e,f\}$, $R = \{(abc, acb), (ade, aed), (fbc, fcb)\}$ and $t = [abcadeafbc]_{\sim_5} = \{abcadeafbc,\ abcadeafcb,\ abcaedafbc,\ abcaedafcb,$

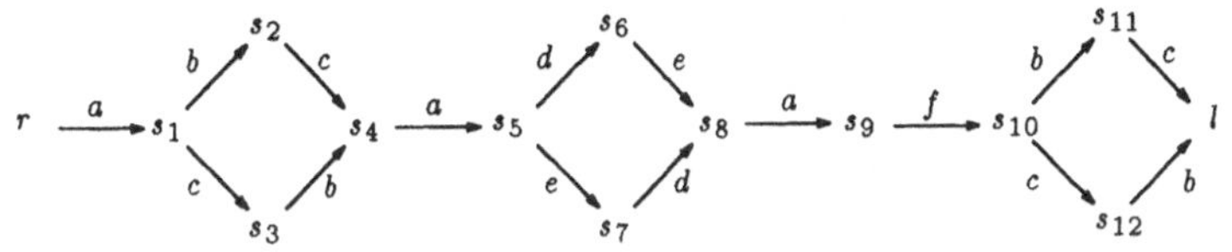

Figure 10: Graph of quasi-prefixes for $t = [abcadeafbc]_{\sim_5}$ (Example 5).

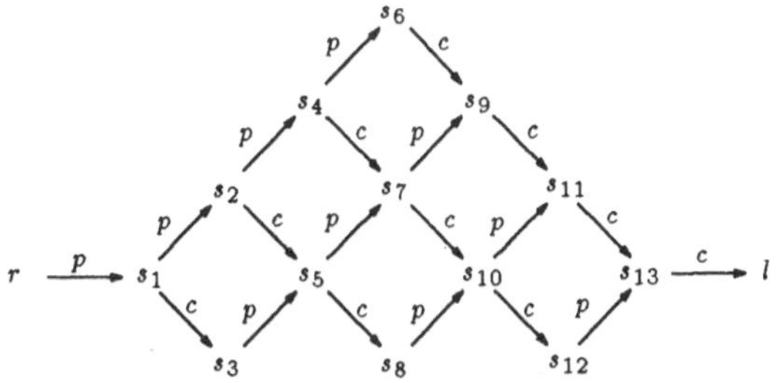

Figure 11: Graph of quasi-prefixes for $t = [ppppcccc]_{\sim_6}$ (Example 6).

acbadeafbc, *acbadeafcb*, *acbaedafbc*, *acbaedafcb*}. The graph of quasi-prefixes of t (up to a renaming of the vertices) is depicted in Figure 10.

The Consumer/Producer paradigm If $R = \{(ppc, pcp)\}$, then, when constructing the graph of quasi-prefixes and adding edges, FD and BD properties are always preserved. In the resulting graph, a vertex has 0, 1 or 2 immediate predecessors or successors; in the latter case, the 2 predecessors (resp. successors) always close the backward (resp. forward) diamond.

Example 6 Consider: $A = \{p, c\}$, $R = \{(ppc, pcp)\}$ and $t = [ppppcccc]_{\sim_6} = \{ppppcccc, pppcpccc, pppccpcc, pppcccpc, ppcppccc, ppcpcpcc, ppcpccpc, ppccppcc, ppccpcpc, pcpppccc, pcppcpcc, pcppccpc, pcpcppcc, pcpcpcpc\}$.

The graph of quasi-prefixes of t (up to a renaming of the vertices) is depicted in Figure 11.

7 Conclusion

Considering a generalized trace defined by an equivalence relation on a monoid of actions, we have studied two graphs: its graphs of quasi-prefixes and the lattice of configurations associated with the intersection of orders between actions. Whereas equal in the classical framework they are in general quite different, but, surprisingly, none of them is included in the other, although the trace is always included in the set of linearizations of the lattice of configurations of its order. Moreover, the trace may be equal to this set of linearizations and admits two distinct such graphs. However, we have shown that there exists only one possible mapping from one of these graphs to the other and have found necessary and sufficient conditions for these graphs to be isomorphic. These conditions lead to a prime event structure representation: this is not surprising, as it is done to fit with the partial order structure. We have now several remarks to point out. First, our conditions are settled with the help of diamond properties: since this kind of properties have been introduced in various ways by several authors they seem to be unavoidable steps to be looked at: further more, FD and BD conditions are easier to test locally than the fact for a lattice to be distributive. Second, note that the initial definition of an equivalence relation on A^* was not formulated in terms of binary concurrency: this arises through diamonds properties and as a consequence of partial order requirements. Does it mean that the ongoing discussion on the adequacy of sequential representations of concurrent systems is not closed? Contributions to that question may be found in [HR91] and [HKT93], but we have the feeling that no definitive answer is actually known. At last, our study has been strictly limited to a single finite trace. A natural first idea is to extend this to a set of traces: we hope it to be possible although the first attempts have not been successful, the notions of events and order being then difficult to formulate. A second idea is to extend to infinite traces: this seems to be probably possible in the case of one single trace, subject to either solve a hard labelling problem or admit an infinite alphabet of actions.

Acknowledgments

Thanks to Serge Bauget, Paul Gastin, Antoine Petit, Laurent Rosaz and other members of the MEP seminar for fruitful discussions on traces and partial orders and to anonymous referees for helpfull comments.

References

[AH93] A. Arnold and J.F. Husson. On an extension of asynchronous automata. Technical Report 93-10-R, Institut de Recherche en Informatique de Toulouse, France, 1993.

[AR88] I.J. Aalbersberg and G. Rozenberg. Theory of traces. *Theoretical Computer Science*, 60(1):1–82, 1988.

[Arn91] A. Arnold. An extension of the notion of traces and of asynchronous automata. *RAIRO, Theoretical Informatics and Applications*, 25(4):335–393, 1991.

[BBR94] S. Bauget, I. Biermann, and B. Rozoy. Generalized traces and partial orders. TR, LRI 95-947, Orsay, France. Presentation at the Caliban meeting (Zaragoza, June 1994), 1994.

[BDK95] F. Bracho, M. Droste, and D. Kuske. Dependence orders for computations of concurrent automata. *Lecture Notes in Computer Science*, 900:467–478, 1995. STACS'95.

[BG95] S. Bauget and P. Gastin. On congruences and partial orders. TR, LITP, Paris, France, to appear in MFCS'95, 1995.

[BR94] I. Biermann and B. Rozoy. Context traces and transition systems. In *ISCIS IX*, Antalya, Turkey, 1994.

[Die90] V. Diekert. Combinatorics on traces. *Lecture Notes in Compuetr Science*, 520, 1990.

[Dro89] M. Droste. Event structures and domains. *Theoretical Computer Science*, 68:37–47, 1989.

[GP92] P. Gastin and A. Petit. Poset properties of complex traces. *Lecture Notes in Computer Science*, 629:255–263, 1992. MFCS'92.

[HKT92] P. Hoogers, H.C.M. Kleijn, and P.S. Thiagarajan. A trace semantics for Petri nets. Technical Report 92-03, Leiden University, 1992. to appear in 1995 in Information and Computation.

[HKT93] P. Hoogers, H.C.M. Kleijn, and P.S. Thiagarajan. Local event structures and Petri nets. *Lecture Notes in Computer Science*, 715:462–476, 1993.

[HR91] H.J. Hoogeboom and G. Rozenberg. Diamond properties of state spaces of elementary net systems. *Fundamenta Informaticae*, XIV:287–300, 1991.

[Maz86] A. Mazurkiewicz. Trace theory. *Lecture Notes in Computer Science*, 255:279–324, 1986. Advanced Course on Petri Nets.

[MMT87] R. McKenzie, G. McNulty, and W. Taylor. *Algebras, Lattices, Varietes*, volume 1. Wadsworth & Brooks/Cole, 1987.

[NPW81] M. Nielsen, G. Plotkin, and G. Winskel. Petri nets, event structures and domains. *Theoretical Computer Science*, 13(1):85–108, 1981.

[Per89] D. Perrin. Partial commutations. *Lecture Notes in Computer Science*, 372:637–651, 1989. ICALP'89.

[Roz92] B. Rozoy. Distributed languages and models for concurency. *Lecture Notes in Computer Science*, 609:267–291, 1992. Advances in Petri Net.

[RT91] B. Rozoy and P.S. Thiagarajan. Trace monoids and event structures. *Theoretical Computer Science*, 91(2):285–313, 1991.

[Sta89] E.W. Stark. Connections between a concrete and an abstract model of concurrent systems. *Lecture Notes in Computer Science*, 442:53–79, 1989. 5th Mathematical Foundations of Programming Semantics, New Orleans, Louisiana, USA.

[Vog91] W. Vogler. A generalization of traces. *RAIRO, Theoretical Informatics and Applications*, 25(2):147–156, 1991.

Orbits, half-frozen tokens and the liveness of weighted circuits*

Piotr Chrząstowski-Wachtel, Marek Raczunas
Institute of Informatics, Warsaw University,
Warszawa, Poland

Abstract

The paper deals with the problem of liveness of weighted conservative circuits. The weight of a marking gives us an important information about its liveness. Some weights correspond only to live markings, some only to dead ones and some to both. The simultaneous presence of live and dead markings with the same weight is associated with the presence of several equivalence classes generated by the solutions of the state equation. We call such classes *orbits*. It is impossible to reach from a given marking a state belonging to a different orbit and this creates an opportunity of coexistence of a dead and a live marking with the same weight. Different orbits are also associated with the presence of a kind of frozen tokens, which we call *half frozen*. A discussion of problems associated with determining liveness for weighted circuits follows and an arithmetical condition to determine whether a given marking is live is presented.

Keywords: Structure theory, Weighted circuits, Diophantine problem of Frobenius, Liveness, Petri nets.

1 Introduction

Circuits form basic subnets. Some problems (like liveness of T-graphs, see e.g. [TCCS 92]) can be reduced to the problem of liveness of circuits. For ordinary nets deciding liveness of circuits is not a problem at all. The problem of determining liveness of a given marking in a weighted circuit is not that easy. Consistence of weighted circuits is equivalent to their conservativeness. Throughout the paper we will talk only about weighted and consistent circuits. From now on we will call them simply circuits. Deciding liveness of circuits can be done behaviourally by firing a T-invariant vector. If this vector is fireable then the circuit is live, otherwise it is not live. A slightly more elaborate algorithm has been presented in [TCCS 92].

There are at least two limitations in this approach to liveness testing. First, such behavioural description is difficult to handle for example when we want to use it in mathematical proofs. This is a general problem of algorithmic characterizations. Second is the complexity problem. In general the size of the minimal T-invariant X may be exponential wrt the number of transitions.

We shall look for some algebraic characterizations of liveness in circuits. In conservative circuits there exists exactly one integer S-vector generating the

*This work has been supported by the ESPRIT Basic Research Project CALIBAN, WG 6067

space of positive S-invariants. We can choose the least S-invariant Y and call it the *weight vector*. So once we choose an initial marking M, we fix the weight $Y^T \cdot M$ of its successors.

Now we can also state an opposite problem: when the weight is fixed, what can be said about the liveness of all markings having this weight. How does one characterize *live weights* — such that every marking of that weight is live? How does one characterize *dead weights* — such that all markings with that weight are not live? Can it be the case that for given weight there exist live and non-live markings at the same time?

The stated problems seem to be not too difficult at the first glance. It was a surprise for the authors to discover how tough these problems can be in general. The paper refers to the solutions to some of the stated questions ([ChR 93],[TCCS 92]), relating them to the generalized Diophantine problem of Frobenius (see e.g. [RCh 92], [Selmer 77], [Brauer 42]).

Maybe the fact of meeting problems in liveness analysis of circuits is not that surprising when we recognize that the attempt of simulation of a weighted circuit by an ordinary net (without weights) leaves in general the class of free choice systems [TCCS 92].

Throughout the paper we adopt standard terminology from the Petri Nets Theory. The reader unfamiliar with the terminology may refer, for instance, to [TCCS 92] for detailed definitions. We assume $\mathcal{C}$ to be a weighted consistent circuit with n *transitions* ($n > 1$) from the set $T = \{t_1, \ldots, t_n\}$ and with n *places* from the set $P = \{p_1, \ldots, p_n\}$. Each transition t_i outputs c_{ii} tokens to the place p_i and takes $c_{(i\ominus 1)i}$ tokens from the place $p_{i\ominus 1}$. The signs $\ominus, \oplus$ denote the subtraction and addition modulo n. By C we denote the *incidence matrix* of $\mathcal{C}$ consisting of c_{ij}'s just defined and of zeros on other coordinates. We put the values c_{ii} with positive signs and the values $c_{i\ominus 1 i}$ with negative signs into C. For convenience we'll use also the notation $a_i = c_{ii}, b_i = -c_{ii\oplus 1}$

A *marking* is a function $M : S \to I\!N$. It defines the distribution of *tokens* on places. Each transition t_i having enough tokens on its input place $p_{i\ominus 1}$ may *fire* taking $b_{i\ominus 1}$ tokens from the place $p_{i\ominus 1}$ and putting a_i tokens on place p_i. The initial marking M is live in $\mathcal{C}$ iff one can fire an infinite sequence of transitions starting from M[1]. The *liveness problem* is the problem of deciding liveness of a given marking.

Consistency of $\mathcal{C}$ means the existence of a positive integer T-vector X such that $CX = \vec{0}$. Conservativeness of $\mathcal{C}$ (equivalent to consistency) means the existence of a positive S-vector Y such that $Y^T C = \vec{0}^T$. Both these conditions hold iff $a_1 \cdots a_n = b_1 \cdots b_n$ [TCCS 92]. Such vectors $X = [x_1, \ldots, x_n]^T$ and $Y = [y_1, \ldots, y_n]^T$ are up to a constant factor uniquely determined by matrix C. We may choose the least positive integer vectors X and Y and call them *T-invariant* and *S-invariant* of $\mathcal{C}$ respectively. The S-invariant Y is called also the *weight vector*. The *weight* of a marking M is the value of the scalar product $Y^T M$. It is a known fact that when transitions are fired, the weight of reachable markings remains unchanged.

We call M' reachable from M iff there exists a sequence of firings transforming M into M'. We call M' *potentially reachable* from M iff the equation $C\vec{z} = M' - M$ has an integer solution. If M is live then the set of reachable markings from M is equal to the set of potentially reachable markings ([TCCS 92]).

[1] Note that this simplified definition of liveness works only for circuits.

For the rest of the paper we assume $\mathcal{C}$ to be a conservative circuit with the incidence matrix $C = \{c_{ij}\}$, the weight (S-invariant) vector $Y = [y_1, \ldots, y_n]^T$ and T-invariant vector $X = [x_1, \ldots, x_n]^T$. Denote by $\mathcal{C}^r$ the reverse circuit of $\mathcal{C}$ (resulting from $\mathcal{C}$ by reversing the arrows). The partial order of markings will be meant componentwise.

2 Weights of live markings

The problem of determining the structural, weight-based characterization of liveness is addressed in this section. We call a given number w a *live weight* iff all markings with weight w are live and there exists at least one such marking. A number d is called a *dead weight* iff no live marking in $\mathcal{C}$ has a weight d. There are weights such that there are both live and dead markings with this weight.
For example consider the net from Fig. 1.

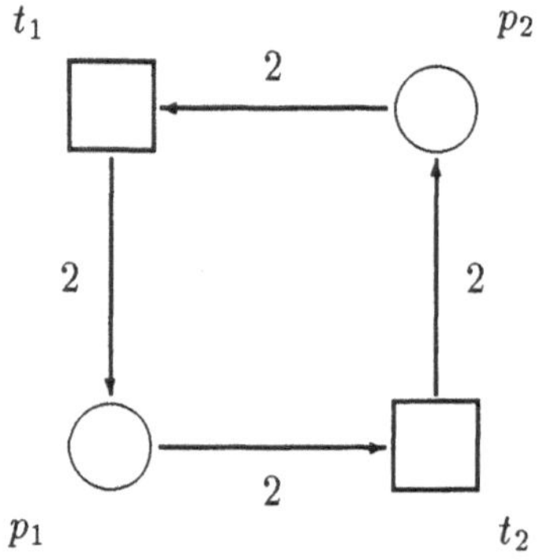

Figure 1: A net for which there exist a dead marking and a live marking with weight 2.

Clearly we have $Y = [1,1]^T$. The marking $M_1 = [2,0]^T$ is live, while $M_2 = [1,1]^T$ is dead; both are of weight 2. The reason for the coexistence of a live and a dead marking with the same weight is the presence of potentially frozen tokens in the presented example (the number of tokens modulo 2 is invariant in both places). For a marking M_0, the place p contains k frozen tokens iff for all markings M reachable from M_0, $M(p) \geq k$, and it contains no frozen tokens, iff $M(p) = 0$ for some reachable marking M.
From the behavioural point of view the odd tokens on both places are frozen, hence superfluous. The first reaction could be to forbid markings with frozen tokens like in M_2 and concentrate on markings, which at every place modulo the greatest common divisor of incoming and outgoing arcs give the value 0. Having two frozen tokens, the marking $M_3 = [3,1]^T$ creates an isomorphic behaviour to that of M_1. After subtracting from $M_3(p_i)$ the remainders $M_3(p_i)$ mod 2, we get M_1. Such a procedure can be generalized, and no frozen tokens are created when this procedure is applied to all places, in case of a live marking.

Proposition 2.1 *Let $d_i = \gcd(a_i, b_i)$. If the marking M is live in C, and for all i, $M(p_i) \bmod d_i = 0$, then there are no frozen tokens in the circuits.*

Proof We show that M is live and $M(p_i) \bmod d_i = 0$, then there exists a follower marking M' such that $M'(p_i) = 0$.
The following sequence M_k, $k = 0, 1, 2, \ldots$ of markings is constructed. $M_0 = M$, and M_{k+1} can be reached from M_k in the following way. We fire at M_k the transition $t_{i\oplus 1}$ as many times as we can. Next we fire $t_{i\oplus 2}$ as many time as we can, and so on, until we fire the transition $t_{i\ominus 1}$ leaving on the place $p_{i\ominus 1}$ the maximum number of tokens that can be put without bothering t_i at all. Next we fire t_i, but only as many times as it is necessary to fire $t_{i\oplus 1}$, so that we stop firing t_i as soon as we recognize that $t_{i\oplus 1}$ is enabled (maybe multiple number of times). Next we again fire $t_{i\oplus 1}$ as many times as we can, resulting in M_{k+1}.
Let $r_k = M_k(p_i)$. Consider two cases

1. $a_i \geq b_i$, so the transition t_i is being fired only once before $t_{i\oplus 1}$ is fired in the last step of the algorithm. In this case the value of r_{k+1} differs from r_k by exactly $a_i \bmod b_i$, because t_i was fired in this case only once. For $k > 0$ the following holds:

$$\begin{aligned} r_k &= (M(p_i) + ka_i) \bmod b_i \\ &= ((M(p_i) \bmod b_i) + ((ka_i) \bmod b_i)) \bmod b_i \end{aligned}$$

   The sequence $(ka_i) \bmod b_i$ consists of all values $0, d_i, 2d_i, \ldots, b_i - d_i$. Since d_i divides $M(p_i)$, the value $-M(p_i) \bmod b_i$ is in the sequence, so there exists such k that $r_k = 0$.

2. $a_i < b_i$, apply the previous proof for the reverse net, where arrows have reverse directions. For live circuits the sets of reachable markings of C and C^r are the same, so this case reduces to the previous one.

□

Observe, that M_2 in the discussed example from Fig.1 is not potentially reachable from M_1 — there are no *integer* solutions of the equation $C\vec{z} = M_2 - M_1$. However, there exist rational solutions for this equation, for instance $[\frac{1}{2}, 0]^T$. So firing "half" of t_1 would transfer M_1 into M_2.
Consider another marking $M_4 = [2, 2]^T$ having the same weight as M_3. In fact M_3 and M_4 (like M_1 and M_2) operate on different *orbits* of (potential) reachability space for a given weight.
Let's make that more precise

Definition 2.1 *Consider the set $\mathbb{Z}_w$ of all integer vectors of space $\mathbb{Z}^n$ such that $Y^T\vec{z} = w$,* for *$\vec{z} \in \mathbb{Z}^n, w \in \mathbb{Z}$. The equivalence relation ρ_w on $\mathbb{Z}_w$ is defined as following: $(\vec{z}_1, \vec{z}_2) \in \rho_w$ iff the equation $C\vec{x} = \vec{z}_1 - \vec{z}_2$ has a solution in integers.*

Definition 2.2 *Any equivalence class of the relation ρ_w is called an* orbit *of weight w.*

Observe that the number of orbits does not depend on the weight w (see [ChR 93] for the proof):

Proposition 2.2 *For each w_1, w_2, the number of orbits of weight w_1 is equal to the number of orbits of weight w_2.*

So the number of orbits is determined by the matrix C itself; we shall denote this number by $Orb(C)$. We call an orbit *nonempty* iff there exists at least one marking belonging to this orbit.
Two markings with the same weight are members of the same orbit iff each of them is potentially reachable from the other one. For instance (Fig.1), $[3,1]^T$ and $[1,3]^T$ do belong to the same orbit of weight 4, while $[3,1]^T$ and $[2,2]^T$ do not. In general, moving frozen tokens from place to place causes changing orbits.
As it will be shown, even in cases where there are no frozen tokens (i.e each place can be emptied), there may coexist live and dead markings with the same weight, clearly belonging to different orbits.

Definition 2.3 *Denote by M_D the marking $[b_1 - 1, \ldots, b_n - 1]^T$.*

Proposition 2.3 ([TCCS 92]) *For every dead marking M_d, $M_d \leq M_D$ componentwise.*

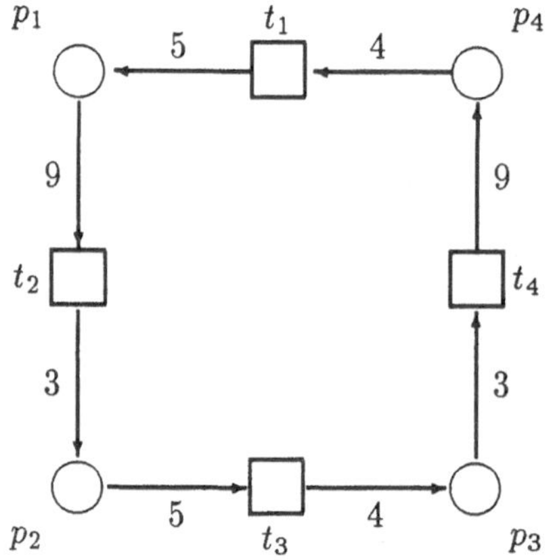

Figure 2: A net without frozen tokens with 3 orbits.

Consider the net from Fig.2, where $Y = [4, 12, 15, 5]^T$ is the weight vector. The biggest dead marking is $M_D = [8, 4, 2, 3]^T$ and $W(M_D) = Y^T M_D = 125$. For this net a live marking exists: $M_L = [8, 4, 3, 0]^T$ with weight equal to 125. While transitions are fired, one can discover no frozen tokens: each place can be emptied. Since the set of potentially reachable markings is equal to the set of reachable markings for live markings in weighted circuits, we may announce the discovery of two different orbits for the considered net (otherwise it would be possible to reach a dead marking from a live one — a contradiction). We may add that there exists another live orbit generated for instance by the marking $M_L' = [8, 4, 1, 6]^T$. And that's all. There are exactly 3 orbits for this net.
Two questions concerning the weight-based characterization of liveness arise:

1. What is the minimal live weight W_L (i.e. such weight that the liveness is guaranteed when a marking weighs W_L)?

2. what is the weight of the least live marking?

The following fact holds:

Proposition 2.4 *The weight w is live iff the equation*

$$Y^T\vec{z} = W(M_D) - w$$

has no nonnegative integer solutions.

Proof This result can be proven following the method used in the proof of theorem 2.5 from [ChR1 93], dealing with the not necessarily least live weight. □

So looking for the least live weight is equivalent to finding the biggest integer g giving no solutions in natural numbers to the equation $Y^T\vec{z} = g$.

Theorem 2.5 *If g is the greatest number having no solution in natural numbers of the equation $Y^T\vec{z} = g$ then $W_L = W(M_D) - g$ is the value of the least live weight.*

Note, that however it is rather clear, and was shown in [TCCS 92], that every marking weighing $W(M_D) - g$ is live, the difficult part of the proof is to show that there exist markings with this weight. The proof can be found in [ChR1 93].
The problem to find such biggest g is known as the Diophantine problem of Frobenius.

Definition 2.4 *Given n relatively prime natural numbers $a_1, \ldots, a_n$. The problem of determining the formula for $g = g(a_1, \ldots, a_n)$ being the greatest number not representable as a linear combination $a_1x_1 + \cdots + a_nx_n$ for some nonnegative integer $x_1, \ldots, x_n$ is called the* Diophantine problem of Frobenius *and the number g — the* Frobenius number *of $a_1, \ldots, a_n$.*

The problem of finding such $g = g(y_1, \ldots, y_n)$ for relatively prime $y_1, \ldots, y_n$ had been posed by Frobenius in his lectures over a century ago. Dozens of papers have been published since then, but no closed formula for g is known up to now. A partial solution to the mentioned problem of Frobenius was presented in [RCh 92]. Though we don't know a general formula for g, several algorithms exist allowing us to calculate the value of g. In many cases the value of g can be easily determined. The paper [Selmer 77] may serve as survey paper about the Frobenius problem.
In fact, as a consequence of Theorem2.5, we can show that the Diophantine problem of Frobenius and the problem of finding the formula for the least live weight in conservative circuits are equivalent.
Let's come back to our example from Fig 2. We can easily verify that $g(Y) = g(4, 5, 12, 15) = g(4, 5) = 11$. Using theorem 2.5 we deduce that the weight $W(M_D) - g = 125 - 11 = 114$ is the least live weight. Is it also the least not dead? No; consider the marking $M_0 = [1, 5, 2, 1]^T$. This is a live marking, and $W(M_0) = 99$.
The reason for this is that there are three orbits in the net. And the condition about nonexistence of solutions of the equation (2.4) should be replaced by the condition of nonexistence of three different solutions ("responsible" for allocating the three orbits). So we come to the following proposition:

Proposition 2.6 *If for some natural b there are m nonempty orbits with weight $W(M_D) - b$ and the equation $Y^T \vec{z} = b$ has at most $m-1$ solutions, then at least one of the orbits is live.*

One can check, that the equation $Y^T \vec{z} = 26$ has only 2 solutions: $26 = 4*4+2*5$ or $26 = 1*4+2*5+1*12$ (while all numbers greater than 26 have more than 3 solutions). So, among all markings of weight $125 - 26 = 99$, only two can be obtained as a result of subtracting 26 weighted tokens from M_D. These two markings belong to two of the three orbits and there is still one orbit left: a live orbit. All three orbits are nonempty: $[1,5,2,1]^T, [4,4,2,1]^T, [7,3,2,1]^T$ are representatives: the first of a live orbit, the other two of dead ones.
The reason why we get several orbits when there are no frozen tokens is that there are still problems with token mobility in the net. When we look at the net we see that the total weighted number of tokens modulo some integer remains invariant in some parts of net. The weighted remainders of tokens are "half-frozen": they can move inside the group of places but are unable to change the group. In particular, if the group of tokens consists of just one place, then half-frozen tokens are just frozen. One can check for instance that the total weighted number of tokens of $\{p_1, p_4\} : y_1 M(p_1) + y_4 M(p_4)$ modulo 9 is constant and different for different orbits (the respective remainders are $0, 3, 6$). So moving the 3 weight units from $\{p_1, p_4\}$ to $\{p_2, p_3\}$ changes the orbit. Note, that moving these 3 weighted tokens cannot be done by transition firing.
This time getting rid of half-frozen tokens is impossible. The lightest live marking, $[1,5,2,1]^T$ occupies the orbit which modulo 9 gives 0 for the sum of weighted tokens on $\{p_1, p_4\}$. Looks nice: this marking does not capture any half-frozen tokens on these two places. However, when other pairs of consecutive places are considered, it becomes clear, that still half-frozen tokens exist there. In general, whenever we encounter a sequence

$$t_i \xrightarrow{a_i} p_i \xrightarrow{b_i} \dots \xrightarrow{a_k} p_k \xrightarrow{b_k} t_{k+1},$$

with $\gcd(a_i y_i, b_k y_k) = d_{i,k}$, then the total number of weighted tokens on places $p_i, \cdots, p_k$ is equal to $\sum_{j=i}^{k} M(p_j) y_j$ and remains constant modulo $d_{i,k}$ for every reachable marking M.
In our example, the corresponding non-one coefficients $d_{i,k}$ are the following: $d_{1,2} = 20, d_{2,3} = 9, d_{3,4} = 20, d_{4,1} = 9$. Hence, for every initial marking M_0 and a follower marking M we have:

$$M_0(p_1)y_1 + M_0(p_2)y_2 \equiv M(p_1)y_1 + M(p_2)y_2 \pmod{20}$$
$$M_0(p_2)y_2 + M_0(p_3)y_3 \equiv M(p_2)y_2 + M(p_3)y_3 \pmod{9}$$
$$M_0(p_3)y_3 + M_0(p_4)y_4 \equiv M(p_3)y_3 + M(p_4)y_4 \pmod{20}$$
$$M_0(p_4)y_4 + M_0(p_1)y_1 \equiv M(p_4)y_4 + M(p_1)y_1 \pmod{9}$$

It is easy to check, that for the initial lightest live marking $[1,5,2,1]^T$, (and for all of its follower markings) not all these congruences give 0 as a result. In fact

$$M_0(p_1)y_1 + M_0(p_2)y_2 \pmod{20} = 4$$
$$M_0(p_2)y_2 + M_0(p_3)y_3 \pmod{9} = 0$$
$$M_0(p_3)y_3 + M_0(p_4)y_4 \pmod{20} = 15$$
$$M_0(p_4)y_4 + M_0(p_1)y_1 \pmod{9} = 0$$

So there are 4 half-frozen weighted tokens on places $\{p_1, p_2\}$ and 15 half frozen tokens on places $\{p_3, p_4\}$ for the marking M_0. Note, that for the other orbits

with the weight 99, which are not live, the total number of frozen tokens is larger. For the orbit of $[7,3,2,1]^T$ the corresponding remainders are $4,3,15,6$, and for the other orbit, of $[4,4,2,1]^T$, they are $4,6,15,3$. If you compare them to the remainders of the live orbit: 4,0,15,0, it looks like if the live orbit trapped less tokens than the dead ones. Is it a general rule, that when you trap less tokens, you have better chances to be live?
Apparently not: consider the 3 orbits of the weight 125. There is one dead, generated by $M_D = [8,4,2,3]^T$ with remainders $0,6,5,2$ and two live orbits generated by the markings $[5,5,2,3]^T$ with remainders $0,0,5,8$ and $[2,6,2,3]^T$ with remainders $0,3,5,5$ consecutively.
Observe, that we cannot prove the condition from proposition 2.6 to be also necessary. Unlike the case of the minimal live weight, we have no guarantee that the solutions of the equation $Y^T\vec{z} = b$ subtracted from M_D will occupy the orbits uniformly. The example from Fig. 3 is an illustration of this phenomenon.

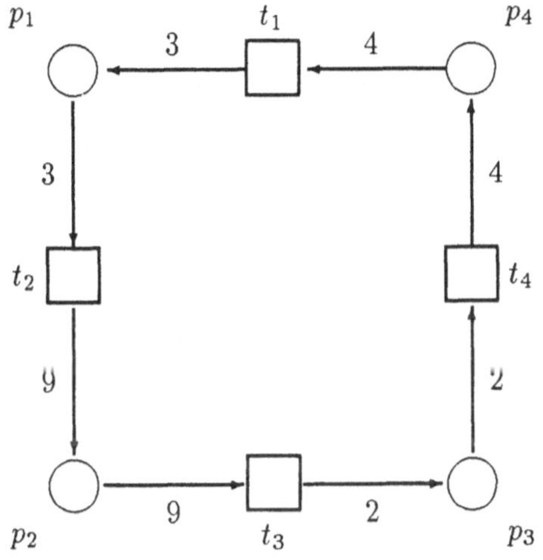

Figure 3: A net exposing that the condition from the proposition 2.5 is not necessary for a weight to be non-dead one. There are 6 orbits and at the same time there are 11 solutions to the equation $Y^T\vec{z} = W(M_D) - 36$, but still one orbit with weight 36 remains live.

For the circuit of Fig. 3 we have $Y = [12,4,18,9]^T$, the 3 tokens on place 1 make the circuit live with the weight 36, which is the minimal non-dead weight. However, the biggest dead marking for this net, M_D equals to $[2,8,1,3]^T$ and it's weight $W(M_D)$ is 101. There are 6 orbits of this net, but the number of the nonnegative solutions of the equation $Y\vec{z} = 101 - 36$ is 11. This means that the 5 orbits with the weight 36 absorb these 11 mortal solutions, while the remaining 6'th orbit survives.
Observe, that it follows also from the corollary 2.4 that all weights W bigger than $W(M_D)$ are live: there is trivially no natural solution for the equation $Y^T\vec{z} = W(M_D) - W$, since the right hand side of the equation is negative.
Some lower bounds for non-dead weights can be presented. One of them is

pretty obvious: we must be able to fire each transition at least once, hence the weight of a live marking must exceed $\max(b_i y_i)$ enabling to collect on each place enough tokens to activate the following transition. Observe, that this is a tight bound, as can be seen from the example of Fig. 3.
We can see now, that all the weights below $\max(b_i y_i)$ are dead, all the weights above $W(M_D)$ are live. The only unsolved case is the segment between these two values.
In the case there is one orbit in the circuit, everything is clear.

Proposition 2.7 *If a circuit has one orbit, then the marking M is live iff the equation*

$$W(M_D) - Y^T z = W(M) \tag{1}$$

has no solution in natural numbers.

Proof To prove this result we will use lemma 2.3 from [ChR1 93], which states, that a weight w is live iff there is no marking with weight $W(M_D) - w$.
If M is live, then, since there is only one orbit, the weight $W(M)$ is live. Hence, the equation (1) has no solution in natural numbers, as it follows from the cited lemma.
And, similarly, if there is no solution to the equation (1), none of the dead markings (each of them componentwise less than M_D) can be reached from M_0: if there were a dead marking of the same weight, it would belong to the same, unique orbit. □

When the number of orbits is greater than one, problems arise. As we have seen, some inevitably half-frozen tokens will be present in the circuit, even when we deal with live markings.
Determining the number of orbits is quite easy:

Theorem 2.8 *Let $C = \{c_{ij}\}_{1 \le i,j \le n}$ be the incidence matrix of a consistent weighted circuit. The number of orbits $Orb(C)$ equals*

$$Orb(C) = \frac{c_{11} \cdots c_{nn}}{lcm(y_1 c_{11}, \ldots, y_n c_{nn})}, \tag{2}$$

The proof can be found in [ChR 93].

3 Arithmetical characterization of liveness

We come back to the question: how to determine whether a given marking is live. We may assume that the weight of the considered marking M lies within the segment of uncertainty $[\max(b_i y_i)..W(M_D)]$, where live and dead markings may coexist and that the number of orbits is greater than one. The main result presented in this section is general, but more difficult to apply when compared to the simplicity of the weight-based characterizations.
This result is based on the following observation. Every transition t_i must be guaranteed at least x_i times to fire in order to assure liveness of M (recall that x_i is the ith entry of the T-invariant X). And once it happens for any of the transitions, it means the initial marking was live. But usually, before

it happens, some transitions (including t_i itself!) have to be fired in order to bring a sufficient amount of tokens on the input place of t_i. Without loss of generality we may concentrate on t_1.

Definition 3.1 *Let $G : \mathbb{N} \times 2^P \rightarrow \mathbb{Z}$ be defined in the following way:*

$$G(x, M) = \lfloor \frac{a_n}{b_n} \lfloor \cdots \lfloor \frac{a_2}{b_2} \lfloor \frac{a_1 x}{b_1} + \frac{M(p_1)}{b_1} \rfloor + \frac{M(p_2)}{b_2} \rfloor + \cdots \rfloor + \frac{M(p_n)}{b_n} \rfloor - x$$

The meaning of the function is the following: if starting from M one can fire the transition t_1 x times, then if we continue firing all transitions but t_1, we'll end up in a marking in which it is possible to fire t_1 exactly $G(x, M)$ times. In other words: after t_1 fires x times starting at M, it is possible to place at least $G(x, M)b_n$, but less than $(G(x, M) + 1)b_n$ tokens on p_n without firing t_1 any more.
The function G reflects how much t_1 can help itself by transferring tokens through itself x times. We concentrated on t_1, but similar functions can be constructed for other transitions as well.
Before we prove the main result of this section, lets make some observations about properties of $G(x, M)$.

Lemma 3.1 $G(0, M) \geq 0$

Proof All numbers used in the formula defining $G(0, M)$ are nonnegative. □

The transition t_2 can be fired at M exactly $\lfloor \frac{M(p_1)}{b_1} \rfloor$ times. After this the transition t_3 can be fired at M exactly $\lfloor a_2 \lfloor \frac{M(p_1)}{b_1} \rfloor + \frac{M(p_2)}{b_2} \rfloor$ times, and so on. After firing consecutively a maximal number of $t_4, \cdots, t_n$ we are able to fire t_1 exactly $G(0, M)$ times. Thus the following lemma holds:

Lemma 3.2 *The transition t_1 cannot be activated by any follower marking of M iff $G(0, M) = 0$.*

Corollary 3.3 *If M is dead then $G(0, M) = 0$.*

Corollary 3.4 *If $G(0, M) = 0$ then M is not live.*

Lemma 3.5 *If $M[t_1^k\rangle M'$, then $G(x, M') = G(x + k, M)$.*

Proof The following equality holds: $M' = M + [ka_1, 0, \cdots, 0, -kb_n]^T$. Substituting the right-hand-side of this equality to the formula for $G(x, M')$ we obtain:

$$\begin{aligned} G(x, M') &= \lfloor \frac{a_n}{b_n} \lfloor \cdots \lfloor \frac{a_1 x}{b_1} + \frac{M(p_1) + ka_1}{b_1} \rfloor + \cdots \rfloor + \frac{M(p_n) - kb_n}{b_n} \rfloor - x \\ &= \lfloor \frac{a_n}{b_n} \lfloor \cdots \lfloor \frac{a_1(x + k)}{b_1} + \frac{M(p_1)}{b_1} \rfloor + \cdots \rfloor + \frac{M(p_n)}{b_n} - k \rfloor - x \\ &= G(x + k, M) \end{aligned}$$

□

Firing transitions other than t_1 does not change the value of G:

Lemma 3.6 *If $M[t_i^k\rangle M'$ and $i > 1$, then $G(x, M) = G(x, M')$.*

Proof Let $M[t_i^k\rangle M'$ for some $i > 1$.
Then $M' = M + [0, \dots, 0, -kb_{i-1}, ka_i, 0, \dots, 0]$. Substituting this value in the formula for $G(x, M')$ we obtain the result in a way analogous to the previous lemma. □

Lemma 3.7 *If $M[w\rangle M'$ and $\#(t_1, w) = k$, then $G(x, M') = G(x + k, M)$, where $\#(t_1, w)$ denotes the number of occurrences of t_1 in $w \in T^*$.*

Proof Direct consequence of 3.5 and 3.6 □

We are ready now to present an arithmetical condition for liveness of a marking.

Theorem 3.8 *The marking M is live iff for all $x \in \mathbb{N} : G(x, M) > 0$.*

Proof ($\Rightarrow$) Let M be live and let $k \geq 0$ be the smallest one for which $G(k, M) = 0$.
Since M is live, there exists M' reachable from M by firing w, such that t_1 was fired exactly k times in w. But then, according to lemma 3.7, $G(0, M') = G(k, M) = 0$, so due to Corollary 3.4, M' is not live. A contradiction.
($\Leftarrow$) Let $\forall k \in \mathbb{N} : G(k, M) > 0$ and M is not live. In this case a sequence $w \in T^*$ exists that transforms M into a dead marking M'. But then $0 = G(0, M')$ (by Corollary 3.4), and at the same time $G(0, M') = G(k, M)$, where $k = \#(t_1, w)$. A contradiction. □

From the above theorem and from lemma 3.2 we obtain immediately:

Corollary 3.9 *The marking M is not live iff $\exists k \in \mathbb{N} : G(k, M) = 0$*

So, finally, we have come out with the arithmetical characterization of liveness in weighted circuits. A natural question arises, how difficult it is to check this arithmetical condition. At first glance it is quite hard, since the condition must hold for all natural numbers. Fortunately we can easily restrict the domain of interest by showing that G is periodical.

Theorem 3.10 *Let $B_i = \frac{b_1 \cdots b_i}{\gcd(b_1 \cdots b_i, a_1 \cdots a_i)}$, and $T = lcm(B_1, \dots, B_n)$. Then $G(x + T, M) = G(x, M)$*

Proof When we concentrate on the ith coordinate in the definition of $G(x + T, M)$, we will be able to rewrite it in the following way:

$$G(x+T, M) = \lfloor \frac{a_n}{b_n} \lfloor \cdots \lfloor \frac{a_i}{b_i} \Big(\frac{T a_1 \cdots a_{i-1}}{b_1 \cdots b_{i-1}} + \lfloor \cdots \lfloor \frac{a_1 x}{b_1} + \frac{M(p_1)}{b_1} \rfloor + \cdots + \rfloor \Big)$$
$$+ \frac{M(p_i)}{b_i} \rfloor + \cdots \rfloor + \frac{M(p_n)}{b_n} \rfloor - x$$

In order to get rid of the coefficient

$$\frac{a_i}{b_i} \frac{T a_1 \cdots a_{i-1}}{b_1 \cdots b_{i-1}} = \frac{T a_1 \cdots a_i}{b_1 \cdots b_i}$$

it must be integer. So T must be divisible by B_i. Eventually we obtain

$$G(x+T,M) = G(x,M) + T\frac{a_1\cdots a_n}{b_1\cdots b_n} = G(x,M) + T$$

So T, as it is defined, is a period of G. From the construction of T we conclude that it must be divisible by all B_i, and that no smaller value can be a period. Hence the least period is the least common multiple of all B_is. □

For instance, for the net of Fig. 2 the consecutive values are:
$b_1 = 9, a_1 = 5, B_1 = 9$
$b_2 = 5, a_2 = 3, B_2 = 3$
$b_3 = 3, a_3 = 4, B_3 = 9$
$b_4 = 4, a_4 = 9, B_4 = 1$

The period of the function G is equal to 9.

4 Conclusions

The problem of determining the liveness of markings in weighted circuits was addressed in the paper. The difficulty of this problem depends heavily on the number of orbits existing in the circuit. In case there is one orbit, determining liveness can be made based on the marking weight only, and reduces to the existence of a solution of a Diophantine linear equation. In case there are more orbits, some problems arise mainly due to the presence of half-frozen tokens. Such tokens can move more or less freely within a subset of places, but cannot leave it. In case of one orbit, the presence of half-frozen tokens may be created by frozen tokens only.
A discussion of cases when the marking weight is not sufficient for determining liveness followed in the paper. A general arithmetical condition equivalent to liveness of a marking was presented. A condition itself, however, is expressed in terms of a function, where taking floors of divisions is necessary, and there is no closed form for it. The expression length defining the function in this condition depends on the circuits length. On the other hand, the function is periodic, and it is enough to check the condition at most within the period.
The condition can simplify the search for the formula for the least weight of a live marking (still an open problem). No firing transitions is necessary and the answers are usually easy to calculate. Although no evidence can be presented here, how quickly the condition can be checked, the authors have used this condition in practice, and in many cases found it easier to use, than any other method.

Acknowledgements

The authors want to express their gratefulness to three anonymous referees for careful reading of the paper and multiple comments that helped much in preparing this version. In particular, the proof of Theorem 3.8 has been significantly simplified, thanks to the suggestions from two referees.

References

[Brauer 42] Brauer A. On a problem of partitions, Amer.J.Math. 1942; 64: 299-312

[ChR 93] Chrząstowski-Wachtel P, Raczunas M. Liveness of Weighted Circuits and the Diophantine problem of Frobenius. Proceedings from the Workshop on Concurrency, Specification and Programming, November 19-21, 1992, Berlin; Fachbereich Informatik der Humboldt-Univeritaet zu Berlin 1993, Informatik-Preprint 22

[ChR1 93] Chrząstowski-Wachtel P, Raczunas M. Liveness of Weighted Circuits and the Diophantine problem of Frobenius. Proceedings from the Fundamentals of Computation Theory conference, Szeged 1993, LNCS 710

[RCh 92] Raczunas M, Chrząstowski-Wachtel P. A Diophantine problem of Frobenius in terms of the least common multiple. To appear in Discrete Mathematics

[Selmer 77] Selmer ES. On the linear Diophantine problem of Frobenius J.reine angew. Math. 1977; 293/294: 1-17

[TCCS 92] Teruel E, Chrząstowski-Wachtel P, Colom JM, Silva M. On weighted T-systems. Advances in Petri Nets, Lecture Notes in Computer Science, 1992; vol.616: 348-367

An Observational Semantics for Linda

Rocco De Nicola
Dipartimento di Scienze dell'Informazione, Università "La Sapienza"
Rome, Italy

Rosario Pugliese
Dipartimento di Scienze dell'Informazione, Università "La Sapienza"
Rome, Italy

Abstract

Linda has just four primitives, all of which are devoted to coordinating the interactions among programs by sharing information maintained in a common data space. To write programs manipulating data, it is necessary to embed Linda in a (functional, imperative, logic, etc.) programming language; this leads to a family of languages based on Linda. We define syntax and semantics for a member of the Linda family, L, that is obtained by embedding Linda in a "simple" sequential language. The semantics of this concurrent programming language is formalized by applying techniques developed within the framework of process algebras. A two-level semantics for L is provided: an operational one in Plotkin's style, based on a *transition system*, and an observational one, based on three *behavioural preorders*, obtained by applying the *testing framework* to L.

1 Introduction

Recently there has been a fresh interest towards global environment models for concurrent languages. Languages such as Linda [1, 2], Concurrent Prolog [3], UNITY [4], Shared Prolog [5], etc., are based on asynchronous, associative communication mechanisms and rely on shared data structures. These differ from the first generation of global environment parallel languages because they offer an explicit control of interactions.

In this paper we want to analyze this communication mechanism by concentrating on Linda, one of the main representatives of the class of new global environment languages listed above. Linda has just four primitives, all designed for coordinating interactions among programs by means of shared informations that are kept in a common data space, known as "Tuple Space". These four primitives can be embedded in a (functional, imperative, logic, etc.) programming language to obtain a concurrent language. This feature renders Linda very general and leads to a family of concurrent languages based on Linda rather than to a single language.

In this paper we provide syntax and operational semantics for a member of the Linda family. Indeed, although pragmatics of Linda have been thoroughly analyzed, little attention has been devoted to formalizing its syntax and semantics. Various papers on Linda propose different syntaxes [6, 1, 7, 2, 8] and little attention has been devoted to equipping it with a formal semantics. Exceptions in this respect are [8, 9, 10, 11], but they deal with the general communication mechanism based on Tuple Space, rather than with a programming language belonging to the Linda family.

We will study the impact of embedding Linda in a simple imperative language (SL) and will study the semantics of the resulting concurrent language that will be called L. We will use semantics prescriptively; some (non-essential) Linda constructs that would lead to involved semantics, or would require introducing additional control structures, will not be considered. L will be equipped with a two–level semantics:

- the first level consists of a *structural operational semantics* (SOS) [12], and defines the *intentional* meaning of each syntactic construct of L;
- the second, more abstract, one is based on *behavioural preorders*, defined by following an approach developed within the framework of Process Algebras [13, 14, 15, 16].

L may be used to describe a system (write a program) with different details depending on the programs that are already available; it is possible to go as far as describing both the desired behaviour of a system (its *specification*), and an actual realization (its *implementation*) as L programs. In general, different programs may represent the same system or may be alternative approaches to tackling a given problem and it could be important to establish their equivalence. This goal can be accomplished by resorting to equivalence notions that are based on the ability of external observers of taking apart programs: two programs will be considered as equivalent if no observer can distinguish them. Thus the observational semantics of L makes it possible to verify equivalence or to establish approximation orderings between distinct descriptions of systems at different levels of abstraction and permits *program verification.*

The rest of the paper is organized as follows. In the next section we informally describe the main features of Linda and provide an abstract syntax for L. In section 3 the operational semantics of L is defined. It is an *interleaving* semantics in that it permits only a single transition at a time. In section 4, three alternative, more abstract semantics for L are defined; they are based on the equivalences generated by three preorders over L programs. All the relations are preserved by the composition operators of the language. In section 5, we briefly discuss related works and suggest future lines of investigation.

2 L: Embedding Linda in an imperative language

2.1 An informal description of Linda

Linda is a coordination language that relies on an asynchronous and associative communication mechanism based on a shared global environment called Tuples Space (TS), a multiset of tuples. A *tuple* is a sequence of *actual fields* (value objects) and *formal fields* (variables) with the constraint that the first field be an actual field, called *logic* name (or *tag*). The basic interaction mechanism is *pattern–matching.* It is used to select a tuple in TS that matches a given tuple t. This operation permits selecting those tuples in TS with the same tag as t and with the same number of fields and such that corresponding fields have matching values or variables. Variables match any value of the same type and two values match only if identical. There are four operations defined over TS:

- **in**(t): the process executing this operation first evaluates t and then looks for a tuple t' matching t in TS . If and when t' is found, it is removed from TS; the corresponding values of t' are assigned to the variables of t and the process continues. If no matching tuple is found, the process is suspended until one is available.

- **read**(t) is similar to **in**(t), but it does not require removal of the matched tuple t' from TS.

- **out**(t): the process executing this operation evaluates t, adds it to TS and proceeds. This is a non–blocking operation.

- **eval**(t) is similar to **out**(t), but rather than forcing evaluation of t, the process executing **eval**(t) creates a new process that will evaluate t and then add the resulting tuple to TS. The tuple will not be available for matching until its evaluation is completed.

It is worth noting that nondeterminism is inherent in the definition of Linda primitives. It arises when more **in/read** operations are suspended waiting for a tuple; when such a tuple becomes available, only one of the suspended operations is nondeterministically selected to proceed. Similarly, when an **in/read** operation has more than one matching tuple one is arbitrarily chosen.

2.2 Abstract Syntax of L

We embed Linda in a sequential imperative language, obtaining the language L whose abstract syntax is defined by the following productions:

$$p ::= \textbf{nil} \mid id := e \mid \textbf{while}\ e\ \textbf{do}\ p \mid p_1; p_2 \mid \textbf{in}(t) \mid \textbf{read}(t) \mid \textbf{out}(t) \mid \textbf{eval}(p)$$

$$t ::= e \mid\ !id \mid t_1, t_2$$

$$e ::= c \mid id \mid e_1\ bop\ e_2$$

$$c ::= v \mid n$$

The syntactical categories Prg, Tpl, Exp and Con (ranged over by p, t, e and c, respectively) defined above describe programs, tuples, expressions and constants of L. The primitive domains of L are: Id, Opbin, Num and Name. As usual, a "primitive" domain is such that the syntactic and the semantic denotation of any element coincide. In L, all constants are natural numbers (Num, ranged over by v) or names (Name, ranged over by n); moreover, we do not completely specify the set of binary operations (OpBin, ranged over by *bop*) over natural numbers; however, we assume that it contains at least the arithmetic operators ($+, -, *, /$ and *mod*) and the relational operators ($=, <, >, \leq, \geq, \neq$). To simplify operational semantics, no value type is associated to variables (Id, ranged over by *id*), so that they can match any constant value, and expressions cannot give rise to ***side effects***. We use the notation $!id$ for formal fields of tuples. Programs cannot be named so it is not possible to write recursive programs or to simultaneously activate two copies of the same program. Finally, we assume that concurrent programs have disjoint variable spaces. This restriction cannot be expressed within our syntax and we assume that it has been (statically) checked separately. This enables us to use a common "global" store for all parallel programs.

For reasons that will be clear later, we let

$$Test^1 \stackrel{def}{=} \{(n, v) \in \text{Tpl} \mid n \in \{\mathbf{END}, \mathbf{START}\}, v \in \text{Num}\}$$

be a set of reserved tuples that programs cannot use and Wv ($\subset$ Id) be a special (infinite) set of identifiers (*working variables*), all beginning with wv.

There are a few syntactical choices which are dictated by the need of providing the language with a simple formal semantics.

The original definition of Linda permits programs to compute values as fields of tuples. In L we have only two syntactical categories: tuples (containing expressions) and programs. Moreover, we only have **eval**(p) where p is a program. This is in fact equally powerful as the original Linda's **eval**(t) where t encompasses both tuples and programs. In our formalism **eval**(**out**(t)) is equivalent to the original **eval**(t). Indeed the two programs would have the same overall effect but the former has the advantage of being explicit.

Also, we did not consider the two predicates **inp** and **readp** that are put forward in some papers on Linda. These predicates have the same functionalities as **in** and **read**, and permit expressions to have "side effects". In fact, if a matching is possible, the evaluation of an **inp**/**readp** returns the boolean value "true" and produces the same effects of an **in**/**read** instruction, both on the store and on TS. Had we introduced the two predicates, the operational semantic of Exp (and that of the other syntactic categories that make use of Exp) would have been more complex and would have required rules with "negative premises". Moreover, **inp** and **readp** represent a sort of statements about the global state of a program and this is not in agreement with the distributed operational semantic we give to L programs.

3 Operational Semantics

3.1 Expressions and Tuples

For defining the meaning of the two syntactical categories Exp and Tpl we define a *denotational semantics*; we are interested in the final result only and not in the evaluation process.

[1]We need *Test* be an infinite set for technical reasons related to the proof of theorem 4.1.

$\text{match}(c, c)$	$c \in (\text{Num} + \text{Name})$
$\text{match}(c, !\,id)$	$c \in (\text{Num} + \text{Name})$ and $id \in \text{Id}$
$\dfrac{\text{match}(t', t'')}{\text{match}(t'', t')}$	
$\dfrac{\text{match}(t'_1, t'_2) \qquad \text{match}(t''_1, t''_2)}{\text{match}((t'_1, t''_1), (t'_2, t''_2))}$	

Table 1: Matching Rules

To evaluate expressions we must know the values of their variables; stores serve this purpose. Stores are modeled by elements (functions) of the domain Store = Id $\rightarrow$ Con + {unbound} ranged over by $\sigma, \sigma', \sigma''$, etc. We use the following standard notations for stores:

$$\Phi = \lambda\, id.\, \text{unbound} \qquad\qquad \sigma[c/id] = \lambda i.[i = id] \rightarrow c, \sigma(i)$$

where the construct $[be] \rightarrow v_1, v_2$ is a form of conditional.

Since expressions do not modify the store, their semantics is defined by means of an *evaluation function* that uses a store to relate expressions with constants. We will write $\mathcal{E}[\![e]\!]\sigma = c$ read as: "given the store σ, the expression e evaluates to the value c". The *expression evaluation function* $\mathcal{E}$: Exp $\rightarrow$ (Store $\rightarrow$ Con) is defined by structural induction via the following clauses:

$$\text{(e1)} \quad \mathcal{E}[\![c]\!]\sigma = c$$

$$\text{(e2)} \quad \mathcal{E}[\![id]\!]\sigma = \sigma(id)$$

$$\text{(e3)} \quad \mathcal{E}[\![e_1 \; bop \; e_2]\!]\sigma = bop(\mathcal{E}[\![e_1]\!]\sigma, \; \mathcal{E}[\![e_2]\!]\sigma)\,.$$

The semantics of tuples is defined similarly. We introduce an auxiliary syntactical category, ETpl (ranged over by *et*), defined by the following productions:

$$et ::= c \mid !\,id \mid et_1, et_2$$

Its elements are tuples whose fields are evaluated. Thus, the *tuple evaluation function* $\mathcal{T}$: Tpl $\rightarrow$ (Store $\rightarrow$ ETpl) is defined by structural induction via the following clauses:

$$\text{(t1)} \quad \mathcal{T}[\![!\,id]\!]\sigma = !\,id$$

$$\text{(t2)} \quad \mathcal{T}[\![e]\!]\sigma = \mathcal{E}[\![e]\!]\sigma$$

$$\text{(t3)} \quad \mathcal{T}[\![t_1, t_2]\!]\sigma = \mathcal{T}[\![t_1]\!]\sigma, \mathcal{T}[\![t_2]\!]\sigma\,.$$

Matching of evaluated tuples is determined via the binary predicate "match" defined in table 1.

3.2 Programs

When considering programs, we are not only interested in their possible final results but also in their interactions with the execution environment. Because of this, we describe how the individual steps of program take place by means of (unlabbelled) *transition systems.*

Definition 3.1 A *Transition System* is a triple $(\mathcal{C}, \mathcal{T}, \rightarrow)$ where:

- $\mathcal{C}$ is the set of configurations;
- $\mathcal{T} \subseteq \mathcal{C}$ is the set of terminal configurations;
- $\rightarrow \subseteq \mathcal{C} \times \mathcal{C}$ is the transition relation defined by the system; it is such that every terminal configuration has no transitions. □

We equip a generic program, say p, with a *structural operational semantics* that is defined in terms of the evolution of quadruples of the form $<p, \sigma, P, T>$ where p is the generic program, σ is the global store, P is the multiset of all parallel programs and T is a multiset of tuples (which models the Tuple Space). From the semantics of programs *in isolation*, the operational semantics of an L system, represented by a set of L programs, is defined in terms of the evolution of triples $<P, \sigma, T>$.

In the following we use $\mathcal{M}$(I) for denoting the set of multisets over a set I, $\{\!| \; i_1, \ldots, i_n \; |\!\}$ for denoting the multiset whose elements are $i_1, \ldots, i_n$ and $\uplus$ for denoting the union operator between multisets.

The following transition system establishes the operational semantics of programs in isolation. We call its configurations "single program configurations"; the transition system describes the evolution of a single component of a (multi) set of concurrent programs.

Configurations: $\mathcal{C} = \text{Prg} \times \text{Store} \times \mathcal{M}(\text{Prg}) \times \mathcal{M}(\text{ETpl})$;

Terminal Configurations: $\mathcal{T} = \{\textbf{nil}\} \times \text{Store} \times \mathcal{M}(\text{Prg}) \times \mathcal{M}(\text{ETpl})$;

Transition Relation: $\rightarrow \subseteq \mathcal{C} \times \mathcal{C}$ is the least relation induced by the rules in tables 2 - 3.

In the following, we will use P and T as generic elements of $\mathcal{M}$(Prg) and $\mathcal{M}$(ETpl) respectively. Note that $\mathcal{M}$(ETpl) is the formalization of TS.

Most of the rules in table 2 are obvious, it is only worth noting that within Rules (p2) and (p3) number 0 is taken as the boolean "false" and any number different from 0 as the boolean "true".

Let us now briefly comment the rules of table 3. Rule (p6) describes the meaning of **eval**; it terminates just after adding the argument program to the multiset of concurrent programs. The direct termination of **eval**, and the possibility of termination of the program executing it, combined with the possibility of nontermination of the created program is a key feature of Linda. It is similar to the operator **new** introduced in APC [18]. This feature of Linda avoids problems of determining distributed termination that are present

(p1)
$$\frac{\mathcal{E}[\![e]\!]\sigma = c}{<id := e, \sigma, P, T> \rightarrow <\mathbf{nil}, \sigma[c/id], P, T>}$$

(p2)
$$\frac{\mathcal{E}[\![e]\!]\sigma = 0}{<\mathbf{while}\ e\ \mathbf{do}\ p, \sigma, P, T> \rightarrow <\mathbf{nil}, \sigma, P, T>}$$

(p3)
$$\frac{\mathcal{E}[\![e]\!]\sigma \neq 0}{<\mathbf{while}\ e\ \mathbf{do}\ p, \sigma, P, T> \rightarrow <p;\mathbf{while}\ e\ \mathbf{do}\ p, \sigma, P, T>}$$

(p4)
$$<\mathbf{nil};p, \sigma, P, T> \rightarrow <p, \sigma, P, T>$$

(p5)
$$\frac{<p_1, \sigma, P, T> \rightarrow <p_1', \sigma', P', T'>}{<p_1; p_2, \sigma, P, T> \rightarrow <p_1'; p_2, \sigma', P', T'>}$$

Table 2: Rules for SL programs

(p6)
$$<\mathbf{eval}(p), \sigma, P, T> \rightarrow <\mathbf{nil}, \sigma, \{\!|\, p \,|\!\} \uplus P, T>$$

(p7)
$$\frac{\mathcal{T}[\![t]\!]\sigma = t'}{<\mathbf{out}(t), \sigma, P, T> \rightarrow <\mathbf{nil}, \sigma, P, \{\!|\, t' \,|\!\} \uplus T>}$$

(p8)
$$\frac{\mathcal{T}[\![t]\!]\sigma = t'}{<\mathbf{read}(t), \sigma, P, T> \rightarrow <\mathbf{read}(t'), \sigma, P, T>}$$

(p9)
$$\frac{\mathcal{T}[\![t]\!]\sigma = t'}{<\mathbf{in}(t), \sigma, P, T> \rightarrow <\mathbf{in}(t'), \sigma, P, T>}$$

(p10)
$$\frac{\mathrm{match}(t_1, t_2)}{<\mathbf{read}(t_1), \sigma, P, \{\!|\, t_2 \,|\!\} \uplus T> \rightarrow <\mathbf{nil}, \sigma[t_1 \Leftarrow t_2], P, \{\!|\, t_2 \,|\!\} \uplus T>}$$

(p11)
$$\frac{\mathrm{match}(t_1, t_2)}{<\mathbf{in}(t_1), \sigma, P, \{\!|\, t_2 \,|\!\} \uplus T> \rightarrow <\mathbf{nil}, \sigma[t_1 \Leftarrow t_2], P, T>}$$

Table 3: Rules for Linda programs

in languages such as CSP [15]. Thus, "a program p is terminated" means that the execution of the program q in "$p; q$" may begin but it does not mean that all programs that have been created by p are terminated too. Rules (p7)–(p11) show that, after the evaluation of its argument tuple, an **out** instruction can always proceed, while an **in/read** instruction may proceed only if a matching tuple is found in T. The notation $\sigma[t_1 \Leftarrow t_2]$, where t_1 and t_2 are evaluated tuples, is defined by the following rules:

$$\sigma[c \Leftarrow c] = \sigma \qquad \sigma[!\,id \Leftarrow c] = \sigma[c/id]$$
$$\sigma[c \Leftarrow !\,id] = \sigma \qquad \sigma[t_1', t_1'' \Leftarrow t_2', t_2''] = \sigma[t_1' \Leftarrow t_2'][t_1'' \Leftarrow t_2'']$$

3.3 Computations and Operational Semantics

The program semantics presented in the previous section does not describe the behaviour of a (multi) set of concurrent programs; indeed, it only describes the evolution of a single program via strictly sequential transitions and the

(int)	$\dfrac{<p,\sigma,P,T> \rightarrow <p',\sigma',P',T'>}{<\{\!\mid p \mid\!\} \uplus P,\sigma,T> \rightarrow_m <\{\!\mid p' \mid\!\} \uplus P',\sigma',T'>}$

Table 4: Axiom for the single–step transition relation

corresponding modifications on the store, on the multiset of the remaining concurrent programs and on the tuple space. By using the transition relation among single program configurations, the transition system that we introduce in this section models the concurrent evolution of a (multi) set of concurrent programs. It is defined as follows:

Configurations: $\mathcal{C}_m = \mathcal{M}(\text{Prg}) \times \text{Store} \times \mathcal{M}(\text{ETpl})$, ranged over by G;

Terminal Configurations: $\mathcal{T}_m = \mathcal{M}(\{\textbf{nil}\}) \times \text{Store} \times \mathcal{M}(\text{ETpl})$;

Transition Relation: $\rightarrow_m \subseteq \mathcal{C}_m \times \mathcal{C}_m$ is the least relation defined by the axiom in table 4.

The resulting relation is a *single–step* transition relation in that it permits only a single transition at a time. In the full paper [17] we define also a *multisteps* transition relation and prove that each multisteps transition can be expressed as a sequence of single–step ones.

Let us briefly comment on axiom (int) of table 4. It states that a transition in the evolution of a multiset of concurrent programs is performed by:

- arbitrarily selecting a program from the multiset to determine a main program and a multiset of (all of the remaining) programs;
- considering a step in the evolution of the chosen program according to the transition relation over programs in isolation;
- rebuilding the multiset of all the concurrent programs by putting together the program and the multiset of programs obtained after the transition.

Note that at each step the selection does not depend on the previous one and that a multiset of concurrent programs is stuck if and only if all of its component programs cannot perform a transition.

The transition system above describes the behaviour of an abstract machine that executes L programs. Once one has specified the set of initial states for beginning the execution of a program p, the operational semantics of the latter may be defined by using the set of transition systems reachable from one of those states.

Indeed we define the operational semantics of a program in terms of a set of computations (maximal sequences of transitions) rather than in terms of transition systems.

Definition 3.2 A *computation from* G_0, the *initial* configuration, is:

- either a finite sequence $\rho = G_0, G_1, G_2, \ldots, G_n$, $n \geq 0$, of configurations such that $G_i \rightarrow_m G_{i+1}$, $0 \leq i < n$, where either $G_n \in \mathcal{T}_m$ (the computation is *terminal*) or $G \notin \mathcal{T}_m$ and $\neg\exists G' \in \mathcal{C}_m : G \rightarrow_m G'$ (the computation is *deadlocked*);
- or an infinite sequence $\rho = G_0, G_1, G_2, \ldots$ of configurations such that $G_i \rightarrow_m G_{i+1}$, $0 \leq i$. □

Note that a computation ρ from G is a *maximal* sequence of configurations, in the sense that it is not "extensible". Comp(G) denotes the *set of computations from G*.

Definition 3.3 The *operational semantics* of a program p is:

$$\mathcal{OS}(p) \stackrel{def}{=} \bigcup_{G \in \mathrm{I}_p} \mathrm{Comp}(G)$$

where the set I_p ($\subset \mathcal{C}_m$) of *initial configurations* for p is:

$$\mathrm{I}_p \stackrel{def}{=} \{<\{\!\!\{ p \}\!\!\}, \sigma, T> \mid \sigma \in \mathrm{Store},\ T \in \mathcal{M}(\mathrm{ETpl})\}.$$ □

Note that for the set I_p other choices were possible; for example, another possibility, rendering I_p independent of a particular store or tuple space, was to set $\mathrm{I}_p \stackrel{def}{=} \{<\{\!\!\{ p \}\!\!\}, \Phi, \emptyset>\}$. However, we prefer considering every initial configuration containing the single program p as an initial configuration for p.

4 Observational Semantics

We cannot use the operational semantics defined in the previous section for comparing programs and verifying their correctness; it is not sufficiently abstract. It gives a different account of any two programs that are syntactically different because it considers also the involved configurations (which, obviously, differ at least in the program component). Then, if we used equality of the operational semantics of programs to establish their equivalence we would essentially differentiate all of them. As an example, let us consider the two programs below:

out(N,5); **out**(M,9).   **out**(M,9); **out**(N,5).

They have different operational semantics; nevertheless they are observationally indistinguishable; they have the same interactions with (add the same two tuples to) the tuple space and do not modify the store. Similarly, the following two programs:

$wv_1 := 3$; $l := 0$; $r := 1$;
while $wv_1 \leq 10$ **do**
  ($wv_1 := wv_1 + 1$; $wv_2 := r$;
  $r := r + l$; $l := wv_2$).

$wv_1 := 10$; $l := 0$; $r := 1$;
while $wv_1 \geq 3$ **do**
  ($wv_1 := wv_1 - 1$; $wv_2 := r$;
  $r := r + l$; $l := wv_2$).

that compute the ninth (l) and the tenth (r) element of Fibonacci's sequence, should be considered as equivalent; they compute the same function between stores, if we ignore the working variables wv_1 and wv_2, and do not require any interaction with the environment.

The problem of lack of abstractness is tackled by resorting to a notion of *semantic equivalence* that permits equating systems that are indistinguishable by external observation. Moreover, the semantic equivalence we define is such that if two semantically equivalent programs p and q are added to a set of concurrent programs, then the resulting sets are semantically equivalent too. Our equivalence relation is a congruence for all the contexts of the language. This can be exploited to obtain modular refinement of programs; it permits replacing a component module (autonomous entity) of a specification with any congruent one, while ensuring that the resulting program is equivalent to the original one.

We formalize the concept of congruence relation for L by means of the notion of "execution context" that is an L program with a "hole", where any other program may be placed.

Definition 4.1 A *legal execution context*, or simply a *context*, is generated by the following productions:

$$\gamma ::= [\,] \mid \textbf{eval}(\gamma) \mid p;\gamma \mid \gamma;p \mid \textbf{while } e \textbf{ do } \gamma$$

We use Γ, ranged over by γ, to denote the set of legal L contexts. □

Let us turn our attention to the working variables. They are the elements of the set Wv defined in section 2.2 and are private, local variables that a program may use in order to control the computation flow. When comparing programs, we abstract from the computations made on them. For this reason, placing a program p in a context γ (we use the notation $\gamma[p]$) could cause the renaming of some of the working variables used by γ in order to avoid variable–clashes with those used by p (which represent different variables).

In the following, we will define three preorders among programs, whose kernels are three equivalences[2]. The basic ingredients of our relations will be two preorders that rely on the ability of programs to respond positively to external tests, and on their inability to respond negatively to them[3]. The third preorder will be obtained by the conjoining of the two basic ones.

4.1 Observers

Observers, like in [19], are essentially programs that have the possibility of making some experiments on the behaviour of a program and eventually reporting their observation. They may use the reserved tuples (**START**,0) and (**END**,0)

[2]A *preorder* $\precsim$ is a transitive and reflexive binary relation that generates an equivalence $\simeq$ in a natural way: $\simeq \,=\, \precsim \cap (\precsim^{-1})$.

[3]Due to the intrinsic nondeterminism of Linda programs, these two notions lead to different preorders.

as argument of the programs **out** and **read** respectively and the other elements of the set *Test* of reserved tuples defined in Section 2.2. In particular, by executing **out**(**START**,0), observers enable the execution of the program they test. By executing **read**(**END**,0), observers are able to start some experiments after the tested program is terminated. By forcing observers to use **read**(**END**,0) instead of **in**(**END**,0) we make sure that if a test on (**END**,0) succeeds then all of them succeed.

Observers may also use the auxiliary program **success-on**(Σ) where Σ is a finite, nonempty subset of Store, to check a condition on the (final) store. Note that no transition rule is associated with such programs as **success-on**(Σ). They may belong to the multiset of concurrent programs of configurations and may appear in the final configuration of a finite computation. This means that the definitions of the sets $\mathcal{C}$ and $\mathcal{T}$ in section 3.2, $\mathcal{C}_m$ (Configurations) and $\mathcal{T}_m$ (Terminal Configurations) in section 3.3 are modified by permitting that such programs as **success-on**(Σ) may belong to the first component of their elements.

Finally, observers may use the variables of the programs they test (this is the only exception to the rule that concurrent programs have disjoint variable spaces) provided that in performing the experiment *o testing p* some of the working variables used by *o* are renamed to avoid variable–clashes with the ones used by *p*.

Definition 4.2 Let $\mathcal{P}_f$(Store), ranged over by Σ, be the set of all finite, nonempty subsets of Store. We say that *o* is an *observer* if it can be generated via the abstract syntax reported below and satisfies the constraint of using the tuple (**START**,0) as argument of **out** only and the tuple (**END**,0) as argument of **read** only.

$$o ::= \mathbf{nil} \mid id := e \mid \mathbf{while}\ e\ \mathbf{do}\ o \mid o_1; o_2 \mid \mathbf{in}(t) \mid \mathbf{read}(t) \mid \mathbf{out}(t) \mid \mathbf{eval}(o) \mid \mathbf{success\text{-}on}(\Sigma) \qquad \square$$

4.2 Preorder and Equivalence Relations

The effect of observers performing tests on programs is formalized by:

Definition 4.3 For a program *p* and observer *o*, the result of the experiment *o testing p* is the following nonempty set of computations:

$$\mathrm{RC}(p, o) = \mathrm{Comp}(<\{\!|\ \mathbf{read}(\mathbf{START},0); p; \mathbf{out}(\mathbf{END},0),\ o\ |\!\},\ \Phi,\ \emptyset>)\,. \qquad \square$$

Note that we could also use **in**(**START**,0) instead of **read**(**START**,0).

To determine whether a program passes a test, we need to state when a computation is successful. Note that an observer can finish with more than one **success-on** program (due to **eval**). The final store σ of a terminal configuration must satisfy each **success-on**, so that we define the success of a computation as the conjunction of the success of every individual **success-on**; a **success-on**(Σ) is successful if there is $\sigma' \in \Sigma$ such that for each of its identifiers not in *Wv* both σ' and σ assume the same value or σ' is undefined.

Definition 4.4 For $P \in \mathcal{M}(\mathrm{Prg})$, let $nt(P) \stackrel{def}{=} \{\!|\, p \in P \mid p \neq \mathbf{nil} \,|\!\}$;

- a terminal configuration $<P, \sigma, T>$ is *successful* if $nt(P) \neq \emptyset$ and $\forall p \in nt(P)$ a finite, nonempty $\Sigma \subset$ Store exists such that $p = \mathbf{success\text{-}on}(\Sigma)$ and $\exists \sigma' \in \Sigma, \forall id \in \mathrm{Id} \setminus Wv$: $\sigma'(id) \neq$ unbound $\Rightarrow \sigma'(id) = \sigma(id)$;
- a configuration is *unsuccessful* if it is not a successful one. □

The notion of successful computation is based on the definition of successful configuration.

Definition 4.5 A finite computation is *successful* if its last configuration is successful; a computation is *unsuccessful* if it is infinite or its last configuration is unsuccessful. □

Due to the relational behaviour of programs, an observer can only establish a sort of "conditional success" of an experiment; that is, an observer delivers a proviso that the final store must satisfy if the computation is successful.

Definition 4.6 For every program p and observer o, we fix:

- $p \; may \; o \stackrel{\text{def}}{\Longleftrightarrow} \exists \rho \in \mathrm{RC}(p, o)$ such that ρ is successful;
- $p \; must \; o \stackrel{\text{def}}{\Longleftrightarrow} \forall \rho \in \mathrm{RC}(p, o) \; \rho$ is successful. □

As in [19], by means of these predicates we define three preorder relations over programs.

Definition 4.7 For a given set $\mathcal{O}$ of observers, we define the following preorders over Prg:

1. $p_1 \sqsubseteq^{\mathcal{O}}_{may} p_2 \stackrel{\text{def}}{\Longleftrightarrow} \forall o \in \mathcal{O}$: $p_1 \; may \; o \Rightarrow p_2 \; may \; o$;
2. $p_1 \sqsubseteq^{\mathcal{O}}_{must} p_2 \stackrel{\text{def}}{\Longleftrightarrow} \forall o \in \mathcal{O}$: $p_1 \; must \; o \Rightarrow p_2 \; must \; o$;
3. $p_1 \sqsubseteq^{\mathcal{O}}_{test} p_2 \stackrel{\text{def}}{\Longleftrightarrow} p_1 \sqsubseteq^{\mathcal{O}}_{may} p_2 \wedge p_1 \sqsubseteq^{\mathcal{O}}_{must} p_2$. □

We will use $\simeq^{\mathcal{O}}_i$ to denote the equivalence relation generated by the preorder $\sqsubseteq^{\mathcal{O}}_i$, $i \in \{may, must, test\}$ (i.e. $\simeq^{\mathcal{O}}_i = \sqsubseteq^{\mathcal{O}}_i \cap (\sqsubseteq^{\mathcal{O}}_i)^{-1}$).

The following theorem permits to perform program verification using the relations defined above because it states that they all are preserved by every composition operator among L programs, i.e. they all are (pre)congruence relations.

Theorem 4.1 Let $\mathcal{E}$ be the set of observers in L; $\forall R \in \{\sqsubseteq^{\mathcal{E}}_{may}, \simeq^{\mathcal{E}}_{may}, \sqsubseteq^{\mathcal{E}}_{must}, \simeq^{\mathcal{E}}_{must}, \sqsubseteq^{\mathcal{E}}_{test}, \simeq^{\mathcal{E}}_{test}\}$: $p \; R \; q \;\Leftrightarrow\; \forall \gamma \in \Gamma: \gamma[p] \; R \; \gamma[q]$. □

Due to lack of space we omit the proof of theorem 4.1; it may be found in the full paper [17].

We can now fully appreciate the role of the reserved tuples *Test*: they are essential for defining (preorder or equivalence) relations that are preserved by the ";" operator. For example, the two programs below cannot be considered as equivalent:

p: **eval**(**in**(M, $!x$));
$y := 5$.

q: **eval**($y := 5$);
in(M, $!x$).

Indeed, only p is able to terminate (in any context). If one considers the context $\gamma = [\,]$;**out**(M,7) the resulting programs $\gamma[p]$ and $\gamma[q]$ behave differently. The first one correctly terminates (the tuple (M,7) is added to TS and, then, **in**(M,$!x$) can terminate by withdrawing it) while the second one deadlocks (the tuple (M,7) cannot be added to TS and thus **in**(M,$!x$) gets stuck). Indeed, the observer:

$$o \;=\; \mathbf{out}(\mathbf{START},0);\mathbf{read}(\mathbf{END},0);\mathbf{out}(\mathrm{M},1);\mathbf{success\text{-}on}(\{\Phi\})$$

distinguishes them: RC(p, o) has a successful computation while RC(q, o) has a deadlocking computation only (o adds the tuple (M,1) to TS if and only if it formerly has observed the termination of the tested program).

5 Related works and further research

We have presented operational and observational semantics of a member of the Linda family obtained by embedding Linda in a very simple imperative language. The operational semantics is described in the SOS style of [12]. The observational semantics follows the paradigm advocated in [19] and there developed for CCS, a model of concurrency based on a synchronous communications paradigm [16]. The presence of full sequential (instead of prefix) composition, global environment and the asynchronous paradigm have rendered the work presented here significantly different.

A few papers [8, 11, 9, 10] have been devoted to formalizing the semantics of the Tuple Space. They rely on different semantic models, such as CCS, Petri Nets, Chemical Abstract Machine or Z notation, and ignore the impact of the host language on the formal definition. More importantly, none of the above mentioned papers tackles the problem of providing a behavioural semantics that permits giving full account of single programs and formally comparing programs for performing program verification.

A work similar in spirit to ours is reported in [20]. There, a compositional semantic model, based on sequences of pairs of states, is defined for a general asynchronous language with partially interpreted actions. Each pair of states encodes the state transformation (initial state - final state) occurred during a transition step of a process. In order to guarantee compositionality, sequences do not necessarily represent connected computations, "gaps" can appear to describe steps performed by the environment. This leads to the explosion of the set of possible sequences to be associated with each process. There would be no problem in defining a similar semantics for L. Our semantics does not require introducing unnecessary dependencies among the distributed components; and these may develop simultaneously exchanging data by means of shared tuples. The different environments in which each program may operate are simulated by the different external observers and gaps are implicit (they are represented by switching to another pivotal program during a computation).

Our approach has already been fruitfully used to provide a two levels observational semantics for VHDL [21], a formalism for describing hardware components [22]. Much work remains to be done in order to fully exploit the semantical characterization we have described in the paper. It would be useful to provide axiomatic characterizations of the defined relations; they would guarantee a fuller understanding of Linda primitives, would permit using a *transformational* method for passing from a program to another and would lay the basis for a set of algebraic laws of asynchronous programming.

By following the approach developed for process algebras it would also be useful to define new, more abstract, operators for composing Linda programs and to introduce scope operators that, by restricting visibility of tuples, would avoid unnecessary matching conflicts. Observers should be extended to report success while asking for more general properties of the final store not just for a finite number of possible values of some of its variables. Also particular subclasses of observers should be defined that fully characterize the (possibly restricted) environments in which programs are going to operate (a sort of "context-depending testing").

In the full paper [17], by means of an example, we show how to use the proposed semantic framework for verifying programs; in particular, we consider L programs for the "prefix sums problem" [23].

Acknowledgments We wish to thank Kees Goossens for the helpful suggestions and the the anonymous referees for their accurate and stimulating comments.

References

[1] Gelernter D. Generative Communication in Linda. ACM Transactions on Programming Languages and Systems 1985; 7(1):80-112

[2] Carriero N, Gelernter D. Linda in Context. Communications of the ACM 1989; 32(4):444-458

[3] Shapiro E. Concurrent Prolog: Collected Papers. The MIT Press, 1987

[4] Chandy KM, Misra J. Parallel Program Design: a Foundation. Addison Wesley, Massachusetts, 1988

[5] Brogi A, Ciancarini P. The concurrent language Shared Prolog. ACM Transactions on Programming Languages and Systems 1991; 13(1):99-123

[6] Gelernter D, Carriero N, Chandran S, et al. Parallel Programming in Linda. In: proc. of the Internatinal Conference on Parallel Programming (St. Charles, Ill., Aug.). IEEE, 1985, pp 255-263

[7] Carriero N, Gelernter D, Leichter J. Distributed Data Structures in Linda. In: proc. of the ACM Symposium on Principles of Programming Languages (St. Petersburg, Fla., Jan.13–15). ACM, New York, 1986, pp 236-242

[8] Ciancarini P, Yanchelevich D. Inside Linda. Technical Report, Yale University, Department of Computer Science, 1990

[9] Butcher P. A behavioural semantics for Linda-2. Software Engineering Journal 1991; 6(4):196-204

[10] Jensen KK. A Formal Linda Definition. Draft, February 1991

[11] Ciancarini P, Jensen KK, Yankelevich D. The Semantics of a Parallel Language based on a Shared Data Space. Technical Report: TR-26/92, Dipartimento di Informatica, Università di Pisa, 1992

[12] Plotkin GD. A Structural Approach to Operational Semantics. Technical Report DAIMI FN-19, Aarhus University, Dep. of Computer Science, Denmark, 1981

[13] Bergstra J, Klop JW. Process Algebra for Synchronous Communication. Information and Control 1984; 60:109-137

[14] Hennessy M. Algebraic Theory of Processes. The MIT Press, 1988

[15] Hoare CAR. Communicating Sequential Processes. Prentice-Hall Int., 1985

[16] Milner R. Communication and Concurrency. Prentice Hall International, 1989

[17] De Nicola R, Pugliese R. Testing Linda: an Observational Semantics for an Asynchronous Language. Research Report: SI/RR - 94/06, Dipartimento di Scienze dell'Informazione, Università di Roma "La Sapienza", 1994

[18] Baeten JCM, Vandrager FM. An Algebra for Process Creation. Acta Informatica 1992; 29(4):303-334

[19] De Nicola R, Hennessy MCB. Testing Equivalence for Processes. Theoretical Computers Science 1984; 34:83-133

[20] de Boer FS, Kok JN, Palamidessi C, et al. The Failure of Failures in a Paradigm for Asynchronous Communication. In: proc. of Concur '91. Springer-Verlag, 1991, pp 111-126 (Lecture Notes in Computer Science no.527)

[21] Goossens KGW. Reasoning about VHDL Using Operational and Observational Semantics. Research Report: SI/RR - 95/06, Dipartimento di Scienze dell'Informazione, Università di Roma "La Sapienza", 1995

[22] The Institute of Electrical and Electronics Engineers. IEEE Standard VHDL Language Reference Manual. IEEE std 1076–1993 edition, 1993

[23] Ladner RE, Fisher MJ. Parallel Prefix Computation. Journal of the ACM 1980; 27(4):831–838

Refinement and Recursion in a High Level Petri Box Calculus

Raymond Devillers
Département d'Informatique, Université Libre de Bruxelles
Boulevard du Triomphe, B-1050 Bruxelles, Belgium
Hanna Klaudel
Laboratoire de Recherche en Informatique, Université Paris Sud
bât. 490, F-91405 Orsay, France

Abstract

The algebra of A-nets, a high level class of labelled Petri nets introduced in the Petri Box Calculus in order to cope with structured data, is extended with a general refinement operator and, based thereon, a general recursion operator; their properties may directly be derived from the corresponding operators for the low level Petri Boxes.

Introduction

The Petri Box Calculus (PBC) is a formalism developed in [BDH92, BDE93] in order to apply Petri net theory to the specification and the verification of concurrent algorithms and also to address the compositional semantics of languages introduced to express them.

The PBC syntax yields *box expressions*, which can be seen as an extension of process terms of CCS [Mil80]; their compositional semantics is given by (classes of) labeled place/transition nets, called *Boxes*. Boxes are nets with two kinds of interface: an entry/exit interface (places) and a communication interface (transitions). Boxes can be composed with each other across these interfaces and are provided with a full algebraic structure which mimics the one for expressions.

The natural domain of application of the Petri Box Calculus is the semantics of parallel languages, and in particular of $B(PN)^2$ (Basic Petri Net Programming Notation) introduced in [BH93] for the specification of concurrent algorithms. A formal low level semantics in terms of Petri Boxes has been proposed in [BH93] by associating a box expression to every meaningful sub-construct of a concurrent $B(PN)^2$ program, and in turn, providing a compositional Petri net semantics [BDH92]. However, the size of the nets obtained is a problem, which has been solved in [BF$^+$95a, BF$^+$95b] by proposing the algebra of M-nets in order to provide a compositional high level semantics for $B(PN)^2$ programs. M-nets are a fairly powerful Petri net model, but the lack of tools to handle complex data structures is a serious drawback if one wants to use it as a semantic domain for a real programming language. Hence, in [KP95], an extension of M-nets, called A-nets, has been introduced with a full abstract data type orientation. However, in the A-net theory, as well as for M-nets, the algebraic structure was not complete, since refinements and (based thereon) recursions were missing.

This paper presents a way to overcome the problem and to define these operators at the A-net level (thus also for the embedded case of M-nets) by using the same devices as for the low level Petri Boxes (labelled trees and sequences). The semantics and unfoldings of A-nets will also be defined for any model of the specification, together with equivalence relations and various properties directly inherited from the low level theory.

1 Petri Boxes

1.1 Low level (Box-like) nets

Petri Boxes, or Boxes for short, are defined as equivalence classes of unmarked labelled place/transition (P/T) Petri Nets. The labels (or status) for the places may be: e (entry places, where tokens may *enter* the net), x (exit places, from where tokens may *leave* the net) and $\emptyset$ (internal places); the transition labels may be *communication* labels (elementary actions from a countable infinite set A_{PBC} or finite multisets of them) or *hierarchical* labels (of the form V, where V is a variable name, liable to be later replaced -possibly recursively- by a whole Box, hence allowing to construct a complex net from simpler ones). We shall denote by $\mathcal{L}$ the set of communication labels and by $\mathcal{V}$ the set of hierarchical labels.

The nets are unmarked but the e-places define a natural initial marking M_e (one token on each e-place and no tokens elsewhere), and the x-places a natural terminating marking M_x (one token on each x-place and no tokens elsewhere), thus allowing to speak about the behaviour of a net; the nets may be weighted (but for the sake of simplicity our examples will not exhibit this feature) and fulfill some constraints:

- each transition has input and output places;
- there are entry and exit places;
- there are neither arcs to entry places nor from exit places.

Various equivalences may be considered to structurally identify nets with the same behaviour. The most elementary one is the *isomorphism* (henceforth denoted $\equiv$) between labelled nets, allowing to change the *names* or *identity* of the nodes.

The most frequently used one is the *duplication* equivalence (henceforth denoted $\equiv_d$), which allows to add/drop duplicate places and (nonhierarchical) transitions (with the same label and weights to successors/predecessors); the fact that the equivalence abstracts from duplicate places allows for instance to consider a unique **stop** Box (without any transition and internal place); the fact that it abstracts from duplicate nonhierarchical transitions allows to get interesting properties for the synchronisation operator [BDH92]. The equivalence between representatives of a same Box may be extended to their coherent markings, i.e., markings which mark equally all the places with the same label and connectivity, and from equivalent coherent markings (like the natural initial ones) the evolutions are equivalent and lead to equivalent coherent markings.

More recently [Dev94], another equivalence was proposed, sometimes called the *step duplication* equivalence (henceforth denoted $\equiv_s$) which additionnally allows to add/drop nonhierarchical transitions whose label and arc weights are the same as for a multiset (step) of other transitions. This may considerably reduce the size of a net, especially when it has been constructed through synchronisation operators.

Our labelled nets will be denoted by $\Sigma = (P, T, W, \lambda)$, where P is the place set, T the transition set, W the arc weights and λ the labelling function on places and transitions; ${}^\bullet\Sigma$ ($\Sigma^\bullet$) will denote the set of entry (exit) places of Σ, and $\ddot{\Sigma} = P \setminus ({}^\bullet\Sigma \cup \Sigma^\bullet)$ the set of internal places of Σ. Figure 1 shows an example of such a labelled net.

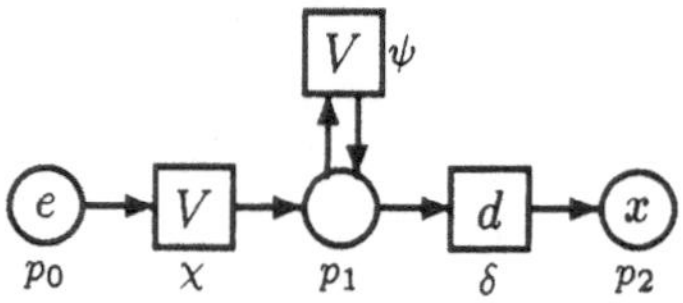

Figure 1: A low level Petri Box net.

1.2 Hierarchical A-nets

A-nets are a special kind of algebraic nets [Vau87, Rei91], mixing the *abstract data type* (shortly adt) features and the Petri Box flavour. Let (S, F, E, X) be a *presentation* of an adt, i.e., a set of *sorts* (type names) S, a *signature* F (a set of function names with their corresponding typed arities $ar : F \to S^+$; the last sort gives the type of the result of the operation, the other ones - if any - give the types of the parameters; if $w \in S^+$, $F_w = ar^{-1}(w)$ will designate the subset of F with arity w), a set of *axioms* E (equations between terms of the same sort, composed from operations and free variables from X, implicitly universally quantified) and a set of typed variables $X = \bigcup_{s \in S} X_s$. From them, we may form as usual the set of terms with variables $T_{F,X}$ (with $(T_{F,X})_s$ denoting the subset of terms of sort s) and the set of *ground* terms T_F (with $(T_F)_s$ denoting the subset of ground terms of sort s).

A *model* of the adt is a many-sorted algebra D, together with a homomorphic application of the adt which satisfies each axiom in E for every sort-preserving assignment $\sigma : X \to D$ (the corresponding evaluation function will be denoted $eval_D^\sigma$). In particular, $(T_F)/_E$, the quotient algebra of T_F modulo the smallest congruence generated by the set of axioms E, is a model of the adt; moreover, it is initial in the class of all models of the adt (see for instance [GTW78]).

Example 1.1
If $S = \{bool, int\}$, $F_{int} = \{1\}$, $F_{int \cdot int \cdot int} = \{+\}$, $F_{bool} = \{true\}$, $F_{bool \cdot bool} = \{not\}$, $F_{int \cdot int \cdot bool} = \{less\}$, $X_{int} = \{x, y, \ldots\}$, and $X_{bool} = \{z, \ldots\}$, then

$$less(+(1, x), 1), \quad not(less(x, y)), \quad not(z), \quad +(1, +(1, 1))$$

are examples of terms with variables; the first three belong to $(T_{F,X})_{bool}$, the last one is ground and belongs to $(T_F)_{int}$. A possible model is given by the usual Boolean and arithmetic domains (if they satisfy the axioms, not given here). ◇

We shall assume that the sort *bool* belongs to S, with the usual constants $true, false \in F_{bool}$; there will also be a special sort, denoted $\bullet$, which will be used to represent a token (control) flow without any further meaning, with a unique constant (also denoted $\bullet$), without any other function symbol nor variable (so that $(T_{F,X})_\bullet = F_\bullet = \{\bullet\}$).

We consider a set of action symbols A with a typed arity function, which will allow to handle parameterized actions, $ar\colon A \rightarrow S^*$; and a *conjugation* bijection $\bar{\ }\colon A \rightarrow A$, such that for all $a \in A$ we have $\bar{a} \neq a$, $\bar{\bar{a}} = a$ and $ar(a) = ar(\bar{a})$. We also define $A_w = ar^{-1}(w)$ with $w \in S^*$, a particular case being the set $A_\emptyset$ of action symbols whose typed arity is the empty word: $A_\emptyset$ is similar to the set A_{PBC} of PBC action names. The action symbols whose arities are nonempty can take parameters which are terms of the appropriate sort from the many sorted term algebra $T_{F,X}$. The set

$$A_{HL} = A_\emptyset \cup \bigcup_{s_1 \cdots s_n \in S^+} \{(a; \vec{u}) \mid a \in A_{s_1 \cdots s_n}, \vec{u} \in (T_{F,X})_{s_1} \times \ldots \times (T_{F,X})_{s_n}\}$$

is the set of *parameterized actions*. Like for the low level nets, we also consider a set $\mathcal{V}$ of *hierarchical actions* (or *net variables*) which are intended to be used in refinements and recursions[1].

Example 1.2 Consider for instance $A_{int} = \{a, \ldots\}$, $A_{bool \cdot int} = \{b, \ldots\}$ and S, F, X as in the previous example, then

$$(a; +(1, x)),\ (a; y),\ (b; not(z), +(1, +(1, 1))),\ (b; less(x, y), +(x, 1))$$

are examples of parameterized actions. The first and second ones take a single parameter of sort *int*, the third and fourth ones take two parameters of sort *bool* and *int*. ◇

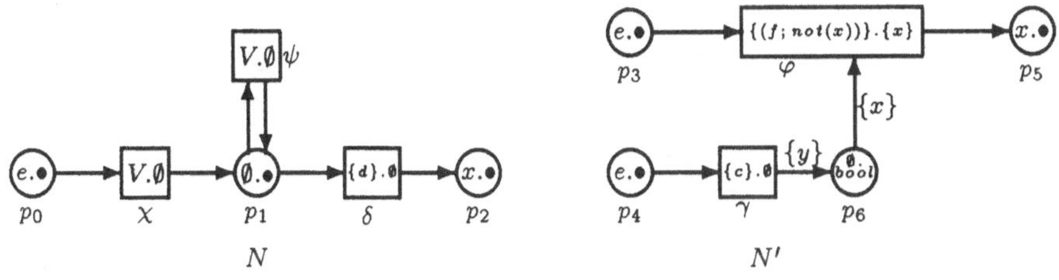

Figure 2: Two A-nets.

[1]This is an addition with respect to the original definition of A-nets in [KP95], which did not consider hierarchical actions since refinements and recursions were out of the theory there.

Definition 1.1 *A-net*

An A-net N is a triple $(P, T; \lambda)$ where P is a set of places, T is a set of transitions, $(P \times T) \cup (T \times P)$ is the set of arcs, and λ is a labeling function on places, transitions and arcs.

1. $\forall p \in P \ : \ \lambda(p) = \alpha_p.\beta_p = \mathit{status.sort}$, where $\alpha_p \in \{e, \emptyset, x\}$ and $\beta_p \in S$; as in the PBC, the places can have one of the three allowed statuses: *entry* (label e), *internal* (label $\emptyset$), or *exit* (label x); the sort specifies the kind of tokens the place can hold.
2. $\forall t \in T \ : \ \lambda(t) = \alpha_t.\beta_t = \mathit{label.guard}$, where α_t is either a finite multiset of parameterized actions (t will then be called a communication transition) or a hierarchical action (t will then be called a hierarchical transition), i.e., $\alpha_t \in \mathcal{M}_f(A_{HL}) \cup \mathcal{V}$, and β_t is a finite set of boolean terms (predicates), i.e., $\beta_t \subseteq (T_{F,X})_{bool} \ \wedge \ |\beta_t| < \infty$, the corresponding guard being intended as the conjunction of all these conditions; as usual, hence, an empty guard will mean *true*, i.e., the condition is always satisfied.
3. $\forall t \in T, \forall p \in P \ : \ \lambda(t,p), \lambda(p,t) \in \mathcal{M}_f((T_{F,X})_{\beta_p})$, i.e., the arc inscriptions are finite multisets of terms with variables of the adequate sort, meaning that a certain amount of tokens with the corresponding semantical contents will be absorbed/produced through them. As usual, arcs with empty labels are not represented in figures (they are considered as nonexistent), ${}^{\bullet}t = \{p \in P \mid \lambda(p,t) \neq \emptyset\}$ and similarly for $t^{\bullet}$, ${}^{\bullet}p$ and $p^{\bullet}$. $\diamond$

Like for the low level Petri Box nets, ${}^{\bullet}N$ ($N^{\bullet}$) will denote the set of entry (exit) places of N, and $\ddot{N} = P \setminus ({}^{\bullet}N \cup N^{\bullet})$ the set of internal places of N.

The concrete semantics of A-nets depends on the choice of a particular model D for the used abstract data type, and we assume that $\top, \bot \in D_{bool}$, $eval_D(true) = \top$ and $eval_D(false) = \bot$. For instance, we could choose as model the initial algebra $(T_F)/_E$, but any coarser model could fit our aim as well. In any case, we may choose $D_{\bullet} = \{\bullet\}$, with the identity for the evaluation function on $(T_{F,X})_{\bullet}$ $(= \{\bullet\})$.

A *marking* of an A-net $N = (P, T, \lambda)$ associates to each place $p \in P$ a finite multiset of model elements of its sort β_p : $M(p) \in \mathcal{M}_f(D_{\beta_p})$. The behaviour of an A-net is then defined by the usual transition rule for algebraic nets [Rei91]:

A transition $t \in T$ is enabled at marking M if there exists an assignment $\sigma : X \rightarrow D$ such that[2] for each $b \in \beta_t$: $eval_D^{\sigma}(b) = \top$ $(= eval_D^{\sigma}(true))$, and for each $p \in P$: $eval_D^{\sigma}(\lambda(p,t)) \subseteq M(p)$. If t is enabled at M for σ, the occurrence of t at M leads (through this σ) to the marking M' defined as follows: $\forall p \in P \ : \ M'(p) = M(p) \ominus eval_D^{\sigma}(\lambda(p,t)) \oplus eval_D^{\sigma}(\lambda(t,p))$. Such an occurrence bears the corresponding label α_t for a hierarchical transition, or $eval_D^{\sigma}(\alpha_t)$ for a communication transition. The definition may also be extended to the concurrent occurrence of a multiset of transitions with a corresponding multiset of assignments.

[2] $eval_D^{\sigma}$ naturally extends to (finite multisets of) terms and parameterized actions.

For our purpose, we shall also restrict the class of A-nets to those which satisfy the usual (for PBC) constraints concerning their structure and labeling of entry/exit (place) interfaces. In particular, it is required that there is at least one entry and at least one exit place, and no incoming arcs to entry places nor outgoing arcs from exit places. The transitions are required to have at least one input place and at least one output place. The motivation of the above is the coherency with PBC requirements (see the unfolding operation below).

The *sort* of entry and exit places must be $\{\bullet\}$, so that the initial (final) marking is unique (like in the PBC again) and consists in putting a single $\bullet$ token in each e-place (x-place) and nothing elsewhere: initial (and final) tokens do not bear any specific semantical meaning, but their mere presence.

Moreover, since it is possible to express constraints between the input and the output of a transition in an algebraic net, while it is not possible to do so across a whole net fragment replacing it (as it will be the case after a refinement or a recursion) without modifying the sorts used by the net, for each hierarchical transition t we shall assume that $\forall p \in {}^\bullet t \cup t^\bullet : \beta_p = \{\bullet\}$, i.e., hierarchical transitions only handle control flow tokens (without any special semantical transformation); as a consequence, each arc to or from t (as well as to an exit place or from an entry place) has a label of the form $n \cdot \bullet$ (n times the unique token expression, n generally being simply 1, in which case the arc label will be omitted in figures), meaning that it may only transport n tokens $\bullet$; and since no true semantical condition may be tested on those tokens, we shall also assume that $\beta_t = \emptyset$, i.e., the condition (guard) always evaluates to true ($\top$). We will see that the resulting theory is general enough in order to get the needed operator synthesis from refinements, and simple enough in order to directly inherit the main operator properties from the low level theory.

2 Unfolding and equivalences

Let D be a model of a given adt (S, F, E, X), with $bool \in S$, $true, false \in F_{bool}$, $F_\bullet = D_\bullet = \{\bullet\}$ and $\top, \bot \in D_{bool}$.

Definition 2.1 *Unfolding*

Given an A-net $N = (P, T, \lambda)$ and a marking M, we define a low level net $U(N) = (P_U, T_U, W_U, \lambda_U)$ and a corresponding marking $U(M)$ as follows :

1. $P_U = \{p_v \mid p \in P \text{ and } v \in D_{\beta_p}\}$, and $\lambda_U(p_v) = \alpha_p$.
2. $T_U = \{t_\sigma \mid t \in T, \sigma : X \to D \text{ and } \forall b \in \beta_t : eval_D^\sigma(b) = \top\}$, and $\lambda_U(t_\sigma) = eval_D^\sigma(\alpha_t)$. We shall not distinguish t_σ and $t_{\sigma'}$ if σ and σ' coincide for all the variables from X occurring in the label of t and in the arc inscriptions to or from t; as a consequence, each high level hierarchical transition will lead to a unique low level one, with the same labelling variable. It may be seen that $A_{PBC} = \{(a; d_1, \ldots, d_n) \mid a \in A, ar(a) = (s_1, \ldots, s_n), \forall i \in \{1, \ldots, n\} : d_i \in D_{s_i}\}$.
3. $W_U(p_v, t_\sigma) = \sum_{u \in eval_D^{\sigma\,-1}(v)} \lambda(p, t)(u)$

4. $U(M)(p_v) = M(p)(v)$. ◇

This is illustrated in Figure 3, which gives the unfolding of the A-net N' of Figure 2 when D_{bool} is the usual Boolean domain.

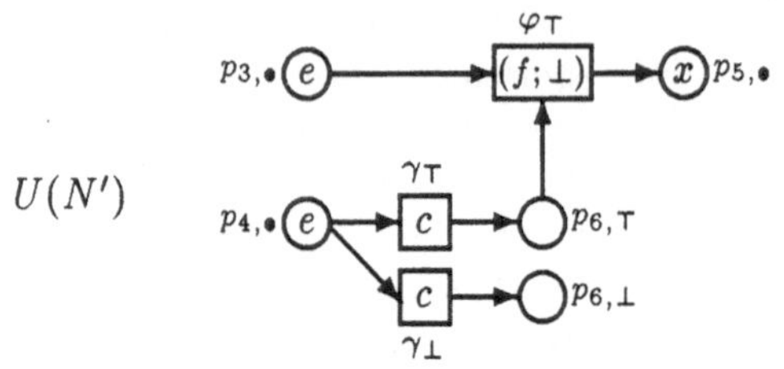

Figure 3: A-net unfolding.

It may easily be checked that

1. $U(N)$ is indeed a low level Petri Box net.

2. $U(M)$ is a marking of $U(N)$, and if M is the natural initial (resp. final) marking of N then $U(M)$ is the natural initial (resp. final) marking of $U(N)$.

3. a transition t is enabled in N at marking M for an assignment σ and leads to the marking M', if and only if t_σ is enabled in $U(N)$ at marking $U(M)$ and leads to the marking $U(M')$.

Such an unfolding also allows to naturally extend to the A-net level the various kinds of equivalence mentioned above: if N_1 and N_2 are two A-nets (for some adt (S, F, E, X) and some model D), and if $\sim$ is any equivalence from $\{\equiv, \equiv_d, \equiv_s\}$,

$$N_1 \sim^D N_2 \text{ iff } U(N_1) \sim U(N_2).$$

While rather powerful, these equivalences are not fully structural however, since they depend on a particular model, and are not always easy to check. The equivalences $\equiv$, $\equiv_d$ and $\equiv_s$ may also be directly defined at the A-net level, but various forms are possible. For instance we may define $N_1 \equiv N_2$ iff there is a label preserving bijection between their nodes, but also $N_1 \equiv' N_2$ iff there is a label preserving bijection between their nodes up to local renaming, i.e., for each (communication) transition it is possible to change the names of the term variables occurring in its label and in the labels of the arcs going in and out of it (i.e., in the area of the transition, following the terminology of [KP95]) since these variables only have a local meaning. It may be observed that $(N_1 \equiv N_2) \Longrightarrow (N_1 \equiv' N_2) \Longrightarrow (N_1 \equiv^D N_2)$. Similar definitions may be introduced for the duplication and step duplication equivalences, but we shall not develop them here.

3 An Algebra of A-nets

In order to support a compositional translation from expressions or languages to net domains, we need to define the same operations at the net level as at the language one. For M-nets [BF+95a] and (nonhierarchical) A-nets [KP95], the most basic operations (i.e., the sequential composition N_1 ; N_2, the parallel composition $N_1 \parallel N_2$, the choice $N_1 \,\square\, N_2$, the iteration $[N_1 \,*\, N_2 \,*\, N_3]$, the synchronization N_1 **sy** a and the restriction operation N_1 **rs** a) were defined individually, in an ad hoc setting. The recursion operator was missing, since it is based on refinements and the latter lacked too. As a further consequence, it was not possible either to synthesize the first four operators (; , $\parallel$, $\square$ and [* *]) in a simple and uniform way from refinements, as it was done in [BDE93] for low level Petri Boxes. We shall now repair this hole in the theory.

3.1 Refinement

The refinement $N[V_i \leftarrow N_i | i \in I]$ means 'N where all V_i-labelled transitions are refined into (i.e., replaced by a copy of) N_i, for each $i \in I$ '. Its definition is slightly technically complex, due to the great generality aimed at in the PBC theory; indeed, refinements are easy to define when the refining nets N_i have a single entry and a single exit place, or when there is a single transition to refine without side loop, but here we want to allow any kind of configuration: any number of refined transitions (possibly infinitely many, due to recursions), any connectivity network, any arc weighting, any number of entry/exit places (possibly continuously infinitely many, due to the cardinality explosion phenomenon [BDH92, BDE93]) compatible with the Box definition.

The definition uses a *labelled tree* device [BDE93] which generalises the kind of multiplicative Cartesian cross product (pre/post places of transitions to be refined with entry/exit places of the refining net) commonly used in the literature [GG89] as the interface places. This setting has not been chosen just for the purpose of treating the general case, but also to get easily the main properties of the refinement operator; with this respect, it has been successfully reused in [Dev93, Dev95, BK95a, BK95b]. In figures, those labelled trees are often represented by the set of all the path labels from the root.

If $V \in \mathcal{V}$ is a hierarchical action and $\mathcal{V}_I = \{V_i \mid i \in I\}$ is a set of such actions, let us define $T^V = \{t \in T \mid \lambda(t) = V.\emptyset\}$ and $T^{\mathcal{V}_I} = \bigcup_{V \in \mathcal{V}_I} T^V$.

Definition 3.1 *General Refinements*

Let $N = (P, T; \lambda)$, $N_i = (P_i, T_i; \lambda_i)$ be A-nets, for each $i \in I$, and for a same adt. $N[V_i \leftarrow N_i | i \in I]$ is defined as the A-net $\tilde{N} = (\tilde{P}, \tilde{T}; \tilde{\lambda})$, where

$\tilde{T} = (T \backslash T^{\mathcal{V}_I}) \cup \bigcup_{i \in I} T^i$

where $T^i = \{t.t_i \mid t \in T^{V_i}, t_i \in T_i\}$ is the set of all (copied) transitions of N_i's refining transitions labelled by V_i's.

$\tilde{P} = \bigcup_{i \in I} P^i \cup \bigcup_{p \in P} P^p$

where $P^i = \{t.p_i \mid t \in T^{V_i}, p_i \in \ddot{N}_i\}$ is the set of all (copied) internal places of N_i's refining transitions labelled by V_i's,
and P^p is the set of all the (isomorphic classes of) labelled trees of the following form:

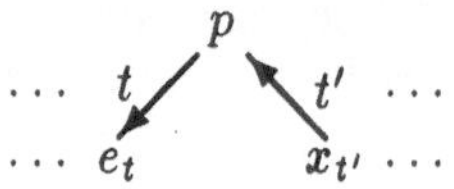

i.e., the root is labelled by p, the arcs are labelled by a transition (in $T^{\mathcal{V}_I} \cap (p^\bullet \cup {}^\bullet p)$) and a direction (up or down); for each $i \in I$ and each (if any) $t \in p^\bullet$ with a label of the form V_i there is an arc labelled t going (down) to (a node labelled by) some (arbitrarily chosen) entry place e_t of N_i and for each $i \in I$ and each (if any) $t' \in {}^\bullet p$ with a label of the form V_i, there is an arc labelled t' coming (up) from (a node labelled by) some (arbitrarily chosen) exit place $x_{t'}$ of N_i; such a tree will also be represented by the sequence set $\{p, p.t.e_t, ..., p.t'.x_{t'}, ...\}$.

$$\tilde{\lambda}(\tilde{t}, \tilde{p}) = \begin{cases} \lambda(t,p) & \text{if } \tilde{t} = t \in (T \backslash T^{\mathcal{V}_I}),\ \tilde{p} \in P^p \\ n \cdot m \cdot \bullet & \text{if } \tilde{t} = t.t_i \in T^i,\ \nwarrow t_{x_i} \text{occurs in } \tilde{p} \in P^p, \\ & \lambda(t,p) = n \cdot \bullet \text{ and } \lambda(t_i, x_i) = m \cdot \bullet \\ \lambda_i(t_i, p_i) & \text{if } \tilde{t} = t.t_i \in T^i,\ \tilde{p} = t.p_i \in P^i \\ \emptyset & \text{otherwise} \end{cases}$$

$\tilde{\lambda}(\tilde{p}, \tilde{t})$ is defined symmetrically

$$\tilde{\lambda}(\tilde{t}) = \begin{cases} \lambda(t) & \text{if } \tilde{t} = t \in (T \backslash T^{\mathcal{V}_I}) \\ \lambda_i(t_i) & \text{if } \tilde{t} = t.t_i, \lambda(t) = V_i.\emptyset \text{ and } t_i \in T_i \end{cases}$$

$$\tilde{\lambda}(\tilde{p}) = \begin{cases} \lambda(p) & \text{if } \tilde{p} \in P^p \\ \emptyset.\beta_{p_i} & \text{if } \tilde{p} = t.p_i \in P^i \end{cases}$$

◇

This definition is illustrated in Figure 4, from the nets in Figure 2. It may be checked that, with the natural initial marking, the behaviour of the refined net corresponds to what may be expected.

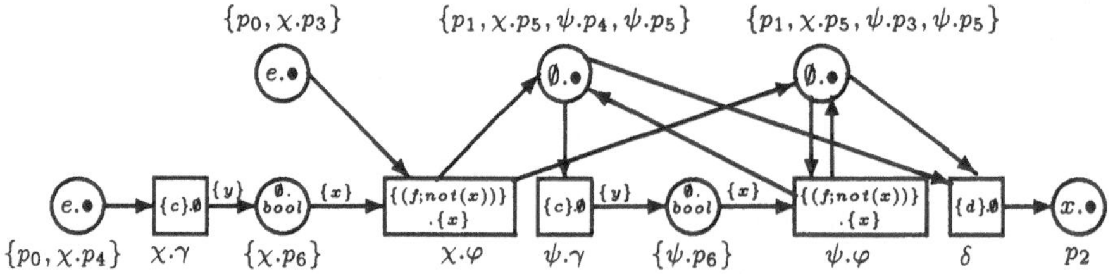

$N[V \leftarrow N']$

Figure 4: A refinement with a side loop.

The result is indeed an A-net; in particular, hierarchical transitions in $\tilde{N}$ inherit correct surroundings from the original nets. It may be observed that this definition mimics faithfully the refinement definition introduced for Petri Box nets in [BDE93, Dev95]. As a consequence, a first basic property of this definition is

Proposition 3.2 *Refinements commute with unfoldings*

For any A-nets N, $\{N_i \mid i \in I\}$, and for any model of their adt

$$U(N[V_i \leftarrow N_i | i \in I]) \equiv U(N)[V_i \leftarrow U(N_i) | i \in I]$$

Proof: This immediately results from the similarity of the definition of refinement between low level Petri Boxes [BDE93] and A-nets, and from the assumptions we made for the latter: since $\bullet$ has a unique model element, each entry place, each exit place and each place connected to a hierarchical transition is unfolded in a single place with the same status; each arc to or from a hierarchical transition has a label of the form $n \cdot \bullet$ and is unfolded in a single arc with weight n. □

For the same reason of definition similarity, all the proofs used in [BDE93] may be reused as such, in order to get the same properties, for instance:

Proposition 3.3 *Expansion Law for Successive Refinements*

$$N[V_i \leftarrow N_i \mid i \in I][W_j \leftarrow N'_j \mid j \in J]$$
$$\equiv N[V_i \leftarrow N_i[W_j \leftarrow N'_j \mid j \in J], W_k \leftarrow N'_k \mid i \in I, k \in K]$$
if $K = \{k \in J \mid W_k \notin V_I\}$.

Also, one immediately gets that,

Proposition 3.4 *Congruence*

If $\sim$ is any equivalence from $\{\equiv, \equiv', \equiv^D, \equiv_d, \equiv'_d, \equiv^D_d, \equiv_s, \equiv'_s, \equiv^D_s\}$ and $N \sim N'$, $N_i \sim N'_i$ for each $i \in I$, then

$$N[V_i \leftarrow N_i \mid i \in I] \sim N'[V_i \leftarrow N'_i \mid i \in I]$$

The usual operators for sequentialisation, choice, parallel composition and iteration, may now be synthesised as claimed above:

Definition 3.5 *Synthesised operators*

Let $N_;$, $N_{\Box}$, $N_{\|}$, N_* and N'_* be the A-nets shown in Figure 5

(i) $N_1; N_2 = N_;[V_1 \leftarrow N_1, V_2 \leftarrow N_2]$ (sequence)

(ii) $N_1 \Box N_2 = N_{\Box}[V_1 \leftarrow N_1, V_2 \leftarrow N_2]$ (choice)

(iii) $N_1 \| N_2 = N_{\|}[V_1 \leftarrow N_1, V_2 \leftarrow N_2]$ (concurrent composition)

(iv) $[N_1 * N_2 * N_3] = N_*[V_1 \leftarrow N_1, V_2 \leftarrow N_2, V_3 \leftarrow N_3]$ (iteration)
or
$[N_1 * N_2 * N_3] = N'_*[V_1 \leftarrow N_1, V_2 \leftarrow N_2, V_3 \leftarrow N_3]$ (1-safe iteration)

◇

From 3.5 and 3.2, one then immediately gets:

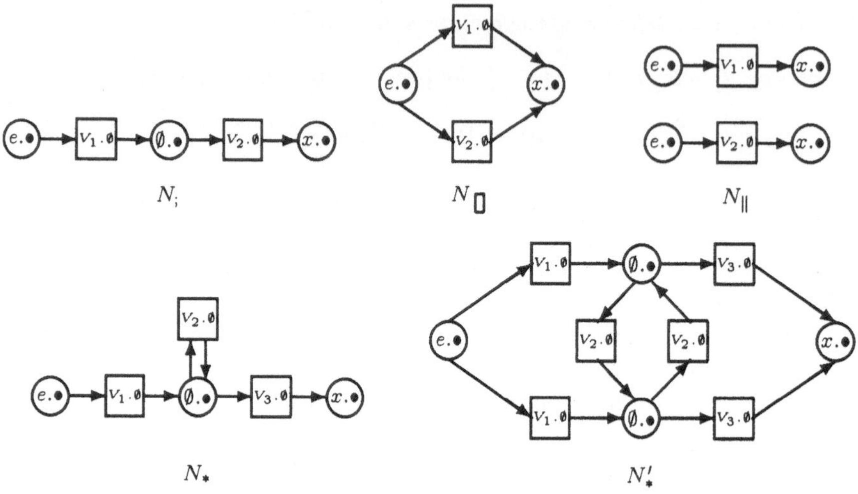

Figure 5: The operative A-nets

Corollary 3.6 *Unfolding commutes with basic operators*

$$U(N_1 \ \mathbf{op}\ N_2) \equiv U(N_1)\ \mathbf{op}\ U(N_2)\ ,\ \textit{for}\ \mathbf{op} \in \{;\,,\,\Box\,,\,\|\}$$
$$U([N_1 * N_2 * N_3]) \equiv [U(N_1) * U(N_2) * U(N_3)]$$

3.2 Recursion

In a similar way, we may now extend the recursion operator defined in [BDE93] (see also [Dev95]) for Petri Box nets, at the A-net level.

In order to define the recursion $\mu\{V_i.N_i | i \in I\}N$, meaning "replace in N all V_i's by N_i's, ad infinitum", we shall introduce some preliminary notations.

Let $\mathcal{V}_I = \{V_i \mid i \in I\}$ be a family of distinct hierarchical actions for an indexing set I, $\{N_i \mid i \in I\}$ be a corresponding family of (not necessarily distinct) A-nets and N be any A-net.

As before, when considering (possibly infinite) labelled trees, we will only be interested in their labels, not in the identity of the various nodes, i.e., we will essentially consider isomorphism classes of such trees. The trees which will be used to define the interface places for the recursion operator, may be constructed from the following tree families, which extend the kind of trees we used for refinements to any (possibly infinite) depth.

Definition 3.7 *A gallery of trees*

(i) basic entry and exit trees:

for each $i \in I$, let τ_i^e be the set of labelled trees where the roots have a label in ${}^\bullet N_i$, the nodes have labels in $\bigcup_{j \in I} {}^\bullet N_j$, and if a node has a label $p \in {}^\bullet N_j$ then for each $k \in I$ and for each $t \in p^\bullet \cap T_j^{V_k}$ (if any) there is an arc labelled by t going (down) to a node labelled by some $p' \in {}^\bullet N_k$. The set τ_i^x of basic exit trees is defined symmetrically.

(ii) basic internal trees :

for each $i \in I$, let $\tau_i^\emptyset$ be the set of labelled trees whose roots have a label of the form $\sigma.p$ where $\sigma \in (\bigcup_{j \in I} T_j^{\mathcal{V}_I})^*$; if σ is empty then $p \in \ddot{N}_i$, otherwise σ starts from a transition in $T_i^{\mathcal{V}_I}$ and, for any pair of successive transitions $t.t'$ in σ, if t has a label $V_j.\emptyset$, then $t' \in T_j^{\mathcal{V}_I}$; moreover, if the last transition of σ has a label $V_j.\emptyset$, then $p \in \ddot{N}_j$; finally, if $p \in \ddot{N}_j$, then for each $k \in I$ and for each $t \in p^\bullet \cap T_j^{V_k}$ (if any) there is an arc labelled by t going to the root of a τ_k^e-tree, and for each $k \in I$ and for each $t \in {}^\bullet p \cap T_j^{V_k}$ (if any) there is an arc labelled by t coming from the root of a τ_k^x-tree.

(iii) place/place trees :

let τ^P be the set of labelled trees where the roots have a label p in P, for each $i \in I$ and for each $t \in p^\bullet \cap T^{V_i}$ (if any) there is an arc labelled by t going to the root of a τ_i^e-tree, and for each $i \in I$ and for each $t \in {}^\bullet p \cap T^{V_i}$ (if any) there is an arc labelled by t coming from the root of a τ_i^x-tree.

(iv) transition/place trees :

let τ^T be the set of labelled trees of the form $t.\tau$ where $t \in T^{\mathcal{V}_I}$ and, if $\lambda(t) = V_k.\emptyset$, $\tau \subset \tau_k^\emptyset$. ◇

All those definitions are schematised in Figure 6; let us also notice that it may happen that τ^T and/or $\tau_i^\emptyset$ are empty, while this may not be the case for τ^P, τ_i^e or τ_i^x.

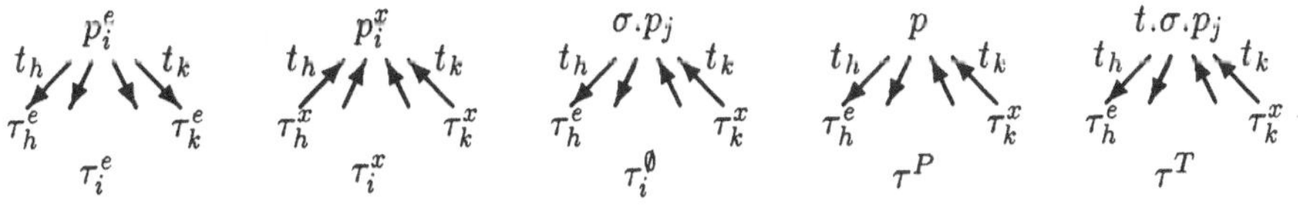

Figure 6: The various (recursive) forms of place-trees

Then, if we interpret the recursion $\mu\{V_i.N_i \mid i \in I\}N$ as a kind of "limit" (for more sound explanations see [BDE93]) of the refinements $N[V_i \leftarrow N_i[V_i \leftarrow N_i[...]]]$ or $N[V_i \leftarrow N_i][V_i \leftarrow N_i]...$, and if we consider the expanded forms[3] of them, we are led to the following, which has strong similarities with the definition for the refinement operators, with some natural additional complexity: transitions will be finite sequences of any length (but with specific constraints) and places will be labelled trees of any depth (possibly infinite).

[3]See [Dev95] for more explanations on tree expansions.

Definition 3.8 *The general simultaneous recursive refinement operator*

$\mu\{V_i.N_i \mid i \in I\}N$, is the A-net $(\tilde{P}, \tilde{T}; \tilde{\lambda})$ defined in the following way :

- $\tilde{P} = \tau^P \cup \tau^T$.
- $\tilde{T}$ is the set of all the finite nonempty transition sequences σ such that
  - the first transition belongs to T, the other ones belong to $\bigcup_{i \in I} T_i$,
  - if a transition in σ has a label $V_i.\emptyset$ $(i \in I)$, then the next transition belongs to T_i; it has a label outside $V_I.\emptyset$ iff it is the last one,
-
  - for any $\tilde{p} \in \tau^T$, $\tilde{\lambda}(\tilde{p}) = \lambda_i(p_i)$, if the root of $\tilde{p}$ has a label $t.\sigma.p_i$ and $p_i \in P_i$ (in fact $\ddot{N}_i$).
  - for any $\tilde{p} \in \tau^P$, $\tilde{\lambda}(\tilde{p}) = \lambda(p)$ if p is the label of the root of $\tilde{p}$.
  - for any $\tilde{t} \in \tilde{T}$, $\tilde{\lambda}(\tilde{t})$ is the label of its last component, i.e., $\lambda(t)$ if $\tilde{t} = t$ and $\lambda_i(t_i)$ if $\tilde{t} = t.\sigma.t_i$ with $t_i \in T_i$.
- a transition $\tilde{t} = \sigma.t'$ is connected to a place $\tilde{p}$ iff $\sigma = \sigma'.\sigma''$, where the root of $\tilde{p}$ is labelled $\sigma'.p$, σ'' is a (possibly empty) transition path (down or up, depending on the direction of the connection) in $\tilde{p}$ from the root to a node labelled p' and t' is connected to p'; the weight is the product of the weights of the corresponding arcs. ◇

We also define $\mu\{V_j.N_j \mid j \in I \setminus \{i\}\}V_i.N_i = \mu\{V_j.N_j \mid j \in I\}N_i$ and $\mu V.N = \mu\{V.N\}N$.

Again the apparent complexity of the definition arises from the great generality which is achieved; in particular, we do not require that the recursion must be guarded, i.e., hierarchical transitions may be connected to entry and/or exit places in any N_i. Also, Definition 3.8 directly gives an explicit form for the result and does not define recursion as a "limit" still to be constructed. However, it is possible to rephrase it denotationally as the fixpoint of a monotonic function, like in [BDE93, BK95b].

The definition is illustrated by the example shown in Figure 7, where the various trees are again represented by their equivalent sequence sets (i.e., the set of all the node labels prefixed by the transition paths to reach them). It may be checked that the behaviour is indeed what could be expected from a recursive net.

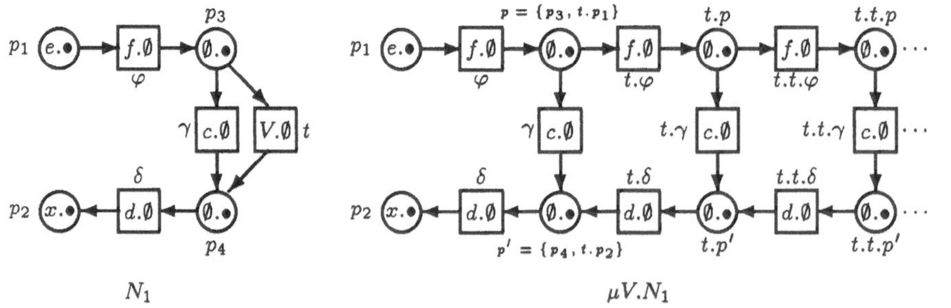

Figure 7: An example of recursion

It may be shown that the result is again an A-net. Moreover, since the definition mimics faithfully the recursion definition introduced for Petri Box nets in [BDE93, Dev95], for the same reasons as for refinement, a first basic property is

Proposition 3.9 *Recursions commute with unfoldings*

For any A-nets N, $\{N_i \mid i \in I\}$, and for any model of their adt

$$U(\mu\{V_i.N_i \mid i \in I\}N) \equiv \mu\{V_i.U(N_i) \mid i \in I\}U(N)$$

And again for the same reason of definition similarity, all the proofs used in [BDE93] may be reused as such, in order to get the same properties, like:

Proposition 3.10 *Some properties of the general recursion operator*

(i) $\mu\{V_i.N_i \mid i \in I\}N \equiv N[V_i \leftarrow \mu\{V_j.N_j \mid j \in I\}N_i \mid i \in I]$ *(substitution property)*
this generalizes the classical fixpoint equation:
$\mu V.N \equiv N[V \leftarrow \mu V.N]$ *(simple recursion)*
and may still be generalized as:

(ii) *Let $\{\mathcal{N}_i \mid i \in I\}$ be the smallest families of nets such that, for any $i \in I$, $\forall J \subseteq I$: $\mu\{V_j.N_j \mid j \in J\}N_i \in \mathcal{N}_i$ and*
if $N_j' \in \mathcal{N}_j$ for each $j \in J \subseteq I$, then $N_i[V_j \leftarrow N_j' \mid j \in J] \in \mathcal{N}_i$,
then if $N_i' \in \mathcal{N}_i$ for each $i \in I$ and $N_j'' \in \mathcal{N}_j$ for each $j \in J \subseteq I$,
$\mu\{V_i.N_i \mid i \in I\}N \equiv \mu\{V_i.N_i' \mid i \in I\}(N[V_j \leftarrow N_j'' \mid j \in J])$.
(expansion law for recursion)

Also, one immediately gets that,

Proposition 3.11 *Congruence*

If $\sim$ is any equivalence from $\{\equiv, \equiv', \equiv^D, \equiv_d, \equiv_d', \equiv_d^D, \equiv_s, \equiv_s', \equiv_s^D\}$ and $N \sim N'$, $N_i \sim N_i'$ for each $i \in I$, then

$$\mu\{V_i.N_i \mid i \in I\}N \sim \mu\{V_i.N_i' \mid i \in I\}N'$$

4 Conclusion

We have provided the A-net domain with the same algebraic structure as the low level Petri Box one, by introducing general simultaneous refinement and recursion operators; the coherence of the two corresponding structures has been exhibited through the unfolding operation, and the properties are inherited from the low level ones; the operators also resist to all the kinds of equivalences which have been considered.

Of course, in order to obtain these results, we had to introduce serious restrictions about the surroundings of the hierarchical transitions (and about

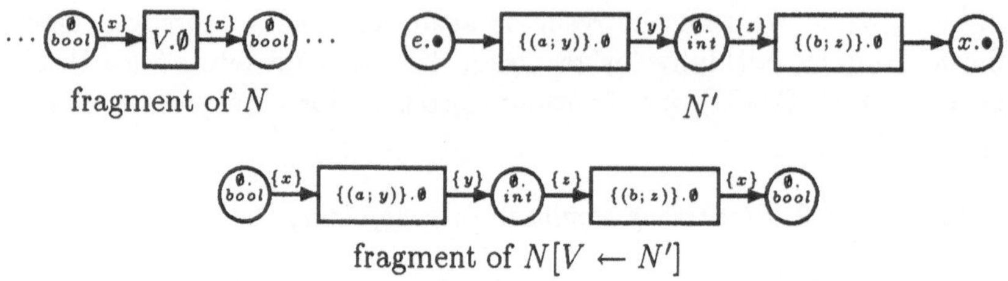

Figure 8: The difficulty of the generalisation

the sort of the entry/exit places, but that was already done in the M-net domain); it does not seem easy to overcome this difficulty, however, as exhibited by Figure 8: the refinement of the A-net fragment given by the first net, when V is replaced by the second A-net, should look like the third fragment, but while in the first one the two x variables were the same (they occur in the surroundings of the same transition) this is no longer the case in the third one, since variables only have a local meaning and may be changed independently around each transition. Hence it should be necessary to *transport* the identity of the variables from the entry of the refined copy to the exit (or the other way round), but presently we do not see any simple and acceptable way to do so; fortunately, this was not necessary in order to be able to synthesize the usual operators from refinement, for instance.

Acknowledgments

This work has been partially supported by the ESPRIT WG 6067 CALIBAN. Most of the work was done while the first author visited the UPVM's Equipe d'Informatique Fondamentale in winter 1994: our thanks go to the Université Paris Val de Marne for the invitation. We also thank the anonymous referees for their inspiring comments.

References

[BDE93] E. Best, R. Devillers, and J. Esparza. General refinement and recursion operators for the Petri Box Calculus. In *STACS'93*, P. Enjalbert et al. (eds.), LNCS 665, pp. 130–140. Springer Verlag, 1993.

[BDH92] E. Best, R. Devillers, and J. Hall. The Box Calculus: a new causal algebra with multi-label communication. *Advances in Petri Nets 1992*, LNCS 609, pp. 21–69, 1992.

[BF+95a] E. Best, H. Fleischhack, W. Fraczak, R.P. Hopkins, H. Klaudel, and E. Pelz. A Class of Composable High Level Petri Nets. To be presented at *ICPN'95*, LNCS, 1995.

[BF+95b] E. Best, H. Fleischhack, W. Fraczak, R.P. Hopkins, H. Klaudel, and E. Pelz. A high level Petri net semantics of $B(PN)^2$. To be presented at *STRICT'95*, LNCS, 1995.

[BH93] E. Best and R.P. Hopkins. $B(PN)^2$ -a Basic Petri Net Programming Notation. *PARLE-93*, LNCS 694, pp. 379–390, 1993.

[BK95a] E. Best and M. Koutny. A refined View of the Box Algebra - a Tutorial- Invited paper at *ICPN'95*, LNCS, 1995.

[BK95b] E. Best and M. Koutny. Solving Recursive Net Equations. Invited paper at *ICALP'95*, LNCS, 1995.

[Dev93] R. Devillers: S-invariant Analysis of General Refined Petri Boxes. In *Computer Science 2: Research and Applications*, pp. 411–428, Plenum Publishing, 1994.

[Dev94] R. Devillers. The Synchronisation Operator Revisited for the Petri Box Calculus. Technical Report LIT-290, Université Libre de Bruxelles, February 1994.

[Dev95] R. Devillers. S-invariant Analysis of General Recursive Petri Boxes. To appear in *Acta Informatica*, 1995.

[GTW78] J.A. Goguen, J.W. Thatcher and E.G. Wagner. An initial algebra approach to the specification, correctness and implementation of abstract data types. *Current Trends in Programming Methodology*, vol. 4, pp. 80-149, Prentice-Hall, 1978.

[GG89] U. Goltz and R.J. van Glabbeek: Refinement of Actions in Causality Based Models. *Proc. of REX Workshop on Stepwise Refinement of Distributed Systems*, LNCS 430, pp. 267–300, 1989.

[KP95] H. Klaudel and E. Pelz. Communication as Unification in the Petri Box Calculus. To be presented at *FCT'95*, LNCS, 1995.

[Mil80] R. Milner. *A Calculus of Communicating Systems*, LNCS 92, 1980.

[Rei91] W. Reisig. Petri nets and algebraic specifications. *TCS 80*, pp. 1–34, 1991.

[Vau87] J. Vautherin. Parallel systems specification with colored Petri nets and algebraic specification. *Advances in Petri Nets 1987*, LNCS 266, pp.293-308, 1987.

Sequentiality by Linear Implication and Universal Quantification

Alessio Guglielmi
Università di Pisa, Dipartimento di Informatica, Corso Italia 40, 56125 Pisa, Italy
e-mail: guglielm@di.unipi.it fax: +39 (50) 887 226

Abstract

In this paper we address the issue of understanding sequential and parallel composition of agents from a logical viewpoint. In particular we use methods of abstract logic programming in linear logic, *i.e.* computations are modeled as proof searches in a suitable fragment of linear logic. While parallel composition has a straightforward treatment in this setting, sequential composition is much more difficult to be obtained. We study a case, directly inspired by Monteiro's distributed logic, in which the causality relation among agents forms a series-parallel order; top agents may be recursively rewritten by series-parallel structures of new agents. We show a very declarative and simple treatment of sequentialization, which smoothly integrates with parallelization, by translating our formal system into linear logic in a complete way. This means that we obtain a full two ways correspondence between proofs and computations; thus we have full correspondence between the two formalisms. Our case study is very general per se, but it should be clear that the methodology adopted should be extensible to orderings more general than the series-parallel ones. The expected outcomes of this research are at least twofold: having some new insights in the design of concurrent languages and formalisms and having a strong starting point for relating linear logic semantics to concurrency semantics.

1 Introduction

Linear logic [4] is a powerful and elegant framework in which many aspects of concurrency, parallelism, non-determinism and synchronization find a natural interpretation. The difficulties of dealing with these issues within classical logic are overcome by the linear logic approach, mainly thanks to the "resource-orientation" of its multiplicative fragment. This roughly amounts to a good treatment of logical formulas as processes, or agents, in a distributed environment [2, 7]. The richness of the calculus and the deep symmetries of its proof theory make it an ideal instrument for purposes such as language design, specification, operational semantics, and it is certainly an interesting starting point for denotational semantics investigations. We are interested here in the "(cut-free) proof search as computation" paradigm, as opposed to the "cut-elimination as computation" one.

While the parallel execution of two agents $A \| A'$ finds a natural understanding as $A \parr A'$ (or $A \otimes A'$ in a symmetrical interpretation), the same cannot be said for their ***sequential*** composition $A\,;A'$. Yet sequential composition is a very important expressive tool and theoretical concept. In fact, we can naively achieve sequential composition in an indirect way, through backchaining. This is not satisfactory for at least two reasons: 1) because it is an unnatural form of encoding, and 2) because backchaining is most naturally thought of, and dealt with, as a non-deterministic tool, while sequential composition is deterministic. A major problem one encounters when trying to express sequentialization is having to make use of "continuations," which are, in our opinion, a concept too distant from a clean, declarative, logical understanding of the subject.

In this paper we offer a methodology, through a simple and natural case study, which deals with sequentiality in a way which certainly does not have the flavor of continuations. Sequentialization is achieved in linear logic by a controlled form of backchaining, whose non-determinism is eliminated by the linearity of the calculus (linear implication) and a declarative way of producing unique identifiers (universal quantification). In our case study these two mechanisms, together with the usual $\parr$ one, are embodied in a translation with a clear declarative meaning.

We introduce the language SMR (Sequential Multiset Rewriting) and give a translation of it into linear logic which is both correct and complete, thus fully relating the two formalisms. Computing in SMR is in the logic programming style: a goal of first order atoms (***agents***) has to be reduced to empty through backchaining by clauses, thus producing a binding for variables. Goals are obtained from agents by freely composing with the two connectives $\diamond$ (***parallel***) and $\triangleleft$ (***sequential***). Every top agent, *i.e.* every agent not preceded by other agents, can give birth to a new subgoal. The declarative meaning of $A \diamond A'$ is that we want to solve problems (to prove) A *and* A'; the meaning of $A \triangleleft A'$ is that we want to solve A *and then* A'. The simplest way to introduce synchronization in this framework is having clauses of the form $A_1, \ldots, A_h \leftarrow G_1, \ldots, G_h$. They state the simultaneous replacement of top agents A_1, $\ldots$, A_h with goals G_1, $\ldots$, G_h, respectively. This framework has been studied by Monteiro, in a more complex framework called distributed logic [10, 11].

It is natural to associate hypergraphs to goals: nodes are agents and hyperarcs express the immediate sequentiality relationship among agents. Thus the hypergraph relative to $G = (a_1 \diamond a_2) \triangleleft a_3 \triangleleft (a_4 \diamond a_5)$ has the two hyperarcs $(\{a_1, a_2\}, \{a_3\})$ and $(\{a_3\}, \{a_4, a_5\})$. Let us associate to every agent a_i the empty agent $\circ_i$, whose declarative meaning is "agent in position i has been solved." A natural interpretation in linear logic of the situation represented in G is given by the formula $(((a_3 \multimap (\circ_1 \parr \circ_2)) \otimes ((a_4 \parr a_5) \multimap \circ_3) \otimes (\circ_4 \parr \circ_5)) \multimap (a_1 \parr a_2))$. Here indices of agents have to be thought of as unique identifiers of the position of the agent in the goal, which is an obvious concern since we have imposed some structure on goals (a partial ordering). Now we need something more: since subgoals appear during the computation as an effect of resolutions, we need a mechanism to "localize" goal descriptions in linear logic, so as to fit them to the contingent goal dynamically. Again, a natural way to do that is describing G as $\forall i_1 i_2 i_3 i_4 i_5 : (((a_{i_3} \multimap (\circ_{i_1} \parr \circ_{i_2})) \otimes ((a_{i_4} \parr a_{i_5}) \multimap \circ_{i_3}) \otimes (\circ_{i_4} \parr \circ_{i_5})) \multimap (a_{i_1} \parr a_{i_2}))$. We do not really need $\otimes$ since $((A_1 \otimes \cdots \otimes A_h) \multimap A) \equiv (A_1 \multimap \cdots \multimap A_h \multimap A)$. It turns out that this very simple-minded idea actually works. Moreover, the $\circ$ goal behaves as a unity for $\diamond$ and $\triangleleft$, as *true* does for *and* in classical logic. Since syntax (and operational semantics) may make somewhat opaque the declarativeness of hypergraphs, which consists essentially of the precedence relations, we shall establish strong bindings between a very declarative notion of normalization for goals and the computations as they are actually performed by the linear logic engine, showing their equivalence.

SMR is a plain generalization of Horn clauses logic programming, using $\diamond$ instead of $\wedge$. As a matter of fact, considering clauses of the form $A \leftarrow A_1 \triangleleft \cdots \triangleleft A_h$, we grasp Prolog's left-to-right selection rule, and of course many more selection rules and much greater control over the order of execution of goals are possible.

In order to link SMR to linear logic we use a fragment of FORUM [8], which is a presentation of linear logic from an abstract logic programming perspective [9]. Its choice is rewarding because FORUM puts under control the large amount of non-determinism of linear logic, which is something in the direction we are pursuing. We refer the reader to the conclusions for a discussion of what we feel is the meaning of this contribution. This paper is rather picky and technical. As a matter of fact, the technique presented works in principle, but the details turned out to be more important than expected. The conference format does not help, so, at least to have some more feeling with the language and its basic mechanisms, the reader is referred to [5] for a more relaxed exposition of an earlier attempt to define and specify SMR. Sect. 2 is devoted to preliminaries and FORUM, in sect. 3 we present SMR, its operational semantics and a study of the normalization properties of goals; then, in sect. 4, the translation into FORUM is shown and correctness and completeness are stated.

2 Basic Notions and Preliminaries

The first subsection fixes the notation for some usual preliminaries. In the second one a brief exposition of the fragment of FORUM we are interested in is given.

2.1 Notation and Basic Syntax

$\mathsf{P}(S)$ stands for the set of subsets of S and $\mathsf{P}_\mathsf{F}(S)$ stands for the set of finite subsets of S.

N is the set of the natural numbers $\{0, 1, 2, \ldots\}$. Given $h \in \mathsf{N}$, indicate with N_h the set $\{h, h+1, h+2, \ldots\}$; given $k \in \mathsf{N}$, indicate with N_h^k the set $\mathsf{N}_h \setminus \mathsf{N}_{k+1}$.

Given $h, k \in \mathsf{N}$, if $h \leqslant k$ then $e|_h^k$ stands for "$e_h, \ldots, e_k$"; if $h > k$ then $e|_h^k$ and $(e|_h^k)$ stand for the empty object "".

Given a set S, indicate with S^+ the set $\bigcup_{i \in \mathsf{N}_1} S^i$ and with $S^\star$ the set $S^+ \cup \{\epsilon_S\}$, where $\epsilon_S \notin S^+$. ϵ_S is the *empty sequence* (*of* S) and we shall write ϵ or nothing instead of ϵ_S. If $s \in S$ then (s) and s denote the same object. On sequences is defined a *concatenation* operator $\|$, with unity ϵ.

$\mathbb{x}$ denotes the set of variables, $\mathbb{p}$ the set of predicates and $\mathbb{A}$ denotes the set of first order atoms.

Given a syntactical object F, $\lceil F \rceil$ denotes the set of free variables in F. For substitutions the usual notation and conventions apply. Let σ denote the set of substitutions, ρ the set of renaming substitutions and let $[\,]$ denote the identity substitution.

2.2 The FORUM$^{\wp \multimap \forall}$ Presentation of a Fragment of Linear Logic

The reader can find in [8] the details missing here. Methods are called this way after [1].

$\mathbb{M}$ is the least set such that: 1) $\mathbb{A} \subset \mathbb{M}$. 2) If $M, M' \in \mathbb{M}$ then $(M \wp M') \in \mathbb{M}$ and $(M \multimap M') \in \mathbb{M}$. 3) If $M \in \mathbb{M}$ and $x \in \mathbb{x}$ then $(\forall x : M) \in \mathbb{M}$. A generic element of $\mathbb{M}$ is a *method* and shall be denoted by M.

$\wp$ associates to the left and $\multimap$ associates to the right. So we shall write $(M \wp M' \wp M'')$ instead of $((M \wp M') \wp M'')$ and $(M \multimap M' \multimap M'')$ instead of $(M \multimap (M' \multimap M''))$. Instead of $(\forall x_1 : (\ldots : (\forall x_h : M)\ldots))$ we shall write $(\forall x_1 \ldots x_h : M)$. Outermost parentheses shall be omitted whenever possible. Given $h \in \mathsf{N}$, $k \in \mathsf{N}_h$ and $f \colon \mathsf{N}_h^k \to \mathbb{M}$, the notation $\bigwp_{i \in \mathsf{N}_h^k} f(i)$ stands for $f(h) \wp \cdots \wp f(k)$; given $g \colon \mathsf{N}_h^k \to \mathbb{x}$, the notation $\bigforall_{i \in \mathsf{N}_h^k} g(i) : M$ stands for $\forall g(h) \ldots g(k) : M$. If $\bar{M} = (M|_1^h) \in \mathbb{M}^+$ then $\bigwp \bar{M}$ stands for $\bigwp_{i \in \mathsf{N}_1^h} M_i$. If $\bar{x} = (x|_1^h) \in \mathbb{x}^\star$ then $\bigforall \bar{x} : M$ stands for $\bigforall_{i \in \mathsf{N}_1^h} x_i : M$ when $h > 0$, and for M when $h = 0$.

We adopt a special kind of sequents, made up from collections of methods with different structures imposed on them: sets, multisets and ordered lists. Sets are used to represent information as in classical logic: this is information which does not change during the computation; a program is represented as a set of methods. Multisets are used to represent the state of the computation, which, of course, changes as the computation goes ahead; here is where linear logic has its main usefulness. Lists of atoms appear in our sequents as a way to limit the choice in the use of right rules; this ordering does not affect correctness and completeness. From the proof theory point of view, sets are places where weakening and contraction rules are allowed, while on multisets and lists these rules are forbidden. In these sequents there is place for one method (which we call "focused") which drives the choice of left inference rules.

A *sequent* is an object of the form $(\Psi; \Gamma \mid M \vdash \Lambda; \Xi)$, where $\Psi \in \mathsf{P}_{\mathsf{F}}(\mathbb{M})$ (the *classical part*), Γ is a finite multiset of methods (the *left linear part*),

Structural rules

$$\mathsf{I}\,\frac{}{\Psi;\{\}_+ \mid A \vdash A;\epsilon} \qquad \mathsf{L}\,\frac{\Psi;\Gamma \vdash \Lambda, A;\Xi}{\Psi;\Gamma \vdash \Lambda;A,\Xi} \qquad \mathsf{D_L}\,\frac{\Psi;\Gamma \mid M \vdash \Lambda;\epsilon}{\Psi;M,\Gamma \vdash \Lambda;\epsilon} \qquad \mathsf{D_C}\,\frac{M,\Psi;\Gamma \mid M \vdash \Lambda;\epsilon}{M,\Psi;\Gamma \vdash \Lambda;\epsilon}$$

Left rules

$$⅋_{\mathsf{L}}\,\frac{\Psi;\Gamma \mid M \vdash \Lambda;\epsilon \quad \Psi;\Gamma' \mid M' \vdash \Lambda';\epsilon}{\Psi;\Gamma,\Gamma' \mid M ⅋ M' \vdash \Lambda \curlyvee \Lambda';\epsilon}$$

$$\multimap_{\mathsf{L}}\,\frac{\Psi;\Gamma \vdash \Lambda;M \quad \Psi;\Gamma' \mid M' \vdash \Lambda';\epsilon}{\Psi;\Gamma,\Gamma' \mid M \multimap M' \vdash \Lambda \curlyvee \Lambda';\epsilon}$$

$$\forall_{\mathsf{L}}\,\frac{\Psi;\Gamma \mid M[t/x] \vdash \Lambda;\epsilon}{\Psi;\Gamma \mid \forall x:M \vdash \Lambda;\epsilon}$$

Right rules

$$⅋_{\mathsf{R}}\,\frac{\Psi;\Gamma \vdash \Lambda;M,M',\Xi}{\Psi;\Gamma \vdash \Lambda;M ⅋ M',\Xi}$$

$$\multimap_{\mathsf{R}}\,\frac{\Psi;M,\Gamma \vdash \Lambda;M',\Xi}{\Psi;\Gamma \vdash \Lambda;M \multimap M',\Xi}$$

$$\forall_{\mathsf{R}}\,\frac{\Psi;\Gamma \vdash \Lambda;M[x/y],\Xi}{\Psi;\Gamma \vdash \Lambda;\forall y:M,\Xi}\,\star$$

*where $x \notin \lceil \Psi;\Gamma \vdash \Lambda;\forall y:M,\Xi \rceil$.

Fig. 1—The FORUM$^{⅋\multimap\forall}$ fragment of FORUM

$M \in \mathbb{M} \cup \{\epsilon_{\mathbb{M}}\}$ (the *focused method*), $\Lambda \in \mathbb{A}^\star$ (the *atomic list*) and $\Xi \in \mathbb{M}^\star$ (the *right linear part*). Instead of $(\Psi;\Gamma|\epsilon_{\mathbb{M}} \vdash \Lambda;\Xi)$ we shall write $(\Psi;\Gamma \vdash \Lambda;\Xi)$. In the following Ψ, Γ, Ξ and Λ shall stand for, respectively, sets, multisets and sequences of methods and sequences of atoms.

With $\Lambda \curlyvee \Lambda'$ we represent any sequence of atoms obtained by an ordered merge of Λ and Λ'.

We outline a sequent presentation of a fragment of the FORUM inference system. FORUM imposes a discipline on the non-deterministic bottom-up construction of proofs, thereby drastically reducing their search space. It turns out that FORUM is equivalent to linear logic, but proofs in FORUM are uniform (see [9]). Since FORUM is much closer to the computations we are interested in, it greatly helped us in finding the way to relate them to linear logic.
The inference system we shall use as an intermediate step from SMR to linear logic is FORUM$^{⅋\multimap\forall}$, meaning that ⅋, $\multimap$ and $\forall$ are the only logical constants this subsystem of FORUM deals with. FORUM$^{⅋\multimap\forall}$ is presented in fig. 1.

The link between FORUM$^{⅋\multimap\forall}$ and linear logic is established by the following proposition, which follows from the result in [8] and the cut-elimination theorem.

Theorem *A sequent* $(M|_1^h;\vdash;M)$ *has a proof in* FORUM$^{⅋\multimap\forall}$ *iff* $(!M_1 \multimap \cdots \multimap\, !M_h \multimap M)$ *has a proof in linear logic.*

3 Syntax and Operational Semantics of SMR

The first subsection deals with the syntax of goals, the second with their "precedence relation" semantics. In the third subsection SMR and its operational semantics are introduced.

3.1 Goals, Contexts and Goal Graphs

We build up the language of goals starting from the empty goal $\circ$ and the set of atoms $\mathbb{A}$, and freely composing with the two connectives $\diamond$ and $\triangleleft$. The connectives have to be thought of as associative and non-idempotent operators; moreover, $\diamond$ is commutative and $\triangleleft$ is not.

The empty goal $\circ$ behaves as a unity for $\diamond$ and $\triangleleft$, like *true* does for the classical logic connective *and*. In the translation from SMR into linear logic it shall be mapped to an atom of a certain class.

Let $\mathbb{A}_G = \mathbb{A} \cup \{\circ\}$. A generic element of $\mathbb{A}_G$ shall be denoted by A.

$\mathbb{G}$ is the least set such that: 1) $\mathbb{A}_G \subset \mathbb{G}$. 2) If $h \in \mathsf{N}_1$ and $G_1, \ldots, G_h \in \mathbb{G}$ then $\diamond(G|_1^h) \in \mathbb{G}$ and $\triangleleft(G|_1^h) \in \mathbb{G}$. A generic element of $\mathbb{G}$ is a *goal* and shall be denoted by G or H. $\diamond$ and $\triangleleft$ are, respectively, the *parallel* and *sequential connective*; goals of the form $\diamond(G|_1^h)$ and $\triangleleft(G|_1^h)$ are, respectively, *parallel* and *sequential goals*. A generic element of $\{\diamond, \triangleleft\}$ shall be denoted by c.

Let us extend the syntax of goals by allowing one or more *holes* _ to appear in place of atoms and of $\circ$. Then we have the set $\mathbb{K}$ of *contexts*, whose generic elements shall be denoted by K. An alternative notation for $c(K|_1^h)$ is $(K_1\, c \cdots c\, K_h)$.

Coordinates uniquely identify occurrences of atoms, empty goals and holes in a context.

Let $\mathbb{k} = \mathsf{N}_1^*$. A generic element of $\mathbb{k}$ is a *coordinate* and shall be denoted by κ.

Let $\mathbb{a} = \mathbb{A}_G \times \mathbb{k}$. A generic element of $\mathbb{a}$ is an *agent*.

Let $\square = \{\_\} \times \mathbb{k}$. A generic element of $\square$ is a *place*. Instead of $(\_, \kappa)$ we shall write $\_{\kappa}$. A generic element of $\mathbb{a} \cup \square$ shall be denoted by a.

As defined below, to every context is associated a hypergraph whose nodes are agents or places.

A *directed hypergraph* is a couple (N, H), where N is a finite set of *nodes* and $H \subseteq (\mathsf{P}(N) \setminus \{\varnothing\})^2$ is a set of *hyperarcs*.

A *context graph* is a directed hypergraph (N, H), where $N \subset \mathbb{a} \cup \square$. Let $\mathbb{Y}$ be the set of context graphs. A generic element of $\mathbb{Y}$ shall be denoted by Y. A context graph (N, H) such that $N \subset \mathbb{a}$ is a *goal graph*.

The "top" and "bottom" of a context graph are, respectively, the sets of agents and places which have no incoming and no outgoing hyperarc.

Define $\mathrm{top}, \mathrm{bot} \colon \mathbb{Y} \to \mathsf{P}(\mathbb{a} \cup \square)$ as $\mathrm{top}(N, H) = \{\, a \in N \mid \forall (N_1, N_2) \in H : a \notin N_2 \,\}$ and $\mathrm{bot}(N, H) = \{\, a \in N \mid \forall (N_1, N_2) \in H : a \notin N_1 \,\}$.

We now want to associate to every context a context graph which represents it.

Contexts are objects recursively made up from inner contexts in two possible ways: as a parallel or as a sequential composition. In the same way a context graph representing a context is made up from the context graphs representing inner contexts. Parallel composition leads to the simple union of the context

graphs; sequential composition introduces a hyperarc for every binary sequential composition.
The coordinate mechanism, which is embedded in the following tricky definition, provides for a constructive way of generating distinct nodes from distinct occurrences of the same atom, empty goal or hole, in a context.
Coordinates are assigned to atoms (or empty goals or holes) in the following way: the position of the atom in the context is a string which records, from left to right, the positions of the contexts which contain the atom, from the outer to the inner. The example which follows the definition shall hopefully clarify this.

For every $\kappa \in \mathbb{K}$ define $[\cdot]_\kappa : \mathbb{K} \to \mathbb{Y}$ as

$$[K]_\kappa = \begin{cases} (\{(K,\kappa)\},\varnothing) & \text{if } K \in \mathbb{A}_G \cup \{\_\} \\ (\bigcup_{i\in\mathbb{N}_1^h} N_i, \bigcup_{i\in\mathbb{N}_1^h} H_i) & \text{if } K = \diamond(K|_1^h) \text{ and } [K_i]_{\kappa\|i} = (N_i,H_i) \\ (\bigcup_{i\in\mathbb{N}_1^h} N_i, \bigcup_{i\in\mathbb{N}_1^h} H_i \cup \bigcup_{i\in\mathbb{N}_2^h}\{(\mathrm{bot}[K_{i-1}]_{\kappa\|(i-1)}, \mathrm{top}[K_i]_{\kappa\|i})\}) & \\ & \text{if } K = \triangleleft(K|_1^h) \text{ and } [K_i]_{\kappa\|i} = (N_i,H_i) \end{cases}.$$

We shall write $[K]$ instead of $[K]_\epsilon$. If $[K]_\kappa = (N,H)$, let $[K]^{\mathrm{N}}_\kappa = N$ and $[K]^{\mathrm{H}}_\kappa = H$.

The context $K = ((A_1 \diamond \_ \diamond \circ) \triangleleft ((\circ \triangleleft \_) \diamond A_6) \triangleleft A_7) \diamond (\_ \triangleleft (A_9 \diamond A_{10}))$, for example, yields

$[K] =$ 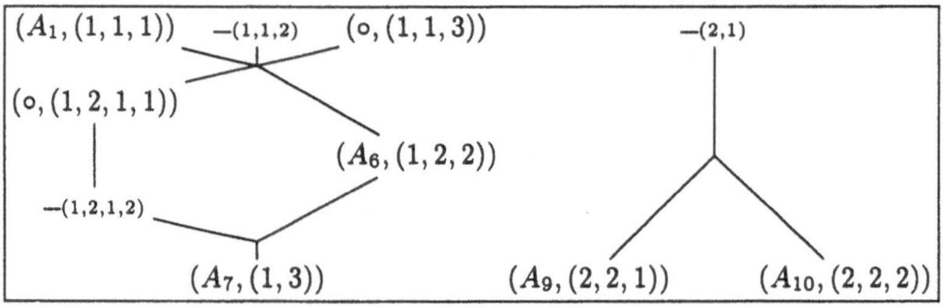

Sometimes coordinates shall not be shown.

We write $a \lessdot K$ or $\__\kappa \lessdot K$ to say that agent a or place $\__\kappa$ appears in the context graph $[K]$.

How well do context graphs represent contexts? The following proposition can be easily proved.

Proposition *For every $\kappa \in \mathbb{K}$ the function $[\cdot]_\kappa : \mathbb{K} \to [\mathbb{K}]_\kappa$ is bijective.*

Given $K, K' \in \mathbb{K}$, $K[\underline{K'}_\kappa]$ stands for K if $\__\kappa \not\lessdot K$ and for the context obtained by K replacing $\__\kappa$ with K' if $\__\kappa \lessdot K$. We shall write $K[\underline{K_1}_{\kappa_1}, \dots, \underline{K_h}_{\kappa_h}]$ instead of $K[\underline{K_1}_{\kappa_1}] \dots [\underline{K_h}_{\kappa_h}]$.

For example $((A_1 \diamond \_ \diamond \circ) \triangleleft ((\circ \triangleleft \_) \diamond A_6) \triangleleft A_7)[\underline{A_2}_{(1,2)}, \underline{\circ}_{(2,1,2)}] = (A_1 \diamond A_2 \diamond \circ) \triangleleft ((\circ \triangleleft \circ) \diamond A_6) \triangleleft A_7$.

3.2 Normalization of Goals

Somewhat orthogonal to the expansion of agents is a notion of normalization (for the terminology refer, for example, to [6]). We introduce a reduction system

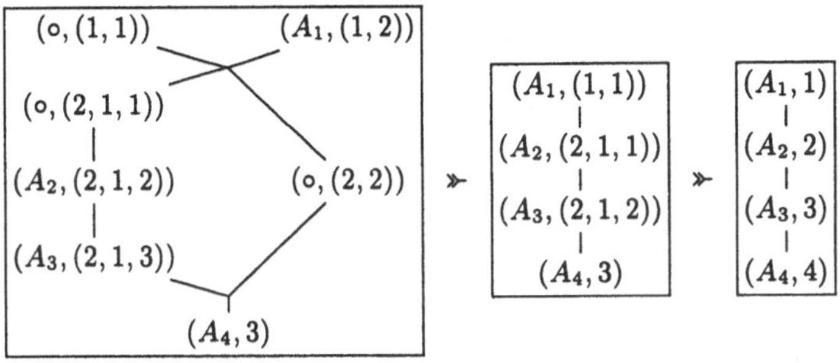

Fig. 2—Elimination of empty goals and of redundant syntax

for goals; intuitively (semantically) the reduction conserves the precedence relations among agents represented in their underlying directed hyperarcs. Since empty goals do not yield further expansions, they are discarded.

Actually there are two conceptually distinct subreductions we consider:

1) Empty goals are discarded while conserving precedence relations among other agents, as in $(\circ \diamond A_1) \triangleleft ((\circ \triangleleft A_2 \triangleleft A_3) \diamond \circ) \triangleleft A_4 \succ\!\!\succ (A_1) \triangleleft ((A_2 \triangleleft A_3)) \triangleleft A_4$. This corresponds to $\circ$ being a unity for $\diamond$ and $\triangleleft$. A reduction of this kind shall be written as $G \succ_{\circ} G'$.
2) Redundant syntax is eliminated, as in $(A_1) \triangleleft ((A_2 \triangleleft A_3)) \triangleleft A_4 \succ\!\!\succ A_1 \triangleleft A_2 \triangleleft A_3 \triangleleft A_4$. It is simply a statement of the associativity of $\diamond$ and $\triangleleft$. A reduction of this kind shall be written as $G \succ_{S} G'$.

The $\succ_{S}$ reduction prevents syntax to go too far away from our semantic requirements, which are better expressed by (the graphical representation of) goal graphs. Notice that the second reduction conserves the shape of the hypergraphs, as could be shown formally. Fig. 2 represents the examples given above.

Define $\succ_{G} = \succ_{\circ} \cup \succ_{S}$ and let $\succ\!\!\succ$ be the transitive reflexive closure of $\succ_{G}$. It can be shown with standard techniques that $\succ\!\!\succ$ is terminating and confluent. So the normal form of a goal G under $\succ\!\!\succ$ is unique, and shall be indicated with nf G.

3.3 Clauses and Operational Semantics of SMR

SMR consists of three components: a set of programs, the set of goals we already defined and a transition relation which nondeterministically transforms goals into goals.

A program is a finite set of clauses. Each clause specifies the synchronous rewriting of some atoms into the same number of goals. Rewriting takes place in the context of a larger goal, in which the rewritten atoms, considered as a multiset, are unifiable with the head of the clause, again considered as a multiset.

The clause specifies also which goal takes the place of which atom (matching one of the atoms in its head), and the usual logic programming mechanism of in-

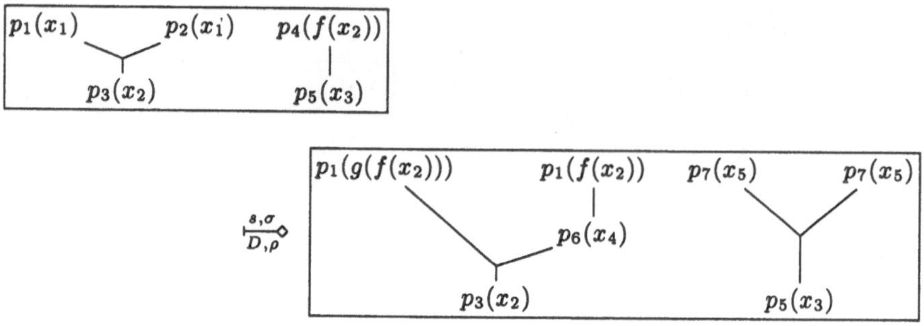

$s = \{(p_2(x_1), (1,1,2)) \mapsto 2,\ (p_4(f(x_2)), (2,1)) \mapsto 1\}$ $\quad \sigma = [g(f(x_2))/x_1, f(x_2)/y]$
$D = (p_4(y), p_2(g(y)) \leftarrow p_7(x_3) \diamond p_7(x_3), p_1(y) \triangleleft p_6(x_4))$ $\quad \rho = [x_5/x_3]$

Fig. 3—Example of resolution (coordinates are not shown)

stantiation with the unifier takes place. We do not insist on the unifiers being mgu's, though this special case can easily be accommodated in our setting.

Let $\mathbb{D} = \{\, (A|_1^h \leftarrow G|_1^h) \mid h \in \mathsf{N}_1,\ A_1, \ldots, A_h \in \mathbb{A}_{\mathsf{G}} \setminus \{\circ\},\ G_1, \ldots, G_h \in \mathbb{G} \,\}$. A generic element of $\mathbb{D}$ is a (*distributed*) *clause* and shall be denoted by D.

Let $\mathbb{P} = \mathsf{P}_{\mathsf{F}}(\mathbb{D})$. A generic element of $\mathbb{P}$ is a *program* and shall be denoted by P.

The following definition needs some explanation. We want to define the set $\mathfrak{s}$ of "selections." Remember that the top of a context goal, and, by extension, of a goal, is the set of agents in that goal which are preceded by no other agent. Every selection associates a unique index in N_1^h to a subset of cardinality h of the top. This is in order to associate to every atom in the head of a clause $A|_1^h \leftarrow G|_1^h$ a corresponding, selected agent in the goal to be rewritten.

Given $G \in \mathbb{G}$ and $h \in \mathsf{N}_1^{|\mathrm{top}\,[G]|}$, let
$\mathfrak{s}_{G,h} = \{\, s \mid s\colon T \to \mathsf{N}_1^h,\ T \subseteq \mathrm{top}\,[G],\ s \text{ is bijective} \,\}$ and
$\mathfrak{s} = \bigcup_{G \in \mathbb{G}} \bigcup_{h \in \mathsf{N}_1^{|\mathrm{top}\,[G]|}} \mathfrak{s}_{G,h}$. A generic element of $\mathfrak{s}$ is a *selection* and shall be denoted by s.

We now define the mechanism by which goals evolve by the action of clauses, in resolutions. It can be informally explained this way, given a goal G and a clause D:
1) Let s be a selection of h agents in the top of G.
2) Rename apart the variables in D.
3) Let σ be a unifier between the selected atoms and the atoms in the head of D (corresponding to them through s).
4) Substitute, in G, the selected atoms with the goals in the body of D which correspond to them through s and the correspondence implied by their order in D. Then apply σ to the goal obtained, and have G'.
Fig. 3 shows an example. Here is the formal definition.

The relation $\mapsto\!\diamond \subset \mathbb{G}^2 \times \mathbbm{s} \times \mathbb{D} \times \mathbbm{p} \times \boldsymbol{\sigma}$ is defined as follows: $(G, G', s, D, \rho, \sigma) \in \mapsto\!\diamond$, where $D = (A|_1^h \leftarrow G|_1^h)$, iff:
1) $s \in \mathbbm{s}_{G,h}$.
2) $\lceil D\rho \rceil \cap \lceil G \rceil = \varnothing$.
3) $\forall (A, \kappa) \in \operatorname{dom} s : A\sigma = A_{s(A,\kappa)}\rho\sigma$.
4) Let K be such that $G = K\left[\underline{A'_1}_{\kappa_1}, \ldots, \underline{A'_h}_{\kappa_h}\right]$, where $\operatorname{dom} s = \{\, (A'_i, \kappa_i) \mid i \in \mathsf{N}_1^h \,\}$; then $G' = \sigma\left(K\left[\underline{G_{s(A'_1,\kappa_1)}\rho}_{\kappa_1}, \ldots, \underline{G_{s(A'_h,\kappa_h)}\rho}_{\kappa_h}\right]\right)$; goals $G_{s(A'_i,\kappa_i)}\rho$ are called *replacing* goals.
Given $P \in \mathbb{P}$ define the relation $\underset{P}{\mapsto\!\diamond} \subset \mathbb{G}^2 \times \boldsymbol{\sigma}$ as $\underset{P}{\mapsto\!\diamond} = \{\, (G, G', \sigma) \mid \exists s \in \mathbbm{s} : \exists D \in P : \exists \rho \in \mathbbm{p} : (G, G', s, D, \rho, \sigma) \in \mapsto\!\diamond \,\}$.
Instead of $(G, G', s, D, \rho, \sigma) \in \mapsto\!\diamond$ and $(G, G', \sigma) \in \underset{P}{\mapsto\!\diamond}$ we shall write $G \overset{s,\sigma}{\underset{D,\rho}{\mapsto\!\diamond}} G'$ and $G \overset{\sigma}{\underset{P}{\mapsto\!\diamond}} G'$.

A goal may evolve either because of a resolution or because of a reduction. The relation $\succ_{\mathsf{G}}$ is not suitable to be translated into FORUM, in particular the problem is with the subreduction $\succ_{\mathsf{O}}$.
Then we introduce the less declarative reduction relation $\succ_{\mathsf{T}}$, where $\succ_{\mathsf{O}}$ is replaced by $\succ_{\mathsf{H}}$. The $\succ_{\mathsf{H}}$ reduction allows collapsing empty goals appearing in the top only. This mechanism can be faithfully represented in $\text{FORUM}^{\wp\multimap\forall}$, whereas with $\succ_{\mathsf{O}}$ this is not possible. With "successful" computations (*i.e.* computations ending in an empty goal) this only has the effect of delaying reduction of non-top empty goals until they eventually reach the top.

The relation $\succ_{\mathsf{H}} \subset \mathbb{G}^2$ is the least set such that:

1) If $G = K\lfloor \underline{\diamond(\overbrace{\circ, \ldots, \circ}^{h})}_{\kappa} \rfloor$, $h \in \mathsf{N}_1$, $\_{\kappa} \lessdot K$ and $\{\, (\circ, \kappa \parallel i) \mid i \in \mathsf{N}_1^h \,\} \subseteq \operatorname{top}\lfloor G \rfloor$ then $G \succ_{\mathsf{H}} K\lfloor \underline{\circ}_{\kappa} \rfloor$.
2) If $G = K[\underline{\triangleleft(\circ, H|_1^h)}_{\kappa}]$, $h \in \mathsf{N}_1$, $\_{\kappa} \lessdot K$ and $(\circ, \kappa \parallel 1) \in \operatorname{top}[G]$ then $G \succ_{\mathsf{H}} K[\underline{\triangleleft(H|_1^h)}_{\kappa}]$.
Let $\succ_{\mathsf{T}} = \succ_{\mathsf{H}} \cup \succ_{\mathsf{S}}$. For $r \in \{\mathsf{O}, \mathsf{H}, \mathsf{S}, \mathsf{G}, \mathsf{T}\}$ let the relation $\succ_r \subset \mathbb{G}^2 \times \boldsymbol{\sigma}$ be defined as $\{\, (G, G', []) \mid G \succ_r G' \,\}$. Clearly $\succ_{\mathsf{T}} \subset \succ_{\mathsf{G}}$.
For every $P \in \mathbb{P}$ and for $r \in \{\mathsf{G}, \mathsf{T}\}$, define the relation $\underset{P}{\to\!\diamond}_r \subset \mathbb{G}^2 \times \boldsymbol{\sigma}$ as $\underset{P}{\to\!\diamond}_r = \succ_r \cup \underset{P}{\mapsto\!\diamond}$. Instead of $(G, G', \sigma) \in \underset{P}{\to\!\diamond}_r$ we shall write $G \overset{\sigma}{\underset{P}{\to\!\diamond}}_r G'$. Clearly $\underset{P}{\to\!\diamond}_{\mathsf{T}} \subset \underset{P}{\to\!\diamond}_{\mathsf{G}}$.

Let SMR be the triple $(\mathbb{P}, \mathbb{G}, \{\, \underset{P}{\to\!\diamond}_{\mathsf{G}} \mid P \in \mathbb{P} \,\})$.

Let $P \in \mathbb{P}$, $r \in \{\mathsf{G}, \mathsf{T}\}$ and $h \in \mathsf{N}$: an object of the form $G_0 \overset{\sigma_1}{\underset{P}{\to\!\diamond}}_r \cdots \overset{\sigma_h}{\underset{P}{\to\!\diamond}}_r G_h$ is a *(r-)computation (by P)*, if $G_h = \circ$ it is a *successful* r-computation *of* G_0 *yielding* $\sigma_1 \cdots \sigma_h$; let $\mathbb{C}_P^r$ be the set of r-computations by P. A generic element of $\mathbb{C}_P^r$ shall be denoted by C.
Let $h \in \mathsf{N}_1$ and $C = (G_0 \overset{\sigma_1}{\underset{P}{\to\!\diamond}}_r \cdots \overset{\sigma_h}{\underset{P}{\to\!\diamond}}_r G_h)$: for $k \in \mathsf{N}_1^h$ every object $G_{k-1} \overset{\sigma_k}{\underset{P}{\to\!\diamond}}_r G_k$ is the kth *step* in C; if $G_{k-1} \succ_r G_k$ it is a *reduction* step ($\succ_r$-*step*), if $G_{k-1} \overset{\sigma}{\underset{P}{\mapsto\!\diamond}} G_k$ it is a *resolution* step ($\mapsto\!\diamond$-*step*).
Define $|\cdot|_{\mathsf{R}} : \mathbb{C}_P^r \to \mathsf{N}$ so that $|C|_{\mathsf{R}}$ is the number of $\mapsto\!\diamond$-steps in C.
Define the relation $\underset{P}{\to\!\diamond\!\diamond}_r \subset \mathbb{G}^2 \times \boldsymbol{\sigma}$ as $\underset{P}{\to\!\diamond\!\diamond}_r = \{\, (G_0, G_h, \sigma_1 \cdots \sigma_h) \mid (G_0 \overset{\sigma_1}{\underset{P}{\to\!\diamond}}_r \cdots \overset{\sigma_h}{\underset{P}{\to\!\diamond}}_r G_h) \in \mathbb{C}_P^r \,\}$. Instead of $(G, G', \sigma) \in \underset{P}{\to\!\diamond\!\diamond}_r$ we shall write $G \overset{\sigma}{\underset{P}{\to\!\diamond\!\diamond}}_r G'$.

The following two theorems are crucial to show, respectively, the correctness and the completeness of the translation from SMR into FORUM. The first is proved by transforming a successful G-computation into an equivalent T-computation, moving to the right (*i.e.* delaying) until possible all occurrences of $\succ_{\mathrm{O}}$-steps and, recursively, all $\succ_{\mathrm{S}}$-steps which depend on each other. The second amounts to an inductive construction of the desired computation.

3.3.1 **Theorem** *If C is a successful* G*-computation of G yielding σ there exists a successful* T*-computation C' of G yielding σ such that $|C|_{\mathrm{R}} = |C'|_{\mathrm{R}}$.*

3.3.2 **Theorem** *If $G \overset{\sigma}{\underset{P}{\Rightarrow}}_{\mathrm{G}} \circ$ then for every $G' \in \{\, H \mid \mathrm{nf}\, H = \mathrm{nf}\, G \,\}$ it holds $G' \overset{\sigma}{\underset{P}{\Rightarrow}}_{\mathrm{G}} \circ$.*

4 SMR and Linear Logic

We first present the translation of SMR into FORUM$^{⅋-\circ\forall}$, then we prove that it is correct and complete wrt linear logic.

4.1 Translation of SMR into FORUM$^{⅋-\circ\forall}$

Let us augment the set of variables by a denumerable set $\mathbb{\pi}$ of *process variables*, which are not allowed to appear in SMR atoms. Agents, *i.e.* atoms decorated by a coordinate, are translated into atoms. The terms inside are left untouched, and the relative position in the goal (coordinate) yields a process variable, which is appended to the resulting atom. Since atoms in SMR do not contain process variables, name clashes are avoided. The empty goal translates into a special atom of the kind $\square\pi$.

Let $\square$ be a distinguished predicate of arity 1.
The function $[\![\cdot]\!] : \mathbb{p} \to \mathbb{p} \setminus \{\square\}$ is chosen such that it is one-one and it holds $\mathrm{ar}[\![p]\!] = \mathrm{ar}\, p + 1$.

While π stands for a generic process variable, object process variables are $\psi_0, \psi_1, \ldots$. Given a coordinate κ, with $\underline{\kappa}$ we shall indicate the unique natural number associated to κ by some bijective function between coordinates and naturals. Given a denumerable set S, we shall indicate with $\langle S\rangle$ the sequence obtained by S by ordering its elements according to a total order of choice. This is alternative to using equivalence classes in the definitions to come, when order of elements is of no importance.

Define $[\![\cdot]\!] : \mathbb{a} \to \mathbb{A}$ as $[\![A, \kappa]\!] = \begin{cases} \square(\psi_{\underline{\kappa}}) & \text{if } A = \circ \\ [\![p]\!](t|_1^h, \psi_{\underline{\kappa}}) & \text{if } A = p(t|_1^h) \end{cases}$. We shall write $A\psi_{\underline{\kappa}}$ instead of $[\![A, \kappa]\!]$ and $\square\pi$ instead of $\square(\pi)$.

We shall call atoms obtained by the translation *agents*, too. In particular, we shall refer to atoms $\square\pi$ as *success agents*.

Let us call *elementary method* a method of the form $(A_1 \parr \cdots \parr A_h) \multimap (A'_1 \parr \cdots \parr A'_{h'})$. To every hyperarc in a goal graph corresponds an elementary method in the translation of the goal relative to the hypergraph. Two auxiliary functions are helpful.

Define $\mathrm{sa}\colon \mathbb{o} \to \mathbb{A}$ as $\mathrm{sa}(A,\kappa) = \Box\psi_{\underline{\kappa}}$.
Define $\mathrm{hm}\colon (\mathsf{P}_{\mathsf{F}}(\mathbb{o}) \setminus \{\varnothing\})^2 \to \mathbb{M}$ as $\mathrm{hm}(N_1, N_2) = (\bigparr\langle [\![N_2]\!]\rangle \multimap \bigparr\langle \mathrm{sa}\, N_1\rangle)$.

Given a goal G and a coordinate κ, the translation $[\![G]\!]_\kappa$ is a method $M_1 \multimap \cdots \multimap M_h \multimap T$, where $M_1, \ldots, M_h$ are the elementary methods obtained by the hyperarcs in $[G]_\kappa$, and T is the translation of $\mathrm{top}[G]_\kappa$. The structure of the hypergraph is kept by process variables (identity of agents), by the $\parr$ connective (parallelism among agents) and by the $\multimap$ connectives which appear in $M_1, \ldots, M_h$ (directionality of hyperarcs). The outer $\multimap$'s do not play a role wrt the structure of the hypergraph. They are used to "load" the linear left context with the structure. Notice that the order $M_1, \ldots, M_h$ is not important.

For every $\kappa \in \mathbb{k}$ define $[\![\cdot]\!]_\kappa\colon \mathbb{G} \to \mathbb{M}$ as $[\![G]\!]_\kappa = (M_1 \multimap \cdots \multimap M_h \multimap \bigparr\langle [\![\mathrm{top}[G]_\kappa]\!]\rangle)$, where $(M|_1^h) = \langle \mathrm{hm}[G]_\kappa^{\mathsf{H}}\rangle$.

Define $[\![\cdot]\!]\colon \mathbb{D} \to \mathbb{M}$ as
$[\![A|_1^h \leftarrow G|_1^h]\!] = \forall\langle\lceil A|_1^h \leftarrow G|_1^h\rceil\rangle : \forall_{i\in\mathsf{N}_1^h}\, \psi_{\underline{i}} : (\bigparr_{i\in\mathsf{N}_1^h} [\![G_i]\!]'_i \multimap \bigparr_{i\in\mathsf{N}_1^h} A_i\psi_{\underline{i}})$,
where, for every $\kappa \in \mathbb{k}$, $[\![\cdot]\!]'_\kappa\colon \mathbb{G} \to \mathbb{M}$ is defined as

$$[\![G]\!]'_\kappa = \begin{cases} G\psi_{\underline{\kappa}} & \text{if } G \in \mathbb{A}_{\mathsf{G}} \\ \forall\langle\{\psi_{\underline{\kappa}} \mid (A,\kappa) \in [G]_\kappa^{\mathsf{N}}\}\rangle : ((\Box\psi_{\underline{\kappa}} \multimap \bigparr\langle \mathrm{sa}\,\mathrm{bot}[G]_\kappa\rangle) \multimap [\![G]\!]_\kappa) . \\ \qquad \text{if } G \notin \mathbb{A}_{\mathsf{G}} \end{cases}$$

4.2 Correctness and Completeness of SMR

Let $\mathsf{R}\,\dfrac{\Psi;\Gamma \vdash \Lambda; M}{\Psi;\Gamma \mid M \multimap \bigparr(M|_1^h) \vdash \Lambda \curlyvee M_1 \curlyvee \cdots \curlyvee M_h;}$ be an inference rule, called *resolution*, defined as a shorthand of the following derivation:

$$\multimap_{\mathsf{L}}\,\dfrac{\Psi;\Gamma \vdash \Lambda; M \qquad \parr_{\mathsf{L}}\,\dfrac{\mathsf{I}\,\dfrac{}{\Psi;\mid M_1 \vdash M_1;} \quad \vdots \quad \mathsf{I}\,\dfrac{}{\Psi;\mid M_h \vdash M_h;}}{\Psi;\mid \bigparr(M|_1^h) \vdash M_1 \curlyvee \cdots \curlyvee M_h;}}{\Psi;\Gamma \mid M \multimap \bigparr(M|_1^h) \vdash \Lambda \curlyvee M_1 \curlyvee \cdots \curlyvee M_h;}.$$

Given $P \in \mathbb{P}$, let $(\!|G|\!)_P = \{\, ([\![P]\!]; \bigparr\langle \mathrm{sa}\,\mathrm{bot}[G]\rangle, \mathrm{hm}[G]^{\mathsf{H}} \vdash A|_1^h;)\rho \mid \rho \in \mathbb{p}_\pi,\ \{A|_1^h\} = [\![\mathrm{top}[G]]\!] \,\}$, where $\mathbb{p}_\pi$ is the set of renaming substitutions on the set of process variables. An element of $(\!|G|\!)_P$ is a *representation* of G.

The following correctness theorem establishes a first connection between SMR and Forum.

Theorem *If $G \overset{\sigma}{\underset{P}{\twoheadrightarrow}}_{\mathsf{G}} \circ$ then for every $\Sigma \in (\!|G\sigma|\!)_P$ there exists a proof Π of* Forum$^{\parr\multimap\forall}$ *with conclusion Σ.*

Sketch of proof Let C be a successful G-computation of G yielding σ. By theorem 3.3.1 there exists a successful T-computation C' of G yielding σ. The proof is by induction on $|C'|_{\mathsf{R}} = |C|_{\mathsf{R}}$. From C' we shall build Π from bottom to top.

.1 If $|C'|_{\mathsf{R}} = 0$ then $C' = (G \succ_{\mathsf{T}} \cdots \succ_{\mathsf{T}} \circ)$. To every $\succ_{\mathsf{S}}$-step corresponds a renaming of some process variables. To every $\succ_{\mathsf{H}}$-step corresponds a sequence, from bottom to top, of a $\mathsf{D_L}$ rule (the focused method is one of the methods corresponding to a "top" hyperarc in $[G]^{\mathsf{H}}$) and of a resolution rule followed by some applications of $\parr_{\mathsf{R}}$ and L. This reduces the problem to finding a proof for $\Sigma' \in (\!|G'|\!)_P$, if $G \succ_{\mathsf{H}} G'$. Proceed inductively. The final step consists in finding a proof of $([\![P]\!]; \parr\langle \mathrm{sa\,bot}[G''] \rangle \vdash \langle \mathrm{sa\,bot}[G''] \rangle;)\rho$, which is made up of $\parr_{\mathsf{L}}$ and I rules, as the right branch in R is; here ρ is a renaming substitution on process variables.

.2 If $|C'|_{\mathsf{R}} > 0$ then $C' = (G \succ_{\mathsf{T}} \cdots \succ_{\mathsf{T}} G' \overset{\sigma'}{\underset{P}{\mapsto\!\diamond}} G'' \underset{P}{\to\!\diamond}_{\mathsf{T}} \cdots \underset{P}{\to\!\diamond}_{\mathsf{T}} \circ)$. The first $\succ_{\mathsf{T}}$-steps are dealt with as in point 1, without the final step. We have to prove that for every $\Sigma' \in (\!|G'\sigma|\!)_P$ there exists a proof Π' of $\mathrm{F{\small ORUM}}^{\parr\multimap\forall}$ such that its conclusion is Σ'. A derivation Δ such that its conclusion is $\Sigma' \in (\!|G'\sigma|\!)_P$, relative to the step $G' \overset{s,\sigma'}{\underset{D,\rho}{\mapsto\!\diamond}} G''$, can be built bottom-up as a sequence of the following rules: $\mathsf{D_C}$ ($[\![D]\!]$ is focused), a sequence of $\forall_{\mathsf{L}}$ (D is instantiated by $\rho\sigma'$), R (the right linear part contains a representation of the replacing goals), then, by means of $\parr_{\mathsf{R}}$, $\forall_{\mathsf{R}}$, $\multimap_{\mathsf{R}}$ and L rules, the right linear part is unloaded and the left linear part is loaded with the representations of the replacing goals. In these representations process variables are either unified with previous "top" process variables (replacing goals are joined to the goal) or created unique by $\forall_{\mathsf{R}}$ (they are relative to inner coordinates in the replacing goals). This guarantees the correspondence between the representation of the new goal and its goal graph.
We have that the only premise of Δ is $\Sigma'' \in (\!|G'''|\!)_P$ and $G''' \ggcurly G''$. Since $G''' \ggcurly G'' \overset{\sigma''}{\underset{P}{\to\!\!\diamond}}_{\mathsf{G}} \circ$, there is, by the induction hypothesis, a proof Π'' such that its conclusion is $\Sigma''\sigma''$, and the theorem is proved.

The other direction of the connection between SMR and FORUM is stated by the following completeness theorem.

Theorem *If for $\Sigma \in (\!|G\sigma|\!)_P$ there is a proof Π of* $\mathrm{F{\small ORUM}}^{\parr\multimap\forall}$ *such that its conclusion is Σ then $G \overset{\sigma}{\underset{P}{\to\!\!\diamond}}_{\mathsf{G}} \circ$.*

Sketch of proof Observe that the application of the rules L, $\parr_{\mathsf{R}}$, $\multimap_{\mathsf{R}}$ and $\forall_{\mathsf{R}}$ is deterministic, in the sense that in a bottom-up construction of a proof every step is uniquely determined. The only choice left is that of a new variable in $\forall_{\mathsf{R}}$: since we build up the proof modulo renaming of process variables, this is not important. We shall show that every possible choice of rules $\mathsf{D_L}$, $\mathsf{D_C}$, $\parr_{\mathsf{L}}$, $\multimap_{\mathsf{L}}$ and $\forall_{\mathsf{L}}$ leads to a proof only if they yield applications of the R rule.
If the $\mathsf{D_C}$ rule is chosen, the focused method becomes the translation $[\![D]\!]$ of a clause D. Then some applications of $\forall_{\mathsf{L}}$ are compulsory. After that, the R scheme is the only possible: it can only be part of a proof if the variables chosen in the $\forall_{\mathsf{L}}$ inferences correspond to a $\mapsto\!\diamond$-step, *i.e.* to a resolution

in SMR. Moreover, the clause must be applicable to the top of the goal, represented in the atoms list, then the $\mathsf{D_L}$ inference should have been wise. After R, all inferences are deterministic again. In this way we have a derivation Δ such that its conclusion is $\Sigma' \in (\!|G\sigma'|\!)_P$ and its only premise is $\Sigma'' \in (\!|G'|\!)_P$. Now it is easy to show that $G' \succ\!\!\!\succ G''$ and $G \overset{\sigma'}{\underset{P}{\mapsto\!\diamond}} G''$. Notice that, by theorem 3.3.2, if $G' \overset{\sigma''}{\underset{P}{\twoheadrightarrow\!\diamond}}_{\mathsf{G}} \circ$ then $G'' \overset{\sigma''}{\underset{P}{\twoheadrightarrow\!\diamond}}_{\mathsf{G}} \circ$.
If the $\mathsf{D_L}$ rule is chosen, the focused method becomes an elementary method relative to a hyperarc. The R scheme must immediately follow: it can only lead to a proof if the elementary method is applicable to the top, *i.e.* the atoms list. This is only possible if, in the top, one or more empty goals appear suitable for a $\succ_{\mathsf{H}}$-step. The exact matching of process variables, *i.e.* coordinates in the goal graph, is ensured both by the translation and the $\forall_{\mathsf{R}}$ rule. Then we obtain a derivation Δ such that its conclusion is $\Sigma' \in (\!|G|\!)_P$ and its only premise is Σ'', where $\Sigma'' \in (\!|G'|\!)_P$ and $G \succ_{\mathsf{T}} G'$.
By considering proofs modulo renaming of process variables, the proof of the theorem is easily obtained by induction on the number of the R rule applications in Π.

Then we can prove the result which tightly links SMR and linear logic:

Theorem $G \overset{\sigma}{\underset{P}{\twoheadrightarrow\!\diamond}}_{\mathsf{G}} \circ$ *iff there is a proof for* $(!M_1 \multimap \cdots \multimap !M_h \multimap \parr\langle \mathrm{sa\,bot}[G]\rangle \multimap M'_1 \multimap \cdots \multimap M'_k \multimap \parr\langle [\![\mathrm{top}[G]]\!]\rangle)$ *in linear logic, where* $(M|_1^h) = \langle [\![P]\!]\rangle$ *and* $(M'|_1^k) = \langle \mathrm{hm}[G]^{\mathsf{H}}\rangle$.

5 Conclusions

We obtained both a declarative and operational understanding of sequencing by associating to every task a couple of statements: 1) that the task i has to be performed by an agent (say a_i) and 2) that when the task is accomplished a signal ($\circ_i$) is issued. The above treatment of sequentiality clearly encompasses paradigms more general than SMR. SMR by itself is a powerful language, as many examples show [10, 5]. We think also that SMR and its methodology are worthy as specification tools, and we are currently investigating their use for the specification of GAMMA [3] and other formalisms.

The translation makes use of the full $\parr$–$\multimap\forall$ fragment of linear logic, thus making full *logical* use of these connectives. This is opposed to, for example, classical logic programming, in which $\Rightarrow$ and $\forall$ are only used in left rules. An important point is that all structural information in SMR goes into the logic, with no need to resort to trickeries with terms. We are also pleased by the correspondence between parts in the sequences of FORUM and our framework: the program in the classical part, the structure of the goal in the left linear part and the top of the goal in the atomic list. The translation is very conservative wrt computational complexity, and FORUM guarantees good operational properties.

If we are satisfied with the translation of SMR, we certainly are not with its *logic*. We think that to fully bring sequentiality to the rank of logic some new logic with a non-commutative connective, together with commutative ones, has to be studied. At least one attempt in this direction exists, pomset logic [12], but until now this logic lacks either a cut-elimination theorem or, equivalently, a sequentialization theorem for its proof nets. Our future work shall go in the direction of investigating that logic with the aim to bring the concept of abstract logic programming [9] in a non-commutative setting, too.

References

[1] J.-M. Andreoli. Logic programming with focusing proofs in linear logic. *Journal of Logic and Computation*, 2(3):297–347, 1992.

[2] J.-M. Andreoli and R. Pareschi. Linear Objects: Logical processes with built-in inheritance. *New Generation Computing*, 9:445–473, 1991.

[3] J.-P. Banâtre and D. Le Métayer. The Gamma model and its discipline of programming. *Science of Computer Programming*, 15(1):55–77, Nov. 1990.

[4] J.-Y. Girard. Linear logic. *Theoretical Computer Science*, 50:1–102, 1987.

[5] A. Guglielmi. Concurrency and plan generation in a logic programming language with a sequential operator. In P. Van Hentenryck, editor, *Logic Programming, 11th International Conference, S. Margherita Ligure, Italy*, pages 240–254. The MIT Press, 1994.

[6] J. W. Klop. Term rewriting systems. In S. Abramsky, D. Gabbay, and T. Maibaum, editors, *Handbook of Logic in Computer Science*, volume 2, pages 1–116. Oxford University Press, 1992.

[7] D. Miller. The π-calculus as a theory in linear logic: Preliminary results. In E. Lamma and P. Mello, editors, *1992 Workshop on Extensions to Logic Programming*, volume 660 of *Lecture Notes in Computer Science*, pages 242–265. Springer-Verlag, 1993.

[8] D. Miller. A multiple-conclusion meta-logic. In S. Abramsky, editor, *Ninth Annual IEEE Symposium on Logic in Computer Science*, pages 272–281, Paris, July 1994.

[9] D. Miller, G. Nadathur, F. Pfenning, and A. Scedrov. Uniform proofs as a foundation for logic programming. *Annals of Pure and Applied Logic*, 51:125–157, 1991.

[10] L. Monteiro. Distributed logic: A logical system for specifying concurrency. Technical Report CIUNL-5/81, Departamento de Informática, Universidade Nova de Lisboa, 1981.

[11] L. Monteiro. Distributed logic: A theory of distributed programming in logic. Technical report, Departamento de Informática, Universidade Nova de Lisboa, 1986.

[12] C. Retoré. Pomset logic. Available by anonymous ftp from cma.cma.fr, Dec. 1993.

Linear Space Algorithm for On-line Detection of Global Predicates*

Roland Jégou
Centre SIMADE, E.N.S. des Mines de Saint-Étienne
Saint-Étienne, France

Raoul Medina, Lhouari Nourine
LIRMM, Université Montpellier II - CNRS UMR C09928
Montpellier, France

Abstract

A fundamental problem in debugging and monitoring distributed computations is to detect whether a state of the system satisfies some predicate. Cooper and Marzullo defined this problem as $Possibly(\Phi)$.

This paper presents the first on-line algorithm using linear space which solves this problem in the general case, improving all existing algorithms both in time and space. It is particularly interesting for the detection of $Possibly(\Phi)$ on potentially infinite computations. To our knowledge, it is also the only algorithm of detection which do not use vectors of timestamps.

The presented algorithm is based on structural properties of the consistent cuts lattice, leading to a new structure which seems promising for the study distributed computations: the consistent cuts tree.

1 Introduction

In this paper we study the detection of $Possibly(\Phi)$ defined by Cooper and Marzullo [7] as follows: $Possibly(\Phi)$ holds on a distributed computation if there exists at least one observation — i.e. a possible interleaving of the events — which reveals the satisfaction of Φ. The problem of detecting $Possibly(\Phi)$ in debugging distributed programs has received great attention these last years [7, 1, 6, 8, 10, 14]. These properties give us informations about errors or good behaviors of an execution [7].

We deal here, with the problem of detecting $Possibly(\Phi)$ where Φ is an arbitrary global predicate. By arbitrary global predicate we mean that no hypothesis is made on the nature of the predicate. The problem consists in detecting if Φ holds on a global state of the execution.

Analyzing a single observation may be sufficient to detect particular predicates such as stable [13] or regular [6] ones, or such as conjunction of local predicates [11, 21]. Many algorithms have been developped in the litterature concerning particular global predicates (eg. conjunctive form predicates [10, 11]

*This work is partially supported by the Région Rhône-Alpes project "Modélisation et Algorithmes massivement parallèles pour les problèmes industriels"

and stable predicates [13]). These algorithms are polynomial since they do not generate all global states: they work directly upon the causal order. The major drawback of these algorithms is that each one is specific to a particular predicate.

Unfortunately, if no hypothesis is made on the nature of the predicate, it may be necessary to analyze all possible observations of the distributed computation. A simple way to achieve this is to consider all the global states. Cooper and Marzullo [7] have given an off-line algorithm for arbitrary global predicates which uses an exponential space in the number of processes. Recently Alagar and Venkatesan [1] have presented an off-line algorithm to detect $Possibly(\Phi)$ which uses $O(n * |E|)$ space memory — where n is the number of processes and $|E|$ is the number of events that have occured in the computation. Both algorithms have an $O(n^3 * i(P))$ time complexity — where $i(P)$ is the total number of global states which may be exponential in the number of processes. Diehl et al [8] and Jard et al [14] have proposed on-line algorithms which runs in $O(n * i(P) + n * |E|^2)$ steps and have exponential space complexities since they store the entire lattice.

Since we make no assumption on the nature of the predicate, we will have to check all the global states of the distributed computation. Thus, the algorithm will be exponential in time complexity. Since time complexity cannot be dragstically reduced in the general case, we focus on the space complexity. We obtain a linear space algorithm. The main advantage of our algorithm is that it is "on–line" and thus may be used to detect $Possibly(\Phi)$ on potentially infinite computations — and this is strengthened by the space complexity used by the algorithm, making it practicable on real applications. Moreover, we improve the time complexity for the detection of $Possibly(\Phi)$ since our algorithm runs in $O(\Delta * i(P) + |U| * log(\Delta))$ where U is the set of arcs in the causality order P, Δ is the maximum *indegree* in P and $i(P)$ is the number of global states of P.

This paper is organized as follows. In Sect. 2 we define the model and the notations we use. We then show the link existing between $Possibly(\Phi)$ detection and the generation of the ideals of an order. The algorithm presented is based on a particular spanning tree of the ideal lattice introduced in [12, 20] and called the *Ideal Tree* or the *Consistent Cuts Tree* of P. This tree and its properties are presented in Sect. 3. It has been already used in distributed system and ordered sets theory [18, 19]. We think that the ideal tree is a pertinent structure to study the ideals of partial orders and thus global states of distributed executions. Indeed it takes less space than the whole lattice and has good algorithmic properties.

In Sect. 4 we present the first on-line algorithm which uses $O(|E| + |U|)$ space and $O(\Delta * i(P) + |U| * log(\Delta))$ time complexity.

Finally, in Sect. 5, we discuss about the detection of $Possibly(\Phi)$ on potentially infinite computations using our algorithm. We show how space storage may be saved for such computations.

2 Model and Notations

A distributed system is a collection $P_1 \cdots P_n$ of processes communicating solely via messages which may suffer from unpredictable delays. We assume that processes do not exhibit faulty behavior. There is no common global time and thus no process can have an instantaneous view of a global state of the system.

A distributed computation is an application running on a distributed system. Each process P_i has a local sequential algorithm determined by internal events and sending or receipt of messages. The local state of a process P_i is determined by its initial local state and by the sequence of events that have occcured on P_i. The set of events occuring on process P_i is denoted by E_i. The set of all events of the distributed computation will be denoted by E.

Events of a distributed computation are ordered by a causality relation — the "happened-before" relation of Lamport [15]. We will denote this relation by $<$. An event e happens before an event f if and only if one of the following conditions holds:

1. e occurs before f on the same process,
2. e is an emission of a message and f the corresponding receive event,
3. There exists an event g such that $e < g < f$.

Events e and e' are said to be *concurrent* (denoted by $e||e'$) if they are not causally related. This relation induces a partial order $P = (E, <)$ called the causal order. The set E_i of events occuring on process P_i is a total order. Then E can be canonically decomposed as the disjoint union of the sets E_i. This relation may be represented by a space-time diagram — see Fig. 1 (a). An *observation* of a distributed computation is a sequence of events respecting the causality relation (this can be interpreted as a valid interleaving of events). To an observation of the causality relation corresponds a *linear extension* L of the order $(E, <)$. Indeed, a linear extension is a sequence of elements denoted by $x_1 <_L \cdots <_L x_{|E|}$ of E such that if $x_i < x_j$ in P then $i < j$.

We suppose that a monitoring process P_0 collects information about each occured event in the distributed computation. To each event is associated the local state of its process after its execution, the identity of its process and the set of its predecessors in the causality relation. The monitoring process will use these informations to check if $Possibly(\Phi)$ holds.

A *consistent cut* C of a distributed computation is a set of events such that if an event e belongs to C and there exists an event f such that $f < e$ then f belongs to C too. A *global state* of a distributed computation is the union of several local states from different processes such that the set of all the sequences of events defines a consistent cut. Thus, a global state may be represented by an n-sized vector where each entry i corresponds to the latest event that have occured on process P_i.

The definition of a consistent cut of a distributed computation and the definition of an ideal of its causal order is exactly the same. Thus the set of all consistent cuts is equal to the set of all ideals. The set of ideals of P ordered by inclusion has a distributive lattice structure. Relevant properties of ideals lattice for our work are described in Sect. 3.

Proposition 1 *Let $(E, <)$ be a distributed computation. Any algorithm which generates all the ideals of $(E, <)$ in a manner such that two consecutive generated ideals differ only by one element may generate all the global states vectors of the distributed computation with the same time complexity.*

Proof: Let I be the first generated ideal. Its global state vector $STATE(I)$ may be computed in $O(|I|)$. Now, in order to generate the next ideal, we simply add or remove an element to I; thus to compute its global state vector it is sufficient to update only the local state of the process where the event has occured. So we may only update the global state vector with the added event or with the predecessor on the same process of the removed event. Since the local state defined by the occurence of an event is stored with this event, retrieving the global state takes constant time. This operation is repeated for each new generated ideal. Thus, computing global states while computing the ideals does not increase the overall time complexity. □.

A consequence of Proposition 1 is that the use of vector of timestamps during the distributed computation is no longer justified for generating the global states if we may generate them as indicated by the proposition. Indeed, previous works on the detection of $Possibly(\Phi)$ [7, 1] used the "vector clock" defined by Fidge and Mattern [9, 16]. Our algorithm generates the ideals as indicated in Proposition 1 and then it only requires the knowledge of the set of *immediate* predecessors in the *transitive reduction* of the order. Droping timestamps vectors is important to save traffic costs on channels (when the number of processes is very large, maintaining such vectors may lead to a congestion of the channels).

Given an event e of the distributed computation, we denote by $PRED(e)$ the set of its predecessors in the causality relation. We suppose that the monitoring process P_0 receives events according to a linear extension of the causal order (this is easy to achieve by delaying the reception of some events). The monitoring process checks if an arbitrary predicate Φ holds for one of the new global states that may appear by the receipt of a new event.

3 A Fundamental Structure: the Ideal Tree

Let $I(P)$ denote the transitive reduction of the ideal lattice of P. This lattice is distributive [3, 16]. Let I and J be two ideals of P. Then (J, I) is an arc of $I(P)$ if and only if $J \setminus I = \{e\}$ where e is an element of P. Thus, (J, I) may be labelled by e. Therefore the set of immediate predecessors of I in $I(P)$ is exactly $IMPRED(I) = \{J \in I(P) \ s.t. \ J = I \setminus \{e\} \ and \ e \in Max(I)\}$. It has been proved that the labels sequence of each path from bottom to top of $I(P)$ corresponds to a linear extension of P, and reciprocally [4].

The fundamental structure used by the algorithm in Sect. 4 is the ideal tree $T(P)$ associated to a linear extension of P. By using the labels of arcs, it gives an efficient way to generate and to encode ideals. The definition of this tree is given in the following theorem.

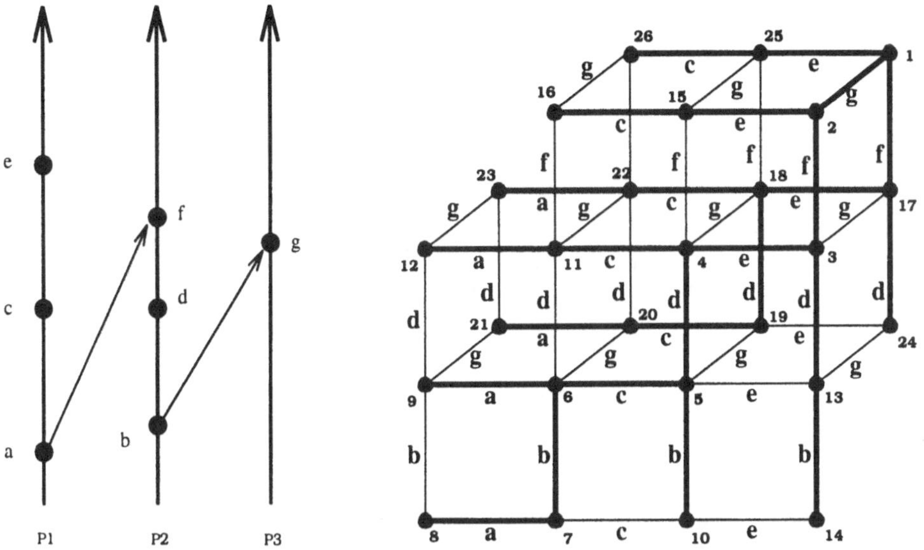

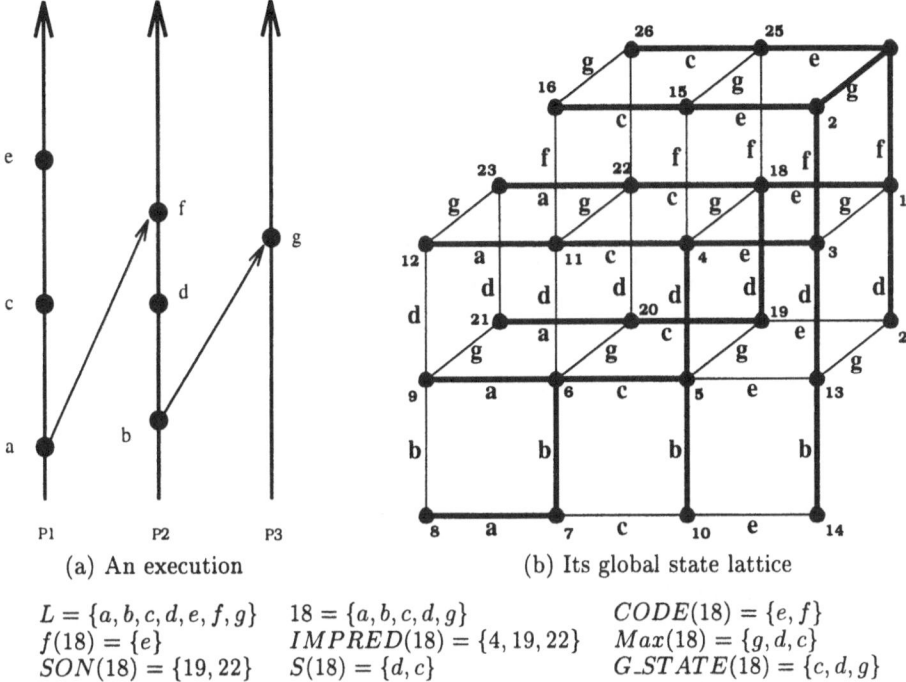

(a) An execution

(b) Its global state lattice

$L = \{a, b, c, d, e, f, g\}$	$18 = \{a, b, c, d, g\}$	$CODE(18) = \{e, f\}$
$f(18) = \{e\}$	$IMPRED(18) = \{4, 19, 22\}$	$Max(18) = \{g, d, c\}$
$SON(18) = \{19, 22\}$	$S(18) = \{d, c\}$	$G\_STATE(18) = \{c, d, g\}$

Figure 1: Example of distributed computation and its global state lattice

Theorem 1 *[12] Let L be any linear extension of P. There exists a unique spanning tree $T(P)$ of $I(P)$ whose root is the top element of $I(P)$ and whose sequence of labels of arcs $e_1, \cdots, e_k$ from the root to any ideal I is such that it satisfies $e_1 >_L \cdots >_L e_k$.*

Remark: It is clear that there also exists an "upward" ideal tree whose root is the bottom of the lattice [19], and whose sequences of labels are ordered according to L. However, this tree is less adapted to on–line algorithms, thus it is not presented here.

Let us denote by $CODE(I)$ the sequence of labels from root to an ideal I. The elements of I are exactly $E \setminus CODE(I)$[1]. In fact, the algorithm handles $CODE(I)$ instead of I. It has been proved by Habib and Nourine [12] that $T(P)$ can be obtained by a unique depth-first search of $I(P)$ which begins from its top.

For an ideal I of $I(P)$, let $f(I)$ be the lowest element of $CODE(I)$ according to L — e.g. e_k in Theorem 1. We denote by $SON(I)$ the set of sons of I in $T(P)$. Let $S(I)$ be the set of labels of the arcs from I to the sons of I.

[1] When appropriate interpretation is clear from context, we use the same notation for a sequence and a set of elements.

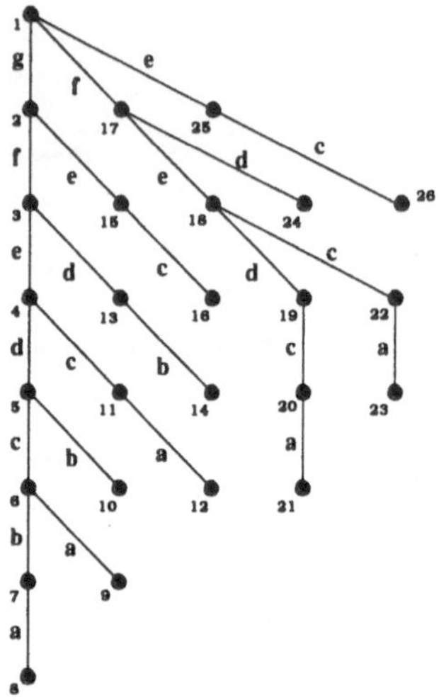

The linear extension used is $L = \{a, b, c, d, e, f, g\}$.

Figure 2: Ideal tree $T(P)$ corresponding to the previous execution

With those notations it can be easely proved that:

1. $S(I) = \{e \in Max(I) \text{ s.t. } e <_L f(I)\}$.
2. $SON(I) = \{J \in I(P) \text{ s.t. } CODE(J) = CODE(I) \cup \{e\}, e \in S(I)\}$.

These properties give a constructive definition of the ideal tree and are the core of our method.

4 An On-Line Algorithm

By *on-line* algorithm for the detection of $Possibly(\Phi)$, we mean that the events are received by the monitoring process according to a linear extension of the final causal order. Notice that this assumption is necessary for on–line verification of predicates since a global state is defined only when all its predecessors have been received. We suppose that at a given step, all the ideals of the current order $P = (E, <)$ have been generated.

The principle of the algorithm is to generate the new ideals created by the occurence of a new element. We suppose that when the event is received, we know the identity $PROC$ of the process where it has occured, and the set $PRED(e)$ of its immediate predecessors in P. It is clear that event e will be maximal in the new order $P \cup \{e\}$. We denote by L the linear extension of the causal order — clearly, it is the order in which the events are received by the monitoring process. We denote by $INC_P(e)$ the set of events concurrent to e. More formally, $INC_P(e) = \{ x \in P \text{ s.t. } x || e\}$. The following theorem gives the idea of the "on-line" algorithm.

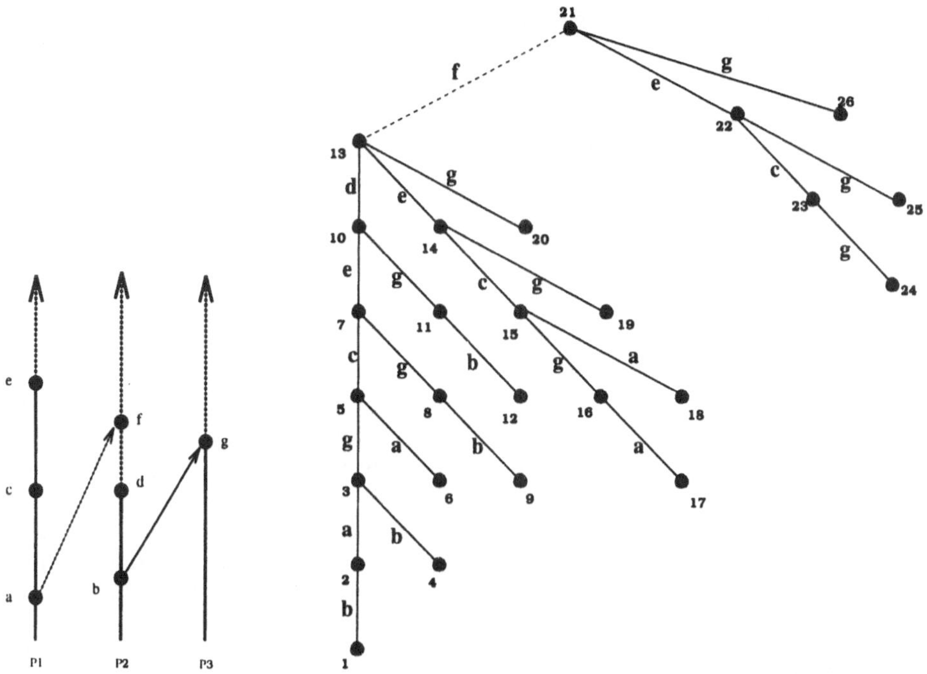

(a) Event f arrives... (b) ... and the new ideals are generated.

Events are received in this order: b, a, g, c, e, d. When f arrives, the ideals from 1 up to 20 have already been checked. The on-line algorithm generates the ideals from 21 to 26.

Figure 3: Example of incremental construction of the ideal lattice

Theorem 2 *Let $P \cup \{e\}$ be an order such that e is maximal in $P \cup \{e\}$.*

$$T(\ P\ \cup\ \{e\}\ )\ =\ T(\ P\ )\ \diamond^e\ T(\ INC_{P\cup\{e\}}(e)\ )$$

where $\diamond^e$ consist in:

- *linking the root of $T(\ P\ )$ as a son of the root of $T(\ INC_{P\cup\{e\}}(e)\ )$ (i.e. the root of $T(\ INC_{P\cup\{e\}}(e)\ )$ becomes the root of $T(\ P\ \cup\ \{e\}\ )$;*
- *labelling the edge between the two roots by e.*

Proof: All ideals of P have already been generated: these ideals do not contain element e. Thus, the remaining ideals to be generated must contain element e, and therefore they must contain all the predecessors of e. We denote by $\downarrow\!e$ the principal ideal of e (i.e. $x \leq e$ if and only if $x \in\downarrow\!e$) corresponding to the causal past of e. For each new ideal to be generated, e will be maximal. Thus, all ideals to be generated may be written as the disjoint union of two sets: $J \cup \downarrow\!e$.

Let us show that $J \cup \downarrow\!e$ is an ideal of $P \cup \{e\}$ if and only if J is an ideal of $INC_{P\cup\{e\}}(e)$.

- **if part**: Suppose that J is not an ideal of $INC_{P\cup\{e\}}(e)$. Then there exists $y \in INC_{P\cup\{e\}}(e)$ and $x \in J$ such that $y < x$ and $y \notin J$. Since $y||e$, y does not belong to $\downarrow\!e$. Thus $J \cup \downarrow\!e$ is not an ideal.

- **only if part**: Suppose that $J \cup \downarrow e$ is not an ideal of $P \cup \{e\}$. Then there exists $y \in P$ and $x \in J \cup \downarrow e$ such that $y < x$ and $y \notin J \cup \downarrow e$. Since $\downarrow e$ contains all the predecessors of e and since e is maximal in $P \cup \{e\}$, we have $y||e$. Thus, y belongs to $INC_{P\cup\{e\}}(e)$. Element x cannot belong to $\downarrow e$ otherwise we would have $y < e$. Thus x belongs to J. Since y does not belongs to J and since we have $y < x$, then J is not an ideal of $INC_{P\cup\{e\}}(e)$.

□.

Thus, for each new event e, we have to compute $INC_{P\cup\{e\}}(e)$ and to search its ideal tree. Let suppose, for the moment, that $INC_{P\cup\{e\}}(e)$ is given. We will search its ideals tree in a depth-first manner, by generating the code of each ideal in the lexicographic order corresponding to the reverse order of L. To achieve this, we must compute for each ideal I the set $S(I)$ containing the labels of the edges to its sons.

Given an ideal I belonging to $INC_{P\cup\{e\}}(e)$, and given $S(I) = \{x_1, \cdots, x_k\}$ — with $x_1 >_L \cdots >_L x_k$ — we denote by $\{I_1, \cdots, I_k\}$ the set containing the sons of I in $T(\ INC_{P\cup\{e\}}(e)\ )$, where $CODE(I_i) = CODE(I) \cup \{x_i\}$.

Lemma 1 *Let $INHERIT\_S(I_i)$ be the set of labels equal to $S(I) \setminus \{x_1, \cdots, x_i\}$. Then $INHERIT\_S(I_i) \subseteq S(I_i)$.*

Proof: By definition, elements in $S(I_i)$ are maximal in I_i, and since we have $CODE(I_i) = CODE(I) \cup \{x_i\}$, all elements in $S(I_i)$ are lower than x_i in the linear extension L. Elements of $S(I)$ are maximal in I, and thus, they are also maximal in I_i. As a consequence, all elements of $S(I)$ which are lower than x_i in L will belong to $S(I_i)$. □.

From Lemma 1 we know a subset of elements of $S(I_i)$. Next lemma give us a way to compute other elements of $S(I_i)$.

Lemma 2 *Let $NEW\_S(I_i)$ be the set of events becoming maximal in $I \setminus \{x_i\}$ — i.e. the immediate predecessors of x_i which become maximal when x_i is removed from I. Then $NEW\_S(I_i) \subseteq S(I_i)$.*

Proof: By definition, elements in $NEW\_S(I_i)$ are maximal in I_i and are lower than x_i in L since they are predecessors of x_i. Thus, $NEW\_S(I_i) \subseteq S(I_i)$. □.

Theorem 3 $S(I_i) = INHERIT\_S(I_i) \cup NEW\_S(I_i)$.

Proof: Let suppose that there exists x belonging to $S(I_i)$ and such that x does not belong to $INHERIT\_S(I_i) \cup NEW\_S(I_i)$. Then x is maximal in I_i and x is lower than x_i in L (by definition).

- Suppose that x is maximal in I. Then x belongs to $S(I)$ since x_i belongs to $S(I)$ and $x <_L x_i$. Thus, x must belong to $INHERIT\_S(I_i)$ since $INHERIT\_S(I_i) = S(I) \setminus \{x_1, \cdots, x_i\}$.

- Suppose that x is not maximal in I. Then x becomes maximal in I_i. Since $I \setminus I_i = \{x_i\}$, x becomes maximal by the removal of x_i. Thus x belongs to $NEW\_S(I_i)$.

So, all elements of $S(I_i)$ belong to $INHERIT\_S(I_i) \cup NEW\_S(I_i)$. Thus $S(I_i) \subseteq INHERIT\_S(I_i) \cup NEW\_S(I_i)$, and by lemma 1 and 2 we obtain the equality. □.

Theorem 3 give a way to compute $S(I_i)$ from the knowledge of $S(I)$. Part of this set is inherited from the father, and new elements are to be computed. The following proposition will give us a method to compute the inherited set in constant time during the depth-first search.

Proposition 2 *Given I an ideal, and $\{I_1, \cdots, I_k\}$ its sons.*

- $INHERIT\_S(I_1) = S(I) \setminus \{x_1\}$
- $INHERIT\_S(I_i) = INHERIT\_S(I_{i-1}) \setminus \{x_i\}, \forall i \in 2..k.$

Proof: Follows directly from the definition of $INHERIT\_S(I_i)$. □.

To compute the elements in $NEW\_S(I_i)$, we use counters associated to each events in P. Those counters give the number of immediate *successors* still present in the order at the current step. Thus, to *virtually* remove an event from the order, it is sufficient to decrease the counter of its immediate predecessors. If the counter reaches 0 then the event becomes maximal in the new current order, and thus it will belong to $NEW\_S(I_i)$.

For each new generated ideal, we compute its corresponding global state from the global state of its father. According to Proposition 1, this may be done in constant time. Then, we may apply the predicate Φ on this global state and check whether or not it holds.

A consequence of Proposition 2 we do not have to store the set $S(I)$ for each generated ideal I: it is sufficient to store the set $INHERIT\_S(I)$ (initially, the set $INHERIT\_S(I)$ is equal to $S(I)$). We may remark that, given I and I_i a son of I, we have $INHERIT\_S(I_i) \subseteq INHERIT\_S(I)$. This means that redundant information is stored. To reduce the space storage required by the algorithm, we merge all the inherited sets in a sorted list S_LIST. To each element of the list, we associate a counter which gives the number of sets $INHERIT\_S$ this element belongs to. When this counter reaches 0, the element is removed from the list. To each ideal I we associate a pointer $pS(I)$ to the first element of S_LIST which belongs to its set $INHERIT\_S(I)$. If I is the current ideal in the depth-first search, then all elements of S_LIST from the element pointed by $pS(I)$ to the end of the list are exactly those of $INHERIT\_S(I)$ (since the S_LIST is sorted according to the reverse order of L).

Algorithm 1 shows the ideals generation algorithm for the set $INC_{P \cup \{e\}}(e)$.

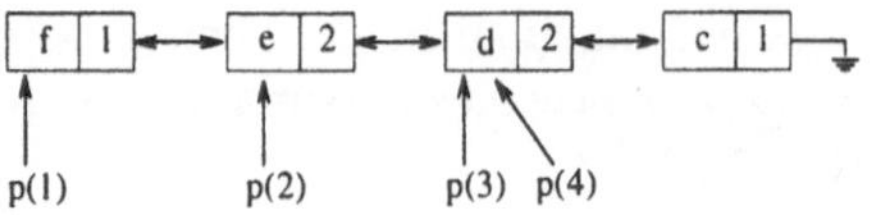

Contents of S_LIST when predicate Φ is checked for ideal 4 in the tree.
$CODE(4) = \{g, f, e\}$ and $G\_STATE(4) = \{c, d, -\}$

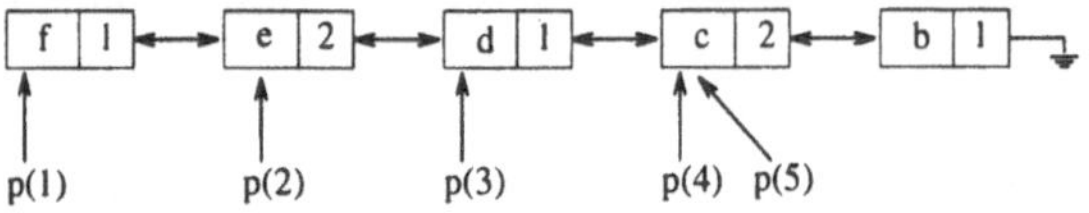

Contents of S_LIST when predicate Φ is checked for ideal 5.
$CODE(5) = \{g, f, e, d\}$ and $G\_STATE(5) = \{c, b, -\}$

Figure 4: Example of S_LIST handling.

Theorem 4 *Algorithm 1 generates all the ideals of the sub-order $INC_{P\cup\{e\}}(e)$ in $O(\Delta.i(INC_{P\cup\{e\}}(e)))$ time complexity and uses linear space storage, where Δ is the maximum in-degree in $INC_{P\cup\{e\}}(e)$ and $i(INC_{P\cup\{e\}}(e))$ is the number of ideals of the sub-order $INC_{P\cup\{e\}}(e)$.*

Proof: For each ideal I, the algorithm computes the set $S(I)$. This set is sorted according to the reverse order of L. The first set $S(I)$ is given as a parameter to the function. Using $S(I)$, the algorithm will generate for each ideal I all its sons in the ideal tree of $INC_{P\cup\{e\}}(e)$. Thus, according to Theorem 1, all ideals of $INC_{P\cup\{e\}}(e)$ will be generated since $T(\ INC_{P\cup\{e\}}(e)\ )$ is a spanning tree of the ideal lattice of $INC_{P\cup\{e\}}(e)$.

Since $S(I)$ is sorted according to the reverse order of L, computing the list $INHERIT\_S(I)$ is done in constant time: we simply remove the first element of $S(I)$. We suppose that the list of predecessors of each event is sorted according to the reverse order of L. Thus, we may compute $NEW\_S(I)$ sorted according to the reverse order of L in Δ steps. The merging of the two sorted list in order to obtain the $S(I)$ of the new generated ideal is done in $|S(I)|$. Thus for each ideal I, the time cost is equal to $\Delta + |S(I)|$.

Each edge (I, J) in the tree is visited twice: the first time downwards from I to J (corresponding to the *if* case in the algorithm), and then upwards from J to I (corresponding to the *else* case in the algorithm). At the first visit, the element corresponding to the label of the edge is removed from the order while computing $NEW\_S(J)$. This costs Δ steps to decrease the counters of the immediate predecessors. When visited upwards, the label is restored in the order. This costs Δ steps too to increase the counters.

We do not take into account the time complexity of checking Φ since this time complexity depends on the nature of Φ.

So, the whole time complexity is bounded by:

$$\sum_{I \in T(\ INC_{P\cup\{e\}}(e)\ )} (2 \times \Delta + |S(I)|) \ = \ 2 \times \Delta + \sum_{I \in T(\ INC_{P\cup\{e\}}(e)\ )} |S(I)|.$$

The overall complexity is in $O(\Delta.i(INC_{P\cup\{e\}}(e)))$ since $|S(I)|$ is the number

Algorithm 1 Generation algorithm for $Possibly(\Phi)$ upon $INC_{P\cup\{e\}}(e)$.

```
Procedure Generate( P, S_LIST, G_STATE ) ;
Data: · Sorted adjacency lists of immediate predecessors of P
      · S_LIST the sorted list of maximal events in P
      · G_STATE the last global state of P
Result: True if Possibly(Φ) holds, False otherwise
begin
    CODE =Empty;
    i = 1;
    pS(1) points on the first element of S_LIST;
    Possibly = Φ( G_STATE );
    while i > 0 and not Possibly do
        if pS(i) ≠ Nil then
            x = Get( S_LIST, pS(i) );
            Push( CODE, x );
            New_S = Decrease( x, G_STATE );
            pS(i + 1) =Merge( S_LIST, pS(i) , New_S );
            i = i + 1;
            Possibly = Φ( G_STATE );
        else
            x = Pop( CODE );
            Increase( x, G_STATE );
            i = i - 1;
    return Possibly;
end
```

of sons of I in $T(\ INC_{P\cup\{e\}}(e)\ )$ and

$$\sum_{I\in T(\ INC_{P\cup\{e\}}(e)\ )} |S(I)| \ = \ i(INC_{P\cup\{e\}}(e)) - 1.$$

Concerning the space complexity, the size of the stack $CODE$ may not exceed $|E|$, since the longest branch in the ideal tree has a length of $|E|$. The list S_LIST may not exceed $|E|$ elements, since each element may appear only once. Thus, the most expensive data structure is the order itself. So the space complexity is linear in the size of the order P. □.

We may notice that in the *if* part of the algorithm, we reach an ideal I by removing an event to the previous ideal (this is done by pushing x in the stack $CODE$). Thus, if the global state of the previous ideal is known, according to Proposition 1 we may compute the global state of I in constant time. The same may apply in the *else* part of the algorithm, since we add an event to I to generate the next ideal (this is done by popping x from the stack $CODE$). Thus, it is sufficient to initialize correctly the global state of the first generated ideal in order to compute the global states of $INC_{P\cup\{e\}}(e)$.

As written previously, the generation may work if the set $INC_{P\cup\{e\}}(e)$ is known. In this case, we must correctly init the algorithm in order to obtain the

Algorithm 2 The "on-line" algorithm for $Possibly(\Phi)$.

```
Data: Events arriving according to a linear extension.
Result: True if Possibly(Φ) holds, False otherwise.
begin
    Possibly =False;
    G_STATE =Null Vector;
    P =Empty;
    MAX =Empty;
    Finished =False;
    while not (Finished Or Possibly) do
        Reception of e with PRED(e);
        G_STATE[PROCe] = e;
        P = P ∪ {e};
        MAX = ({e} + MAX \ PRED(e));
        S_LIST = MAX \ {e};
        pS(1) points to the first element of S_LIST;
        Possibly =Generate( P, S_LIST, G_STATE);
        L = L + {e};
        Sort PRED(e) according to the reverse order of L;
        Finished =(e is the last event);
    Return Possibly;
end
```

first ideal $ROOT$. We have seen that $S(I) = INHERIT\_S(I) \cup NEW\_S(I)$, for any ideal I. Initially, the set $INHERIT\_S(ROOT)$ is empty since the $ROOT$ has no father. So, $S(ROOT)$ is equal to the maximal elements in $INC_{P\cup\{e\}}(e)$. Thus, to apply the algorithm upon $INC_{P\cup\{e\}}(e)$ it is sufficient to know the elements which are maximal in this sub–order. Obviously, the maximal elements are the maximal elements of P which are not predecessors of the new element e. So, we can apply this algorithm directly upon P by computing only the maximal elements of $INC_{P\cup\{e\}}(e)$. By the way, we may notice that the global state of the first generated ideal for $INC_{P\cup\{e\}}(e)$ will be the final global state of P updated with event e. This will initialize correctly Algorithm 1. So, at each incremental step we may only store the last global state of the current order.

Algorithm 2 summarizes the on–line procedure to detect $Possibly(\Phi)$.

Theorem 5 *Given an order P and L a linear extension of P, Algorithm 2 generates all the ideals of P in an "on-line" manner in $O(\Delta.i(P)+|U|*log(\Delta))$ time complexity and uses linear space, where $i(P)$ is the number of ideals of P and $|U|$ is the number of edges in P.*

Proof: According to Theorem 2, the algorithm generates all the ideals in an "on–line" manner, since for each new event arrived it calls the function `Generate()`. No supplementary data structure is introduced, so the space complexity is the same as for Algorithm 1.

From Theorem 4 we deduce that the cost of generating all the sub–trees is:

$$\sum_{e\in P}(\Delta.i(INC_{P\cup\{e\}}(e))) \;=\; i(P)$$

where $i(P)$ is the total number of ideals of P. For each new element e, the algorithm inserts e in the order by increasing the counters of its immediate predecessors. This costs $|PRED(e)|$ steps. The algorithm then computes the maximal elements in $INC_{P\cup\{e\}}(e)$: this is done in $|PRED(e)|$ steps too, by using an array indexed by elements of E where each entry either nil or a pointer on the position of the element in the list of the maximal elements of P. The list of immediate predecessors of e is sorted according to the reverse order of L: this can be done in $O(|PRED(e)| \times log|PRED(e)|)$. The other operations take constant time.

Since $|PRED(e)|$ is bounded by Δ, we have:

$$\sum_{e\in P}(|PRED(e)| \times log|PRED(e)|) \leq log\Delta \times \sum_{e\in P}|PRED(e)|.$$

And since —PRED(e)— is the number of edges arriving to e in P, we have:

$$\sum_{e\in P}|PRED(e)| \leq |U|.$$

Thus, the total time complexity is in $O(\Delta.i(P) + |U|.log\Delta)$. □.

5 Conclusion and Discussions

In this paper, we have presented an on-line algorithm which detects $Possibly(\Phi)$. It uses $O(|E| + |U|)$ space and $O(\Delta.i(P) + |U|.log(\Delta))$ time complexities. The algorithm relies on structural properties of the ideal lattice, and particularly on a special spanning tree of this lattice.

We can notice that Δ is the maximal in–degree of the considered order. If the monitoring process receives P as its transitive reduction (or compute its transitive reduction on–line), then Δ will be optimum.

It is also noticeable that this algorithm may be used to detect $Possibly(\Phi)$ on potentially infinite computations. To our knowledge, this is the first algorithm in this category. For such computations, it is important to reduce the amount of stored data. We may use an incremental garbage collector such as the one presented in [17]. Indeed, some elements will never be used again by the generation algorithm for new events. The basic idea is to remove from the order all the elements which will be predecessors of any new upcoming event. This is formalized by the following proposition.

Proposition 3 *Let $e_1, \cdots, e_n$ be the last events received respectively from processes $P_1, \cdots, P_n$. Let e be an upcoming event. Then, for any x belonging to $\bigcap_{i\in 1..n} \downarrow e_i$, x will not belong to $INC_{P\cup\{e\}}(e)$.*

An algorithm which uses such property may be found in [17].

We may remark that if we consider an off-line analysis, the time complexity will be $O(\Delta.i(P))$ since the adjacency lists of predecessors in the order P may be sorted during a preprocessing [19]. This improves the ideals generation algorithms proposed by Steiner [22] and Bordat [5]. Steiner's algorithm runs in $O(|E|.i(P))$ and requires linear space storage, while Bordat's one runs in $O(\omega(P).i(P))$, where $\omega(P)$ is the width of P (bounded by the number of processes in our case), and requires $O(|E|^2)$ space. Both algorithms are not adapted for on–line generation. To our knowledge, the only other algorithm which detects $Possibly(\Phi)$ in an on–line manner is due to Diehl et al. [8]. However, they store the whole lattice and thus have an exponential space complexity.

Generating the ideals of a partial order has many applications in computer science, discrete optimization and operations research. Dynamic programming algorithms, for precedence constrained scheduling problems, for project scheduling and assembly line balancing, all use ideal generation procedures for partial orders [22]. Computing reachability in reliability networks also requires the generation of ideals [2]. According to the remark concerning Theorem 1, an off–line algorithm which searches the "upward" tree may be derived from the presented algorithm [19]. This off–line algorithm seems to be well–suited for the analysis of possible runs when dealing with conflicts, such as in event structures.

Acknowledgements

We are indebted to the anonymous referees for their insightful comments, which greatly improved the presentation of the paper. We also wish to thank Michel Habib for its comment on earlier versions of this work, and Brigitte Rozoy for having explained us the connection existing between our algorithm and event structures problems.

References

[1] S. Alagar and S. Venkatesan. Techniques to tackle state explosion in global predicate detection. In *Proc. IEEE International Conference on Parallel and Distributed Systems*, pp 412–417, Taiwan, December 1994.

[2] M.O. Ball and J.S. Provan. Calculating bounds on reachability and connectedness in stochastic networks. *Networks*, 13(1), 253–278, 1983.

[3] G. Birkhoff. *Lattice Theory*, volume 25 of *Coll. Publ. XXV*. American Mathematical Society, Providence, 3rd edition, 1967.

[4] R. Bonnet and M. Pouzet. Extensions et stratifications d'ensembles dispersés. *C.R. Acad. Sci.*, t. 268-Série A:1512–1515, 1969.

[5] J.P. Bordat. Calcul des idéaux d'un ordonné fini. *Recherche Opérationnelle / Operations Research*, 25(3):265–275, 1991.

[6] B. Charron-Bost, C. Delporte-Gallet, and H. Fauconnier. Local and temporal predicates in distributed systems. Technical Report LITP 92.36, Institut Blaise Pascal, Université Paris 7, France, April 1992.

[7] R. Cooper and K. Marzullo. Consistent detection of global predicates. In *Proc. ACM/ONR Workshop on Parallel and Distributed Debugging*, pp 163–173, Santa Cruz, California, May 1991.

[8] C. Diehl, C. Jard, and J.-X. Rampon. Reachability analysis on distributed executions. In Jouannaud Gaudel (ed), *TAPSOFT*, Springer-Verlag, Orsay (France), 1993, pp 629–643 (LNCS 688).

[9] C.J. Fidge. Timestamps in message-passing systems that preserve the partial ordering. In *Proc. 11th Australian Computer Science Conference*, pp 55–66, University of Queensland, Australia, February 1988.

[10] V.K. Garg and B. Waldecker. Detection of unstable predicates in distributed programs. In *Proc. of 12th Conference on the Foundations of Software Technology & Theoretical Computer Science*, 1992, Springer Verlag, pp 253–264, (LNCS 652).

[11] V.K. Garg and B. Waldecker. Detection of weak unstable predicates in distributed programs. *IEEE Transactions on Parallel and Distributed Systems*, To appear.

[12] M. Habib and L. Nourine. Tree structures for distributive lattice and its application. Research report, LIRMM, Montpellier, France, October 1993.

[13] J.-M. Helary and M. Raynal. Towards the construction of distributed detection programs with an application to distributed termination. Research Report 1460, INRIA, Rennes, France, Juin 1991.

[14] C. Jard, G.-V. Jourdan, and J.-X. Rampon. Some 'on-line' computations of the ideal lattice of posets. Internal Publication PI-773, IRISA, Rennes, France, October 1993.

[15] L. Lamport. Time, clocks, and the ordering of events in a distributed system. *Communications of the ACM*, 21(7):558–565, July 1978.

[16] F. Mattern. Virtual time and global states of distributed systems. In M. Cosnard et al. (ed), *Parallel and Distributed Algorithms*, pp 215–226. Elsevier / North-Holland, 1989.

[17] R. Medina. Incremental Garbage Collection for Causal Relationship Computation in Distributed Systems. In *Proc. 5th IEEE Symposium on Parallel and Distributed Processing*, pp 650–655, Dallas (Texas), 1993.

[18] R. Medina and L. Nourine. Growing and pruning the consistent cuts tree. Research Report 93-074, LIRMM, Montpellier, France, December 1993.

[19] R. Medina and L. Nourine. Algorithme efficace de génération des idéaux d'un ensemble ordonné. *C.R. Acad. Sci.*, t. 319-Série I:1115–1120, 1994.

[20] L. Nourine. *Quelques propriétés algorithmiques des treillis*. PhD thesis, Université Montpellier II, Montpellier, France, June 1993.

[21] R. Schwarz and F. Mattern. Detecting causal relationships in distributed computations - in search of the holy grail. *Distributed Computing*, 7(3):149–174, 1994.

[22] G. Steiner. An algorithm for generating the ideals of a partial order. *Operations Res. Lett.*, 5:317–320, 1986.

A Simple Decision Method for the Linear Time Mu-calculus

Roope Kaivola*
Laboratory for Foundations of Computer Science†
University of Edinburgh, United Kingdom

Abstract

The linear time mu-calculus νTL is a language extending standard linear time temporal logic with fixpoint operators. We present a method for deciding whether a given νTL-formula is satisfiable, and give a direct proof of its completeness. Although simpler than the existing methods, it gives rise to an algorithm working in the same $2^{O(n^2 \log n)}$ time as these, or alternatively, to a polynomial space, singly exponential time algorithm. What is more important, the method allows us to devise a tableau system to support manual or computer-aided (as opposed to fully automated) satisfiability checking.

1 Introduction

One of the most widely accepted and established temporal or modal logic based methods for specifying properties of concurrent systems is the linear time temporal logic TL, built around *nexttime* and *until* operators. For surveys of the area, see [6, 13]. However, in the course of research it has become apparent that standard TL is not expressive enough for all aspects of this task. In [17] it was shown that TL cannot express such a simple property as *ϕ holds at every other moment*. More importantly, in [10] it was shown that the inability to express such properties makes TL inadequate for modular verification.

To remedy this problem, a formalism called ETL, extending TL with an infinite set of temporal operators corresponding to regular grammars was introduced in [16, 17]. An alternative approach, applied already earlier in the context of branching time [7, 8], is extending TL with maximal and minimal fixpoint operators ν and μ [3]. Although the resulting language νTL is expressively equivalent to ETL, it is syntactically much more concise, containing only boolean operators, fixpoints and the *nexttime* temporal operator.

In this paper we examine the decision problem for νTL. Despite the syntactic simplicity of νTL, this problem has turned out to be rather intricate. The easiest way of establishing decidability of νTL by mapping it to the monadic second-order theory of one successor $S1S$, is for any practical purpose useless, since the decision procedure for $S1S$ [4] is non-elementary.

The usual starting point of the more practical approaches to deciding the satisfiability of a νTL-formula ϕ, is constructing a graph, called a *tableau* or *Hintikka structure*, that takes care of the satisfaction of ϕ as far as propositional

*e-mail: Roope.Kaivola@dcs.ed.ac.uk, tel: +44-131-650 5997, fax: +44-131-667 7209
†address: The King's Buildings, Edinburgh EH9 3JZ

connectives, unwinding of fixpoints, and consistency between successive states is concerned. The intricate part of the problem is then deciding whether it is possible to read a model from this Hintikka structure in such a way that no minimal fixpoint is infinitely regenerated. There have been basically two approaches to this problem.

First, one can reduce the satisfiability problem to the emptiness of automata on infinite strings. A Hintikka structure can be viewed as an automaton Al_ϕ recognising structures that are models of ϕ as far as local aspects are concerned. Intersecting the local automaton Al_ϕ with a global automaton Ag_ϕ preventing infinite regeneration of minimal fixpoints yields an automaton A_ϕ recognising the models of ϕ [14, 15]. Checking A_ϕ for emptiness answers then the question of satisfiability of ϕ. The global automaton Ag_ϕ is normally obtained by complementing an automaton recognising infinite regeneration of a minimal fixpoint. An application of this approach to νTL, leading to a deterministic $2^{O(n^3)}$ (where $n = |\phi|$) time algorithm is described in [15]. Speeding this algorithm up by Safra's construction [11] yields a better, $2^{O(n^2 \log n)}$ time bound.

Another approach is using special graph marking algorithms to extract from the Hintikka structure a path where no minimal fixpoints are infinitely regenerated [2, 9]. In [2] this leads to a decision method taking singly exponential time and space (no accurate bounds presented). In [9] an algorithm inspired by Safra's construction [11] working in $2^{O(n^2 \log n)}$ time is described.

A drawback in these approaches is the non-transparency caused by the rather complex automata constructions (complementation of Ag_ϕ) or graph marking algorithms. This means that although they certainly give an answer to the question of whether ϕ is satisfiable, they do not give an insight into why or how this happens. Moreover, it is difficult for the verifier to take advantage of her intuitions about the formula to ease the task, by using a computer-aided as opposed to fully automated approach.

In this paper we describe a new and, in our opinion, substantially clearer decision method for νTL. The approach is based on decorating Hintikka structures with an explicit dependency relation bearing witness to the fact that no minimal fixpoints are infinitely regenerated. This gives rise to a clear and natural decision method. One way of looking at the difference between the current and earlier work is that here we separate the previously intertwined issues of what kind of structure can be exhibited to witness the satisfiability of a formula, and how such a structure can be found.

From an almost trivial nondeterministic algorithm guessing a structure witnessing the satisfiability of a given formula, we can easily read a deterministic algorithm working in $2^{O(n^2 \log n)}$ time, i.e. matching the time complexity of the rather more complicated algorithms above. Using a different determinisation construction, we also obtain an algorithm working in polynomial space and singly exponential time. What is more important, in contrast to the earlier decision procedures, the clarity of the method presented here allows us to derive a complete set of top-down tableau rules suited for manual or computer-aided satisfiability checking.

A technical point that may have some interest on its own is the direct proof of completeness of the decision method and the finite model property for νTL. Unlike earlier proofs, it does not rely on automata theory [15] or mapping νTL to some other calculus, such as $S1S$ or ETL [2].

2 Preliminaries

Definition 2.1 Σ^* (Σ^ω) is the set of finite (infinite) strings of elements of Σ, $|s|$ is the length of s, and ϵ the empty string. If $s = a_1 \ldots a_n \ldots$, s_i is the element a_i. For all $n \in N$, $[n] = \{i \in N \mid 1 \leq i \leq n\}$. Notation $\bar{q}$ denotes a vector $\bar{q} = (q_1, \ldots, q_n)$, and $\bar{q}_i = q_i$ its elements. If R and R' are relations, $R \circ R'$ denotes their composition, i.e. $R \circ R' = \{(a, c) \mid \exists b : aRb \wedge bR'c\}$, and R^* the reflexive, transitive closure of R. We use Ord to denote the set of ordinals, and $\preceq$ their standard ordering. □

Let us recall definitions of linear mu-calculus syntax and semantics, and introduce some related notation. The language is built from propositions, boolean connectives, the minimal and maximal fixpoint operators μ and ν, and two temporal operators, the *strong nexttime* $\oplus$ and the *weak nexttime* $\ominus$. Intuitively, $\oplus\phi$ means *there is a next moment in time and ϕ is true at that moment*, whereas $\ominus\phi$ means *if there is a next moment in time, then ϕ is true at that moment.*

Definition 2.2 Let us fix two disjoint countable sets, $\mathcal{Z}_c$, the set of *propositions*, and $\mathcal{Z}_v$, the set of *variables*, and define $\mathcal{Z} = \mathcal{Z}_c \cup \mathcal{Z}_v$. The formulae of νTL are defined by the abstract syntax:

$$\phi ::= a \mid \neg a \mid z \mid \phi_1 \wedge \phi_2 \mid \phi_1 \vee \phi_2 \mid \oplus\phi \mid \ominus\phi \mid \mu z.\phi \mid \nu z.\phi$$

where a varies over $\mathcal{Z}_c$ and z over $\mathcal{Z}_v$. Symbol $\odot$ refers to both $\oplus$ and $\ominus$, and σ to both ν and μ. An occurrence of a variable z in ϕ is *bound* iff it is within a subformula $\sigma z.\phi'$ of ϕ and *free* otherwise. A formula ϕ without any free variables is a *sentence*. If z a variable, $\phi[\phi'/z]$ is the result of simultaneously substituting ϕ' for all free occurrences of z in ϕ. □

We suppose above that the formulae discussed are already in the positive normal form as defined e.g. in [13], consequently omitting the negation operator from the language. Furthermore, in any νTL-formula ϕ, we assume that all the bound variables are distinct, and that all occurrences of bound variables are guarded, i.e. that each occurrence of variable z in $\sigma z.\phi$ is in a subformula of the type $\odot\phi'$. These restrictions are immaterial as any formula can be mechanically transformed into an equivalent one fulfilling them [2, 13].

Definition 2.3 A *model* M is a finite or infinite sequence of sets of propositions, $M \in (2^{\mathcal{Z}})^* \cup (2^{\mathcal{Z}})^\omega$. The set of states of model M is $\mathrm{st}(M) = \{i \in N \mid i \leq |M|\}$ if M is finite, and $\mathrm{st}(M) = N$ otherwise. □

Definition 2.4 Let M be a model and ϕ a νTL-formula. The set of states of M satisfying ϕ, denoted $||\phi||_M$, is defined for $\phi = a, \neg a, z, \phi_1 \wedge \phi_2, \phi_1 \vee \phi_2$ by the obvious interpretation, and by $||\oplus\phi||_M = \{i \in \mathrm{st}(M) \mid i+1 \in ||\phi||_M\}$, $||\ominus\phi||_M = \{i \in \mathrm{st}(M) \mid \text{if } i{+}1 \in \mathrm{st}(M) \text{ then } i{+}1 \in ||\phi||_M\}$, $||\mu z.\phi||_M = \bigcap\{W \subseteq \mathrm{st}(M) \mid ||\phi||_{M[W/z]} \subseteq W\}$, $||\nu z.\phi||_M = \bigcup\{W \subseteq \mathrm{st}(M) \mid W \subseteq ||\phi||_{M[W/z]}\}$, where $M[W/z]$ is defined by: $M[W/z]_i = M_i \cup \{z\}$ if $i \in W$, $M[W/z]_i = M_i \setminus \{z\}$ if $i \in \mathrm{st}(M) \setminus W$.

We say that ϕ is *true* at state i of M, and write $M, i \models \phi$ iff $i \in ||\phi||_M$. We say that ϕ is *satisfiable* iff there exists a model M such that $M, 1 \models \phi$. □

Definition 2.5 If a variable z is bound in ϕ_0 by a ν (μ) fixpoint, we call it a ν (μ) variable, respectively, and write νz (μz) for the unique subformula $\nu z.\phi$ ($\mu z.\phi$) of ϕ_0 bound by the variable z. We use $\mathrm{var}\nu(\phi_0)$ ($\mathrm{var}\mu(\phi_0)$) to denote the set of ν (μ) variables bound by fixpoints in ϕ_0, and define $\mathrm{var}(\phi_0) = \mathrm{var}\nu(\phi_0) \cup \mathrm{var}\mu(\phi_0)$. The subformula relation induces a partial ordering on $\mathrm{var}(\phi_0)$ by: $z \trianglelefteq z'$ iff σz is a subformula of $\sigma z'$. Extend this partial ordering to a total ordering $\trianglelefteq$ of $\mathrm{var}(\phi_0)$ in some fixed, arbitrary way. □

We routinely omit explicit reference to the ϕ_0 on which the notation of the previous definition depends, as it can be deduced from the context.

Definition 2.6 For all ordinals $\alpha \in \mathrm{Ord}$, the *fixpoint approximants* $\mu^\alpha z.\phi$ and $\nu^\alpha z.\phi$ are defined by: $\mu^0 z.\phi = \bot$, $\nu^0 z.\phi = \top$, $\sigma^{\alpha+1} z.\phi = \phi[\sigma^\alpha z.\phi/z]$, $\mu^\lambda z.\phi = \bigvee_{\alpha \prec \lambda} \mu^\alpha z.\phi$ and $\nu^\lambda z.\phi = \bigwedge_{\alpha \prec \lambda} \nu^\alpha z.\phi$, where λ is a limit ordinal. □

Proposition 2.7 [Knaster-Tarski] $\mu z.\phi = \bigvee_\alpha \mu^\alpha z.\phi$, $\nu z.\phi = \bigwedge_\alpha \nu^\alpha z.\phi$. □

3 Hintikka structures

In this section we define the structures used as a basis of the decision method.

Definition 3.1 Let ϕ_0 be a νTL-formula. The *closure of* ϕ_0, denoted $\mathrm{cl}(\phi_0)$, is the minimal set of formulae containing ϕ_0 and fulfilling

- if $\neg\phi \in \mathrm{cl}(\phi_0)$ or $\odot\phi \in \mathrm{cl}(\phi_0)$ then $\phi \in \mathrm{cl}(\phi_0)$
- if $\phi \vee \phi' \in \mathrm{cl}(\phi_0)$ or $\phi \wedge \phi' \in \mathrm{cl}(\phi_0)$ then $\phi \in \mathrm{cl}(\phi_0)$ and $\phi' \in \mathrm{cl}(\phi_0)$
- if $\sigma z.\phi \in \mathrm{cl}(\phi_0)$ then $\phi[\sigma z.\phi/z] \in \mathrm{cl}(\phi_0)$ □

Definition 3.2 Let ϕ_0 be a νTL-formula. An *atom* A of ϕ_0 is a subset of the closure of ϕ_0, $A \subseteq \mathrm{cl}(\phi_0)$, such that

- if $\neg a \in \mathrm{cl}(\phi_0)$ then $\neg a \in A$ iff $a \notin A$
- if $\phi \vee \phi' \in \mathrm{cl}(\phi_0)$ then $\phi \vee \phi' \in A$ iff $\phi \in A$ or $\phi' \in A$
- if $\phi \wedge \phi' \in \mathrm{cl}(\phi_0)$ then $\phi \wedge \phi' \in A$ iff $\phi \in A$ and $\phi' \in A$
- if $\sigma z.\phi \in \mathrm{cl}(\phi_0)$ then $\sigma z.\phi \in A$ iff $\phi[\sigma z.\phi/z] \in A$ □

Definition 3.3 Let A and B be atoms of ϕ_0. We say that B is a *successor* of A iff for all $\phi \in A$, if ϕ is of the form $\phi = \odot\phi'$, then $\phi' \in B$. □

Definition 3.4 Let A be an atom of ϕ_0. We say that A is *terminal* iff there is no $\phi \in A$ of the form $\phi = \oplus\phi'$. □

The structures used for the decision method are essentially finite strings of atoms, equipped with a relation witnessing that no minimal fixpoints are infinitely regenerated. To this purpose we decorate atoms with *dependency relations* $\rightarrow$. Intuitively, $\phi \rightarrow \phi'$ means *ϕ' being true justifies ϕ being true.*

Definition 3.5 Let ϕ_0 be a νTL-formula. An *adorned atom* is a pair $(A, \rightarrow)$, where A is an atom of ϕ_0 and $\rightarrow$, a *dependency relation*, is a minimal relation $\rightarrow\ \subseteq A \times A$ fulfilling:

- if $\phi \vee \phi' \in A$, then either $\phi \vee \phi' \rightarrow \phi$ or $\phi \vee \phi' \rightarrow \phi'$
- if $\phi \wedge \phi' \in A$, then both $\phi \wedge \phi' \rightarrow \phi$ and $\phi \wedge \phi' \rightarrow \phi'$ □

Definition 3.6 A *Hintikka structure* H for a νTL-formula ϕ_0 is a pair

$$H = (A_1 \ldots A_k, (B_1, \rightarrow_1) \ldots (B_m, \rightarrow_m))$$

where

- the *initial part* $A_1 \ldots A_k$ is a nonempty sequence of atoms of ϕ_0, and
- the *recurring part* $(B_1, \rightarrow_1) \ldots (B_m, \rightarrow_m)$ is a sequence of adorned atoms of ϕ_0

such that

- $\phi_0 \in A_1$
- A_{i+1} is a successor of A_i for every $1 \leq i < k$
- if $m = 0$, then A_k is terminal
- if $m > 0$, then B_1 is a successor of A_k, B_{i+1} is a successor of B_i for every $1 \leq i < m$, and B_1 is a successor of B_m. □

Next, we want to characterise the property that there is no minimal fixpoint formula $\mu z.\phi$ such that $\mu z.\phi$ being true at a state of the structure is justified by itself being true at the same state, i.e. that there is no vicious circle causing infinite regeneration. For this, we join the dependency relations in the adorned atoms to a global relation.

Definition 3.7 Let H be a Hintikka structure as in 3.6. Define a binary relation $\rightarrow$ relating formula, index pairs by:

$$(\phi, i) \quad \rightarrow \quad (\phi', i) \qquad \text{iff} \quad i \in [m] \text{ and } \phi \rightarrow_i \phi'$$

For every variable $z \in \text{var}(\phi_0)$, define a similar relation $\xrightarrow{z}$ by:

$$(\sigma z.\phi, i) \quad \xrightarrow{z} \quad (\phi[\sigma z.\phi/z], i) \quad \text{iff} \quad i \in [m] \text{ and } \sigma z.\phi \in B_i$$

Furthermore, define relations $\xrightarrow{\odot}$ and $\xhookrightarrow{\odot}$ by:

$$\begin{array}{llll} (\odot\phi, i) & \xrightarrow{\odot} & (\phi, i+1) & \text{iff} \quad i \in [m-1] \text{ and } \odot\phi \in B_i \\ (\odot\phi, m) & \xhookrightarrow{\odot} & (\phi, 1) & \text{iff} \quad \odot\phi \in B_m \end{array}$$

Define then

$$\begin{array}{lll} \xrightarrow{\trianglelefteq z} & = & \bigcup_{x \trianglelefteq z} \xrightarrow{x} \\ \rightarrow^* & = & (\rightarrow \cup \xrightarrow{\odot})^* \\ \xrightarrow{\trianglelefteq z *} & = & (\xrightarrow{\trianglelefteq z} \cup \rightarrow \cup \xrightarrow{\odot})^* \\ \xrightarrow{z *} & = & \xrightarrow{\trianglelefteq z *} \circ \xrightarrow{z} \circ \xrightarrow{\trianglelefteq z *} \\ \xhookrightarrow{\trianglelefteq z *} & = & (\xrightarrow{\trianglelefteq z} \cup \rightarrow \cup \xrightarrow{\odot} \cup \xhookrightarrow{\odot})^* \\ \xhookrightarrow{z *} & = & \xhookrightarrow{\trianglelefteq z *} \circ \xrightarrow{z} \circ \xhookrightarrow{\trianglelefteq z *} \end{array}$$

□

The general idea of all these relations is that $(\phi, i) \dot{\rightarrow} (\psi, j)$ means that ϕ in B_i depends on ψ in B_j. The definition of $\rightarrow$ says simply that if ϕ and ψ are connected by the local dependency relation $\rightarrow_i$, then ϕ at B_i depends on ψ at the same B_i. The relations $\xrightarrow{z}$ reflect dependencies related to unwinding of fixpoints: every $\sigma z.\phi$ in B_i depends on $\phi[\sigma z.\phi/z]$ in B_i in a way that requires unwinding the fixpoint σz. The relations $\xrightarrow{\odot}$ and $\xhookrightarrow{\odot}$ express the idea that every $\odot\phi$ in B_i depends on ϕ in B_{i+1} if $i < m$, and every $\odot\phi$ in B_m depends on ϕ in B_1, reflecting the intended looping back from B_m to B_1.

These immediate dependency relations $\rightarrow$, $\xrightarrow{z}$, $\xrightarrow{\odot}$ and $\xhookrightarrow{\odot}$ are then joined to global ones. Intuitively, $(\phi, i) \xrightarrow{\trianglelefteq z}{}^{*} (\psi, j)$ holds iff there is a path of immediate dependencies (not using the looping back from B_m to B_1) from ϕ in B_i to ψ in B_j so that no fixpoints outside the scope of z are unwound in this path. The relation $(\phi, i) \xrightarrow{z}{}^{*} (\psi, j)$ is similar, but it is required that the fixpoint σz is unwound at least once along the corresponding path of immediate dependencies. The relations $\xhookrightarrow{\trianglelefteq z}{}^{*}$ and $\xhookrightarrow{z}{}^{*}$ are analogous, but with the looping back from B_m to B_1 allowed.

Definition 3.8 Let ϕ_0 and H be as in 3.6. We say that H is *proper* iff there is no $i \in [m]$, $\phi \in B_i$ and μ-variable $z \in \mathrm{var}\mu(\phi_0)$ such that $(\phi, i) \xhookrightarrow{z}{}^{*} (\phi, i)$. □

Intuitively, this states that there is no looping path of dependencies in the structure in which a minimal fixpoint μz would be unwound at least once and no fixpoints outside the scope of μz would. Notice that as all formulae are assumed to be guarded here, the above is equivalent to requiring that there is no $\phi \in B_1$ and $z \in \mathrm{var}\mu(\phi_0)$ such that $(\phi, 1) \xhookrightarrow{z}{}^{*} (\phi, 1)$.

Example 3.9 Let $\phi_0 = \nu z.\mu y.(a \wedge \oplus z \vee \oplus y)$, expressing the property *infinitely often a*. Define $\phi_z = \phi_0$, $\phi_y = \mu y.(a \wedge \oplus\phi_z \vee \oplus y)$, $\phi_\vee = a \wedge \oplus\phi_z \vee \oplus\phi_y$. Then $\mathrm{cl}(\phi_0) = \{\phi_z, \phi_y, \phi_\vee, a \wedge \oplus\phi_z, a, \oplus\phi_z, \oplus\phi_y\}$.

One Hintikka structure H for ϕ_0 is $H = (A_1, (B_1, \rightarrow_1)(B_2, \rightarrow_2))$ where

$$
\begin{array}{ll}
A_1 = \{\phi_z, \phi_y, \phi_\vee, a \wedge \phi_z, a, \phi_z\} & \\
B_1 = \{\phi_z, \phi_y, \phi_\vee, \oplus\phi_y\}, & \{\phi_\vee \rightarrow_1 \oplus \phi_y\} \\
B_2 = \mathrm{cl}(\phi_0), & \{\phi_\vee \rightarrow_2 \oplus \phi_y, a \wedge \oplus\phi_z \rightarrow_2 a, a \wedge \oplus\phi_z \rightarrow_2 \oplus \phi_z\}
\end{array}
$$

In H, $(\phi_z, 1) \xrightarrow{z} (\phi_y, 1)$, $(\phi_z, 2) \xrightarrow{z} (\phi_y, 2)$, $(\phi_y, 1) \xrightarrow{y} (\phi_\vee, 1)$, $(\phi_y, 2) \xrightarrow{y} (\phi_\vee, 2)$, $(\oplus\phi_y, 1) \xrightarrow{\odot} (\phi_y, 2)$, $(\oplus\phi_y, 2) \xhookrightarrow{\odot} (\phi_y, 1)$ and $(\oplus\phi_z, 2) \xhookrightarrow{\odot} (\phi_z, 1)$.

The structure is not proper, since $(\phi_y, 1) \xhookrightarrow{y}{}^{*} (\phi_y, 1)$ as

$$
(\phi_y, 1) \xrightarrow{y} (\phi_\vee, 1) \rightarrow (\oplus\phi_y, 1) \xrightarrow{\odot} (\phi_y, 2) \xrightarrow{y} (\phi_\vee, 2) \rightarrow (\oplus\phi_y, 2) \xhookrightarrow{\odot} (\phi_y, 1)
$$

If we define a Hintikka structure H' as H but with

$$
\rightarrow_2 = \{\phi_\vee \rightarrow_2 a \wedge \oplus\phi_z, a \wedge \oplus\phi_z \rightarrow_2 a, a \wedge \oplus\phi_z \rightarrow_2 \oplus \phi_z\}
$$

the structure H' is proper. □

4 Completeness

In this section we show that a sentence is satisfiable exactly when there exists a proper Hintikka structure for it. The easy half is showing that the existence of a proper Hintikka structure guarantees the satisfiability of the sentence.

Theorem 4.1 [Soundness] Let ϕ_0 be a νTL-sentence. If there exists a proper Hintikka structure H for ϕ_0, then ϕ_0 is satisfiable.

Proof: By projecting every state A_i (B_i) of H to the set $A_i \cap \mathcal{Z}$ ($B_i \cap \mathcal{Z}$), we get a model M. An easy induction shows that $M, 1 \models \phi_0$. □

For the other direction, we first need some concepts concerning ordinal signatures. The discussion here follows closely to [14, 9].

Definition 4.2 Let X be a set, and $\leq \subseteq X \times X$ a binary relation on X. We say that $\leq$ is a *total ordering* of X iff it is reflexive, transitive and total. Define: $x < x'$ iff $x \leq x'$ and not $x' \leq x$. We say that $<$ is *well-founded* iff there are no infinite descending chains $x_1 > x_2 > x_3 > \ldots$ of elements of X. □

Definition 4.3 A *signature* is a finite sequence of ordinals. The set of signatures is denoted by $\mathrm{Sig} = \mathrm{Ord}^*$. Let us extend the ordering $\preceq$ of ordinals to signatures as the usual lexicographic ordering of strings, based on $\preceq$. Furthermore, for each $k \in N$, define $\preceq_k \subseteq \mathrm{Sig} \times \mathrm{Sig}$ by: $\alpha_1 \ldots \alpha_n \preceq_k \beta_1 \ldots \beta_m$ iff $\alpha_1 \ldots \alpha_{\min(k,n)} \preceq \beta_1 \ldots \beta_{\min(k,m)}$, and $\prec_k \subseteq \mathrm{Sig} \times \mathrm{Sig}$ by: $\alpha_1 \ldots \alpha_n \prec_k \beta_1 \ldots \beta_m$ iff $\alpha_1 \ldots \alpha_{\min(k,n)} \prec \beta_1 \ldots \beta_{\min(k,m)}$. □

Let us remember that the standard ordering $\preceq$ of ordinals is total and $\prec$ well-founded, and that the same holds of its lexicographic extension to signatures. It is also easy to see that the orderings $\preceq_k$ are total although not antisymmetric, and $\prec_k$ well-founded,

Definition 4.4 Let ϕ be a νTL-sentence. The *μ-height* of ϕ, denoted $\mu h(\phi)$, is the depth of nesting of μ-subsentences of ϕ. □

For example, μ-height of $\mu x.(a \vee \oplus \mu y.(x \vee \oplus y))$ is 1, since $\mu y.(x \vee \oplus y)$ is not a sentence.

Definition 4.5 Let ϕ be a νTL-sentence of μ-height n, and $s = \alpha_1 \ldots \alpha_n \in \mathrm{Sig}$ a signature of length n. We denote by $\phi : s$ the sentence obtained from ϕ by replacing each μ-subsentence $\mu z.\phi'$ of ϕ by $\mu^{\alpha_i} z.\phi'$, where $i = \mu h(\mu z.\phi')$. □

Proposition 4.6 Let ϕ be a νTL-sentence, M a model, and $i \in \mathrm{st}(M)$ a state of M, such that $M, i \models \phi$. There exists a signature $s \in \mathrm{Sig}$ such that $M, i \models \phi : s$.

Proof: By the Knaster-Tarski fixpoint theorem [14, 9]. □

This fact justifies the following definition.

Definition 4.7 Let ϕ, M and i be as in 4.6. The *signature of ϕ* at state i of M is $\mathrm{sig}(\phi, i) = \min_{\preceq}\{s \in \mathrm{Sig} \mid M, i \models \phi : s\}$. □

Lemma 4.8 Let M be a model and $i \in \text{st}(M)$ a state of M.

1 If $M, i \models \phi \vee \phi'$, then $\text{sig}(\phi \vee \phi', i) = \text{sig}(\phi, i)$ or $\text{sig}(\phi \vee \phi', i) = \text{sig}(\phi', i)$.
2 If $M, i \models \phi \wedge \phi'$, then $\text{sig}(\phi \wedge \phi', i) \succeq \text{sig}(\phi, i)$ and $\text{sig}(\phi \wedge \phi', i) \succeq \text{sig}(\phi', i)$.
3 If $M, i \models \oplus\phi$, then $\text{sig}(\oplus\phi, i) = \text{sig}(\phi, i+1)$.
4 If $M, i \models \ominus\phi$ and $i+1 \in \text{st}(M)$, then $\text{sig}(\ominus\phi, i) = \text{sig}(\phi, i+1)$.
5 If $M, i \models \nu z.\phi$ and $\mu h(\nu z.\phi) = k$, $\text{sig}(\nu z.\phi, i) \succeq_k \text{sig}(\phi[\nu z.\phi/z], i)$.
6 If $M, i \models \mu z.\phi$ and $\mu h(\mu z.\phi) = k$, $\text{sig}(\mu z.\phi, i) \succ_k \text{sig}(\phi[\mu z.\phi/z], i)$.

Proof: [14, lemma 3.5] □

The following technical lemma allows us to take an infinite sequence of signatures corresponding to a sequence of states in an infinite model, and extract a prefix of this sequence in such a manner that the corresponding prefix of the model can be used to build a proper Hintikka structure.

Lemma 4.9 Let X be a set, $\leq$ a well-founded total ordering of X, and $\bar{x}^1\bar{x}^2 \ldots \in (X^n)^\omega$ an infinite sequence of n-vectors of elements of X. There are indices h and h' such that $h < h'$ and $\bar{x}_k^h \leq \bar{x}_k^{h'}$ for all $1 \leq k \leq n$.

Proof: This is a formulation of Dickson's lemma, originally in [5]. □

Theorem 4.10 [Completeness] Let ϕ_0 be a νTL-sentence. If ϕ_0 is satisfiable, then there exists a proper Hintikka structure H for ϕ_0.

Proof: Suppose that $M, 1 \models \phi_0$. For each $i \in \text{st}(M)$, let $A_i = \{\phi \in \text{cl}(\phi_0) \mid M, i \models \phi\}$. If M is finite, $H = (A_1 \ldots A_{|M|}, \epsilon)$ provides the required, trivially proper Hintikka structure for ϕ_0. Suppose then that M is infinite.

For each A_i, define a relation $\to'_i$ by:

- if $\phi_1 \wedge \phi_2 \in A_i$, then $\phi_1 \wedge \phi_1 \to'_i \phi_1$, $\phi_1 \wedge \phi_1 \to'_i \phi_2$
- if $\phi_1 \vee \phi_2 \in A_i$, then $\phi_1 \vee \phi_2 \to'_i \phi_1$ if $\phi_1 \in A_i$ and $\text{sig}(\phi_1 \vee \phi_2, i) = \text{sig}(\phi_1, i)$, and otherwise $\phi_1 \vee \phi_2 \to'_i \phi_2$.

Since there are only finitely many atoms of ϕ_0, there is an infinite sequence of indices $m_1 < m_2 < \ldots$ such that $A = A_{m_1} = A_{m_2} = \ldots$ Fix some enumeration $\phi_1, \ldots, \phi_n$ of elements of A, and define a sequence $\bar{x}^1\bar{x}^2 \ldots \in (\text{Sig}^n)^\omega$ by $\bar{x}^i = (\text{sig}(\phi_1, m_i), \ldots, \text{sig}(\phi_n, m_i))$. By 4.9, there are h and h' such that $h < h'$ and for all $1 \leq i \leq n$, $\bar{x}_i^h \preceq \bar{x}_i^{h'}$, i.e. for all $\phi \in A$, $\text{sig}(\phi, m_h) \preceq \text{sig}(\phi, m_{h'})$.

Define a Hintikka structure H by $H = (A_1 \ldots A_k, (B_1, \to_1) \ldots (B_m, \to_m))$ where $k = m_h - 1$, $m = m'_h - m_h$, $B_i = A_{k+i}$ and $\to_i = \to'_{k+i}$ for every $i \in [m]$. Let $\to$, $\xrightarrow{z}$ etc. be as in 3.7. By lemma 4.8 we know that

- if $(\phi, i) \to (\psi, j)$ then $\text{sig}(\phi, i+k) \succeq \text{sig}(\psi, j+k)$ (by choice of $\to'$)
- if $(\phi, i) \xrightarrow{z} (\psi, j)$ for ν-variable z and $h = \mu h(\phi)$, $\text{sig}(\phi, i+k) \succeq_h \text{sig}(\psi, j+k)$
- if $(\phi, i) \xrightarrow{z} (\psi, j)$ for μ-variable z and $h = \mu h(\phi)$, $\text{sig}(\phi, i+k) \succ_h \text{sig}(\psi, j+k)$
- if $(\phi, i) \xrightarrow{\odot} (\psi, j)$, then $\text{sig}(\phi, i+k) = \text{sig}(\psi, j+k)$
- if $(\phi, i) \xhookrightarrow{\odot} (\psi, j)$, then $\text{sig}(\phi, i+k) \succeq \text{sig}(\psi, j+k)$ (by choice of k, m and the fact that $\text{sig}(\phi, m_{h'}) \succeq \text{sig}(\phi, m_h)$)

These imply that

- if $(\phi, i) \xhookrightarrow{\trianglelefteq z *} (\psi, j)$ and the μ-height of every formula in the corresponding sequence of dependencies is at least h, then $\text{sig}(\phi, i+k) \succeq_h \text{sig}(\psi, j+k)$

Suppose then that H was not proper, i.e. that for some $i \in [m]$, $\phi \in B_i$ and $z \in \text{var}\mu(\phi_0)$, $(\phi, i) \overset{z}{\hookrightarrow}{}^*(\phi, i)$. Take the sequence of dependencies corresponding to $(\phi, i) \overset{z}{\hookrightarrow}{}^*(\phi, i)$. Somewhere along this sequence there is a dependency $(\mu z.\psi, j) \overset{z}{\rightarrow}(\psi[z/\mu z.\psi], j')$. Define $h = \mu h(\mu z.\psi)$. As z is the outermost variable unwound along this sequence of dependencies and the sequence forms a loop, every formula along the sequence must contain $\mu z.\psi$ as a subsentence, i.e. the μ-height of every formula along the sequence is at least h. But then by the above $\text{sig}(\phi, i+k) \succ_h \text{sig}(\phi, i+k)$, clearly a contradiction. □

As an easy corollary, we obtain the finite model property for νTL.

Definition 4.11 An infinite model M is *ultimately periodic* iff there are $k \geq 0$ and $m > 0$ such that $M_i = M_{i+m}$ for all $i > k$. We call the minimal such k and m the *initial length* and the *period length* of M, respectively. □

Corollary 4.12 [Finite models] A νTL-sentence is satisfiable iff it has a finite or ultimately periodic model. □

5 Decision method

Given a Hintikka structure H for ϕ_0, determining whether it is proper or not can be done in one pass through the structure.

Definition 5.1 Let H be as in 3.6. For notational convenience, extend $\text{var}(\phi_0)$ by a new element 0, write $\overset{0}{\rightarrow}{}^*$ for $\rightarrow^*$, and extend $\trianglelefteq$ by: $0 \trianglelefteq z$ for all $z \in \text{var}(\phi_0)$.

For every $i \in [m]$, define $D(i)$, the set of dependencies between B_1 and B_i by: $D(i) = \{(\phi, \psi, z) \mid (\phi, 1) \overset{z}{\rightarrow}{}^*(\psi, i)\}$.

Define also DL, the set of dependencies from B_1 looping back to B_1 by: $DL = \{(\phi, \psi, z) \mid (\phi, 1) \overset{z}{\hookrightarrow}{}^*(\psi, 1)\}$. □

It is easy to see that $D(1)$ is determined by $(B_1, \rightarrow_1)$, every $D(i+1)$ is determined by $D(i)$ and $(B_{i+1}, \rightarrow_{i+1})$, and DL is determined by $D(m)$. What is more, H is proper iff there is no ϕ and μ-variable z such that $(\phi, \phi, z) \in DL$. If we can find a bound so that if there is any proper Hintikka structure for ϕ_0, then there is one having size smaller than the bound, the decision problem for νTL can be solved by the naive nondeterministic algorithm in figure 1.

To derive such a size bound, we take a Hintikka structure and remove from it portions that do not have a bearing on whether the structure as a whole is proper or not. For example, if there are $i < j \leq k$ such that $A_i = A_j$, we can remove the sequence from A_{i+1} to A_j. Similarly, if there are $i < j \leq m$ such that $B_i = B_j$ and $D(i) = D(j)$, i.e. for all ϕ, ψ, z, $(\phi, 1) \overset{z}{\rightarrow}{}^*(\psi, i)$ iff $(\phi, 1) \overset{z}{\rightarrow}{}^*(\psi, j)$, then the sequence from B_{i+1} to B_j can be removed and the result is proper if the original was. These observations alone would be enough to establish a size bound. However, by analysing the situation more carefully we can obtain a tighter bound.

```
input: a νTL-formula φ0
begin
        /* Guess the initial part of a Hintikka structure. */
        guess k and m obeying the size bound (a function of |φ0|)
        guess A1, an atom of φ0, such that φ0 ∈ A1
        for i := 2 to k
                guess Ai, a successor of Ai-1
        if m = 0 then
                if Ak is terminal, then accept, else reject
        /* Guess the recurring part of a Hintikka structure */
        guess (B1, →1), an adorned atom of φ0, such that B1 is a successor of Ak
        compute D(1) on the basis of (B1, →1)
        for i := 2 to m
                guess (Bi, →i), such that Bi is a successor of Bi-1
                compute D(i) on the basis of D(i - 1) and (Bi, →i)
        if B1 is not a successor of Bm, reject
        /* Check that the Hintikka structure is proper. */
        compute DL from D(m)
        if there are φ ∈ B1, z ∈ varμ(φ0) such that (φ, φ, z) ∈ DL then reject
        accept
end
```

Figure 1: A nondeterministic decision algorithm Alg_{ND1} for νTL

Definition 5.2 Let ϕ_0 and H be as in 3.6. We say that H is *redundant* iff

1 there are $1 \leq i < j \leq k$ such that $A_i = A_j$, or

2 there are $1 \leq i < j \leq m$ such that $B_i = B_j$ and for all $\phi \in B_1$, $\psi \in B_i$

- if $(\phi, 1) \xrightarrow{z}{}^{*} (\psi, i)$ for $z \in \mathrm{var}(\phi_0) \cup \{0\}$, then there is $x \in \mathrm{var}(\phi_0) \cup \{0\}$ such that $x \trianglelefteq z$ and $(\phi, 1) \xrightarrow{x}{}^{*} (\psi, j)$
- if $(\phi, 1) \xrightarrow{z}{}^{*} (\psi, i)$ for $z \in \mathrm{var}\mu(\phi_0)$, then there is $x \in \mathrm{var}\mu(\phi_0)$ such that $z \trianglelefteq x$ and $(\phi, 1) \xrightarrow{x}{}^{*} (\psi, j)$ □

Proposition 5.3 Let ϕ_0 be a νTL-formula. If there is a proper Hintikka structure for ϕ_0, there is a proper non-redundant Hintikka structure for ϕ_0.

Proof: Let us show that for any redundant proper Hintikka structure H for ϕ_0, there is a strictly smaller proper Hintikka structure H' for ϕ_0. Let H be a proper redundant Hintikka structure for ϕ_0. If H fulfils clause 1 of 5.2, removing the sequence from A_{i+1} to A_j gives a proper H' such that $|H'| < |H|$. If H fulfils clause 2 of 5.2, removing the sequence from $(B_{i+1}, \rightarrow_{i+1})$ to $(B_j, \rightarrow_j)$ gives a Hintikka structure H' such that $|H'| < |H|$.

Suppose then that H' is not proper, i.e. that there are $\phi \in B_1$, $z \in \mathrm{var}\mu(\phi_0)$ such that $(\phi, 1) \overset{z}{\hookrightarrow}{}^{*} (\phi, 1)$ in H'. The corresponding sequence of dependencies can be divided to segments from some $\psi \in B_1$ to some $\psi' \in B_i$, and from some $\psi \in B_i = B_j$ (in H) to some $\psi' \in B_1$. Take any segment of the first type, and the outermost variable x unwound along this segment.

- If $x = z \in \mathrm{var}\mu(\phi_0)$, there is a $z' \in \mathrm{var}\mu(\phi_0)$ such that $z \trianglelefteq z'$ and $(\psi, 1) \xrightarrow{z'}{}^{*} (\psi', j)$ in H,
- If $x \neq z$, i.e. $x \triangleleft z$, there is some $x' \in \mathrm{var}(\phi_0) \cup \{0\}$ such that $x' \trianglelefteq x$ and $(\psi, 1) \xrightarrow{x'}{}^{*} (\psi', j)$ in H.

In both cases we can replace the segment in H' by a segment in H so that the resulting sequence of dependencies in H corresponds to $(\phi, 1) \overset{z'}{\hookrightarrow}{}^{*} (\phi, 1)$ in H for some μ-variable $z' \in \mathrm{var}\mu(\phi_0)$. Consequently, H is not proper, contrary to the assumption. □

Definition 5.4 Let ϕ_0 and H be as in 3.6. For every $i \in [m]$, define

$$\begin{aligned} D^{\min}(i) &= \{(\phi, \psi, z) \in D(i) \mid \text{there is no } x \triangleleft z \text{ such that } (\phi, \psi, x) \in D(i)\} \\ D^{\mu\max}(i) &= \{(\phi, \psi, z) \in D(i) \mid z \in \mathrm{var}\mu(\phi_0) \text{ and there is no } z \triangleleft x \text{ such} \\ &\qquad \text{that } x \in \mathrm{var}\mu(\phi_0) \text{ and } (\phi, \psi, x) \in D(i)\} \end{aligned}$$

Define also

$$\begin{aligned} DL^{\mu\max} &= \{(\phi, \psi, z) \in DL \mid z \in \mathrm{var}\mu(\phi_0) \text{ and there is no } z \triangleleft x \text{ such} \\ &\qquad \text{that } x \in \mathrm{var}\mu(\phi_0) \text{ and } (\phi, \psi, x) \in DL\} \end{aligned}$$

□

Proposition 5.5 Let ϕ_0 and H be as in 3.6. If H is non-redundant, then $k \leq 2^n$ and $m \leq 2^{3n^2 \log n}$, where $n = |\phi_0|$.

Proof: As $|\mathrm{cl}(\phi_0)| \leq |\phi_0| = n$, there are fewer than 2^n distinct atoms for ϕ_0. As no atom can occur twice in $A_1 \ldots A_k$, $k \leq 2^n$.

If for some $1 \leq i < j \leq m$, $B_i = B_j$, $D^{\min}(i) = D^{\min}(j)$, and $D^{\mu\max}(i) = D^{\mu\max}(j)$, then H is redundant. As for every $(\phi, \psi) \in \mathrm{cl}(\phi_0)^2$ there is at most one $z \in \mathrm{var}(\phi_0)$ such that $(\phi, \psi, z) \in D^{\min}(i)$, and as $|\mathrm{var}(\phi_0)| \leq |\phi_0| = n$, there are only $n^{(n^2)} = 2^{n^2 \log n}$ different choices for $D^{\min}(i)$. As the same holds for $D^{\mu\min}(i)$, there are at most $2^{2n^2 \log n}$ distinct choices for $D^{\mu\max}(i)$ and $D^{\min}(i)$, and there are at most $2^n \cdot 2^{2n^2 \log n} \leq 2^{3n^2 \log n}$ distinct situations that can be encountered in the recurring part without the structure being redundant. □

Corollary 5.6 [Small models] A νTL-formula ϕ_0 is satisfiable iff it either has a finite model of length $\leq 2^n$, or an ultimately periodic model with initial length $\leq 2^n$ and period length $\leq 2^{3n^2 \log n}$, where $n = |\phi_0|$. □

We have now established the bound required by the nondeterministic decision algorithm Alg_{ND1} in figure 1. With some simple modifications we can make the algorithm more efficient.

Proposition 5.7 There is a nondeterministic decision algorithm Alg_{ND2} for νTL working in $O(n^2 \log n)$ space.

Proof: Observe the following facts about Alg_{ND1}:

- After computing $D(i)$, $D(i-1)$ is never referred to again, and the space used for storing $D(i-1)$ may be reused for storing $D(i+1)$
- After processing A_i, or $(B_i, \rightarrow_i)$ for $i > 2$, A_{i-1} or $(B_{i-1}, \rightarrow_{i-1})$ is never referred to again, and storage space may be similarly reused.

Therefore, in addition to k, m, i, taking $O(n^2 \log n)$ space ($n = |\phi_0|$), we only need to store a fixed number of (adorned) atoms, each taking $O(n)$ space, and a fixed number of dependency sets, each taking $O(n^3)$ space, i.e. Alg_{ND1} can be modified to work in $O(n^3)$ space.

However, we can still do better. Notice that

- H is proper iff there is no $\phi \in B_1$ and $z \in \mathrm{var}\mu(\phi_0)$ such that $(\phi, \phi, z) \in DL^{\mu\max}$.
- $DL^{\mu\max}$ is determined by $D^{\mu\max}(m)$ and $D^{\min}(m)$
- $D^{\mu\max}(i+1)$ and $D^{\min}(i+1)$ are determined by $D^{\mu\max}(i)$, $D^{\min}(i)$, $\rightarrow_{i+1}$, and $D^{\mu\max}(1)$ and $D^{\min}(1)$ are determined by $\rightarrow_1$

Consequently, to check that H is proper we can just store $D^{\mu\max}(i)$ and $D^{\min}(i)$ instead of $D(i)$. Both $D^{\mu\max}(i)$ and $D^{\min}(i)$ can be stored in $O(n^2 \log n)$ space. This results in an algorithm working in $O(n^2 \log n)$ space. □

Applying some elementary complexity theory gives us deterministic algorithms.

Proposition 5.8 There is a deterministic decision algorithm for νTL, working in $2^{O(n^2 \log n)}$ space and time.

Proof: Apply the transformation of [1, theorem 2.8 (g)] to Alg_{ND2}. □

In principle, this determinisation works by writing down all the possible execution states of Alg_{ND2} on the given input ϕ_0 as a graph, and by looking for a path from the initial to an accepting state in this graph. In a practical implementation, the search for such a path would be done as a depth-first search, creating new nodes of the graph when necessary as the search progresses.

Alternatively, by a time-space tradeoff, we can obtain a deterministic algorithm working in singly exponential time but in polynomial space.

Proposition 5.9 There is a deterministic decision algorithm for νTL, working in $O(n^4 \log^2 n)$ space and $2^{O(n^4 \log^2 n)}$ time.

Proof: Apply Savitch's theorem ([12], also [1, cor. 2.1]) to transform Alg_{ND2} into a deterministic algorithm working in $O((n^2 \log n)^2) = O(n^4 \log^2 n)$ space. □

6 Tableaux

The decision method presented in the previous sections works in a 'bottom-up' fashion, by starting from the consistent sets of subformulae of the given formula. In practice, however, it is often easier to work in a 'top-down' manner, starting from just the formula itself and dissecting it into subformulae only when this becomes necessary. In this section we show how to reformulate the decision procedure as a set of tableau rules for this purpose.

Tableau rules			
name	application	condition	dependencies
$\wedge$	$\dfrac{\Gamma, \phi \wedge \phi'}{\Gamma, \phi, \phi'}$		$\phi \wedge \phi' \rightarrow \phi$ $\phi \wedge \phi' \rightarrow \phi'$
$\vee$-left	$\dfrac{\Gamma, \phi \vee \phi'}{\Gamma, \phi}$		$\phi \vee \phi' \rightarrow \phi$
$\vee$-right	$\dfrac{\Gamma, \phi \vee \phi'}{\Gamma, \phi'}$		$\phi \vee \phi' \rightarrow \phi'$
σ	$\dfrac{\Gamma, \sigma z.\phi}{\Gamma, \phi[\sigma z.\phi/z]}$		$\sigma z.\phi \xrightarrow{z} \phi[\sigma z.\phi/z]$
$\odot$	$\dfrac{\odot\phi_1, \ldots, \odot\phi_n,\ a_1, \ldots, a_m,\ \neg a'_1, \ldots \neg a'_{m'}}{\phi_1, \ldots, \phi_n}$	$a_i, a'_i \in \mathcal{Z}_c$	$\odot\phi_i \rightarrow \phi_i$

Note: In the first four rules, the set Γ is disjoint from $\phi \wedge \phi'$ etc.
In addition to the dependencies above, $\psi \rightarrow \psi$ for each $\psi \in \Gamma$.

Figure 2: Tableau rules

Definition 6.1 Let ϕ_0 be a νTL-formula. A *tableau* T for ϕ_0 consists a finite sequence $\Gamma_1\Gamma_2 \ldots \Gamma_{n-1}\Gamma_n$ and an index $k \in N$, the *recurrence point* of T, where

- every Γ_i is a set of νTL-formulae
- every Γ_{i+1} is derived from Γ_i by applying one of the rules in figure 2,
- $\Gamma_1 = \{\phi_0\}$ and
- Γ_n is a leaf (see 6.2) and no Γ_i for $i < n$ is a leaf

We say that i is a $\odot$-*point* of T iff Γ_i is of the form that the $\odot$-rule can be applied to it, i.e. every $\phi \in \Gamma_i$ is either of the form $\phi = \odot\phi'$, or a (negated) atomic proposition.

For every i, the rule applied at point i of T induces dependency relations $\rightarrow$, $\xrightarrow{z}$ between formulae in Γ_i and Γ_{i+1} as indicated in figure 2. We write $(\phi, i) \xrightarrow{\trianglelefteq z}{}^*(\psi, j)$ iff there is a path of dependencies from $\phi \in \Gamma_i$ to $\psi \in \Gamma_j$ such that no fixpoints outside the scope of σz are unwound along this path, and $(\phi, i) \xrightarrow{z}{}^*(\psi, j)$ iff there is a path of dependencies from $\phi \in \Gamma_i$ to $\psi \in \Gamma_j$ such that the fixpoint σz is unwould along this path and no fixpoints outside the scope of σz are. □

Definition 6.2 Let T be as in 6.1. Γ_n is a *leaf* iff n is a $\odot$-point of T and one of the following holds:

1 There is some $a \in \mathcal{Z}_c$ such that $\{a, \neg a\} \subseteq \Gamma_n$.

2 There is a point m such that $m < n \leq k$ and $\Gamma_m = \Gamma_n$

3 There is a point m such that $k < m < n$, $\Gamma_m = \Gamma_n$ and for all $\phi \in \Gamma_k$, $\psi \in \Gamma_n$:

- if $(\phi, k) \xrightarrow{z}{}^*(\psi, m)$ for $z \in \text{var}(\phi_0) \cup \{0\}$,
  there is $x \in \text{var}(\phi_0) \cup \{0\}$ such that $x \trianglelefteq z$ and $(\phi, k) \xrightarrow{x}{}^*(\psi, n)$

- if $(\phi, k) \xrightarrow{z}{}^{*} (\psi, m)$ for $z \in \mathrm{var}\mu(\phi_0)$,
  there is $x \in \mathrm{var}\mu(\phi_0)$ such that $z \trianglelefteq x$ and $(\phi, k) \xrightarrow{x}{}^{*} (\psi, n)$

4 There is no formula of the form $\oplus\phi \in \Gamma_n$, and case 1 does not hold.

5 The recurrence point $k < n$, $\Gamma_k = \Gamma_n$ and there is no $z \in \mathrm{var}\mu(\phi_0)$ and $\phi_1, \ldots, \phi_h \in \Gamma_n$ such that $\phi_1 = \phi_h$, $(\phi_i, k) \xrightarrow{\trianglelefteq z}{}^{*} (\phi_{i+1}, n)$ for every $i < h$, and $(\phi_i, k) \xrightarrow{z}{}^{*} (\phi_{i+1}, n)$ for some $i < h$.

A leaf is *successful* in cases 4 and 5, and *unsuccessful* otherwise. A tableau is successful iff its leaf Γ_n is successful. □

The conditions on an unsuccessful leaf of type 2 correspond to clause 1 of the definition 5.2 of a Hintikka structure being redundant, and those on a leaf of type 3 correspond to clause 2 of 5.2. Intuitively, an unsuccessful leaf of these types can be seen as saying: 'There is already an earlier situation Γ_m that is similar to the current one Γ_n as far as dependencies from the recurrence point Γ_k are concerned. If we want to find a successful tableau for ϕ_0, we could just as well go on building it from Γ_m onwards instead of Γ_n.'

Proposition 6.3 Let ϕ_0 be a νTL-formula. There is a non-redundant proper Hintikka structure for ϕ_0 iff there is a successful tableau for ϕ_0.

Proof: From a non-redundant proper Hintikka structure we can read a successful tableau, and from a successful tableau a proper Hintikka structure. □

Corollary 6.4 A νTL-formula is satisfiable iff it has a successful tableau. □

We can view a tableau as a sequence of choices in a systematic search to find a model for ϕ_0. It is easy to see that such a search always reaches a leaf after a finite number of steps. Conversely, if wanting to show that ϕ_0 is not satisfiable, we can do this by constructing a tree-like structure where each branch is a tableau, and each node either has one son, justified by the rules $\wedge$, σ or $\odot$, or two sons, one of which is justified by the rule **∨-left** and the other by **∨-right**, both applied to the same formula. If each leaf in such a tree is unsuccessful, then ϕ_0 is not satisfiable.

Acknowledgements

I would like to thank Colin Stirling for many useful discussions, and the anonymous referees for their constructive comments. This work has been funded by the Academy of Finland, the British Council, the CVCP Overseas Research Students Awards Scheme and the Finnish Cultural Foundation.

References

[1] Balcázar, J. L. & Díaz, J. & Gabarró. J.: *Structural Complexity I*, EATCS Monographs on Theoretical Computer Science vol. 11, Springer-Verlag, 1988

[2] Banieqbal, B. & Barringer, H.: Temporal Logic with Fixed Points, in *Temporal Logic in Specification*, LNCS vol. 398, Springer-Verlag, 1989, pp. 62-74

[3] Barringer, H. & Kuiper, R. & Pnueli, A.: A Really Abstract Concurrent Model and its Temporal Logic, in *Proc. of the 13th ACM POPL*, 1986

[4] Büchi, J. R.: On a Decision Method in Restricted Second-Order Arithmetics, in *Proc. of the 1960 International Congress on Logic, Methodology and Philosophy of Science*, Stanford University Press, 1962, pp. 1-12

[5] Dickson, L. E.: Finiteness of the Odd Perfect and Primitive Abundant Numbers with Distinct Prime Factors, in *Americal Journal of Mathematics*, vol. 35, 1913, pp. 413-422

[6] Emerson, E. A.: Temporal and Modal Logic, in van Leeuwen, J. (ed.): *Handbook of Theoretical Computer Science*, Elsevier/North-Holland, 1990, pp. 997-1072

[7] Emerson, E. A. & Clarke, E. M.: Characterising Correctness Properties of Parallel Programs using Fixpoints, in *Proc. of the 7th ICALP*, LNCS vol. 85, Springer-Verlag, 1980, pp. 169-181

[8] Kozen, D.: Results on the Propositional μ-calculus, in *Theoretical Computer Science*, vol. 27, 1983, pp. 333-354

[9] Lichtenstein, O.: *Decidability, Completeness, and Extensions of Linear Time Temporal Logic*, PhD thesis, The Weizmann Institute of Science, Rehovot, Israel, 1991

[10] Lichtenstein, O. & Pnueli, A. & Zuck, L.: The Glory of the Past, in *Proc. of Workshop on Logics of Programs*, LNCS vol. 193, Springer-Verlag, 1985, pp. 97-107

[11] Safra, S.: On the Complexity of ω-Automata, in *Proceedings of the 29th Symposium on Foundations of Computer Science*, 1988, pp. 319-327

[12] Savitch, W. J.: Relationships Between Nondeterministic and Deterministic Tape Complexities, in *Journal of Computer and System Sci.*, vol. 4, 1970, pp. 177-192

[13] Stirling, C.: Modal and Temporal Logics, in Abramsky, S. & al. (eds.): *Handbook of Logic in Computer Science*, Oxford University Press, 1992, pp. 477-563

[14] Streett, R. S. & Emerson, E. A.: An Automata Theoretic Decision Procedure for the Propositional Mu-Calculus, in *Information and Computation*, vol. 81, 1989, pp. 249-264

[15] Vardi, M. Y.: A Temporal Fixpoint Calculus, in *Proceedings of the 15th ACM Symposium on Priciples of Programming Languages*, 1988, pp. 250-259

[16] Wolper, P.: *Synthesis of Communicating Processes from Temporal Logic Specifications*, PhD thesis, Stanford University, 1982

[17] Wolper, P.: Temporal Logic can be More Expressive, in *Information and Control*, vol 56, 1983, pp. 72-99

Message passing mutex*

Ekkart Kindler, Rolf Walter
Humboldt-Universität zu Berlin, Institut für Informatik
10099 Berlin, Germany

Abstract

We present a new solution of the mutual exclusion problem, which is modelled as a Petri net. In order to present a concise model of the algorithm, we extend Petri nets by the concepts of *progress* and *non-progress transitions* and *fair arcs*.

Moreover, we introduce a simple temporal logic in order to express and verify properties of distributed algorithms. The verification rules allow for rigorous reasoning close to the arguments of an informal proof.

The verification method is applied to the new solution as well as to Peterson's solution. This allows for a comparison of the two algorithms.

1 Introduction

The mutual exclusion problem (mutex, for short) mirrors typical phenomena in distributed computing in a nice and simple setting. Dijkstra [7] states the mutex problem as follows:

> We consider N cyclic processes. In each cycle a so-called "critical section" occurs. Outside the critical section each process may remain forever or it may request access to its critical section, spontaneously. A solution to this problem must guarantee that at any moment only one process is in its critical section and that each process requesting access to the critical section reaches its critical section eventually. The use of a static priority is not allowed. Reading and writing of shared variables is indivisible.

In the beginning sixties the solvability of mutex was an open question [7]. The first two proposals to solve the problem were wrong (cf. [9]). A precise specification and a rigorous way of reasoning turned out to be necessary to avoid mistakes, even if only two processes are considered. Many contributions which solve mutex [10, 8, 13], compare or verify solutions [14, 19, 6, 11, 18, 2] demonstrate the ongoing academic significance of the problem. Mutex became a paradigm for distributed algorithms.

There are mainly two aspects of distributed programming that constitute the challenge of mutex. Firstly, the assumed granularity of an *atomic action* justifies the complexity of each solution. For example, the use of a semaphore easily solves the mutex problem. But it contradicts the requirement, that reading and writing of variables are assumed to be the only indivisible operations.

*supported by the Deutsche Forschungs Gemeinschaft SFB 342, TP YE1, the ESPRIT Basic Research WG 6067 Caliban and the DFG-Project: Distributed Algorithms

We choose a message passing based solution, such that no particular granularity of atomic actions has to be defined.

The second aspect results from the requirement that a process is never forced to request an access to the critical section. As a consequence every formal correctness proof of a mutex solution must include a priority or fairness argument, when a process is allowed to stop outside the critical section. Dijkstra used the *program statement* "remainder of the cycle in which stopping is allowed" to focus on this aspect. We slightly extend the Petri net model to cope with these particular aspects of distributed algorithms: we use *weak* transitions (graphically inscribed with "w") to denote, that these transitions may occur, but must not. And we introduce the notion of fair behaviour of a transition w.r.t. a place in its preset. This is graphically indicated by a white arrowhead of the connecting arc. Formal means to reason about these Petri net models are introduced.

In Section 2 a new solution to the mutex problem for two processes is modelled as a system net, which we call *message passing mutex.* Section 3 formally introduces system nets and its semantics. Section 4 introduces formulas to describe *invariant* and *liveness* properties of system nets. Section 5 introduces proof rules for the derivation of such properties. Section 6 proves message passing mutex correct. Section 7 applies the proof techniques to Peterson's mutex algorithm and compares the two algorithms.

2 The message passing mutex algorithm

Figure 1 shows the net model of a solution to the mutex problem. There are

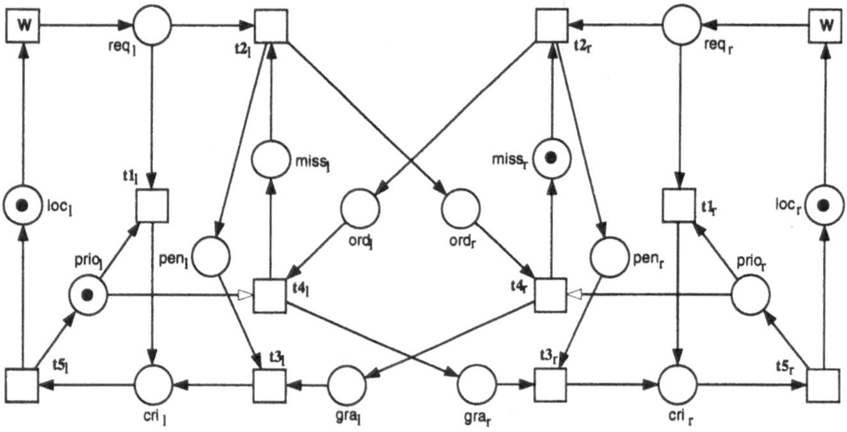

Figure 1: A message passing algorithm for mutual exclusion

two agents (called *left* and *right*), which cyclically adopt the states *loc*al, *req*uest and *cri*tical. The algorithm guarantees that *left* and *right* are never *cri*tical at the same time. This property is called *mutex property.* Moreover, a *req*uesting agent will eventually enter its *cri*tical section. This property is called *evolution property*.

The algorithm itself works as follows: In the beginning, *left* has *prio*rity. It can immediately enter its *cri*tical section, when access is *req*uested. In contrast, *right misses* the priority. When *right req*uests access to the critical section, it *ord*ers the priority from *left*. *Right* is *pen*ding until *left grants* priority for *right*. Then *right* immediately enters its *cri*tical section. Now, *right* has priority and the roles of *left* and the *right* are exchanged.

The mutex property is denoted by the invariant formula $\Box\neg(cri_l \wedge cri_r)$, the temporal formulas $req_l \leadsto cri_l$ and $req_r \leadsto cri_r$ denote the evolution property. Validity of the latter depends on a fair behaviour of transitions t_4. Note, that an agent cannot prevent the other agent from becoming *req*uesting. On the other hand, an agent may remain *loc*al forever; it cannot be forced to become *req*uesting. This is indicated by the inscription "w" in the corresponding transitions.

A proof of the above properties is shown in Section 5.

3 Basic definitions

In this section we introduce *net systems* for modelling distributed algorithms. Essentially, a net system Σ is a T-restricted contact-free elementary net system [17] with the following three extensions:

1. transitions with loops may fire when the corresponding place is marked;
2. we distinguish two types of transitions, called *non-progress transitions* T_w (graphically indicated by a label "w") and *progress transitions* $T \setminus T_w$;
3. we distinguish *fair* arcs F_f, graphically represented by a white arrowhead.

A *run* ρ of a net system Σ is a process [5] of the net with two additional requirements concerning the progress transitions and the fair arcs (see Def. 7).

3.1 Net Systems

Definition 1 (Net, Marking)

A triple $N = (S, T; F)$ is a *net* iff S and T are disjoint sets and $F \subseteq (S \times T) \cup (T \times S)$. A subset $M \subseteq S$ is called a *marking* of the net N.

As usual, the elements of S, T, and F are called *places*, *transitions*, and *arcs* of the net and are represented by circles, squares, and arrows, respectively. A marking is represented by black dots (*token*) in the corresponding places.

Notation 2 (Pre- and Postset, Minimal and Maximal Elements)

Let $N = (S, T; F)$ be a net and $x \in S \cup T$ an element of N. The *preset* ${}^\bullet x$ and the *postset* $x^\bullet$ of x are defined by ${}^\bullet x = \{y \in S \cup T \mid (y, x) \in F\}$ and $x^\bullet = \{y \in S \cup T \mid (x, y) \in F\}$. For a set $X \subseteq S \cup T$ we define ${}^\bullet X = \bigcup_{x \in X} {}^\bullet x$ and $X^\bullet = \bigcup_{x \in X} x^\bullet$.

The *minimal elements* ${}^\circ N$ and the *maximal elements* N° of a net N are defined by ${}^\circ N = \{x \in S \cup T \mid {}^\bullet x = \emptyset\}$ and $N^\circ = \{x \in S \cup T \mid x^\bullet = \emptyset\}$.

A net $N = (S, T; F)$ is *T-restricted* iff ${}^\circ N \cup N^\circ \subseteq S$.

The occurrence rule is defined such that transitions with a loop may fire, when the corresponding place is marked.

Definition 3 (Occurrence Rule, Reachable Markings)

Let $N = (S, T; F)$ be a net and M be a marking of N. A transition $t \in T$ is *enabled at* M iff $t^\bullet \cap M \subseteq {}^\bullet t \subseteq M$.

The *effect* M^t of a transition at marking M is defined by $M^t = (M \setminus {}^\bullet t) \cup t^\bullet$.

A transition t that is enabled at M may *occur* resulting in the *follower marking* M^t.

The set of markings $[M\rangle$ that are *reachable* from a marking M is defined as the least set such that

- $M \in [M\rangle$ and
- if $M_1 \in [M\rangle$ and $t \in T$ is enabled at M_1 then $M_1^t \in [M\rangle$.

Note, that the set M^t is defined for every transition t, even when t is not enabled at marking M.

A net system is a T-restricted, finite net together with an initial marking M_0, a set of distinguished *non-progress* transitions T_w, and a set of distinguished *fair* arcs F_f. Moreover, we require that the initial marking M_0 is *contact-free* for N, i.e. for each marking $M \in [M_0\rangle$ a transition t is enabled at M iff ${}^\bullet t \subseteq M$.

Definition 4 (Net System)

Let $N = (S, T; F)$ be a T-restricted finite net, M_0 a contact-free marking of N and $T_w \subseteq T$ and $F_f \subseteq F$. Then $\Sigma = (N, M_0, T_w, F_f)$ is a *net system*.

3.2 Runs

A run of a net system is a particularly labelled *occurrence net* (cf. [5]).

Definition 5 (Occurrence Net)

A net $K = (B, E; \lessdot)$ is an occurrence net iff

1. K is T-restricted,
2. the transitive closure of $\lessdot$ (denoted by $<$ in the sequel) is acyclic,
3. for each $b \in B$ the conditions $|\, {}^\bullet b \,| \leq 1$ and $|\, b^\bullet \,| \leq 1$ are satisfied, and
4. for each element $x \in B \cup E$ the set of its predecessors $\{y \in B \cup E \mid y < x\}$ is finite.

Definition 6 (co-sets and slices)

Let $K = (B, E; \lessdot)$ be an occurrence net.

1. A set $Q \subseteq B$ is a *co-set* of K iff for any two $b, b' \in Q$ neither $b < b'$ nor $b' < b$ holds true.
2. A co-set $Q \subseteq B$ is a *slice* of K iff there exists no co-set Q' of K that properly contains Q.

A slice of an occurrence net corresponds to a global state. Since a slice of K is a marking of K, the reachability of a slice Q from another slice Q' is well-defined (cf. Def. 3).

A run of a net system is now defined as a labelling function of an occurrence net. The conditions 1–4 (cf. [5]) are standard; these conditions imply that slices of the run correspond to markings of the system. Conditions 5 and 6 add the progress and fairness condition and will be explained below.

Definition 7 (Run of a Net System)
Let $\Sigma = (N, M_0, T_w, F_f)$ be a net system with $N = (S, T; F)$, $K = (B, E; <)$ an occurrence net, and $\rho : B \cup E \rightarrow S \cup T$ a labelling function such that

1. $\rho(B) \subseteq S$ and $\rho(E) \subseteq T$,
2. for each slice Q of K the restricted mapping $\rho\,|_Q$ is injective,
3. $\rho({}^\circ K) = M_0$,
4. for each $e \in E$ the conditions ${}^\bullet\rho(e) = \rho({}^\bullet e)$ and $\rho(e)^\bullet = \rho(e^\bullet)$ hold,
5. for each $t \in T$ with ${}^\bullet t \subseteq \rho(K^\circ)$ holds $t \in T_w$, and
6. for each $t \in T$ and each $s \in S$ such that ${}^\bullet t \subseteq \rho(K^\circ) \cup \{s\}$ holds: if for each slice Q of K there exists a slice Q' reachable from Q with $s \in \rho(Q)$, then $(s,t) \notin F_f$.

Then the pair (K, ρ) is a *run* of Σ.

We write ρ instead of (K, ρ), when K is clear form the context.

Condition 5 requires that "at the end of the run" no progress transition t of the net system is enabled (the contact-freeness guarantees that the condition ${}^\bullet t \subseteq \rho(K^\circ)$ is sufficient). Condition 6 is similar to the progress requirement. Assume that s occurs infinitely often in (labellings of) slices of the run and only s is missing to enable transition t "at the end of the run" $\rho(K^\circ)$. Then the run is unfair with respect to the arc (s,t).

Thus, the progress requirement guarantees that a run does not stop, when a progress transition is activated. The fairness requirement guarantees that an repeatedly enabled transition will eventually occur. This fairness requirement is weaker than many other notions of fairness; but, it turned out to be sufficient for modelling most distributed algorithms.

4 Properties of Distributed Algorithms

Now, we introduce a temporal logic in order to express properties of distributed algorithms. The temporal logic is based on a propositional logic with propositional variables S, which correspond to the places of the net system, and the usual logical operators $\wedge$, $\vee$, $\Rightarrow$, etc. and $\neg$. The propositional formulas over S are denoted by $\mathcal{P}(S)$. The propositional formulas are interpreted on markings $M \subseteq S$ in the usual way: we assign *true* to a propositional variable $s \in M$ and *false* to a propositional variable $s \notin M$. When $\varphi \in \mathcal{P}(S)$ evaluates to true according to that assignment we write $M \models \varphi$.

Thus, the propositional formulas express properties of markings of net systems. For example $\neg(cri_l \wedge cri_r)$ is true for those markings that do not contain both places cri_l and cri_r (mutual exclusion).

This notion can be lifted to the slices of a run, because slices of a run correspond to markings of the system net. For a formula $\varphi \in \mathcal{P}(S)$ a slice Q of (K, ρ) is called *φ-slice* iff $\rho(Q) \models \varphi$.

Based on the propositional formulas we introduce two types of unnested temporal formulas: $\Box\varphi$ (pronounced "always" φ) to express *invariant properties* and $\varphi \leadsto \psi$ (pronounced φ "leadsto" ψ) to express *liveness properties* of a distributed algorithm.

The validity of a temporal formula is defined on runs first. A formula is valid for a net system, iff it is valid for each run of the net system.

Definition 8 (Temporal Operators)

Let $\varphi, \psi \in \mathcal{P}(S)$. Then $\Box\varphi$ and $\varphi \leadsto \psi$ are *temporal formulas* over S.

Let Σ be a net system with places S and (K, ρ) a run of Σ. Then $\Box\varphi$ is *valid in run ρ* (denoted by $\rho \models \Box\varphi$), iff each slice of K is a φ-slice. $\varphi \leadsto \psi$ is *valid in run ρ*, iff from each φ-slice a ψ-slice is reachable.

A temporal formula φ is *valid in a net system* Σ (denoted $\Sigma \models \varphi$) iff φ is valid in each run of the net system.

Many case studies have shown, that these simple temporal formulas are sufficient for expressing properties of typical distributed algorithms. For instance, $\Box\neg(cri_l \wedge cri_r)$ expresses the mutual exclusion property and the formulas $req_l \leadsto cri_l$ and $req_r \leadsto cri_r$ express the evolution property.

5 Proof rules for net systems

In this section we introduce rules for verifying properties of distributed algorithms. These rules can be partitioned into *pick-up* and *combination* rules. The pick-up rules reflect typical Petri net arguments. For instance, invariant properties $\Box\varphi$ are derived from *traps* and *S-invariants* of the net. The combination rules are classical in the field of temporal logic. For instance, liveness properties are verified by a *proof graph*, which was introduced as proof lattice by Owicki and Lamport [12].

5.1 Invariant properties

An S-invariant assigns a weight to each place, such that for all reachable markings the total number of tokens in each place (in our case 0 or 1) multiplied by the corresponding weight is the same. The significance of S-invariants originates from the fact, that S-invariants can be easily checked by methods of linear algebra. We syntactically represent an S-invariant of a system net Σ by

$$\Sigma \models w_1 \cdot s_1 + \ldots + w_n \cdot s_n = w$$

where the w_i are non-zero integers representing the weight for the distinct places $s_1, \ldots, s_n$ of Σ and w is an integer representing the weight function applied to the initial marking. For example,

$$prio_l + miss_l - req_l - loc_l = 0$$

is an S-invariant of the net system shown in Fig. 1.

A trap of a net system is a set of places $S' = \{s_1, \ldots, s_n\}$, such that for any marking M that contains a place of S' each marking M' that is reachable from M contains a place of S', too. We syntactically represent an initially marked (i.e. $M_0 \cap S' \neq \emptyset$) trap of Σ by

$$\Sigma \models s_1 + \ldots + s_n \geq 1$$

The trap property can be easily checked for a given net. For a more detailed information on S-invariants and traps we refer to [16].

From S-invariants and traps, we conclude some invariant properties:

Rule 1 (S-invariants, Traps)

1. From $\Sigma \models s_1 + \ldots + s_n = 1$ we conclude $\Sigma \models \Box[s_j \Rightarrow (\neg s'_1 \wedge \ldots \wedge \neg s'_k)]$ for $s_j \notin \{s'_1, \ldots, s'_k\} \subseteq \{s_1, \ldots, s_n\}$.
2. From $\Sigma \models s_1 + \ldots + s_n \geq 1$ or $\Sigma \models s_1 + \ldots + s_n = 1$ we conclude $\Sigma \models \Box[s_1 \vee \ldots \vee s_n]$.
3. From $\Sigma \models w_1 \cdot s_1 + \ldots + w_k \cdot s_k - w_{k+1} \cdot s_{k+1} - \ldots - w_n \cdot s_n = 0$, where $w_i \geq 1$ for $1 \leq i \leq n$, we conclude $\Sigma \models \Box[(s'_1 \vee \ldots \vee s'_l) \Rightarrow (s_1 \vee \ldots \vee s_k)]$ for $\{s'_1, \ldots, s'_l\} \subseteq \{s_{k+1}, \ldots, s_n\}$.

The invariant properties justified by S-invariants and traps can be combined by other rules. We can *substitute* a formula by a propositionally equivalent formula in any temporal formula. For instance, we (implicitly) substitute $(\neg s_1 \vee \neg s_2)$ or $\neg(s_1 \wedge s_2)$ for $s_1 \Rightarrow \neg s_2$. Moreover, each propositional tautology is an invariant of a system. These rules will be used implicitly in this paper.

In addition, we will use the following obvious rules:

Rule 2 (Conjunction, Weakening, and Disjunction)

1. From $\Sigma \models \Box\varphi$ and $\Sigma \models \Box\psi$ we conclude $\Sigma \models \Box\varphi \wedge \psi$.
2. From $\Sigma \models \Box\varphi$ and $\Sigma \models \Box\varphi \Rightarrow \psi$ we conclude $\Sigma \models \Box\psi$.
3. From $\Sigma \models \Box\varphi \Rightarrow \psi$ and $\Sigma \models \Box\varphi' \Rightarrow \psi'$ we conclude $\Sigma \models \Box(\varphi \vee \varphi') \Rightarrow (\psi \vee \psi')$

5.2 Liveness properties

The *progress rule* derives a leadsto property for a progress transition t of the net system, which is enabled at a marking $P \supseteq {}^\bullet t$. Then, either this transition or a conflicting transition will eventually occur. Some of the conflicting transitions can be excluded if an invariant property states that these transitions are not enabled when P is valid.

For a set $S' = \{s_1, \ldots, s_n\} \subseteq S$ we use S' as an abbreviation for the formula $s_1 \wedge \ldots \wedge s_n$.

Rule 3 (Progress)

Let Σ be a net system, t a progress transition of Σ, $T_1, T_2 \subseteq T$, $P \subseteq S$, and $A(t') \subseteq S$ for each $t' \in T_1$. If

1. ${}^\bullet t \subseteq P$,
2. $T_1 \cup T_2 = P^\bullet$,
3. for each $t' \in T_2$ holds $\Sigma \models \Box P \Rightarrow \neg {}^\bullet t'$, and
4. $A(t') \subseteq P^{t'}$ for each $t' \in T_1$

hold true, then we conclude $\Sigma \models P \leadsto \bigvee_{t' \in T_1} A(t')$.

The next pick-up rule proves a liveness property guaranteed by a fair arc (s,t). This rule requires a liveness property, which guarantees that s repeatedly is true in the run.

Rule 4 (Fairness)

Let Σ be a net system, (s,t) a fair arc of Σ, and $P = {}^\bullet t \setminus \{s\}$. If

1. $P^\bullet = \{t\}$ and
2. $\Sigma \models P \leadsto s$

hold true, then we conclude $\Sigma \models P \leadsto t^\bullet$.

Liveness properties justified by these pick-up rules can be combined by other rules. The notion of proof graphs unifies many classical rules for leadsto properties (like transitivity, and disjunction). Moreover, proof graphs present the relevant behaviour of a net system in an appealing graphical way.

Definition 9 (Proof Graph)

Let $\Phi \subseteq \mathcal{P}(S)$ and $G = (\Phi, \rightarrow)$ be a finite acyclic graph.

The set of successors of a node $\varphi \in \Phi$ is denoted by $\varphi^\bullet = \{\psi \in \Phi \mid \varphi \rightarrow \psi\}$.

G is a *proof graph* for a net system iff for each node $\varphi \in \Phi$ with $\varphi^\bullet \neq \emptyset$

$$\Sigma \models \varphi \leadsto \bigvee_{\psi \in \varphi^\bullet} \psi$$

From a proof graph, which is justified by many (simple) liveness properties, we conclude more complex liveness properties by the following rule.

Rule 5 (Proof Graph) Let $G = (\Phi, \rightarrow)$ be a proof graph for a net system Σ and G° be the set of *nodes without successors* (w.r.t.. $\rightarrow$).

Then we conclude for any node $\varphi \in \Phi$

$$\Sigma \models \varphi \leadsto \bigvee_{\psi \in G^\circ} \psi$$

The use of proof graphs is particularly appealing, when the nodes are conjunctions[1] of places of the system.

The proofs of the properties $\Sigma \models \varphi \leadsto \bigvee_{\psi \in \varphi^\bullet} \psi$ are called the *justifications* of the proof graph $G = (\Phi, \rightarrow)$. This justifications are either applications of the progress or the fairness rule or are applications of other proof graphs.

Moreover, from $\Sigma \models \Box[\varphi \Rightarrow (\psi \vee \chi)]$, we can derive $\Sigma \models \varphi \leadsto (\varphi \wedge \psi \vee \varphi \wedge \psi)$. Therefore, we allow such invariant properties in justifications of a proof graph without explicitly converting them to leadsto properties.

6 Correctness of message passing mutex

Now, we will apply the proof rules to verify that the message passing mutex algorithm shown in Fig. 1 satisfies the formal requirements for a mutex solution.

$$\begin{array}{ll} I1: & prio_l + cri_l + gra_l + gra_r + cri_r + prio_r = 1 \\ I2: & prio_l + miss_l - req_l - loc_l = 0 \\ I3: & gra_r + cri_r + prio_r - ord_r - miss_l = 0 \\ I4: & pen_r - ord_l - gra_r = 0 \\ I4': & pen_l - ord_r - gra_l = 0 \\ I5: & pen_l + cri_l + prio_l + miss_l = 1 \end{array}$$

Table 1: S-Invariants of Σ

The mutex property is an immediate consequence of the S-invariant $I1$ depicted in Table 1. The formal application of the rule is denoted as follows:

$$(0) \quad \Box\neg(cri_l \wedge cri_r) \quad \text{by S-inv. } I1 \ (Rule 1.1)$$

The proven formula is preceded by a number in parentheses, which will be used for references. The formula is followed by the rule that is applied for proving the formula; in this case it is a reference to an S-invariant $I1$. In the proof of the evolution property, we will use the progress rule with references to the corresponding transitions. A list of excluded conflicting transitions together with the excluding invariant properties follows. Similarly we use the fairness rule with a reference to the transition and the used liveness property.

The formal proof of the evolution property $req_l \leadsto cri_l$ is shown in Table 2. It is an justification of the proof graph of Fig. 2. The number of a node in this proof graph is a reference to the justification in Table 2.

In the following, we give an informal proof with references to the formal proof of Table 2. When req_l holds true, either $prio_l$ or $miss_l$ is true (1). Transition $t1_l$ or $t4_l$ will, therefore, eventually occur, which results in cri_l or $req_l \wedge miss_l$ (3). When *left* does not immediately enter its critical section but hands over its priority to *right*, it will eventually order (ord_r) the priority from *right* (4).

For proving that the order ord_r will eventually be accepted by *right* we first prove $ord_r \leadsto prio_r$ by the proof graph depicted in Fig. 3: by S-invariant $I3$ either gra_r, cri_r, or $prio_r$ are true, when ord_r holds (5). When the priority

[1] Remember, that a conjunction of places $S' \subseteq S$ is abbreviated by S'.

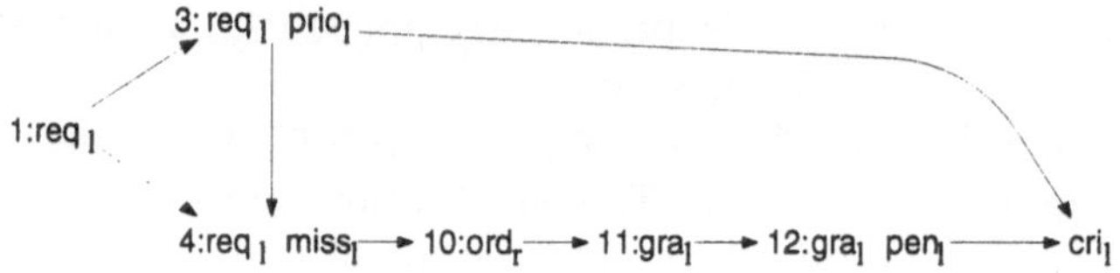

Figure 2: Proof graph for $req_l \leadsto cri_l$

is *gra*nted to *right*, *right* is pending (6). Therefore, it will eventually enter its critical section cri_r (7) and leave it again; then $prio_r$ holds (8). Thus, we have $ord_r \leadsto prio_r$.

Since ord_r is made false only, when the right agent returns the priority to *left* (transition $t4_r$), the fairness assumption and the above leadsto property guarantee that $t4_r$ will occur. Then we have gra_l (10). Symmetrically to $gra_r \leadsto cri_r$ we prove $gra_l \leadsto cri_l$ (12).

(1)	$\Box req_l \Rightarrow prio_l \vee miss_l$	by S-inv. $I2$ (Rule 1.3)
(2)	$\Box \neg prio_l \vee \neg miss_l$	by S-inv. $I5$ (Rule 1.1) and subst.
(3)	$req_l \wedge prio_l \leadsto cri_l \vee (req_l \wedge miss_l)$	progress with $t1_l$ or $t4_l$; $t2_l$ excluded by (2)
(4)	$req_l \wedge miss_l \leadsto ord_r$	progress with $t2_l$; $t1_l$ excluded by (2)
(5)	$\Box ord_r \Rightarrow (gra_r \vee cri_r \vee prio_r)$	by S-inv. $I3$ (Rule 1.3)
(6)	$\Box gra_r \Rightarrow pen_r$	by S-inv. $I4$ (Rule 1.3)
(7)	$gra_r \wedge pen_r \leadsto cri_r$	progress with $t3_r$;
(8)	$cri_r \leadsto prio_r$	progress with $t5_r$;
(9)	$ord_r \leadsto prio_r$	by proof graph of Fig. 3 with (5)–(8)
(10)	$ord_r \leadsto gra_l$	fairness with $t4_r$ and (9)
(11)	$\Box gra_l \Rightarrow pen_l$	by S-inv. $I4'$ (Rule 1.3)
(12)	$gra_l \wedge pen_l \leadsto cri_l$	progress with $t3_l$;
(13)	$req_l \leadsto cri_l$	proof graph of Fig. 2 with (1)–(12)

Table 2: Proof of the evolution property for Σ

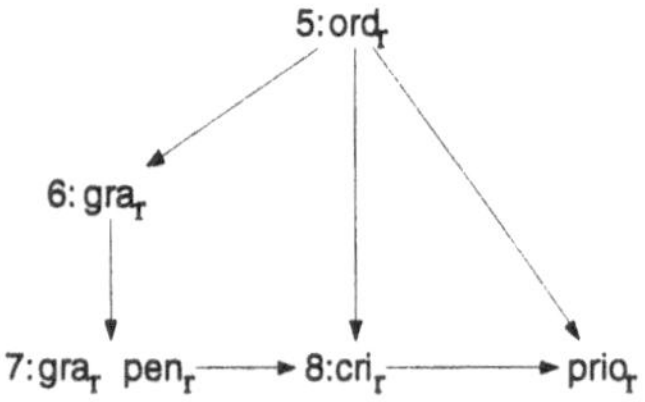

Figure 3: Proof graph $ord_r \leadsto prio_r$

Altogether, we have shown that Fig. 2 is a proof graph for the liveness property $req_l \leadsto cri_l$ (13). The proof of $req_r \leadsto cri_r$ is symmetric to this one: we just exchange the indices l and r.

7 Peterson's mutex algorithm

Peterson [13] proposed an elegant solution to the mutex problem. Nevertheless, formal proofs of Peterson's algorithm which avoid behavioural reasoning or handwaving are rare. For instance, [1] proves only the mutex property and [2] proves only the evolution property. Both properties of Peterson's algorithms are proven e.g. in [11, 4].

In this section we apply the introduced proof rules to the net model[2] of Peterson's algorithm, depicted in Figure 4. Afterwards, we compare the two presented algorithm.

7.1 Modelling and correctness

Initially, both agents are *loc*al and "*out* of the game". In the beginning, *right*

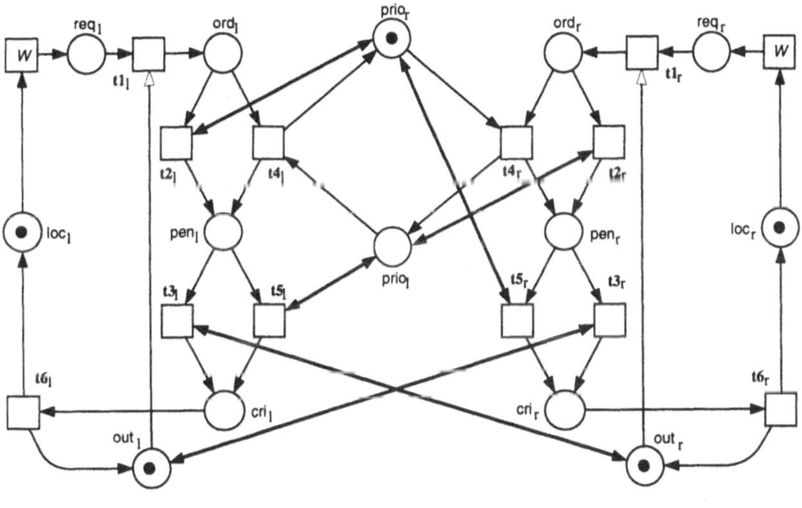

Figure 4: Σ_{Pet}: Peterson's mutex algorithm

has *prio*rity. The place *out* can be tested (loops are depicted as boldfaced single arcs with two arrowheads) to determine, if an agent is outside the critical section. Independently, each agent may request access to the critical section.

A *req*uesting agent becomes *ord*ering (t_1). Then, it is no longer out of the game . An ordering agent may become *pen*ding by occurrence of transition t_2 or t_4. Immediately after the occurrence, the other agent has *prio*rity . There are also two ways to become *cri*tical: either the other agent is actually out

[2]The Petri net models used in [4, 15] are quite similar. Best, however, does not model weak transitions, and Reisig does not consider fairness arguments in the proof.

of the game (t_3), or the priority mechanism guarantees access to the critical section (t_5).

We are going to prove the mutex and evolution property using the same techniques as before. Table 3 summarizes some traps and S-invariants of Σ_{Pet}.

$$\begin{array}{ll}
T: & prio_l + pen_l + out_r + ord_r \geq 1 \\
T': & prio_r + pen_r + out_l + ord_l \geq 1 \\
I1: & ord_l + pen_l + cri_l + out_l = 1 \\
I1': & ord_r + pen_r + cri_r + out_r = 1 \\
I2: & prio_l + prio_r = 1 \\
I3: & out_l - loc_l - req_l = 0 \\
I3': & out_r - loc_r - req_r = 0
\end{array}$$

Table 3: S-invariants and traps of Σ_{Pet}

In contrast to the message passing solution both, S-invariants and traps, are necessary to prove the mutex property. This combined method to prove invariant properties of distributed algorithms was introduced in [3].

(A) $\Box\neg prio_l \Rightarrow (pen_l \vee out_r \vee ord_r)$	by Trap T (Rule 1.2)
(B) $\Box\neg prio_r \Rightarrow (pen_r \vee out_l \vee ord_l)$	by Trap T' (Rule 1.2)
(C) $\Box(\neg prio_l \vee \neg prio_r) \Rightarrow (pen_l \vee out_r \vee ord_r) \vee (pen_r \vee out_l \vee ord_l)$	(Disj with A,B)
(D) $\Box\neg prio_l \vee \neg prio_r$	by S-inv. $I2$ (Rule 1.1),
(E) $\Box(pen_l \vee out_r \vee ord_r \vee pen_r \vee out_l \vee ord_l)$	(Weakening with D,C)
(F) $\Box(pen_l \vee out_l \vee ord_l) \Rightarrow \neg cri_l$	by S-inv. $I1$ (Rule 1.1)
(G) $\Box(pen_r \vee out_r \vee ord_r) \Rightarrow \neg cri_r$	by S-inv. $I1'$ (Rule 1.1)
(H) $\Box(pen_l \vee out_l \vee ord_l \vee pen_r \vee out_r \vee ord_r) \Rightarrow (\neg cri_l \vee \neg cri_r)$	(Disj with F,G)
(I) $\Box(\neg cri_l \vee \neg cri_r)$	(Weakening with E,H)

Table 4: $\Sigma_{Pet} \models \Box\neg(cri_l \wedge cri_r)$

Table 4 formally proves the mutex property: trap T guarantees that one place of $prio_l$, pen_l, out_r and ord_r is marked. If $prio_l$ is not marked, one of the remaining three places is marked. The same holds for $prio_r$ w.r.t. the initially marked trap T'. Since always either $prio_l$ or $prio_r$ does not hold, at least one of the six remaining places of both traps is marked (E). From the S-invariants $I1, I1'$ we conclude that always at least one agent is not in its critical section.

The arguments that prove evolution for *left* are summarized in the proof graph depicted in Fig. 5. Its justification in Table 5 omits the justified $\leadsto$-formulas, since they are deducible from the proof graph itself. The proof graph formalizes the following behavioural reasoning: If *left* requests access to the critical section it is always out of the game (2).
Now we must apply a fairness argument, since *right* may test out_l infinitely often (by repeated occurrence of the transition sequence $t2_r, t3_r, t6_r, t1_r$). The fair arc $(out_l, t1_l)$ assures that *left* eventually orders its access (3).

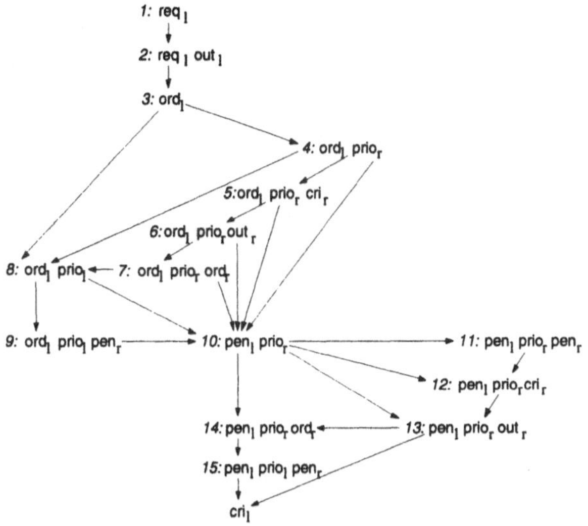

Figure 5: A proof graph for $\Sigma_{Pet} \models req_l \leadsto cri_l$

(1) $\Box\ req_l \Rightarrow out_l$ by S-inv. I3 (Rule 1.3)
(2) fairness with $t1_l$ and $req_l \leadsto out_l$, shown with (1)
(3) $\Box\ prio_l \vee prio_r$ by S-inv. $I2$ (Rule 1.2)
(4) progress with $t2_l$ or $t4_l, t4_r, t5_r$;
(J) $\Box cri_r \Rightarrow (\neg pen_r \wedge \neg ord_r)$ by S-inv. $I1'$ (Rule 1.1)
(5) progress with $t2_l$ or $t4_l, t6_r$; $t4_r, t5_r$ excluded by (J)
(K) $\Box out_r \Rightarrow \neg pen_r$ by S-inv. $I1'$ with (Rule 1.1)
(L) $\Box ord_l \Rightarrow (\neg pen_l \wedge \neg out_l)$ by S-inv. $I1$ with (Rule 1.1)
(6) progress with $t2_l$ or $t4_l, t1_r$; $t3_l, t4_r, t5_r$ excluded by (K, L),
(M) $\Box ord_r \Rightarrow (\neg pen_r \wedge \neg out_r)$ by S-inv. $I1'$ with (Rule 1.1)
(7) progress with $t2_l$ or $t4_l, t2_r, t4_r$; $t5_r$ excluded by (M)
(N) $\Box pen_l \Rightarrow (\neg ord_l \wedge \neg out_l)$ by S-inv. $I1$ with (Rule 1.1)
(8) progress with $t4_l$ or $t2_l, t2_r$; $t5_l$ excluded by (N),
(O) $\Box pen_r \Rightarrow (\neg ord_r \wedge \neg out_r)$ by S-inv. $I1'$ with (Rule 1.1)
(9) progress with $t4_l$ or $t2_l$; $t5_l, t2_r, t3_r, t5_r$ excluded by $(L), (O), (D)$
(10) $\Box\ pen_l \vee cri_r \vee out_l \vee ord_r$ by (F)
(11) progress with $t5_r$ or $t3_r$; $t2_l, t3_l, t4_l, t5_l, t4_r$ excluded by $(N), (O), (D)$
(12) progress with $t6_r$; $t2_l, t3_l, t5_l, t4_r, t5_r$ excluded by $(N), (J), (D)$
(13) progress with $t3_l$ or $t5_l, t1_r$; $t2_l, t4_r, t5_r$ excluded by (K)
(14) progress with $t4_r$ or $t2_r$; $t2_l, t3_l, t5_l, t5_r$ excluded by $(N), (M), (D)$
(15) progress with $t5_l$ or $t3_l$; $t4_l, t2_r, t3_r, t5_r$ excluded by $(N), (O), (D)$.

Table 5: Justification for the proof graph of Fig. 5

In a second step each ordering agent becomes pending (10). As long as *left* has priority (8), *right* may only become pending (9). Then *right* gets blocked and *left* is guaranteed to become pending, too. Otherwise *right* has priority (4). *Left* may also become pending now. But we can not exclude, that *right* eventually turns the priority before (4)–(8).

In a final step we argue, that *left* eventually becomes critical when it is pending. If *right* is or will become out of the game (11)–(13), $t3_l$ may occur. If *right* gets its request ordered before, than priority eventually changes (15) and both agents are pending. Since *left* has priority now, *left* succeeds.

The formal proof of $req_r \leadsto cri_r$ is symmetric.

7.2 Comparison

Compared to other solutions the interaction in Peterson's algorithm is very simple. The elegance of this solution is due to this fact. Compared to the proof of Peterson's algorithms, the proof of the message passing mutex is simple. The reason might be that the message passing solution is not the sum of two sequential processes. When *left* is local and has priority it must be ready to grant an order of *right*. Concurrently, it might require access to the critical section. Peterson's algorithm and most other algorithms we know in the literature are the combination of two sequential components. This explains, perhaps, why this solution has never been published before.

A consequence of a message passing based solution is that the priority has to be handed over, repeatedly. The algorithm avoids a livelock, since an agent immediately enters its critical section, when the priority is handed over. If we look closely to the proof graph of Peterson's algorithm we note that the priority mechanism is more intrinsic to avoid live- and deadlocks. The priority may change three times before an ordering agent enters the critical section.

Finally, the fairness arguments needed in both proofs are similar. The simplicity of the applied fairness notion shows, that a fair treatment of a single transition for each agent is sufficient in each algorithm. For the remaining transitions no fairness argument is required.

8 Conclusion

The mutex problem extracts some interesting phenomena of many distributed algorithms — even when only two agents are concerned. Therefore, good solutions and adequate correctness proofs are still a great challenge. In particular, it is interesting to apply new proof methods to that problem.

We introduced a formal model for distributed algorithms. The two mutex solutions show that we can capture typical requirements of distributed algorithms in this model. To this end, we introduced non-progress transitions and fairness. There are more general approaches to formalize "weak transitions" in the literature, which use the concept of an environment. In UNITY [6], for example, weak transition are modelled as a transition of the environment which satisfies an unless-property. We prefer the concept of non-progress transitions because of its simplicity.

As fairness concept we introduced fair arcs, which impose additional requirements on the runs of a system. This fairness concept is weaker than most other fairness requirements. Therefore, it is easier to realize the fairness requirement in an "implementation" of the net model.

Moreover, we introduced a temporal logic to express properties of distributed algorithms. The proof rules allow for a rigorous proof of these properties. Proof graphs nicely formalize behavioural arguments.

Acknowledgements We thank J. Desel, W. Reisig, W. Volger and various anonymous referees for helpful discussions and comments.

References

1. K. R. Apt and E. R. Olderog. *Programmverifikation.* Springer-Verlag, 1994.
2. M. Ben-Ari. *Mathematical Logic for Computer Science.* Prentice Hall, 1993.
3. E. Best. Representing a program invariant as a linear invariant in a Petri net. *EATCS Bulletin*, (17):2–11, June 1982.
4. E. Best. *Semantik.* Vieweg-Verlag, 1995.
5. E. Best and C. Fernández. *Nonsequential Processes*, volume 13 of *EATCS Monographs on Theoretical Computer Science.* Springer-Verlag, 1988.
6. K. M. Chandy and J. Misra. *Parallel Program Design: A Foundation.* Addison-Wesley, 1988.
7. E. Dijkstra. Solution of a problem in concurrent programming control. *Communication of the ACM*, 8(9):569, 1965.
8. R. W. Doran and L. K. Thomas. Variants of the sofware solutions to mutual exclusion. *Information Processing Letters*, 10(4,5):206–208, 1980.
9. D.E. Knuth. Additional comments on a problem in concurrent programming control. *Communications of the ACM*, 9(5):321–322, 1966.
10. L. Lamport. A new solution to Dijkstra's concurrent programming problem. *Communications of the ACM*, 21(7):453–455, 1974.
11. Z. Manna and A. Pnueli. A temporal proof methodology for reactive systems. In M. Broy, editor, *Programm Design Calculi*, volume 118 of *NATO ASI Series F*, pages 287–323, 1992.
12. S. Owicki and L. Lamport. Proving liveness properties of concurrent programs. *ACM Transactions on Programming Languages and Systems*, 4(3):455–495, July 1982.
13. G.L. Peterson. Myths about the mutual exclusion problem. *Information Processing Letters*, 12(3):115–116, 1981.
14. M. Raynal. *Algorithms for Mutual Exclusion.* The MIT Press, 1986.
15. W. Reisig. Correctness proofs of distributed algorithms. Informatik-Bericht 38, Humboldt-University of Berlin, 1994.
16. W. Reisig. *Petri Nets*, volume 4 of *EATCS Monographs on Theoretical Computer Science.* Springer-Verlag, 1982.
17. P.S. Thiagarajan. Elementary net systems. In W. Brauer, W. Reisig, and G. Rozenberg, editors, *Petri Nets: Central Models and Their Properties*, volume 254 of *LNCS*, pages 26–59. Springer-Verlag, September 1986.
18. J. L. A. van de Snepscheut. *What Computing is all about.* Springer-Verlag, 1993.
19. D. J. Walker. Automated analysis of mutual exclusion algorithms using CCS. *Formal Aspects of Computing*, (1):273–292, 1989.

Possible and Guaranteed Concurrency in CSP

Marta Kwiatkowska*
School of Computer Science, University of Birmingham
Edgbaston, Birmingham B15 2TT, UK

Iain Phillips†
Department of Computing, Imperial College
180 Queens Gate, London SW7 2BZ, UK

Abstract

As part of an effort to give a "truly concurrent" semantics to process algebra, we propose a framework of refinements of the failures model for CSP with concurrency, conflict and causality relations on traces. These relations are defined by induction over syntax of CSP processes. We study in detail two new semantics: the *possible concurrency* (where two traces are said to be concurrent if they *may* be observations of the same concurrent run) and the *possible conflict* (two traces are said to be in conflict if they *may* be observations of two different runs). The *guaranteed concurrency* is obtained from the possible conflict semantics. Although the expansion law is necessarily weakened to an inequality, we show that most of the CSP laws are preserved, the exception being the idempotency of choice for the possible conflict refinement. Finally, we show that our semantics is well-founded by demonstrating a strong connection with the existing event structures semantics for CSP. The latter results show that, in a certain sense, concurrency distinctions can be made at the level of syntax, without resorting to reasoning about event occurrences.

1 Introduction

This paper is part of an effort by many authors to give a "truly concurrent" semantics to process algebra. Many authors have previously contributed in this area, see *e.g.* [20, 17, 8, 18, 4, 9, 13, 1, 15, 2], and the methods have included translation into Petri nets, deriving an event structure, and enhancing the labels in labelled transition systems, often leading to descriptions of low-level character. We propose starting at the high level of abstraction usually associated with process algebras, and define a refinement of the existing failures model directly from the syntax of CSP terms. The virtues of abstraction have long been recognised in software engineering and specification, and this is where this approach may be utilised.

*This research was started when the author was Visiting Academic at the Department of Computing, Imperial College supported by the Nuffield Science Foundation grant (SCI/124/528/G).

†Supported by SERC grant GR/F72475.

As the touchstone for whether a semantics is truly concurrent, as usual we take the example processes $a \parallel b$ and $a{\cdot}b+b{\cdot}a$. These two are of course equated in the standard interleaving semantics of process algebras such as CCS, CSP, ACP due to the expansion law. However, we would expect them to be distinguished by a truly concurrent semantics, since $a \| b$ exhibits concurrency (in that actions a and b are causally independent), whereas $a{\cdot}b + b{\cdot}a$ does not.

Trace theory [14] provides one solution to the example. The sequences of possible actions are just ε, a, b, ab, ba in each case. In the case of $a \parallel b$, since a, b are independent, denoted $a \; \iota \; b$, we say that ab and ba are equivalent, since they are the same except that independent actions are permuted. However, in the case of $a{\cdot}b + b{\cdot}a$ the sequences ab, ba are not equivalent, and so we have distinguished the processes.

The disadvantage of trace theory is that it does not handle non-determinism and deadlock. For instance the processes $a{\cdot}b$ and $a + a{\cdot}b$ have the same sequences, namely ε, a, ab, and there is no independence between actions in either case. However, the second process may deadlock after performing a, so that a distinction should be made. It is for this reason that Milner was led to consider a semantics for processes more refined than that used in formal language theory. He proposed observation equivalence (bisimulation), which takes us a long way from traces, and is generally considered the most refined behavioural equivalence. Brookes, Hoare and Roscoe [5] took a different approach, in which they refined the traces[1] model with deadlock information telling us which actions may be refused after some sequence of actions, giving the so-called *failures* model.

We aim to obtain a semantics for CSP-like languages capable of making concurrency distinctions as in Mazurkiewicz trace theory, while, at the same time, capable of handling non-determinism and deadlock in the process algebra style. An extension of trace theory to cater for Petri nets appeared in [11]. We propose a new framework of *refinements* of the classical, failures semantics with relations on traces capturing the notions of *possible*, and their duals, *guaranteed*, relations of concurrency, conflict and causality. These relations are defined by induction over syntax of CSP processes. We discuss two refinements of the failures semantics (with possible concurrency and conflict) and the corresponding laws of the calculi. Finally, we demonstrate a strong connection with the event structures semantics for CSP.

This paper had its origins in a preliminary report presented at an internal workshop [12], where the idea of a conflict relation on traces was introduced. In this paper, we propose a whole new framework as outlined above, including material from [12] for the sake of completeness. A true concurrency semantics for CSP, without the inequational version of the expansion law, appeared in [18]. It would be interesting to compare the two approaches.

The paper is organised as follows. Section 2 introduces the central idea, while Section 3 contains an overview of CSP. In Section 4 we define two main

[1] Note that in the CSP world "trace" means a sequence of actions performed by a process, whereas for trace theory, it means an equivalence class of such sequences under the equivalence generated by the independence relation.

refinements of the failures model and summarise the corresponding laws. Finally, Section 5 states a formal connection of our abstract semantics with event structures.

2 Discussion: "may" or "must" concurrency?

The equational version of the expansion law is a natural consequence of the statement that concurrency is not observable. As a result, concurrency is only structural, and not behavioural, since the expansion law acts as a "rewrite" rule on syntactic forms. This also means that syntactic distinctions are ignored at the behavioural level. We do not argue with the view that concurrency is not observable. Instead, we propose to view non-interleaving theories as based on an element of knowledge about the internal structure of the system. One may say that the issue is approached from a different perspective, that of a *system designer*, rather than a *system observer*.

The aim of this paper is to obtain a semantic model for CSP-like languages which

- is *non-interleaving*, in the sense that it does not satisfy the expansion law, but instead it satisfies its weaker, inequational version;
- is *simple* and *abstract*, *i.e.* makes sense of processes such as $a \parallel b + (a \cdot b)$ without resorting to relabelling;
- perturbs *classical semantics* as little as possible, *i.e.* it preserves as many laws as possible and does not reject classical semantics in favour of existing non-interleaving models.

Much has been published on the subject of non-interleaving semantics for process algebras. The advantage of models such as Petri nets, event structures and pomsets is that they are concrete, intuitive and graphical, and can therefore serve as design tools. However, they can be difficult to reason about, and not sufficiently abstract. Often two levels are introduced, one for event occurrences and one for abstract actions. This is cumbersome to use, as reasoning about a system must involve both levels. Furthermore, there is too much dependence on how particular nets or event structures are manufactured, and too much arbitrariness in the way the maps on processes are defined, possibly leading to inconsistencies with operational intuitions.

We feel that, alongside models such as event structures, there is much to gain from complementing existing interleaving theories with abstract, non-interleaving ones. As we are hoping to show in this paper, non-interleaving *can* be handled at a high level of abstraction, and can, therefore, support *meaningful specifications*. The motivation for using non-interleaving is that it adequately represents phenomena present in hardware (*e.g.* at the logic gate level), and also leads to more efficient verification of certain classes of properties.

The central idea of this paper is to *refine* the existing classical semantics, *e.g.* failures in the case of CSP, with additional concurrency information. Our intention is that this is defined by induction over syntax of process expressions, in the same way as the failures themselves can be defined directly from syntax (see Section 4.1). The idea is as follows. Consider a process algebra language. We think of traces of a given process P as *observations*. We propose to refine the existing semantics with a binary *concurrency* relation (denoted *conc*) on traces meaning two traces are in the relation iff they are observations of the same concurrent run. For example, we would have ab *conc* ba in $a \parallel b$, but not in $ab + ba$. Note that this is more general than the approach used, say, in Mazurkiewicz trace theory, where the concurrency relation is on actions, because it can handle context dependency. However, there is a difficulty with non-determinism. Consider the process $(a \parallel b) + (a{\cdot}b + b{\cdot}a)$. Clearly, in the subprocess on the left ab *conc* ba, while in the subprocess on the right this is *not* the case. Should ab *conc* ba hold or not in *the whole* process? One could rename or index the actions to distinguish between "the a on the left" and "the a on the right", but this introduces two levels of reasoning, which we believe is unnecessary. Instead we propose to make a choice between interpreting the concurrency relation either as:

> it is *possible* for ab to be concurrent with ba, in the sense that *there exists* a concurrent run such that both traces can be observed from it;

or

> it is *guaranteed* that ab is concurrent with ba, in the sense that *for every* concurrent run both traces can be observed.

We shall refer to the first interpretation above as the *possible* (also "*may*") concurrency (denoted *co*()), while the latter will be called *guaranteed* (also "*must*") concurrency and will be denoted *gco*().

Formally, given a process expression P, we shall distinguish an indexed family of relations $co(P) \subseteq tr(P)^2$ defined by induction over syntax, where $tr(P)$ denotes the set of traces of P.

We would anticipate that the two interpretations above will correspond to different sets of laws. The decision regarding which interpretation to use would depend on application; we feel that guaranteed concurrency would correspond to measuring the amount of concurrency, while possible concurrency to describing low-level behavioural phenomena. This can be explained as follows. In the guaranteed concurrency semantics we obtain $a{\cdot}b + b{\cdot}a \leq a \parallel b$, which can be understood as "the more concurrent the process, the higher it is in the ordering". On the other hand, the possible concurrency semantics yields $a \parallel b \leq a{\cdot}b + b{\cdot}a$: the process lower in the ordering has more complex behaviour (a, b can be seen as either ordered in time or in indeterminate order) than the process higher in the ordering. Thus, "the more abstract (less complex) the process, the higher it is in the ordering".

We also note that, independently of deriving the (possible or guaranteed) concurrency, one may also consider relations on traces corresponding to the (possible or guaranteed) *conflict* or *causality* relations. Roughly speaking, two traces would be in a *possible conflict* relation if they *may* be observed of two different runs, originating from either side of the choice operator. The *guaranteed conflict* would be the obvious dual. The definition of the causality relations would be similar, and would correspond to the sequential composition of processes.

These relations can be combined with the classical semantics to give models of varying powers of distinction. Concurrency and conflict will also be related under complement – some approximation of, say, guaranteed concurrency can be obtained from possible conflict, as we shall show later.

In this paper, we shall define refinements of the failures semantics for CSP with possible concurrency and possible conflict relations. We believe, however, that the above ideas are applicable to a range of process algebras, although we should emphasise that the particular definitions and results presented here are dependent on the semantics of the CSP mixed parallel.

3 Preliminaries

3.1 Notation

We shall use the following notation conventions throughout the paper. *Act* is a set of action symbols[2]. The set of finite strings (also called *traces*) over the set *Act* is denoted by $Act^\star$; ε is the empty trace. We use a, b, c to range over *Act*, and s, t, u to range over $Act^\star$. Given $a \in Act$ and traces $s, t \in Act^\star$, we write as, sa and st for the respective concatenations. Given a set L, the set of all finite subsets of L will be denoted by $\mathcal{P}_f(L)$, and the set of all its subsets by $\mathcal{P}(L)$. We use A, B, X, Y to range over $\mathcal{P}_f(Act)$. $s \backslash d$, where $d \in Act$, denotes *hiding*, *i.e.* the string s with all occurrences of d deleted. Given $A \in \mathcal{P}_f(Act)$, $s \lceil A$ denotes *restriction*, *i.e.* the string s with all occurrences of actions outside the set A deleted.

3.2 The language of CSP

We use a subset of Communicating Sequential Processes (CSP) as described in [6]. This is essentially an alphabet-free version of CSP as introduced in [10]. In this paper only a summary of the language and the model is presented. We refer the reader to [10, 5, 6] for more details.

The syntax of CSP process expressions is as follows:

$$P ::= 0 \mid a{\cdot}P \mid P \sqcap Q \mid P \,\square\, Q \mid P_A \parallel Q_B \mid P \backslash a \mid f(P) \mid x \mid \mu x.P$$

where P, Q ranges over (syntactic) process expressions, a is an action ranging over the set *Act*, A, B are (finite) sets of actions, and x is a process variable. 0 represents a deadlocked process (usually denoted STOP). $a{\cdot}P$ denotes prefixing the process P with the action a (usually $a \rightarrow P$). We shall often elide $\cdot$ and 0; *e.g.*

[2]We prefer to use the term "action" instead of "event" so as to distinguish between actions and their occurrences; the latter are usually referred to as "events".

the process $a{\cdot}b{\cdot}0$ will simply be denoted by ab. $P \sqcap Q$ is the non-deterministic choice (internal), while $P \,\Box\, Q$ the deterministic choice (external). $P_A \parallel P_B$ is the mixed parallel operator of [6]; here A is the named alphabet for the process P, B the named alphabet for the process Q, and $P_A \| Q_B$ requires simultaneous participation of P and Q on any actions belonging to $A \cap B$, but the process P may progress independently on actions belonging to $A - B$ (and symmetrically for Q). This operator resembles the parallel operator introduced by Hoare in [10], where alphabets of processes were used in place of arbitrary sets A and B. If it is clear from the context, the alphabets will be omitted. $P\backslash a$ denotes the hiding operator, while $f(P)$ renaming. We assume that $f : Act \longrightarrow Act$ is a 1-1 function. This restriction guarantees no auto-concurrency, and is convenient when considering the relationship with event structures. Finally, recursion is handled by terms $\mu x.P$, where μ is a variable binding operator.

We shall denote the set of terms defined by the above syntax by CSP.

4 Semantics

In this section we define the *possible conflict* and *possible concurrency* semantics for CSP. We begin by reviewing the failures semantics. We also study the notion of *guaranteed concurrency* via the notion of possible conflict, in a sense that will be made precise in Section 5.

4.1 Failures semantics

The failures semantic model [5] views a process as a set of *failures*, that is, pairs (s, X) where $s \in Act^*$ is a trace which the process *may* perform and X is a finite set of actions which the process *may* refuse after having done s. Formally, a failure set is a subset $F \subseteq Act^* \times \mathcal{P}_f(Act)$ satisfying the following four conditions:

(F1) $(\varepsilon, \emptyset) \in F$;
(F2) $(st, \emptyset) \in F \Rightarrow (s, \emptyset) \in F$;
(F3) $(s, X) \in F \wedge Y \subseteq X \Rightarrow (s, Y) \in F$;
(F4) $(s, X) \in F \wedge (sc, \emptyset) \notin F \Rightarrow (s, X \cup \{c\}) \in F$.

For any set F of failures we define

$$tr(F) = \{s \mid \exists X.(s, X) \in F\}.$$

The set of failures is denoted by $\mathbb{F}$ and the ordering is as follows. We say $F \sqsubseteq F'$ (F is more non-deterministic than F') iff $F \supseteq F'$.

There exists a refinement of the failures model with divergences [6] needed to ensure that the law $P \,\|\, \mathrm{CHAOS} = \mathrm{CHAOS}$ is satisfied. For reasons of space we shall not consider divergences in this paper, but an extension to include them is straightforward.

CSP can be given denotational semantics by means of a compositional mapping $fail(\cdot)$ from process expressions to the failure sets [5, 10] (we omit the details for reasons of space). The treatment of recursion is standard and depends on the operators being monotone and continuous in the domain $(\mathbb{F}, \sqsubseteq)$.

4.2 The refined semantics

We shall add an extra component to the model, in the shape of a binary relation C on $tr(F)$. This will be either the possible concurrency relation or the possible conflict relation, which we define below.

Let $\mathbb{FC} = \{\langle F, C\rangle \mid F$ is a failure set, C a binary relation on $tr(F)\}$. An ordering is defined on $\mathbb{FC}$ as follows:

$$\langle F, C\rangle \sqsubseteq \langle F', C'\rangle \iff F \supseteq F' \text{ and } C \supseteq C'.$$

This is clearly a refinement of the usual failures ordering.

We may think of the pairs $\langle F, C\rangle$ as the *possibilities* of a process. Going up in the ordering corresponds to discarding possibilities.

As far as using $\mathbb{FC}$ as a semantic domain is concerned, it will be enough to assert that $(\mathbb{FC}, \sqsubseteq)$ is a dcpo, with directed joins given by set intersection. In fact, of course, we can say more, but our intention is to place further restrictions on the relations C to refine our model in the form of axioms similar to (F1)-(F4) above. Some ideas on how this may be done emerge from Section 5, where connections with event structures are explored.

We are going to define two alternative semantic maps from CSP terms to the domain $\mathbb{FC}$. The first corresponds to possible concurrency, and the second to possible conflict. We give them component-wise, so that they are

$$P \mapsto \langle fail(P), co(P)\rangle; P \mapsto \langle fail(P), cf(P)\rangle$$

The maps appear in Tables 1 and 2.

As explained in Section 2, two traces s, t of a process P are possibly concurrent ($s\ co(P)\ t$) if they may be observations of the same run of P. They possibly conflict ($s\ cf(P)\ t$) if they may be observations of two conflicting runs of P. The style of our tables is to define $co(P)$ and $cf(P)$ by induction on syntax. Recall that we have $co(P), cf(P) \subseteq tr(P)^2$ as background conditions. The rules (sym) and (refl) are used for convenience, to avoid complicating the definition.

The crucial rule for $co()$ is (L-R-$\|$), which introduces concurrency. Any valid trace is concurrent with itself by the rule (refl). The other rules preserve it through the syntactic structure. As far as $cf()$ is concerned, the crucial rules are (L-R-$\Box$) and (L-R-$\sqcap$), which introduce conflict. Again, the other rules preserve it through the syntactic structure.

Example 4.1 *Consider $a \| b$. $ab\lceil\{a\}\ co(a)\ ba\lceil\{a\}$ and $ab\lceil\{b\}\ co(b)\ ba\lceil\{b\}$ by rule (refl), and so $ab\ co(a \| b)\ ba$.*

Proposition 4.2 *With both the $co()$ and $cf()$ semantics, all the operations on $\mathbb{FC}$ are monotone and continuous.*

Table 1: The possible concurrency relation

(refl)	$\dfrac{s \in tr(P)}{s\ co(P)\ s}$	(sym)	$\dfrac{s\ co(P)\ s'}{s'\ co(P)\ s}$
(pref)	$\dfrac{s\ co(P)\ s'}{as\ co(a{\cdot}P)\ as'}$	(hide)	$\dfrac{s\ co(P)\ s'}{s\backslash a\ co(P\backslash a)\ s'\backslash a}$
(ren)	$\dfrac{s\ co(P)\ s'}{f(s)\ co(f(P))\ f(s')}$	(L-R-$\parallel$)	$\dfrac{s\lceil A\ co(P)\ s'\lceil A \quad s\lceil B\ co(Q)\ s'\lceil B}{s\ co(P_A \parallel Q_B)\ s'}$
(L-$\sqcap$)	$\dfrac{s\ co(P)\ s'}{s\ co(P \sqcap Q)\ s'}$	(R-$\sqcap$)	$\dfrac{s\ co(Q)\ s'}{s\ co(P \sqcap Q)\ s'}$
(L-$\Box$)	$\dfrac{s\ co(P)\ s'}{s\ co(P \Box Q)\ s'}$	(R-$\Box$)	$\dfrac{s\ co(Q)\ s'}{s\ co(P \Box Q)\ s'}$

The proposition allows recursive processes to be handled by taking least fixed points in the standard way.

We shall show in Section 5 that the intuitive explanations of our semantic maps we have given in terms of runs can be justified in terms of event structures.

4.3 The laws

We have given two alternative semantics for the CSP operators. We now investigate to what extent the usual laws of CSP are preserved. The interleaving law $a \parallel b = ab \Box ba$ is where the differences appear most sharply. Of course, $fail(a \parallel b) = fail(ab \Box ba)$. However, $co(a \parallel b) \supseteq co(ab \Box ba)$, so that under the possible concurrency ($co()$) semantics

$$a \parallel b \sqsubseteq ab \Box ba.$$

On the other hand, $cf(a \| b) \subseteq cf(ab \Box ba)$, so that under the possible conflict $cf()$ semantics

$$a \parallel b \sqsupseteq ab \parallel ba.$$

Tables 3 and 4 show the laws which we have verified as sound for the $co()$ semantics. Tables 3 and 5 summarize the laws for the $cf()$ semantics. Notice that the idempotency of choice fails. Clearly $cf(P \sqcap P) = tr(P)^2$, so that $P \sqcap P$ has maximal conflict.

Table 2: The possible conflict relation

(pref)	$\dfrac{s \;\; cf(P) \;\; s'}{as \;\; cf(a\cdot P) \;\; as'}$	(sym)	$\dfrac{s \;\; cf(P) \;\; s'}{s' \;\; cf(P) \;\; s}$
(hide)	$\dfrac{s \;\; cf(P) \;\; s'}{s\backslash a \;\; cf(P\backslash a) \;\; s'\backslash a}$	(ren)	$\dfrac{s \;\; cf(P) \;\; s'}{f(s) \;\; cf(f(P)) \;\; f(s')}$
(L-$\Vert$)	$\dfrac{s\lceil A \;\; cf(P) \;\; s'\lceil A}{s \;\; cf(P_A \Vert Q_B) \;\; s'}$	(R-$\Vert$)	$\dfrac{s\lceil B \;\; cf(Q) \;\; s'\lceil B}{s \;\; cf(P_A \Vert Q_B) \;\; s'}$
(L-$\sqcap$)	$\dfrac{s \;\; cf(P) \;\; s'}{s \;\; cf(P \sqcap Q) \;\; s'}$	(R-$\sqcap$)	$\dfrac{s \;\; cf(Q) \;\; s'}{s \;\; cf(P \sqcap Q) \;\; s'}$
(L-R-$\sqcap$)	$\dfrac{s \in tr(P) \;\; s' \in tr(Q)}{s \;\; cf(P \sqcap Q) \;\; s'}$		
(L-$\square$)	$\dfrac{s \;\; cf(P) \;\; s'}{s \;\; cf(P \square Q) \;\; s'}$	(R-$\square$)	$\dfrac{s \;\; cf(Q) \;\; s'}{s \;\; cf(P \square Q) \;\; s'}$
(L-R-$\square$)	$\dfrac{s \in tr(P) \;\; s' \in tr(Q)}{s \;\; cf(P \square Q) \;\; s'} \;\; s, s' \neq \varepsilon$		

Table 3: The laws of CSP with possible conflict and possible concurrency

$P \square Q$	$=$	$Q \square P$	
$P \square (Q \square R)$	$=$	$(P \square Q) \square R$	
$P \square 0$	$=$	P	
$a(P \sqcap Q)$	$=$	$aP \square aQ$	
$P \sqcap Q$	$\sqsubseteq$	P	
$P \sqcap Q$	$=$	$Q \sqcap P$	
$P \sqcap (Q \sqcap R)$	$=$	$(P \sqcap Q) \sqcap R$	
$P \sqcap Q$	$\sqsubseteq$	$P \square Q$	
$P_A \Vert Q_B$	$=$	$Q_B \Vert P_A$	
$P_A \Vert (Q_B \Vert R_C)_{B\cup C}$	$=$	$(P_A \Vert Q_B)_{A\cup B} \Vert R_C$	
$P_A \Vert 0_B$	$=$	P_A	if $A \cap B = \emptyset$
$P_A \Vert 0_A$	$=$	0	
$(aP)\backslash a$	$=$	$P\backslash a$	
$(aP)\backslash b$	$=$	$a(P\backslash b)$	if $a \neq b$
$(P \sqcap Q)\backslash a$	$=$	$(P\backslash a) \sqcap (Q\backslash a)$	

Table 4: Additional laws of CSP with possible concurrency

$$
\begin{array}{rcll}
P \Box P & = & P & \\
P \sqcap P & = & P & \\
P \Box (Q \sqcap R) & = & (P \Box Q) \sqcap (P \Box R) & \\
aP_A \parallel bQ_B & \sqsubseteq & a(P_A \parallel bQ_B) \Box b(aP_A \parallel Q_B) & \text{where } A \cap B = \emptyset, \\
 & & & a \in A, b \in B
\end{array}
$$

Table 5: Additional laws of CSP with possible conflict

$$
\begin{array}{rcll}
P \Box P & \sqsubseteq & P & \\
P \Box (Q \sqcap R) & \sqsupseteq & (P \Box Q) \sqcap (P \Box R) & \\
P \sqcap P \sqcap P & = & P \sqcap P & \\
aP_A \parallel bQ_B & \sqsupseteq & a(P_A \parallel bQ_B) \Box b(aP_A \parallel Q_B) & \text{where } A \cap B = \emptyset, \\
 & & & a \in A, b \in B
\end{array}
$$

5 Connection with event structures

We adopt the modelling of CSP in prime event structures described in [13]. Let $\mathcal{E}(P)$ denote the event structure associated with CSP process P. We shall describe how we can give a concurrency / conflict semantics to event structures in a simple and intuitive way. Let $co(E)$, resp. $cf(E)$, denote the possible concurrency, resp. conflict, semantics of an event structure E. Then we shall establish

$s\ co(P)\ t$ iff $s\ co(\mathcal{E}(P))\ t$

$s\ cf(P)\ t$ iff $s\ cf(\mathcal{E}(P))\ t$

As well as providing evidence that our two semantics are well-founded, this will enable us to study the properties of our semantic domains more easily.

We start by defining the event structures which we shall be using. A *partially labelled prime event structure* $(E, \leq, \#, \ell)$ is a set E of events equipped with a partial order $\leq$ (causation), an irreflexive symmetric conflict relation $\#$, and a partial labelling function $\ell : E \longrightarrow Act$, satisfying

1. (axiom of finite causes) for every $e \in E, \{e' \mid e' \leq e\}$ is finite;
2. (forward propagation of conflict) if $e \# e' \leq e''$ then $e \# e''$.

Here *Act* is the set of actions which we earlier adopted for CSP. The labelling is partial because hidden events will be unlabelled. We shall refer to partially labelled prime event structure simply as event structures. We let $E, \ldots$ range over event structures, and $e, \ldots$ over events (members of E). Prime event structures were defined in [16].

We say that events e, e' of E are *concurrent* (notation $e \; co \; e'$) if none of $e \leq e'$, $e' \leq e$, $e \# e'$ holds. E is said to be *non-autoconcurrent* (nac) if concurrent events cannot have the same label, i.e., if $e, e' \in E$ are concurrent, then it is not the case that $\ell(e)$, $\ell(e')$ are both defined and equal to each other.

A *run* (or configuration) r of an event structure E is a conflict-free, downwards-closed subset of E. In other words:

1. if $e, e' \in r$ then not $e \# e'$;

2. if $e \in r$ and $e' < e$ then $e' \in r$.

We say that runs r, r' of E are *compatible* (denoted $r \; cp \; r'$) if $r \cup r'$ is a run. This amounts to seeing whether $r \cup r'$ is conflict-free, since downwards closure is automatic. If $r \cup r'$ is not a run then we say that r and r' *conflict* (denoted $r \# r'$).

Now we define what it means for a trace t to be an *observation* of a run r of an event structure E. A trace is a member of Act^*. It is useful to regard such strings as also being linearly ordered multisets, in the obvious way. Let $obs_E(t, r)$ iff there is a partial bijection $f : r \to t$ such that

1. if $e \in r$ and $\ell(e)\downarrow$ then $f(e)\downarrow$ and $f(e) = \ell(e)$;

2. if $e, e' \in r$, $e \leq e'$ and $\ell(e), \ell(e')\downarrow$, then $f(e) \leq f(e')$

where $g(e)\downarrow$ means $g(e)$ is defined. This amounts to saying that t is a linearisation of the visible events in r. We shall drop the subscript E if no confusion will arise. If E is nac then f is unique.

We define the *traces* of E, denoted $tr(E)$, to be the set of all t such that $obs(t, r)$ for some run r of E. For $t \in tr(E)$ let $runs(t)$ denote the set of all runs r of E such that $obs(t, r)$.

Definition 5.1 *Let E be an event structure and r a run of E.*

1. *r is* live *for an event $e \in E$ if $e \notin r$ and $r \cup \{e\}$ is a run. Denote the set of such events by $live(r)$.*

2. *r is* stable *if it is maximal with respect to hidden events, that is, $live(r)$ contains no hidden events.*

3. *r* refuses *a set $X \subseteq Act$ if r is stable and $X \cap \{\ell(e) \mid e \in live(r)\} = \emptyset$.*

4. *For $t \in Act^*, X \in \mathcal{P}_f(Act)$, $(t, X) \in fail(E)$ iff there exists a run r such that $obs(t, r)$ and r refuses X.*

Proposition 5.2 *Let P be a finite CSP process. Then $fail(P) = fail(\mathcal{E}(P))$.*

Now we turn our attention to our concurrency and conflict semantics. We define the counterpart of the $co()$ and $cf()$ relations, as well as additional relations $cp()$ of *compatibility*, $gco()$ of *guaranteed concurrency*, and $pre()$ of *causal precedence*. We then show that there exist connections between them. The complement of the relation $cf()$ is denoted by $\overline{cf}()$.

We shall need the following auxiliary notions. String prefix order is denoted by $\leq$. We say s *perm* t iff the trace s is a permutation of t. $s \subseteq t$ denotes the multiset order; more formally, $s \subseteq t$ iff $\exists s'.s \; perm \; s' \leq t$. For $s, t \in Act^\star$, their *string difference* is defined inductively by: $\varepsilon - s = \varepsilon$, $s - \varepsilon = s$, $a - a = \varepsilon$, and

$$as - t \quad = \quad \begin{cases} s - (t - a) & \text{if } a \in t \\ a(s - t) & \text{otherwise} \end{cases}$$

The *leftmost difference* $\mathrm{ld}(s,t)$ of traces $s, t \in Act^\star$ is defined to be $head(s - t)$ if $s - t \neq \varepsilon$, and undefined otherwise.

Definition 5.3 *Let E be an event structure and $s, t \in tr(E)$.*

1. *$s \; cp(E) \; t$ iff there are $r \in runs(s), r' \in runs(t)$ such that $r \; cp \; r'$;*
2. *$s \; cf(E) \; t$ iff there are $r \in runs(s)$, $r' \in runs(t)$ such that $r \# r'$;*
3. *$s \; co(E) \; t$ iff there is $r \in runs(s) \cap runs(t)$;*
4. *$s \; gco(E) \; t$ iff $runs(s) = runs(t)$;*
5. *$s \; pre(E) \; t$ iff $s \subseteq t$ and $s \; \overline{cf}(E) \; t$.*

Lemma 5.4 *Let E be nac. If $s, t \in tr(E)$, r, r' are runs of E such that $s \subseteq t$, $obs(s, r)$, $obs(t, r')$ and $r \; cp \; r'$, then $obs(t, r \cup r')$.*

We can now show that the $co()$ and $gco()$ relations are the natural counterparts of the Mazurkiewicz trace equivalence since they relate two different observations of the same run.

Proposition 5.5 *Let E be nac. For any $s, t \in tr(E)$:*

1. *$s \; co(E) \; t$ iff $s \; cp(E) \; t$ and s perm t;*
2. *$s \; cp(E) \; t$ iff $s(t - s) \; co(E) \; t(s - t)$;*
3. *$s \; gco(E) \; t$ iff $s \; \overline{cf}(E) \; t$ and s perm t.*

Proof: Follows from Lemma 5.4. □

These last two results mean that the $cp(E)$, resp. the $gco(E)$, semantics may be derived from our $co(E)$, resp. $cf(E)$, semantics. Furthermore, the next result shows that the relation $pre()$ is a counterpart of the causal ordering on E, as well as the Mazurkiewicz trace preorder. Part (1) clearly corresponds to forward propagation of conflict.

Proposition 5.6 *Let E be nac.*

1. *If $s \; cf(E) \; t$ and $t \; pre(E) \; u$ then $s \; cf(E) \; u$.*
2. *$pre(E)$ is a preorder.*

Proof: (1) is by Lemma 5.4. (2) comes from (1). □

Now, if $s\ gco(E)\ t$ then s and t should be in a sense indistinguishable:

Proposition 5.7 *Let E be nac.*

1. *Suppose* $u\ cf(E)\ sv$ *and* $s\ gco(E)\ t$. *Then* $u\ cf(E)\ tv$.
2. *Suppose* $(su, X) \in fail(E)$ *and* $s\ gco(E)\ t$. *Then* $(tu, X) \in fail(E)$.

The following result shows that there is a connection between conflict and failure semantics.

Proposition 5.8 *Let E be nac.*

1. *Suppose* $(s, X), (s, Y) \in fail(E)$ *and* $s\ \overline{cf}(E)\ s$. *Then* $(s, X \cup Y) \in fail(E)$.
2. *Suppose* $(s, \{a\}) \in fail(E)$, $t \in tr(E)$ *and let* $a = \mathrm{ld}(s,t)$. *Then* $s\ cf(E)\ t$.

Part (1) states that if a process is deterministic in the sense of not having self-conflict, then it is deterministic in the failures semantics (*i.e.* does not have essentially different refusals). A simple case of part (2) would be where $(b, \{a\}) \in fail(E)$ and $ba \in tr(E)$. We expect that there must be two conflicting runs, where a is either performed or refused. The result guarantees that $b\ cf(E)\ ba$ in accordance with this expectation.

Now we wish to establish that our two semantics ($co()$ and $cf()$) are sound under the event structure interpretation. This is shown by structural induction on finite processes using the definitions of the CSP operators on event structures as in [13].

We do not repeat the definition of the map from CSP to event structures given in [13]. The most difficult part of this is the definition of the parallel composition of two event structures. We adopt a slightly different version, which we denote $E_{1A} \parallel E_{2B}$ (or $E_1 \parallel E_2$ where A and B are understood). It is similar to those given in [19] and [13]. Essentially the runs of $E_1 \parallel E_2$ are sets of pairs of events from E_1, E_2. If r is a run of $E_1 \parallel E_2$ then we can project r in a standard way to get runs $\pi_1(r)$, $\pi_2(r)$ of E_1, E_2 respectively.

Non-autoconcurrency is a particularly useful property, and so we would like the image of a CSP process to be a nac event structure. Various authors have argued against autoconcurrency, see *e.g.* [3]; for example, it is disallowed in trace theory. It turns out that the mixed parallel is the right one to consider since it gives rise to non-autoconcurrent event structures.

Proposition 5.9 *Let P be a finite CSP process. Then* $\mathcal{E}(P)$ *is nac.*

The only difficulty with the above result is in showing that if E_1, E_2 are nac then so is $E_1 \parallel E_2$. It is important to note that nac would fail if we allowed many-one renaming.

Lemma 5.10 *Let E_1, E_2 be nac. Let r, r' be runs of $E = E_1 \parallel E_2$. Then*

$$r \# r' \quad \textit{iff} \quad \pi_1(r) \#_1 \pi_1(r') \quad \textit{or} \quad \pi_2(r) \#_2 \pi_2(r')$$

Proof: (Remarks.) The "if" is straight from the definition of parallel composition. For "only if" we need nac. □

The nac assumption is essential in the above result. For a counterexample, let $E_1 = \{e\}, E_2 = \{e', e''\}$, with $e' \ co_2 \ e''$. Let all events be labelled with a. So E_2 exhibits autoconcurrency. Now $\{(e, e')\}$ and $\{(e, e'')\}$ are both runs of $E_1 \parallel E_2$ and their projections do not conflict. However, $\{(e, e'), (e, e'')\}$ is not a run of $E_1 \parallel E_2$.

Proposition 5.11 *If P is a finite CSP process, then $cf(P)=cf(\mathcal{E}(P))$.*

Proof: By structural induction on P. The case for parallel composition is handled by Lemma 5.10. We omit the checks for the other operators. □

Lemma 5.12 *Let E_1, E_2 be event structures, let $E = E_{1A} \parallel E_{2B}$, and let $s, s' \in tr(E)$. If $s \lceil A \ co(E_1) \ s' \lceil A$ and $s \lceil B \ co(E_2) \ s' \lceil B$ then $s \ co(E) \ s'$.*

Proof: (Idea.) We take common runs for $s \lceil A$, $s' \lceil A$ and $s \lceil B$, $s' \lceil B$ and "knit" them together to form runs for E. □

Proposition 5.13 *If P is a finite CSP process, then $co(P)=co(\mathcal{E}(P))$.*

The import of Propositions 5.11 and 5.13, together with Proposition 5.2, is that our semantics, while simpler and more abstract, is "as good as" the event structure semantics. Thus, concurrency distinctions can be made at a high level of abstraction, without resorting to reasoning about event occurrences.

6 Further work

A further refinement of the semantics would be by adding causality information. The definition is similar to that of the possible concurrency relation $co()$.

Another direction is to adopt a more discriminating notion of an observation, such as failures, instead of traces. Observations as failures are well motivated, since they can be explained in terms of experiments on machines. A generalization of the concurrency relation to work on failures instead of traces is straightforward, and it would lead to finer distinctions.

Finally, we would like to derive a complete axiomatization of the failure–concurrency, resp. failure–conflict, pairs analogous to the axioms for failure–divergence pairs [5, 6]. It is relatively well known that any finite CSP process is equivalent to one obtained using 0, prefixing, $\sqcap$ and $\Box$ [7]. The interesting point is that the parallel operator $\parallel$ is not used. We would anticipate that in our model, if there is a satisfactory axiomatization, the parallel operator would play an essential role.

7 Conclusion

We have proposed a framework of refinements of the standard failures model for CSP, the purpose of which was to weaken the expansion law to an inequality. We have shown that the additional information about concurrency, conflict and causality can be derived systematically from the syntax of processes, without the need for additional labelling of actions, as long as the relations are interpreted as representing the *possibility* (or, dually, the *guarantee*) of the given phenomenon. We have carried out the construction for a large subset of CSP, which includes hiding, with respect to the relations of possible concurrency and conflict. We have found that the laws of the "possible concurrency" model are essentially the same as the original CSP laws, the exception being the expansion law. The "possible conflict" (or "guaranteed concurrency") calculus is less well behaved, as it also fails the idempotency of choice. The resulting inequational calculi are only slightly more complex than the original failures calculus.

We believe that this framework is, to a large extent, independent of the choice of the language and model. However, we should point out that the CSP parallel is the right one to make the construction work.

Acknowledgements

We would like to thank Bill Roscoe, Rob van Glabbeek, Frits Vaandrager, Samson Abramsky and Mike Shields for helpful discussions on the subject. Bill Roscoe's contribution in particular was invaluable, in that he directed our attention to the significance and potential technical advantages of possible concurrency, an aspect we had rather neglected, being hampered at the time by a belief that guaranteed concurrency would be of more interest in applications.

References

[1] Luca Aceto and Uffe Engberg. Failure semantics for a simple process language with refinement. In *Foundations of Software Technology and Theoretical Computer Science*, pages 89–108. Springer-Verlag, 1991.

[2] J.C.M. Baeten and J.A. Bergstra. Non interleaving process algebra. In *CONCUR 93*, volume 715 of *Lecture Notes in Computer Science*, pages 308–323. Springer-Verlag, 1993.

[3] Bard Bloom and Marta Z. Kwiatkowska. Trade-offs in true concurrency: Pomsets and Mazurkiewicz traces. In S. Brookes, M. Main, A. Melton, M. Mislove, and D. Schmidt, editors, *Mathematical Foundations of Programming Semantics*, volume 598 of *Lecture Notes in Computer Science*. Springer-Verlag, 1992.

[4] Gerard Boudol and Ilaria Castellani. Permutation of transitions: an event structure semantics for CCS and SCCS. In *Linear Time, Branching Time and Partial Order in Logics and Models for Concurrency*, volume 354 of *Lecture Notes in Computer Science*, pages 411–427. Springer-Verlag, 1989.

[5] S. D. Brookes, C. A. R. Hoare, and A. W. Roscoe. A theory of communicating sequential processes. *Journal of the ACM*, 31(3):560–599, 1984.

[6] S. D. Brookes and A. W. Roscoe. An improved failures model for communicating processes. In *Seminar on Concurrency*, volume 197 of *Lecture Notes in Computer Science*, pages 281–305. Springer-Verlag, 1985.

[7] Stephen Brookes. *A Model for Communicating Sequential Processes*. PhD thesis, Oxford University, 1983.

[8] P. Degano and U. Montanari. Concurrent histories: A basis for observing distributed systems. *Journal of Computer and System Sciences*, 34:422–461, 1987.

[9] M. Hennessy. Concurrent testing of processes. Technical Report 11/91, University of Sussex, 1991.

[10] C.A.R. Hoare. *Communicating Sequential Processes*. Prentice-Hall International, 1985.

[11] P.W. Hoogers, H. Kleijn, and P.S. Thiagarajan. A trace semantics for petri nets. In *ICALP'92*, volume 623 of *Lecture Notes in Computer Science*. Springer-Verlag, 1992.

[12] Marta Z. Kwiatkowska and Iain C.C. Phillips. Concurrency and conflict in CSP. In G.L. Burn, S.J. Gay, and M.D. Ryan, editors, *Theory and Formal Methods 1993: Proceedings of the First Imperial College, Department of Computing, Workshop on Theory and Formal Methods*, Workshops in Computer Science, pages 209–225. Springer-Verlag, 1993.

[13] R. Loogen and U. Goltz. Modelling nondeterministic concurrent processes with event structures. *Fundamenta Informaticae*, XIV:39–74, 1991.

[14] Antoni Mazurkiewicz. Basic notions of trace theory. In *Linear Time, Branching Time and Partial Order in Logics and Models for Concurrency*, volume 354 of *Lecture Notes in Computer Science*, pages 25–34. Springer-Verlag, 1989.

[15] M. Mukund and M. Nielsen. CCS, locations and asynchronous transition systems. In R. Shyamasundar, editor, *Foundations of Software Technology and Theoretical Computer Science*, volume 652 of *Lecture Notes in Computer Science*. Springer-Verlag, 1992.

[16] M. Nielsen, G. Winskel, and G. Plotkin. Petri nets, event structures and domains. *Theoretical Computer Science*, 13:85–108, 1981.

[17] M.W. Shields. Deterministic asynchronous automata. In *Formal Methods in Programming*. North Holland, 1985.

[18] D. Taubner and W. Vogler. The step failures semantics. In *STACS'87*, volume 247 of *Lecture Notes in Computer Science*, pages 348–359. Springer-Verlag, 1987.

[19] F. Vaandrager. A simple definition for parallel composition of prime event structures. Technical Report CS-R8903, CWI, 1989.

[20] Glynn Winskel. Event structure semantics for CCS and related languages. In *Automata, Languages and Programming*, volume 140 of *Lecture Notes in Computer Science*, pages 561–567. Springer-Verlag, 1982.

Metric completion versus ideal completion

Mila E. Majster-Cederbaum, Christel Baier
Fakultät für Mathematik und Informatik, Universität Mannheim
Germany

Abstract

Complete partial orders have been used for a long time for defining semantics of programming languages. In the context of concurrency de Bakker and Zucker [4] proposed a metric setting for handling concurrency, recursion and nontermination, which has proved to be very successful in many applications. Starting with a semantic domain D for 'finite behaviour' we investigate the relation between the ideal completion $Idl(D)$ and the metric completion which are both suitable to model recursion and infinite behaviour. We also consider the properties of semantic operators.

1 Introduction

In order to provide denotational semantics to programming languages complete partial orders have been successfully used to model recursive or infinite behaviour of programs. In the context of concurrency de Bakker and Zucker [4] (going back to ideas of M. Nivat) proposed to use complete metric spaces in order to model the behaviour of recursive or infinite concurrent systems. Some semantic domains for modelling concurrent systems, e.g. event structures, trees, pomsets and strings, can be endowed with both a metric and a partial order structure. One way of looking at defining semantics is that one first provides a semantic domain for 'finite behaviour' and secondly uses a completion technique to obtain a domain for 'infinite behaviour'. In this paper we investigate the connection between metric completion and ideal completion techniques. These results are related to our previous investigations [1, 2, 3, 9] and shed light on the question of the influence of the choice of mathematical discipline on semantics. We also discuss similar work which has been done in [5]. Other attempts to 'reconcile' the metric and order approach can be found e.g. in [10, 13, 14].

We assume that D is a semantic domain for nonrecursive programs of a CCS-like language as finite strings or (labelled) trees of finite height. $\sqsubseteq$ is a partial order on D such that D has a bottom element $\perp$ which either can be the meaning of the *nil* program (the program which does not perform any action) or which represents a totally undefined process. If we have semantic operators on D which are monotone w.r.t. $\sqsubseteq$ then the ideal completion $Idl(D)$ can be used as semantic domain for a denotational cpo semantics which extends the semantics on D for recursive programs. On the other hand if D is endowed with a metric such that the semantic operators are non-distance-increasing resp. contracting we get a denotational metric semantics on the metric completion $\overline{D}$. The question arises in which way the metric and ideal completion are related and how the denotational semantics on $Idl(D)$ resp. $\overline{D}$ are connected.

In this paper we answer these questions under the assumption that $(D, \sqsubseteq)$ can be endowed with a finite length. This length induces a metric on D. By a finite length we mean a function which assigns the maximal number of atomic steps to each element x of D which are needed for the execution of x. Here the elements of D are considered as processes. E.g. the length of a finite string is its usual length, the length of a tree is its height. The distance $d(x, y)$ induced by a length counts the maximal number n of steps on which the executions of x and y coincide (and then $d(x, y) = 1/2^n$).

The paper is organized as follows: Section 2 presents the concept of a length and a weight on a pointed poset. The relationship between the metric and ideal completion of a pointed poset with a finite length is shown in section 3. We show that the metric d on D can be lifted to a metric on $Idl(D)$. In section 3.1 we present conditions for the completeness of $Idl(D)$ as a metric space and we show that then there exists an isometric embedding

$$\mathcal{I} : \overline{D} \to Idl(D).$$

Section 3.2 deals with the question when $\mathcal{I}$ is an isometry, i.e. when the metric completion can be identified with the ideal completion. Section 3.3 shows that under certain conditions the topology induced by the metric induced by a length coincides with the Lawson topology on $Idl(D)$. In section 4 we discuss the connection between the canonical extensions of monotone and non-distance-increasing operators on the ideal and the metric completions (section 4.1) and we get a consistency result for denotational semantics Me_{cms} on $\overline{D}$ and Me_{cpo} on $Idl(D)$ of the form

$$\mathcal{I} \circ Me_{\text{cms}} = Me_{\text{cpo}}$$

(section 4.2). Section 5 presents conditions for the equivalence of the metric induced by a length and the metric of [5].

2 Pointed posets with a length or weight

Definition 2.1 *Let $(D, \sqsubseteq)$ be a pointed poset (i.e. a partially order set with a bottom element which we denote by $\bot_D$ or shortly $\bot$.) A* length *on $(D, \sqsubseteq)$ is a function $\rho : D \to \mathbb{N}_0 \cup \{\infty\}$ such that for all $x, y \in D$:*
(i) $\rho(x) = 0 \iff x = \bot_D$
(ii) $x \sqsubseteq y \implies \rho(x) \leq \rho(y)$
Let $Fin(D, \rho)$ denote the collection of all $y \in D$ such that $\rho(y) < \infty$. For all $x \in D$ we define:

$$\downarrow^n (x) = \{y \in D : y \sqsubseteq x,\ \rho(y) \leq n\}, \quad \downarrow^{\text{fin}}(x) = \bigcup_{n \in \mathbb{N}_0} \downarrow^n_\rho (x).$$

ρ is called finite *iff $Fin(D, \rho) = D$, i.e. $\rho(x) < \infty$ for all $x \in D$. An element $x \in D$ is called* approximable *(w.r.t. ρ) iff x is the least upper bound of $\downarrow^{\text{fin}}(x)$. $\mathcal{M}(D, \sqsubseteq, \rho)$ denotes the set of approximable elements.*

Lemma 2.2 *Let $(D, \sqsubseteq)$ be a pointed poset and ρ a length on $(D, \sqsubseteq)$. Then*

$$d[\rho](x, y) = \inf \left\{ \frac{1}{2^n} : \downarrow^n (x) = \downarrow^n (y) \right\}$$

is a pseudo ultrametric on D and an ultrametric on $\mathcal{M}(D, \sqsubseteq, \rho)$. $Fin(D, \rho)$ is a subspace of $\mathcal{M}(D, \sqsubseteq, \rho)$.

In particular: If ρ is a finite length on a pointed poset $(D, \sqsubseteq)$ then $d[\rho]$ is an ultrametric on D. In general the induced metric space $\mathcal{M}(D, \sqsubseteq, \rho)$ is not complete. In order to ensure the completeness of $\mathcal{M}(D, \sqsubseteq, \rho)$ we need additional assumptions.

Definition 2.3 *Let $(D, \sqsubseteq)$ be a pointed poset. A* weight *on $(D, \sqsubseteq)$ is a length ρ on $(D, \sqsubseteq)$ such that for all $x \in D$ and $n \geq 0$ the set $\downarrow^n (x)$ has a greatest element which we denote by $x[n]$. $x[n]$ is called the n*-cut *of x w.r.t. ρ.*

Let $(D, \sqsubseteq)$ be a cpo. A continuous weight *on $(D, \sqsubseteq)$ is a weight ρ on $(D, \sqsubseteq)$ such that for each $n \geq 0$ the function $D \to D$, $x \mapsto x[n]$, is continuous.*

Lemma 2.4 *Let ρ be a continuous weight on a cpo $(D, \sqsubseteq)$. Then the induced ultrametric space $\mathcal{M}(D, \sqsubseteq, \rho)$ is complete and*

$$\lim_{n \to \infty} x_n = \bigsqcup_{n \geq 0} x_n$$

for each monotone Cauchy sequence (x_n) in $\mathcal{M}(D, \sqsubseteq, \rho)$.

The concept of a finite weight can be realized on various domains, e.g.

- finite strings over some alphabet A (endowed with the prefixing ordering and the weight $\rho(x) = |x|$ where $|x|$ means the usual length of the string x)
- trees of finite height endowed with Winskels partial order [16] and the height as underlying weight
- prime event structures of finite depth with Winskels partial order [15] and the depth as underlying weight.

In order to give an example for a finite length which is not a weight we consider the pointed poset of traces in the sense of [11].

Example 2.5 In [8] the concept of (finite) traces is generalized to traces of length up to ω by dealing with the order resp. metric completion. We recall the basic notions of trace theory a la [8, 11]: Let (A, ι) be a **concurrent alphabet**, i.e. A is a set of actions and ι an irreflexive and symmetric relation on A (called **independency**). A **trace** is an equivalence class $[x]$ of a finite string x over A where the underlying equivalence relation $\equiv$ is the reflexive, transitive closure of $\equiv'$ which is given by:

$$x \equiv' y \quad :\Longleftrightarrow \quad \exists \alpha, \beta \in A,\ z, w \in A^* \ : \ \alpha \,\iota\, \beta \ \wedge \ x = z\alpha\beta w \ \wedge \ y = z\beta\alpha w$$

If $\sigma = [x]$ is a trace then $|\sigma| = |x|$ where $|x|$ means the usual length of x. In the following *Traces* denotes the set of traces w.r.t. a fixed concurrent alphabet (A, ι) and $\sqsubseteq$ means the lifting of the prefixing ordering on A^* to traces. I.e.

$$[x] \sqsubseteq [y] \quad :\Longleftrightarrow \quad \exists x', y', z \in A^* \ : \ x' \equiv x \ \wedge \ y' \equiv y \ \wedge \ y' = x'z$$

If $n \in I\!N$ then

$$\sigma^{(n)} = \{ \sigma' \in D : |\sigma'| \leq n, \sigma' \sqsubseteq \sigma \}.$$

[8] considers the following metric d on traces:

$$d(\sigma, \tau) = \inf \left\{ \frac{1}{2^n} : \sigma^{(n)} = \tau^{(n)} \right\},$$

This metric metric d coincides with the metric $d[\rho]$ where the length ρ is given by:

$$\rho : D \to I\!N_0, \ \rho(\sigma) = |\sigma|$$

If $\iota \neq \emptyset$ then ρ is not a weight, e.g. if $\alpha, \beta \in A$, $\alpha \ \iota \ \beta$ then

$$\downarrow^1 ([\alpha\beta]) = \{\bot, [\alpha], [\beta]\}$$

does not contain a greatest element since $[\alpha]$, $[\beta]$ are incomparable.

3 The ideal completion as a metric space

Given a finite length on a pointed poset D the set of left-closed subsets $\mathcal{P}_\downarrow(D)$ can be endowed with a continuous weight. Hence $\mathcal{P}_\downarrow(D)$ turns into a complete metric space. Then the ideal completion $Idl(D)$ as a subspace of $\mathcal{P}_\downarrow(D)$ is also a metric space. We present conditions which ensure the completeness of $Idl(D)$ and we show that if $Idl(D)$ is complete then the metric completion of D can be embedded into the ideal completion. We also present conditions for the isometry of the metric and ideal completion.

Let $(D, \sqsubseteq)$ be a pointed poset and $X \subseteq D$. Then

$$X \downarrow = \{ y \in D : y \sqsubseteq x \text{ for some } x \in X \}$$

if $X \neq \emptyset$ and $\emptyset \downarrow = \{\bot\}$. If $x \in D$ then $x \downarrow = \{x\} \downarrow$. X is called `left-closed` iff $X \downarrow = X$. Let $\mathcal{P}_\downarrow(D)$ denote the set of left-closed subsets of D.

X is called `directed` iff for all $x, y \in X$ there exists an upper bound of x and y in X, i.e. there exists $z \in X$ with $x \sqsubseteq z$ and $y \sqsubseteq z$. X is called an `ideal` iff X is left-closed and directed.

$$Idl(D) = \{ X \subseteq D : X \text{ is an ideal} \}$$

denotes the set of ideals of $(D, \sqsubseteq)$. $(\mathcal{P}_\downarrow(D), \subseteq)$ and $(Idl(D), \subseteq)$ are cpo's. The later is called the `ideal completion` of $(D, \sqsubseteq)$.

If ρ is a length on a pointed poset $(D, \sqsubseteq)$ then ρ induces a continuous weight (also called ρ) on $\mathcal{P}_\downarrow(D)$ which is given by:

$$\rho(X) = \sup \{ \rho(x) : x \in X \}$$

Then the n-cut of $X \in \mathcal{P}_\downarrow(D)$ is

$$X[n] = \{ x \in X : \rho(x) \leq n \}.$$

If ρ is finite then all elements of $\mathcal{P}_\downarrow(D)$ are approximable. By Lemma 2.4:

Lemma 3.1 *Let ρ be a finite length on a pointed poset $(D, \sqsubseteq)$. Then $(\mathcal{P}_\downarrow(D), d[\rho])$ is a complete metric space. The metric $d[\rho]$ on $\mathcal{P}_\downarrow(D)$ is given by the formula:*

$$d[\rho](X,Y) = \inf \left\{ \frac{1}{2^n} : X[n] = Y[n] \right\}$$

Remark 3.2 Let ρ be a finite length on a pointed poset $(D, \sqsubseteq)$. If (X_n) is a Cauchy sequence in $(\mathcal{P}_\downarrow(D), d[\rho])$ then

$$\lim_{n\to\infty} X_n = \bigcup_{k\geq 0} X_{n_k}[k]$$

where (X_{n_k}) is a subsequence of (X_n) with $d[\rho](X_n, X_m) \leq 1/2^k$ for all n, $m \geq n_k$.

The restriction of ρ on $Idl(D)$ is a length on $(Idl(D), \subseteq)$, but in general not a weight.

Notation 3.3 *Let (M, d) be a metric space. Then the metric completion of (M, d) is denoted by $(\overline{M}, \overline{d})$. We assume that $M \subseteq \overline{M}$ and that d is the restriction of $\overline{d}$ on M.*

If (M, d) and (N, d') are metric spaces and $f : M \to N$ a non-distance-increasing function then $\overline{f}$ denotes the unique non-distance-increasing function $\overline{M} \to \overline{N}$ with $\overline{f}(x) = f(x)$ for all $x \in M$.

If ρ be a finite length on a pointed poset $(D, \sqsubseteq)$, $d = d[\rho]$, then

$$\imath : D \to Idl(D), \ \imath(x) = x\downarrow$$

is an isometric embedding of the metric space D into the metric space $Idl(D)$. Hence the canonical extension $\overline{\imath} : \overline{D} \to \overline{Idl(D)}$ of $\imath$ is an isometric embedding. If ρ is a finite length on $(D, \sqsubseteq)$ such that $(Idl(D), d[\rho])$ is a complete metric space then the metric completion of $(D, d[\rho])$ can be embedded into the ideal completion of $(D, \sqsubseteq)$.

3.1 The completeness of the ideal completion as a metric space

We present conditions which ensure the completeness of $(Idl(D), d[\rho])$.

Theorem 3.4 *Let ρ be a finite length on a pointed poset $(D, \sqsubseteq)$, $d = d[\rho]$, such that for all x, $y \in D$: If $\{x, y\}$ is bounded in D (i.e. there exists $z \in D$ with $x \sqsubseteq z$ and $y \sqsubseteq z$) then the least upper bound $x \sqcup y$ of x and y exists in D. Then $(Idl(D), d[\rho])$ is a complete metric space and $\overline{\imath} : \overline{D} \to Idl(D)$ is an isometric embedding.*

Proof: Let (X_n) be a Cauchy sequence of ideals. W.l.o.g. $d[\rho](X_n, X_m) \leq 1/2^n$ for all $m \geq n \geq 0$. (Otherwise we deal with a subsequence (X_{n_k}) of (X_n).) Then $X_n[n] = X_m[n] \subseteq X_m[m]$ for all $m \geq n \geq 0$. By Remark 3.2 $X = \bigcup X_n[n]$

is the limit of (X_n) in $(\mathcal{P}_\downarrow(D), d[\rho])$. We have to show that X is an ideal. (Then $X = \lim X_n$ in $(Idl(D), d[\rho])$.)

X is left-closed (since the sets $X_n[n]$ are left-closed). Now we show that X is directed: Let x, $y \in X$. Then there exists $n \geq 0$ with x, $y \in X_n[n]$. Since X_n is directed there exists $w \in X_n$ with x, $y \sqsubseteq w$. By assumption $z = x \sqcup y$ exists.

If $m \geq n$ then x, $y \in X_n[n] \subseteq X_m[m]$. Since X_m is directed there exists $w_m \in X_m$ with x, $y \sqsubseteq w_m$. Then:

$$z \sqsubseteq w_m \quad \forall\, m \geq n$$

Let $m = \max\{n, \rho(z)\}$. Then $z \in X_m[m] \subseteq X$. □

Equivalent descriptions for the condition in Theorem 3.4 are presented in the following remark:

Remark 3.5 Let $(D, \sqsubseteq)$ be a pointed poset. Then the following are equivalent:
(I) Whenever x, $y \in D$ have an upper bound in D then $x \sqcup y$ exists.
(II) $(D, \sqsubseteq)$ is finite bounded complete. I.e. for each finite subset of D which is bounded the least upper bound exists.
(III) $(Idl(D), \subseteq)$ is bounded complete. I.e. for each bounded subset of $Idl(D)$ the least upper bound exists.

Lemma 3.6 *Let ρ be a finite length on a pointed poset $(D, \sqsubseteq)$ such that for each ideal I and each natural number n the set $I[n]$ is directed. Then*
(a) ρ is a continuous weight on $(Idl(D), \subseteq)$.
(b) $(Idl(D), d[\rho])$ is a complete metric space and $\bar{\imath} : \overline{D} \to Idl(D)$ is an isometric embedding.

Proof: $I[n]$ is the n-cut of $I \in Idl(D)$ and $I = \bigcup I[n]$. The functions $I \mapsto I[n]$ are continuous. Hence ρ is a continuous weight on $Idl(D)$. Then we apply Lemma 2.4. □

Theorem 3.7 *Let ρ be a finite weight on a pointed poset $(D, \sqsubseteq)$. Then $(Idl(D), d[\rho])$ is a complete metric space and $\bar{\imath} : \overline{D} \to Idl(D)$ is an isometric embedding.*

Proof: For each ideal I the set $I[n]$ is directed. Then we apply Lemma 3.6. □

3.2 Metric and ideal completion

We show that under the assumptions of Theorem 3.4 resp. 3.7 and the additional assumption that the sets $I[n]$ are finite the metric completion of D and the ideal completion of D are isometric.

Theorem 3.8 *Let ρ be a finite length on a pointed poset $(D, \sqsubseteq)$ such that the following conditions (i) and (ii) are satisfied:*
(i) $(D, \sqsubseteq)$ is finite bounded complete.
(ii) For all $I \in Idl(D)$ the set $I[n]$ is finite.
Then $\bar{\imath} : \overline{D} \to Idl(D)$ is an isometry.

Proof (Sketch): By Theorem 3.4 it is sufficient to show that $\bar{\imath}$ is surjective. Let I be an ideal. By assumption (ii) for all $n \geq 0$ the n-cut $I[n]$ of I is finite. Since I is directed there exists an upper bound $x_n \in I$ of $I[n]$, i.e. $y \sqsubseteq x_n$ for all $y \in I[n]$. Then

$$\downarrow^n (x_n) \ = \ (x_n \downarrow)[n] \ = \ I[n].$$

We define by induction on k a subsequence (x_{n_k}) of (x_n) as follows: In the basis of induction ($k = 0$) we define $n_0 \ = \ 0$. In the induction step $k \Longrightarrow k+1$ we define

$$n_{k+1} \ = \ 1 \ + \ \max \ \{ \ \rho(x_{n_k}), \ n_k \ \}.$$

It can be shown that (x_{n_k}) is a Cauchy sequence in D and $\bar{\imath}(\lim x_{n_k}) \ = \ I$. □

Theorem 3.9 *Let ρ be a finite weight on a pointed poset $(D, \sqsubseteq)$ such that for all $I \in Idl(D)$ the n-cut $I[n]$ is a finite set. Then $\bar{\imath} : \overline{D} \to Idl(D)$ is an isometry.*

Proof (Sketch): By Theorem 3.7 it is sufficient to show that $\bar{\imath}$ is surjective. It can be shown that for each ideal I there exists a sequence (x_n) in I with $I[n] \ = \ x_n \downarrow$. Then (x_n) is a Cauchy sequence and $\bar{\imath}(\lim x_n) \ = \ I$. □

3.3 The $d[\rho]$-topology and the Lawson topology on $Idl(D)$

We show that under certain conditions the topology induced by $d[\rho]$ equals the Lawson topoloy on $Idl(D)$. We omit the general definitions of the topologies induced by a metric resp. the Lawson topology. They can be found e.g. in [6, 7]. We only specify the open sets w.r.t. the $d[\rho]$-topology and the Lawson topolgy.

- A subset U of $Idl(D)$ is open w.r.t. the $d[\rho]$-topology if and only if U can be written as union of balls

  $$B_n(I) \ = \ \{ \ J \in Idl(D) \ : \ I[n] = J[n] \ \}$$

  where $n \in \mathbb{N}_0$ and $I \in Idl(D)$.

- Let $(D, \sqsubseteq)$ be a pointed poset such that D is countable. Then a subset U of $Idl(D)$ is open w.r.t. the Lawson topoloy if and only if

  $$U \ = \ \bigcup_{i \in A} \left(\bigcap_{j \in B(i)} W_{i,j} \cap \bigcap_{j \in C(i)} V_{i,j} \right)$$

  where A is an arbitrary indexing set, for all $i \in A$ the sets $B(i)$, $C(i)$ are finite sets and for all $j \in B(i)$ there exists $x \in D$ with $W_{i,j} = W_x$, for all $j \in C(i)$ there exists $x \in D$ with $V_{i,j} = V_x$.

Here:

$$W_x \ = \ \{I \in Idl(D) : x \in I\}, \quad V_x \ = \ \{I \in Idl(D) : x \notin I\}$$

Theorem 3.10 *Let ρ be a finite length on a countable pointed poset $(D, \sqsubseteq)$ such that for all $n \geq 0$ the set*

$$\{x \in D : \rho(x) \leq n\}$$

is finite. Then the $d[\rho]$-topology coincides with the Lawson topology. I.e. if U is a subset of $Idl(D)$ then U is open w.r.t. the $d[\rho]$-topology if and only if U is open w.r.t. the Lawson topology.

Proof: Since the union and finite intersection of open sets is always open it is sufficient to show that the sets W_x, V_x are open w.r.t. the $d[\rho]$-topology and that the balls $B_n(I)$ are open w.r.t. the Lawson topology. If $x \in D$ then

$$W_x = \bigcup \{ B_n(I) : I \in W_x \}, \quad V_x = \{ B_n(I) : I \in V_x \}$$

are open w.r.t. the $d[\rho]$-topology. If $I \in Idl(D)$ and $n \geq 0$ then Let $B_n(I) = W \cap V$ where

$$W = \bigcap_{x \in I[n]} W_x, \quad V = \bigcap_{x \in C} V_x$$

and where $C = \{x \in D \setminus I : \rho(x) \leq n\}$. Since the sets $I[n]$ and C are finite we get: $B_n(I)$ is open w.r.t. the Lawson topology. □

Example 3.11 [8] considers the metric

$$d^*(I, J) = \inf \left\{ \frac{1}{2^n} : I[n] = J[n] \right\}$$

on the ideal completion of *Traces* (cf. Example 2.5). This is the metric induced by the length ρ on *Idl(Traces)* where ρ is an in Example 2.5. [8] shows:
(I) If A is countable then the ideal completion of *Traces* is a complete metric space.
(II) If A is finite then the metric completion of *Traces* and the ideal completion (endowed with the metric d^*) are isometric.
(III) The d^*-topology and the Lawson topology on the ideal completion of *Traces* are the same.
These results are special cases of our results: (I) follows by Theorem 3.4, (II) by Theorem 3.8, (III) by Theorem 3.10.

4 Denotational semantics on the metric and ideal completion

In order to compare a cpo denotational semantics Me_{cpo} on $Idl(D)$ with a metric denotational semantics Me_{cms} on $\overline{D}$ we assume that the semantic operators on $Idl(D)$ resp. $\overline{D}$ are the canonical extensions of semantic operators on D. Now we discuss the relationship between the canonical extensions and we present conditions for a consistency result of the form $\bar{\imath} \circ Me_{\text{cms}} = Me_{\text{cpo}}$.

4.1 The canonical extensions of operators on the metric and ideal completion

Let ρ be a finite length on a pointed poset $(D, \sqsubseteq)$. If $f : D \to D$ is a function which is monotone w.r.t. $\sqsubseteq$ and non-distance-increasing w.r.t. $d[\rho]$ then the question arises in which way the canonical extension

$$f^* : Idl(D) \to Idl(D), \quad f^*(I) = f(I) \downarrow$$

of f on the ideal completion and the canonical extension

$$\overline{f} : \overline{D} \to \overline{D}, \ \overline{f}\left(\lim_{n\to\infty} x_n \right) = \lim_{n\to\infty} f(x_n)$$

of f on the metric completion are related.

Lemma 4.1 *Let ρ be a finite weight on a pointed poset D such that $(Idl(D), d[\rho])$ is complete. If $f : D \to D$ is monotone and non-distance-increasing then*

$$\overline{\imath} \circ \overline{f} = f^* \circ \overline{\imath}$$

and f^* is non-distance-increasing w.r.t. $d[\rho]$. If f is contracting then also f^* is contracting.

In the general case (i.e. ρ is a length but not a weight) we cannot establish the result of Lemma 4.1. Nevertheless we have the following:

Lemma 4.2 *Let ρ be a finite length on a pointed poset D such that $(Idl(D), d[\rho])$ is complete. Let $f : D \to D$ be a function which is monotone and non-distance-increasing. Then $\overline{f}$ is contracting with contracting constant $1/2$ and*

$$\overline{\imath}(\ fix(\overline{f})\ ) = lfp(f^*).$$

Here $fix(\overline{f})$ means the unique fixed point of $\overline{f}$ and $lfp(f^*)$ the least fixed point of f^*.

4.2 The consistency of denotational semantics on the metric and ideal completion

Our aim is to establish a consistency result for denotational semantics on the complete metric space $\overline{D}$ and the cpo $Idl(D)$. Here we assume that D is a semantic domain for nonrecursive programs and that ρ is a finite length on D. We consider a language where recursion is modelled by declarations, i.e. a program is a pair $< s, \sigma >$ where s is a statement (which is built from operator symbols like prefixing or sequential composition, nondeterministic choice, parallelism, etc. and process variables) and a declaration σ (i.e. σ is a function which assigns a statement $\sigma(x)$ to each process variable x). $\mathcal{L}$ denotes the set of statements.

For each operator symbol ω in $\mathcal{L}$ let ω_D be a semantic operator on D which is monotone w.r.t. $\sqsubseteq$ and non-distance-increasing/contracting w.r.t. $d[\rho]$. For a fixed declaration σ we may define a mapping

$$F : (\mathcal{L} \to D) \ \to \ (\mathcal{L} \to D)$$

by structural induction on $s \in \mathcal{L}$:

$$F(f)(a) = a_D \qquad \text{for each constant symbol } a \text{ in } \mathcal{L}$$

$$F(f)(x) = f(\sigma(x)) \quad \text{for each process variable } x$$

$$F(f)(\omega(s_1,\ldots,s_n)) = \omega_D(F(f)(s_1),\ldots,F(f)(s_n))$$

$$\text{for each } n\text{-ary operator symbol } \omega \text{ in } \mathcal{L}$$

Similary we get mappings

$$F_{\mathrm{cms}} : (\mathcal{L} \to \overline{D}) \to (\mathcal{L} \to \overline{D})$$

$$F_{\mathrm{cpo}} : (\mathcal{L} \to Idl(D)) \to (\mathcal{L} \to Idl(D))$$

where we use the canonical extensions $\overline{\omega_D}$ resp. ω_D^* as semantic operators. Since F_{cpo} is continuous we have a denotational cpo semantics on $Idl(D)$:

$$Me_{\mathrm{cpo}} : \mathcal{L} \to Idl(D), \quad Me_{\mathrm{cpo}} = lfp(F_{\mathrm{cpo}}).$$

Under certain conditions (e.g. the guardedness of the statements $\sigma(x)$ in the sense of [12]) the function F_{cms} is contracting and hence has a unique fixed point. We get a metric denotational semantics on $\overline{D}$:

$$Me_{\mathrm{cms}} : \mathcal{L} \to \overline{D}, \quad Me_{\mathrm{cms}} = fix(F_{\mathrm{cms}}).$$

We establish the following consistency result:

Theorem 4.3 *Let ρ be a finite length on a pointed poset $(D, \sqsubseteq)$ such that $(Idl(D), d[\rho])$ is complete. Then*

$$\bar{\imath} \circ Me_{\mathrm{cms}} = Me_{\mathrm{cpo}}.$$

Proof (Sketch): By structural induction on $s \in \mathcal{L}$ it can be shown that

$$(\imath \circ F(f))(s) = F_{\mathrm{cpo}}(\imath \circ f)(s)$$

for all functions $f : \mathcal{L} \to D$. Now we assume that s is a fixed statement. By Banach's and Tarski's fixed point theorem:

$$Me_{\mathrm{cms}}(s) = \lim_{n\to\infty} f_n(s), \quad Me_{\mathrm{cpo}}(s) = \bigcup_{n\geq 0} g_n(s)$$

where $f_0 = \lambda t.\bot$, $f_{n+1} = F_{\mathrm{cms}}(f_n)$ and $g_0 = \lambda t.\{\bot\}$, $g_{n+1} = F_{\mathrm{cpo}}(g_n)$. By induction on n it can be shown that $\imath \circ f_n = g_n$. Then

$$\bar{\imath}(\, Me_{\mathrm{cms}}(s)\,) = \lim_{n\to\infty} f_n(s)\downarrow = \bigcup_{n\geq 0} \downarrow^n (f_n(s)).$$

and

$$Me_{\mathrm{cpo}}(s) = \bigcup_{n\geq 0} f_n(s)\downarrow$$

By the monotonicity of F_{cpo} it can be concluded that $\bar{\imath}(Me_{\mathrm{cms}}(s)) = Me_{\mathrm{cpo}}(s)$.
□

5 The connection between $d[\rho]$ and the metric d_ϕ of Comyn and Dauchet

We show the connection of the metric $d[\rho]$ on $Idl(D)$ and the metric d_ϕ of [5]. In contrast to our approach [5] only deal with countable posets. If $(D, \sqsubseteq)$ is a poset and $\phi : \mathbb{N}_0 \to D$ a surjective mapping then the metric d_ϕ of [5] is given by:

$$d_\phi(I, J) = \frac{1}{1 + \min\{n : \phi(n) \in I \Delta J\}}$$

where

$$I \Delta J = I \setminus J \cup J \setminus I$$

Our aim is to show the equivalence of d_ϕ and $d[\rho]$. Equivalence of two metrics d, d' on a set M means that d and d' induce the same topology on M. This is the case if and only if the Cauchy sequences in (M, d) and the Cauchy sequences in (M, d') are exactly the same. Given two equivalent metrics d and d' on a set M the completions of (M, d) and (M', d') are the same (more precisely: the underlying set is the same and the metrics are equivalent). In particular (M, d) is complete if and only if (M, d') is complete. I.e. when the equivalence of d_ϕ and $d[\rho]$ is shown then the results of the earlier sections also hold for d_ϕ instead of $d[\rho]$: We assume that D is endowed with the metric

$$d[\phi](x, y) = d_\phi(\, x\downarrow,\, y\downarrow).$$

We show that $\imath : D \to Idl(D)$ is an isometric embedding of the metric space $(D, d[\phi])$ into the metric space $(Idl(D), d_\phi)$. Since d_ϕ and $d[\rho]$ are equivalent $d[\rho]$ and $d[\phi]$ are equivalent metrics on D. If $(\overline{D}, \overline{d})$ is the completion of $(D, d[\rho])$ then there exists an equivalent metric $\overline{d_\phi}$ on $\overline{D}$ such that $(\overline{D}, \overline{d_\phi})$ is the completion of $(D, d[\phi])$. $(Idl(D), d[\rho])$ is complete if and only if $(Idl(D), d_\phi)$ is complete. In this case: The canonical extension $\overline{\imath} : \overline{D} \to Idl(D)$ of $\imath$ w.r.t. $d[\rho]$ and $d[\rho]$ coincides with the canonical extension of $\imath$ w.r.t. $d[\phi]$ and d_ϕ. By our results of section 3: If $(Idl(D), d_\phi)$ is complete then $\overline{\imath} : (\overline{D}, \overline{d_\phi}) \to (Idl(D), d_\phi)$ is an isometric embedding.

If D is finite then also $Idl(D)$ is finite. Since the topology induced by a metric on a finite set is always the discrete metric we get in the finite case that all metrics on $Idl(D)$ are equivalent. Therefore in the following we only consider the case that D is infinite and countable.

Instead of the metric d_ϕ we consider the following metric d^*_ϕ which is equivalent to d_ϕ:

$$d^*_\phi(I, J) = \frac{1}{2^{\min\{n:\phi(n) \in I \Delta J\}}}$$

Remark 5.1 In [5] it is shown that whenever ϕ, $\psi : \mathbb{N}_0 \to D$ are surjective mappings then the induced metrics d_ϕ and d_ψ are equivalent. Hence d^*_ϕ and d^*_ψ are equivalent. Since D is infinite there exists a bijective mapping $\psi : \mathbb{N}_0 \to D$. Therefore we may assume that the underlying mapping $\phi : \mathbb{N}_0 \to D$ is bijective.

Lemma 5.2 *Let ρ be a finite length on a pointed poset $(D, \sqsubseteq)$ and $\phi : \mathbb{N} \to D$ a surjective mapping. Then: Each Cauchy sequence in $(Idl(D), d[\rho])$ is a Cauchy sequence in $(Idl(D), d^*_\phi)$.*

Proof (Sketch): Let (I_n) be a Cauchy sequence in $(Idl(D), d[\rho])$. Let $\varepsilon > 0$ and $k_0 \geq 1$ such that $1/2^{k_0} < \varepsilon$. Let $m_0 = \max\{\rho(\phi(i)) : 0 \leq i < k_0\}$ and let $n_0 \geq 0$ such that

$$d[\rho](I_n, I_m) \leq \frac{1}{2^{m_0}} \quad \forall\, m, n \geq n_0.$$

Let $N = \max\{n_0, m_0\}$. Then it can be shown that for all n, $m \geq N$:

$$\min\{\, l : \phi(l) \in I_n \Delta I_m \,\} \geq k_0$$

Therefore $d_\phi^*(I_n, I_m) \leq 1/2^{k_0} < \varepsilon$ for all n, $m \geq N$. □

Lemma 5.3 *Let ρ be a finite length on a pointed poset $(D, \sqsubseteq)$ such that for all $n \geq 0$ the set*

$$\{x \in D : \rho(x) \leq n\}$$

is finite. Let $\phi : \mathbb{N}_0 \to D$ be a surjective mapping. Then: Each Cauchy sequence in $(Idl(D), d_\phi^*)$ is a Cauchy sequence in $(Idl(D), d[\rho])$.

Proof (Sketch): By Remark 5.1 we may assume that ϕ is bijective. Let $\sigma : \mathbb{N} \to \mathbb{N}_0$ be given by:

$$\sigma(n) = \min\{\, \rho(\phi(k)) : k > n \,\}$$

Let $\varepsilon > 0$ and $k_0 \geq 1$ with $1/2^{k_0 - 1} < \varepsilon$ and let $A = \{x \in D : \rho(x) < k_0\}$. By assumption A is a finite set. Since ϕ is bijective $\phi^{-1}(A)$ is finite. Let

$$r = \max \phi^{-1}(A)$$

and let $N \geq 0$ such that

$$d_\phi^*(I_n, I_m) < \frac{1}{2^r} \quad \forall\, n, m \geq N.$$

It can be shown that $d[\rho](I_n, I_m) \leq 1/2^{k_0 - 1} < \varepsilon$ for all n, $m \geq N$. □

By Lemma 5.2 and 5.3 and the equivalence of d_ϕ and d_ϕ^* we get:

Theorem 5.4 *Let ρ be a finite length on a pointed poset $(D, \sqsubseteq)$ such that for all $n \geq 0$ the set*

$$\{x \in D : \rho(x) \leq n\}$$

is finite. Let $\phi : \mathbb{N}_0 \to D$ be a surjectice mapping. Then $d[\rho]$ and d_ϕ are equivalent.

If $(D, \sqsubseteq)$ is a pointed poset and $\phi : \mathbb{N}_0 \to D$ a surjective and monotone mapping then

$$\rho = \rho_\phi : D \to \mathbb{N}_0, \quad \rho(x) = \min\{\, n \geq 0 : \phi(n) = x \,\}$$

is a finite length on $(D, \sqsubseteq)$ which satisfies the conditions of Theorem 5.4. Hence the metric d_ϕ on $Idl(D)$ in the sense of [5] is equivalent to the metric $d[\rho]$. I.e. in this case the approach of [5] is a special case of our approach. But in general there does not exist a finite length ρ such that the sets $\{x \in D : \rho(x) \leq n\}$ are finite, e.g. when there exists $x \in D$ with $x\downarrow$ infinite.

References

[1] C. Baier, M.E. Majster-Cederbaum: Denotational semantics in the cpo and metric approach, Theoretical Computer Science, Vol. 135, 1994.

[2] C. Baier, M.E. Majster-Cederbaum: Construction of a cms on a given cpo, Techn. Report 11/94, Universität Mannheim, 1994.

[3] C. Baier, M.E. Majster-Cederbaum: The connection between initial and unique solutions of domain equations in the partial order and metric approach, in print.

[4] J.W. de Bakker, J.I.Zucker: Processes and the Denotational Semantics of Concurrency, Information and Control, Vol.54, No. 1/2, pp 70-120, 1982.

[5] G. Comyn, M. Dauchet: Metric approximations in ordered domains, Algebraic Methods in Semantics, Cambridge University Press, 1985.

[6] G. Gierz, H. Hofmann, K. Keimel, J. Lawson, M. Mislove, D. Scott: A Compendium of Continuous Lattices, Springer-Verlag, 1980.

[7] K. Kuratowski: Toplogy, Academic Press, 1966.

[8] M.Z. Kwiatkowska: On three constructions of infinite traces, Techn. Report No. CSD-48, University of Leicester, 1991.

[9] M. Majster-Cederbaum, F. Zetzsche: The comparison of a cpo-based with a cms-based semantics for CSP, Theoretical Computer Science, Vol. 124, 1994.

[10] S. Matthews: The Cycle Contracting Mapping Theorem, Research Report 228, Department of Computer Science, University of Warwick, 1992.

[11] A. Mazurkiewicz: Basic notions of trace theory, in Linear Time, Branching Time and Partial Order in Logics and Models for Concurrency, Lecture Notes in Computer Science 354, 1989.

[12] R. Milner: Communication and Concurrency, Prentice Hall, 1989.

[13] M.B. Smyth: Quasi-uniformities: Reconciling Domains with Metric Spaces, Proc. of the Third Workshop on the Mathematical Foundations of Pragramming Language Semantics, Lecture Notes in Computer Science 298, 1988.

[14] K. Weihrauch, U. Schreiber: Metric spaces defined by weighted algebraic cpo's, Fundamentals of Computation Theory, FCT '79, Proceedings of the Conference on Algebraic, Arithmetric, and Categorial Methods in Computation Theory, Akademie-Verlag, 1979.

[15] G. Winskel: Event Structure Semantics for CCS and Related Languages, Proc. ICALP 82, Lecture Notes in Computer Science 140, Springer-Verlag, pp 561-576, 1982.

[16] G. Winskel: Synchronisation trees, Theoretical Computer Science, Vol. 34, 1984.

Keeping Track of the Latest Gossip in Message-Passing Systems

Madhavan Mukund* K. Narayan Kumar† Milind Sohoni‡

Abstract

Consider a distributed system in which processes exchange information by passing messages. The *gossip problem* is the following: Whenever a process q receives a message from another process p, q must be able to decide which of p and q has more recent information about r, for every other process r in the system. With this data, q is in a position to update its knowledge about the global state of the system.

We propose a solution where each message between processes carries information about the current state of knowledge of the sender. This information is *uniformly* bounded if we make reasonable assumptions about the number of undelivered messages present at any time in the system. This means that the overhead of maintaining the latest gossip is a constant, independent of the length of the underlying computation.

Introduction

We tackle a natural problem from distributed computing, involving time-stamps. Let $\mathcal{P}$ be a set of computing agents or processes which exchange information by passing messages. The *gossip problem* is the following: Whenever a process q receives a message from another process p, q must be able to decide which of p and q has more recent information about r, for every other process r.

By keeping track of the latest gossip about other agents, each process can consistently update its knowledge about the global state of the system whenever it receives some new information from another process. Computing global information about the system from local information is, of course, a central issue in distributed computing. So, a solution to the gossip problem would be useful in a wide variety of applications involving distributed systems.

The gossip problem has been investigated in [5, 9] for systems where processes synchronize periodically and exchange information. We extend the solution proposed in [5, 9] to a general message-passing model, where processes communicate by sending messages on point-to-point channels. We assume that communication is guaranteed—all messages in the system are eventually delivered. However, we permit indefinite delays in transit. Also, messages need not be received in the order in which they were sent.

In our solution to the gossip problem, processes exchange only a bounded amount of gossip information with each message they send. At the heart of our

*School of Mathematics, SPIC Science Foundation, 92 G.N. Chetty Road, T. Nagar, Madras 600 017, India. E-mail: madhavan@ssf.ernet.in

†Computer Science Group, Tata Institute of Fundamental Research, Homi Bhabha Road, Colaba, Bombay 400 005, India. E-mail: kumar@tcs.tifr.res.in

‡Department of Computer Science and Engineering, Indian Institute of Technology, Bombay 400 076, India, E-mail: sohoni@cse.iitb.ernet.in

solution is a protocol for time-stamping messages in the system using a finite set of labels.

Time-stamping is a well-established technique for ordering events in a distributed setting [6, 7]. Time-stamping protocols which use only a bounded set of labels to tag events have attracted a fair amount of attention in recent years. Protocols have been exhibited for systems in which processes communicate via a shared memory [3, 4], as also for systems where processes synchronize periodically and exchange information [1, 2, 9]. However, no such protocol seems to exist for message-passing systems.

It is not difficult to see that in the completely general message-passing model described above, bounded time-stamping is not possible. In order to make the problem tractable, we have to restrict the model by placing a bound on the number of unacknowledged messages that can be present in the system at any time. We believe that this restriction is a natural one and that all "reasonable" distributed algorithms for message-passing systems would actually conform to the paradigm we propose.

An important feature of our solution to the gossip problem is that it does not introduce any additional messages—it just adds some additional data to each message of the underlying computation. This additional data is guaranteed to be uniformly bounded. So, given any distributed algorithm which conforms to the restricted model we work with, we can enhance the algorithm to also keep track of the latest gossip with only a constant overhead in message complexity.

The paper is organized as follows. In the next section we introduce our model of computation and formulate the gossip problem in terms of a natural partial order on the events in the system. Section 2 describes ideals, which capture the notion of a partial view of a distributed computation. Sections 3 and 4 describe a protocol to solve the gossip problem for *fifo* computations, where messages are delivered in the order they were sent. Each process maintains what we call *primary information* about the computation, using potentially unbounded labels to distinguish messages in the system. In Section 5 we show how to convert this protocol to one which uses bounded time-stamps, thereby establishing a solution to the gossip problem in the fifo case. In the final section, we give a quick sketch of how to extend this solution to the case where messages are not necessarily delivered in the order in which they were sent.

1 The Model

Let $\mathcal{P} = \{p_1, p_2, \ldots, p_N\}$ be a set of processes which communicate with each other through messages. We assume that messages are never lost—that is, the communication medium is reliable. However, there may be an arbitrary delay between the sending of a message and its receipt. Further, messages need not be received in the order in which they were sent.

We assume that communication is point-to-point. So, each message is addressed to a specific process and is not seen by any of the other processes in the system. Thus, each transmission of a message from a process p to a process q consists of two distinct actions; the action $s_{p\to q}$ corresponds to the sending of the message from p to q and the action $r_{q\leftarrow p}$ corresponds to its receipt by q.

We can regard a computation of the system as a word over the alphabet $\mathcal{C} = \mathcal{C}_S \cup \mathcal{C}_R$ where $\mathcal{C}_S = \{s_{p\to q} \mid p, q \in \mathcal{P}\}$ is the set of *send actions* and

$\mathcal{C}_R = \{r_{p \leftarrow q} \mid p, q \in \mathcal{P}\}$ is the set of *receive actions.* Each action in $\mathcal{C}$ is "executed" by a single process. So, we can also partition $\mathcal{C}$ across processes—for each process p, $\mathcal{C}_p = \{s_{p \rightarrow q} \mid q \in \mathcal{P}\} \cup \{r_{p \leftarrow q} \mid q \in \mathcal{P}\}$ is the set of *p-actions* that p participates in directly.

We shall regard a word $u \in \mathcal{C}^*$ of length m as a function $u : [1..m] \rightarrow \mathcal{C}$, where $[1..m]$ denotes the set $\{1, 2, \ldots, m\}$ if $m \geq 1$ and is $\emptyset$ if $m = 0$. For $u \in \mathcal{C}^*$ and $c \in \mathcal{C}$, $\#_c(u)$ denotes the number of occurrences of c in u. We can extend this to subsets $X \subseteq \mathcal{C}$: $\#_X(u) = \sum_{c \in X} \#_c(u)$.

Not every word corresponds to a valid computation—for instance, we must insist that messages are received only after they are sent. In addition, since messages need not be received in the order they were sent, to completely specify a computation we need to match each receive event to the corresponding send event. With this in mind, we define computations as follows:

Computations A *computation* over $\mathcal{C}$ is a pair (u, ϕ) where $u : [1..m] \rightarrow \mathcal{C}$ is a word and $\phi : [1..m] \rightarrow [1..m]$ is a partial function such that:

(i) The domain of ϕ, $dom(\phi)$ is $\{i \mid u(i) \in \mathcal{C}_R\}$.

(ii) ϕ is injective over $dom(\phi)$—for each $i, j \in dom(\phi)$, $i \neq j \Rightarrow \phi(i) \neq \phi(j)$.

(iii) For each $i \in dom(\phi)$, $\phi(i) < i$.

(iv) If $u(i) = r_{q \leftarrow p}$ then $u(\phi(i)) = s_{p \rightarrow q}$.

If $\phi(i) = j$, then $u(i)$ is a receive action whose corresponding send action is $u(j)$. By condition (ii), we may also refer to i unambiguously as $\phi^{-1}(j)$.

EXAMPLE: *Let $\mathcal{P} = \{p, q\}$ and let u be the string $s_{p \rightarrow q} s_{p \rightarrow q} s_{p \rightarrow q} r_{q \leftarrow p} r_{q \leftarrow p}$ and ϕ be the function where $\phi(4) = 3$ and $\phi(5) = 1$. So, in this computation, the message sent from p to q at $u(1)$ is overtaken by the message sent at $u(3)$. Moreover, the message sent at $u(2)$ has not reached q at all.*

Events and causality The word u imposes a total, temporal order on the actions observed during a computation (u, ϕ). However, in order to analyze the flow of information between processes, we need a more accurate description of the cause and effect relationship between the different actions in u.

Let (u, ϕ) be a computation, where $u : [1..m] \rightarrow \mathcal{C}$. We associate with (u, ϕ) a set of events $\mathcal{E}_u = \{(i, u(i)) \mid i \in [1..m]\}$.

Let $e = (i, u(i))$ be an event in $\mathcal{E}_u$. When there is no ambiguity, we shall use e to denote both i and $u(i)$. For instance, $e \in \mathcal{C}_p$ denotes that $u(i) \in \mathcal{C}_p$—in other words, e is a *p-event.* Similarly, if we say $f = \phi(e)$ we mean that $f = (j, u(j))$ is an event such that $\phi(i) = j$. We shall also use $\mathcal{E}_u$ and u interchangeably in expressions such as $\#_c(\mathcal{E}_u)$, which denotes $\#_c(u)$.

As we mentioned earlier, u imposes a total, temporal order on the events in $\mathcal{E}_u$. Let $e, f \in \mathcal{E}_u$. Then $e < f$ provided $e = (i, u(i))$, $f = (j, u(j))$ and $i < j$. As usual $e \leq f$ if $e < f$ or $e = f$.

Messages introduce causality across processes. For each pair $(p, q) \in \mathcal{P} \times \mathcal{P}$ such that $p \neq q$, define $\lessdot_{pq}$ to be the ordering

$$e \lessdot_{pq} f \triangleq e \in \mathcal{C}_p, f \in \mathcal{C}_q \text{ and } \phi(f) = e.$$

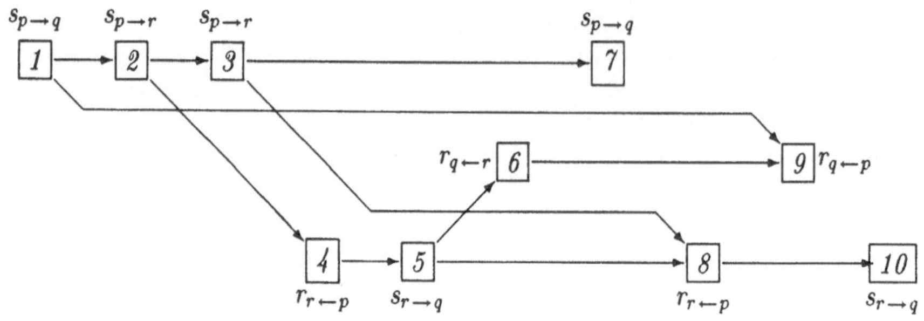

Figure 1: An example

In addition, each process p orders the events it participates in. Define $\lessdot_{pp}$ to be the strict ordering

$$e \lessdot_{pp} f \triangleq e < f,\ e \in \mathcal{C}_p,\ f \in \mathcal{C}_p \text{ and for all } e < g < f,\ g \notin \mathcal{C}_p.$$

The set of all p-events in $\mathcal{E}_u$ is totally ordered by $\lessdot^*_{pp}$, the reflexive, transitive closure of $\lessdot_{pp}$.

Define $e \sqsubset f$ if for some $p, q \in \mathcal{P}$ (where p and q need not be distinct), $e \lessdot_{pq} f$ and $e \sqsubseteq f$ if $e = f$ or $e \sqsubset f$. Let $\sqsubseteq^*$ denote the transitive closure of $\sqsubseteq$. If $e \sqsubseteq^* f$ then we say that e is *below* f. The partial order $\sqsubseteq^*$ records the information we require about causality and independence between events in $\mathcal{E}_u$.

Let $e \in \mathcal{E}_u$ be a p-event. The set of events below e is $e\!\downarrow = \{f \mid f \sqsubseteq^* e\}$. These represent the only actions which are known to p when e occurs.

Latest information Consider a computation (u, ϕ) and its associated set of events $\mathcal{E}_u$. The $\sqsubseteq^*$-maximum p-event in $\mathcal{E}_u$ is denoted $max_p(\mathcal{E}_u)$—this is the last event in $\mathcal{E}_u$ in which p has taken part. This quantity is well defined whenever $\#_{\mathcal{C}_p}(\mathcal{E}_u) > 0$, since all p-events in $\mathcal{E}_u$ are totally ordered by $\sqsubseteq^*$. (Recall that $\mathcal{C}_p \subseteq \mathcal{C}$ is the set of p-actions, so $\#_{\mathcal{C}_p}(\mathcal{E}_u)$ denotes, by convention, the number of p-actions mentioned in the string u.)

Let $p, q \in \mathcal{P}$. If $\#_{\mathcal{C}_p}(\mathcal{E}_u) > 0$, the latest information p has about q in $\mathcal{E}_u$ corresponds to the $\sqsubseteq^*$-maximum q-event in the set $max_p(\mathcal{E}_u)\!\downarrow$, provided the set of q-events below $max_p(\mathcal{E}_u)$ is not empty. We denote this event by $latest_{p\leftarrow q}(\mathcal{E}_u)$. (If there are no q-events in $max_p(\mathcal{E}_u)\!\downarrow$, then $latest_{p\leftarrow q}(\mathcal{E}_u)$ is undefined.)

EXAMPLE: *Let $\mathcal{P} = \{p, q, r\}$. Consider the computation (u, ϕ), where $u = s_{p\to q} s_{p\to r} s_{p\to r} r_{r\leftarrow p} s_{r\to q} r_{q\leftarrow r} s_{p\to q} r_{r\leftarrow p} r_{q\leftarrow p} s_{r\to q}$, $\phi(4) = 2$, $\phi(6) = 5$, $\phi(8) = 3$ and $\phi(9) = 1$. Figure 1 is a picture of $(\mathcal{E}_u, \sqsubseteq^*)$. The arrows in the figure correspond to the basic relations $\lessdot_{pq}$, which generate $\sqsubseteq^*$.*

In this computation, $max_q(\mathcal{E}_u) = (9, r_{q\leftarrow p})$. Though at $max_q(\mathcal{E}_u)$, process q hears from process p, $latest_{q\leftarrow p}(\mathcal{E}_u)$ does not correpond to $\phi(max_q(\mathcal{E}_u))$. Instead, $latest_{q\leftarrow p}(\mathcal{E}_u) = (2, s_{p\to r})$—process q hears this information indirectly, via process r.

The gossip problem

Let p, q and r be processes such that $latest_{p \leftarrow r}(\mathcal{E}_u)$ and $latest_{q \leftarrow r}(\mathcal{E}_u)$ are both defined. Since both of these are r-events, they must be ordered by $\lhd^*_{rr}$. So, the latest information that p and q have about r will always be comparable.

The gossip problem is the following.

> *Whenever a process q receives a message from another process p, q must be able to decide which of p and q has heard more recently from r, for every other process r in the system.*

One way to resolve this problem is as follows. As the computation progresses, each action is assigned a label by the process involved in that action. These labels allow processes to refer to events in an unambiguous manner. Each process then maintains the labels corresponding to its latest information. These labels are passed on with each communication in such a way that the process receiving the message can consistently update its own latest information.

The labels which are assigned to events during a computation are essentially *time-stamps.* A trivial solution to the time-stamping problem is for each process to maintain a local counter and assign strictly increasing counter values to the actions it executes. Then, two processes p and q can compare their latest information about r by checking which of their "latest" r labels is larger.

This scheme has the following drawback: As the computation progresses, the time-stamps assigned to events grow without bound. As a result, processes need to send longer and longer messages to transfer the labels corresponding to their latest information.

We seek a solution to the gossip problem where message lengths are bounded. This will ensure that the overhead of maintaining gossip information remains a constant, regardless of the length of the underlying computation.

To achieve this, we need to devise a scheme for labelling events using a bounded set of time-stamps. This means that, eventually, the same time-stamp will be assigned to more than one event. We need to ensure that time-stamps are reused in such a way that the update of latest information is not affected.

In principle, this should be possible. Let $\mathcal{E}_u$ be the events corresponding to the computation (u, ϕ) and let the number of processes in the system be N. Regardless of the number of events in $\mathcal{E}_u$, at most N^2 of them are relevant for solving the gossip problem—we only need to be able to compare the labels of events of the form $latest_{p \leftarrow q}(\mathcal{E}_u)$ for each pair $p, q \in \mathcal{P}$. So, in effect, at most N^2 of the events in $\mathcal{E}_u$ constitute "current" gossip. Moreover, once an event becomes "obsolete" its time-stamp can be safely reused—an "obsolete" event can never become "current" at a later stage in the computation.

However, in the completely general model we have considered so far, it is impossible to achieve our goal. Since messages can be delayed indefinitely, a process p may send unboundedly many messages to q without knowing whether any or all of them have reached. Until p receives some confirmation from q that a particular message has reached, that message's time-stamp cannot be reused. Thus, p will potentially need to use an unbounded number of time-stamps to label its messages to q. (Notice that this problem arises even if messages are delivered in the order in which they were sent—the main source of difficulty is the fact that there is no bound on the delay in delivering a particular message.)

B-bounded computations To overcome this problem, we need to restrict the class of computations we permit. Intuitively, we must bound the number of unacknowledged messages between any pair of processes. More formally, we say that (u, ϕ) is a *B-bounded computation* provided the following holds:

$$\text{For each event } e = (i, s_{p\to q}) \text{ in } \mathcal{E}_u,\ \#_{s_{p\to q}}(e\downarrow) - \#_{r_{q\leftarrow p}}(e\downarrow) \leq B.$$

In other words, p can send a fresh message to q only if, as far as it knows, the number of messages which it has already sent to q and which are yet to be delivered is less than B. Notice that p need not get direct acknowledgments from q. For instance, q could tell r that it has received a particular message m from p and r, in turn, could pass on this information to p.

Even for B-bounded computations, it is not immediate that the gossip problem has a solution. Suppose process p sends process q a message m. Though p is guaranteed to receive an acknowledgment for this message by the time it sends its next B messages to q, it cannot naïvely reuse m's time-stamp once m is acknowledged. In between, q may have passed on the information in m to another process r, in which case the message m would still constitute "current" gossip for r. So, p has to have some means of recording which of its time-stamps are "in use" in the system at any given time.

Fifo computations Though we eventually present a solution to the gossip problem for all B-bounded computations, it will be convenient to first look at the problem in a restricted setting, where messages are delivered in the order in which they were sent. Formally, we say that a computation (u, ϕ) is *fifo* if the following holds: For every pair of processes p and q and for any two receive events e and f of the form $r_{p\leftarrow q}$, $e \sqsubset^+ f$ implies $\phi(e) \sqsubset^+ \phi(f)$, where $\sqsubset^+$ denotes the transitive closure of the relation $\sqsubset$.

Though messages between processes may not overtake each other in a fifo computation, gossip information along one route may overtake gossip information on another route. For instance, suppose p sends a message m to q and then sends a message m' to r, following which r sends a message m'' to q which reaches q before m does. Then, by the time m reaches q, the information that m contains about the state of p is obsolete. So, even in a fifo computation, each process has to have a way of resolving whether the message it has just received has already been superseded by earlier, indirect information.

For the next four sections, we assume that every computation we deal with is B-bounded and fifo.

2 Ideals

Let us fix a computation (u, ϕ), where $u : [1..m] \to \mathcal{C}$, and the corresponding set of events $\mathcal{E}_u$, which we shall denote as just $\mathcal{E}$ from now on, for convenience.

The main source of difficulty in solving the gossip problem is the fact that the processes in $\mathcal{P}$ need to compute global information about the computation (u, ϕ) while each process only has access to a local, "partial" view of u. Although partial views of (u, ϕ) correspond to subsets of $\mathcal{E}$, not every subset of $\mathcal{E}$ arises from such a partial view. Those subsets of $\mathcal{E}$ which do correspond to partial views of (u, ϕ) are called ideals.

Ideals A set of events $I \subseteq \mathcal{E}$ is called an *order ideal* if I is closed with respect to $\sqsubseteq^*$—i.e., $e \in I$ and $f \sqsubseteq^* e$ implies $f \in I$ as well. We shall always refer to order ideals as just *ideals*.

The requirement that an ideal be closed with respect to $\sqsubseteq^*$ guarantees that the observation it represents is "consistent"—whenever an event e has been observed, so have all the events in the computation which necessarily precede e. Clearly the entire set $\mathcal{E}$ is an ideal, as is $e\!\downarrow$ for any $e \in \mathcal{E}$. It is easy to see that if I and J are ideals, so are $I \cup J$ and $I \cap J$.

EXAMPLE: *In Figure 1, the set* $I = \{(1, s_{p\to q}), (2, s_{p\to r}), (4, r_{r\leftarrow p}), (5, s_{r\to q}), (6, r_{q\leftarrow r})\}$ *is an ideal. However, the set* $J = \{(1, s_{p\to q}), (2, s_{p\to r}), (3, s_{p\to r}), (5, s_{r\to q}), (6, r_{q\leftarrow r})\}$ *is not an ideal, since* $(4, r_{r\leftarrow p}) \sqsubseteq^* (5, s_{r\to q})$ *but* $(4, r_{r\leftarrow p}) \notin J$.

We need to generalize the notion of $max_p(\mathcal{E})$, the maximum p-event in $\mathcal{E}$, to all ideals $I \subseteq \mathcal{E}$.

p-views For an ideal I, the $\sqsubseteq^*$-maximum p-event in I is denoted $max_p(I)$, provided $\#_{c_p}(I) > 0$. The p-view of I is the set $I_p = max_p(I)\!\downarrow$. So, I_p is the set of all events in I which p can "see". (By convention, if $max_p(I)$ is undefined—i.e., if there is no p-event in I—the p-view I_p is the empty set.)

EXAMPLE: *In Figure 1, consider the ideal* $I = \{(1, s_{p\to q}), (2, s_{p\to r}), (4, r_{r\leftarrow p}), (5, s_{r\to q}), (6, r_{q\leftarrow r})\}$ *Then* I_p*, the p-view of I is* $\{(1, s_{p\to q}), (2, s_{p\to r})\}$ *whereas* I_q*, the q-view of I, is the entire ideal I.*

3 Primary information

For processes $p, q \in \mathcal{P}$, we have already defined $latest_{p\leftarrow q}(\mathcal{E})$, the latest information that p has about q after (u, ϕ). We can extend this definition to arbitrary ideals.

Let $I \subseteq \mathcal{E}$ be an ideal and $p, q \in \mathcal{P}$. Then $latest_{p\leftarrow q}(I)$ denotes the $\sqsubseteq^*$-maximum q-event in I_p, provided $\#_{c_q}(I_p) > 0$. So, $latest_{p\leftarrow q}(I)$ is the latest q-event in I that p knows about. (As usual, if there is no q-event in I_p, the quantity $latest_{p\leftarrow q}(I)$ is undefined.)

It is clear that $latest_{p\leftarrow q}(I)$ will always correspond to a send action from C_q. However $latest_{p\leftarrow q}(I)$ need not be of the form $s_{q\to p}$; the latest information that p has about q in I may have been obtained indirectly.

To maintain and update the latest information of processes, we need to expand the set of events that each process keeps track of. This expanded set will be called primary information. The primary information of a process contains not only its latest information about every other process but also information about acknowledged and unacknowledged messages in the system.

Message acknowledgments Let $I \subseteq \mathcal{E}$ be an ideal and $e \in I$ an event of the form $s_{p\to q}$. Then, e is said to have been *acknowledged* in I if $\phi^{-1}(e)$ exists and, moreover, belongs to I_p. Otherwise, e is said to be *unacknowledged* in I.

Notice that it is not enough for a message to have been received in I to deem it to be acknowledged. We demand that the event corresponding to the receipt of the message be "visible" to the sending process.

For an ideal I and a pair of processes p, q, let $received_{q \leftarrow p}(I)$ be the most recent message from p which q has received in I. In other words, $received_{q \leftarrow p}$ is the $\sqsubseteq^*$-maximum $s_{p \to q}$ event e in I such that $\phi^{-1}(e)$ belongs to I. Since (u, ϕ) is a fifo computation, all messages corresponding to $s_{p \to q}$ events below e must have also been received by q.

Let $unack_{p \to q}(I)$ be the set of unacknowledged $s_{p \to q}$ events in I. By definition, this set consists of all $s_{p \to q}$ events in I which lie above $received_{q \leftarrow p}(I_p)$. Since the underlying computation (u, ϕ) is B-bounded, $unack_{p \to q}(I)$ never contains more than B events.

It will be convenient to define the notion of primary information with respect to ideals rather than processes.

Primary information Let $I \subseteq \mathcal{E}$ be an ideal. The *primary information* of I, $primary(I)$, consists of the following events in I:

- The set $latest(I) = \{max_p(I) \mid p \in \mathcal{P}\}$.
- The collection of sets $unack(I) = \{unack_{p \to q}(I) \mid p, q \in \mathcal{P}\}$.
- The set $received(I) = \{received_{q \leftarrow p}(I) \mid p, q \in \mathcal{P}\}$.

Observe that we can treat $unack(I)$ to be just a set of events, though we have defined it as a collection of sets. For each event $e = (i, u(i))$ in $unack(I)$, the action $u(i)$ immediately determines the particular set $unack_{p \to q}(I)$ to which e belongs—$e \in unack_{p \to q}(I)$ iff $u(i) = s_{p \to q}$. Similarly, we can regard $latest(I)$ and $received(I)$ as just sets of events. For each event $e = (i, u(i))$ in $latest(I)$, $e = max_p(I)$ iff $u(i) \in C_p$. In the same way, for each event $(i, u(i))$ in $received(I)$, $e = received_{q \leftarrow p}(I)$ iff $u(i) = s_{p \to q}$.

On the other hand, $primary(I)$ is an *indexed* set of events. If e belongs to $primary(I)$ we also need to record which of the three components e belongs to. However, often we shall abuse notation and treat $primary(I)$ as just a set of events. It is clear that $primary(I)$ contains only a bounded number of events.

Let $I \subseteq \mathcal{E}$ be an ideal and p a process such that $I_p \neq \emptyset$. Then, $primary(I_p)$ denotes the primary information of p in I—i.e., p's primary information is just the primary information of the p-view of I. Clearly, the "latest information" of p after I is contained in its primary information—for every process q, $latest_{p \leftarrow q}(I)$ is just $max_q(I_p)$.

To compare and update primary information, processes also need to remember how their primary events are ordered by $\sqsubseteq^*$.

Primary graph Let $I \subseteq \mathcal{E}$. The *primary graph* of I, $primary\text{-}graph(I)$ is the directed graph (V, E) where:

- $V = primary(I)$ (where $primary(I)$) is represented as an indexed set of events.)
- For $v_1, v_2 \in V$, let e_1 and e_2 be the corresponding events from I. Then, $(v_1, v_2) \in E$ iff $e_1 \sqsubseteq^* e_2$.

As with primary information, the primary graph of a process p in I is just the graph *primary-graph*(I_p).

For the moment we shall ignore the issue of assigning bounded time-stamps to events and assume that events are assigned unambiguous labels by some mechanism. For instance, as we mentioned earlier, each process could maintain a local counter and assign an increasing sequence of unique time-stamps to the events that it participates in.

Our first goal is to exhibit a procedure by which processes update their primary graphs *without* relying on the temporal order of events. Processes will only utilise the information about causality recorded in the primary graphs. All comparisons and updates of primary information will based purely on equality of event labels. This feature will allow us to extend the algorithm smoothly to the case where processes reuse labels.

4 Comparing primary information

Let $\mathcal{E}$ be the events corresponding to a computation (u, ϕ). Then, each ideal $I \subseteq \mathcal{E}$ corresponds to a possible partial computation of (u, ϕ). Let us assume that at the end of any partial computation I, each process maintains the information *primary-graph*(I_p).

In general, for an ideal $I \subseteq \mathcal{E}$, each pair of processes p and q will have incomparable information about I. The events known to both p and q lie in the ideal $I_p \cap I_q$. Events lying "above" the intersection are known to only one of the processes.

Suppose that q receives a message from p during the computation. Then, we have an event $e_p \in \mathcal{E}$ of the form $s_{p \to q}$ and a corresponding event $e_q \in \mathcal{E}$ of the form $r_{q \leftarrow p}$ such that $\phi(e_q) = e_p$.

Let $e'_q = max_q(e_q{\downarrow} - \{e_q\})$—i.e., e'_q is the maximum q-event strictly below e_q. So, $e'_q{\downarrow}$ represents the state of q's knowledge before receiving this message from p. Let I be the ideal $e_p{\downarrow} \cup e'_q{\downarrow}$.

For the moment, let us assume that $max_p(e'_q{\downarrow}) \sqsubset^+ e_p$—i.e., the message sent at e_p has "new" information for q about the state of p. Then, $e_p{\downarrow} = I_p$ and $e'_q{\downarrow} = I_q$. The message sent by p at e_p contains the primary graph *primary-graph*(I_p). On the other hand, before receiving this message, q's primary graph corresponds to *primary-graph*(I_q).

Our first observation is that if q knows both *primary-graph*(I_p) and *primary-graph*(I_q), it can determine which events in the two primary graphs lie within $I_p \cap I_q$ and which lie outside this intersection.

Lemma 1 *Let $I \subseteq \mathcal{E}$ be an ideal and p, q a pair of processes. Then, for each maximal event e in $I_p \cap I_q$, either $e \in$ latest$(I_p) \cap$ unack(I_q) or $e \in$ unack$(I_p) \cap$ latest(I_q).*

Proof Let e be a maximal event in $I_p \cap I_q$. Suppose it is an r-event, for some $r \in \mathcal{P}$. The event e must have $\sqsubset$-successors in both I_p and I_q. However, observe that any event f in $\mathcal{E}$ can have at most two immediate $\sqsubset$-successors—one "internal" successor within the process and, if f is a send event, one "external" successor corresponding to the matching receive event.

Thus, the maximal event e will have a $\lessdot_{rr}$ successor e_r and a $\lessdot_{rs}$ successor e_s, corresponding to some $s \in \mathcal{P}$. Assume that $e_r \in I_q - I_p$ and $e_s \in I_p - I_q$. Since the r-successor of e is outside I_p, $e = max_r(I_p)$. So e belongs to $latest(I_p)$. On the other hand, e is an unacknowledged $s_{r \to s}$ event in I_q. So, $e \in unack_{r \to s}(I_q)$, which is part of $unack(I_q)$.

Symmetrically, if $e_r \in I_p - I_q$ and $e_s \in I_q - I_p$, we find that e belongs to $unack(I_p) \cap latest(I_q)$. □

Thus, when q receives p's primary graph, q can collect together in a set M all the events that lie in $latest(I_p) \cap unack(I_q)$ and $unack(I_p) \cap latest(I_q)$. By the preceding lemma, the events in M subsume the maximal events in the intersection $I_p \cap I_q$. (It is easy to see that those events in M which are not actually maximal still lie within the intersection.)

The process q can use M to check whether a primary event $e \in primary(I_p) \cup primary(I_q)$ lies inside or outside the intersection—e lies inside the intersection iff it lies below one of the elements in M. These comparisons can be made using the edge information in the graphs $primary\text{-}graph(I_p)$ and $primary\text{-}graph(I_q)$.

Now, it is easy for q to compare the events in $latest(I_p)$ with those in $latest(I_q)$ to determine which of p and q have more recent information about every other process r.

Lemma 2 *Let $I \subseteq \mathcal{E}$ be an ideal and p, q a pair of processes. Let $e = max_r(I_p)$ and $f = max_r(I_q)$ such that $e \neq f$. Then, $e \sqsubset^+ f$ iff $f \in I_q - I_p$.*

Proof ($\Rightarrow$) Suppose that $e \sqsubset^+ f$. If f belongs to I_p then $e \neq max_r(I_p)$, which is a contradiction. On the other hand, f clearly belongs to I_q, so $f \in I_q - I_p$.

($\Leftarrow$) Suppose that $f \in I_q - I_p$. If $f \sqsubset^+ e$, then $f \in I_p$ since I_p is an ideal. This is a contradiction. So it is not the case that $f \sqsubset^+ e$. Since $f \neq e$ and all r-events are totally ordered by $\sqsubseteq^*$, we must have $e \sqsubset^+ f$. □

Once q has compared all events of the form $max_r(I_p)$ and $max_r(I_q)$, it can easily update its sets $unack_{r \to s}(I_q)$, where $r \neq q$. The process which has better information about r also has better information about unacknowledged events of the form $s_{r \to s}$ in I. In other words, q inherits the sets $unack_{r \to s}(I_p)$ for every process r such that $max_r(I_p)$ is more recent than $max_r(I_q)$. On the other hand, if $max_r(I_p)$ is older than $max_r(I_q)$, then q ignores p's sets $unack_{r \to s}(I_p)$ since it already has better information about these events.

At this stage, using the data in $primary\text{-}graph(I_p)$ and $primary\text{-}graph(I_q)$, q has updated all of $primary(e_q{\downarrow})$ except for the sets $\{unack_{q \to r}(e_q{\downarrow})\}_{r \in \mathcal{P}}$ and $received(e_q{\downarrow})$. (Recall that e_q was the event where q received p's message sent at e_p.)

We first describe how to construct $received(e_q{\downarrow})$. Clearly $received_{q \leftarrow p}(e_q{\downarrow}) = e_p$. To fill in the rest of $received(e_q{\downarrow})$, we need the following observation, which we state without proof.

Proposition 3 *For every pair of processes (r, s) such that $(r, s) \neq (p, q)$,*

$$received_{s \leftarrow r}(e_q{\downarrow}) = \begin{cases} received_{s \leftarrow r}(I_p) & \text{if } max_s(I_q) \sqsubset^+ max_s(I_p) \\ received_{s \leftarrow r}(I_q) & \text{otherwise} \end{cases}$$

So, once q has compared the events in $latest(I_p)$ and $latest(I_q)$, it can decide which of the events in $received(e_q\downarrow)$ are inherited from $received(I_p)$ and which are retained from $received(I_q)$.

Process q can now use the information in $received(e_q\downarrow)$ to update the sets $\{unack_{q\to r}(I_q)\}_{r\in\mathcal{P}}$ by purging acknowledged events from these lists. Formally, for every process r, $unack_{q\to r}(e_q\downarrow)$ consists of all the events in $unack_{q\to r}(I_q)$ which lie above $received_{r\leftarrow q}(e_q\downarrow)$. This update can be made with the edge information in $primary\text{-}graph(I_q)$.

Having constructed the sets $latest(e_q\downarrow)$, $unack(e_q\downarrow)$ and $received(e_q\downarrow)$, we need to extend this set to the graph $primary\text{-}graph(e_q\downarrow)$.

Let $f_1, f_2 \in primary(e_q\downarrow)$. If both f_1 and f_2 came from $primary(I_p)$, then we add an edge from f_1 to f_2 in $primary\text{-}graph(e_q\downarrow)$ iff a corresponding edge existed in $primary\text{-}graph(I_p)$. A symmetric situation applies if both f_1 and f_2 were contributed by $primary(I_q)$.

So, the only interesting case is when f_1 and f_2 originally came from different processes. Without loss of generality, suppose that f_1 came from $primary(I_p)$ and f_2 from $primary(I_q)$. Then, from the method which we used to compare events, we know that f_1 must have been in $I_p - I_q$ and f_2 must have been in $I_q - I_p$. So, it is clear that f_1 and f_2 are unordered in $\mathcal{E}$ and there is therefore no edge between them in $primary\text{-}graph(e_q\downarrow)$.

So far, we have shown how to construct $primary\text{-}graph(e_q\downarrow)$ assuming that $max_p(e'_q\downarrow) \sqsubset^+ e_p$. If q already has better information about p—i.e., $e_p \sqsubseteq^* max_p(e'_q\downarrow)$—it is easy to see that e_p must in fact lie strictly below $max_p(e'_q\downarrow)$, since q's knowledge of e_p must have come from a message sent by p *after* e_p to some other process r. In this case, q can no longer use Lemma 1 to compute the maximal elements in the intersection $e_p\downarrow \cap e'_q\downarrow$ since the p-view of the ideal $I = e_p\downarrow \cup e'_q\downarrow$ is *not* $e_p\downarrow$ but $max_p(e'_q\downarrow)\downarrow$.

However, it is easy to see that if $e_p \sqsubset^+ max_p(e'_q\downarrow)$ then $e_p\downarrow \subseteq e'_q\downarrow$. So, for every other process r, $max_r(e_p\downarrow) \sqsubseteq^* max_r(e'_q\downarrow)$. It then follows that the only update that q has to make to $primary\text{-}graph(e'_q\downarrow)$ is to set $received_{q\leftarrow p}(e_q\downarrow)$ to e_p. The rest of $primary\text{-}graph(e_q\downarrow)$ is inherited from $primary\text{-}graph(e'_q\downarrow)$.

All we need is a means for q to detect whether e_p is a "new" message. Based on this, q can decide whether to ignore the information in $primary\text{-}graph(e_p\downarrow)$. This is not very difficult because of the following observation.

Proposition 4 *Let e_p be a $s_{p\to q}$ event such that $\phi^{-1}(e_p) = e_q$ and let $e'_q = max_q(e_q\downarrow - \{e_q\})$. Then $e_p \sqsubseteq^* max_p(e'_q\downarrow)$ iff $e_p \in unack_{p\to q}(e'_q\downarrow)$.*

In other words, to check if the message is old, all q has to do is to see if e_p is already in its set $unack(e'_q\downarrow)$. This leads to the following general statement which summarizes the results of this section.

Lemma 5 *Let e_p be a $s_{p\to q}$ event in $\mathcal{E}$ such that $\phi^{-1}(e_p) = e_q$. Let $e'_q = max_q(e_q\downarrow - \{e_q\})$. Then, q can construct $primary\text{-}graph(e_q\downarrow)$ from the graphs $primary\text{-}graph(e_p\downarrow)$ and $primary\text{-}graph(e'_q\downarrow)$.*

5 Bounded time-stamps

To make the protocol described in the previous section effective, we have to bound the amount of information recorded in the primary graph of each process by limiting the size of the labels used to identify events.

Notice that the procedure for updating primary graphs only checks the labels of events which actually lie in the primary graphs of the sending and receiving processes. Call an event e "current" in I if e belongs to $primary(I_p)$ for some process p.

Let N be the number of processes in the system. Since the underlying computation is B-bounded, we know that there are at most $N + (B+1)N^2$ distinct events in $primary(I_p)$ for process p—there are at most N events in $latest(I_p)$, at most B events in each of the sets $unack_{q \to r}(I_p)$ and at most N^2 events overall in $received(I_p)$. So, at any given time, the number of events across the system which are current is bounded by $N(N + (B+1)N^2)$.

From the way primary information is defined, it is clear that all current events are actually send events. Each send event $s_{p \to q}$ begins by being current—the moment the message is sent, the event is added to the list of unacknowledged messages from p to q. Eventually, q acknowledges this message and p gets rid of it from its unacknowledged list. Meanwhile, as the computation progresses, this event may get added to the primary information of other processes. However, gradually it recedes into the past, until it drops out of the primary information of *all* processes. At this time, the label assigned to this event can be reused—the old event with the same label can *never* become current again.

Processes can keep track of which events in the system are current by maintaining one additional level of data, called secondary information.

Secondary information Let I be an ideal. The *secondary information* of I is the collection of (indexed) sets $primary(e{\downarrow})$ for each event e in $primary(I)$. This collection of sets is denoted $secondary(I)$.

The following lemma says that the only p-events which can be current in the system are those which occur in p's secondary information.

Lemma 6 *Let $I \subseteq \mathcal{E}$ be an ideal and e a p-event which belongs to $primary(I_q)$ for some process q. Then, $e \in secondary(I_p)$.*

A proof of this lemma can be found in the full paper [8]. We will use the preceding result in the following form.

Corollary 7 *Let e be a p-event such that $e \notin secondary(I_p)$. Then $e \notin primary(I_q)$ for any $q \in \mathcal{P}$.*

Our update procedure does not rely on the temporal order of events. So long as all processes which refer to the same label in their primary information are actually talking about the same event, reusing labels should cause no confusion. Therefore, if p knows that no p-event labelled ℓ is currently part of the primary information of any process in the system, it can safely use ℓ to time-stamp the next message which it sends.

Secondary information can be updated in a straightforward manner when we update primary information—if q inherits an event e from p's primary information, it also inherits the secondary information $primary(e{\downarrow})$ associated with e. Notice that it suffices to maintain secondary information as an indexed set—we do not need to maintain secondary *graphs* as we do primary graphs.

This at once gives us a protocol which solves the gossip problem for B-bounded fifo computations.

The gossip protocol

Let $\mathcal{L}$ be a finite set of labels of such that $|\mathcal{L}| > (N^2+(B+1)N^3)$. All processes use the set $\mathcal{L}$ to time-stamp messages. Each process p maintains its primary graph $\textit{primary-graph}_p = (V_p, E_p)$, where V_p consists of the (indexed) sets of labels $\textit{latest}_p$, $\textit{unack}_p$ and $\textit{received}_p$.

A typical element of $\textit{latest}_p$ is a pair of the form (ℓ, q)—this will mean that the maximum q-event known to p is time-stamped ℓ. Similarly, elements of $\textit{unack}_p$ and $\textit{received}_p$ are triples. An entry (ℓ, q, r) in $\textit{unack}_p$ signifies that, as far as p knows, the $s_{q\to r}$ event labelled ℓ has not been acknowledged. In the same vein, a typical entry (ℓ, q, r) in $\textit{received}_p$ denotes that, as far as p knows, the most recent message from q to r which has actually been delivered is time-stamped ℓ.

Finally, the process p maintains its secondary information $\textit{secondary}_p$ as an indexed set of labels. If $\bar{e}$ is a tuple from $\textit{latest}_p \cup \textit{unack}_p \cup \textit{received}_p$, then an event in $\textit{latest}(\bar{e}\downarrow)$ will be represented as $(\ell', r, \bar{e})$, indicating that the maximum r-event in $\bar{e}\downarrow$ is time-stamped ℓ'. In a similar manner, an entry $(\ell', r, s, \bar{e})$ in $\textit{unack}(\bar{e}\downarrow)$ signifies that there is a $s_{r\to s}$ event time-stamped ℓ' which is unacknowledged within $\bar{e}\downarrow$. And, finally, an entry $(\ell', r, s, \bar{e})$ in $\textit{received}(\bar{e}\downarrow)$ means that the most recent message from r to s which has been delivered within $e\downarrow$ is timestamped ℓ'.

Initially, for each p, $\textit{latest}_p$, $\textit{unack}_p$, $\textit{received}_p$ and $\textit{secondary}_p$ are empty.

Sending a message When p sends a message to q it does the following:

- Choose a label ℓ from $\mathcal{L}$ which does not appear as the first component of any tuple in $\textit{latest}_p \cup \textit{unack}_p \cup \textit{received}_p \cup \textit{secondary}_p$.
- Remove the old event (ℓ', p) from $\textit{latest}_p$, if it exists. Also remove all associated events from $\textit{secondary}_p$—i.e., tuples of the form (ℓ'', p', ℓ', p) and $(\ell'', p', p'', \ell', p)$.
- Add (ℓ, p, q) to $\textit{unack}_p$ and (ℓ, p) to $\textit{latest}_p$. Add an edge in E_p from each tuple in $\textit{latest}_p \cup \textit{unack}_p \cup \textit{received}_p$ to the new tuples $(\ell, p, q) \in \textit{unack}_p$ and $(\ell, p) \in \textit{latest}_p$.
- For each pair (ℓ', p') in $\textit{latest}_p$, add (ℓ', p', ℓ, p) to $\textit{secondary}_p$. Similarly, for each triple (ℓ', p', p'') in $\textit{unack}_p \cup \textit{received}_p$, add $(\ell', p', p'', \ell, p)$ to $\textit{secondary}_p$.
- Send $\textit{primary-graph}_p$ and $\textit{secondary}_p$ to q.

Receiving a message On receiving a message from p, q does the following:

- Extract the label ℓ of the new message.
- Replace the current triple (ℓ', p, q) in $\textit{received}_q$, if it exists, by (ℓ, p, q).
- If (ℓ, p, q) does not already belong to $\textit{unack}_q$ then update $\textit{primary-graph}_q$ and $\textit{secondary}_q$ by comparing $\textit{primary-graph}_p$ and $\textit{secondary}_p$ in the message with $\textit{primary-graph}_q$ and $\textit{secondary}_q$ currently maintained by q.

So, on sending a message, p chooses a label ℓ which is not currently in use in the system and uses ℓ to time-stamp the message. It then replaces the latest p label in $latest_p$ by ℓ and also adds ℓ to the list of unacknowledged $s_{p \to q}$ events. Finally, it places the new event at the "top" of its primary graph and sets the secondary information with respect to the new event to be its overall primary information. It then sends its new data structures $primary\text{-}graph_p$ and $secondary_p$ to q.

When q receives the message labelled ℓ, it first records in $received_q$ that the most recently received message from p is labelled ℓ. (Extracting the label of the message can be done by looking, for instance, for the E_p^*-maximal event in $primary\text{-}graph_p$). It then checks whether the message is new. If so, it updates its primary graph and secondary information following the results in Lemmas 1 and 2 and Proposition 3.

Message complexity

Lemma 8 *Let N be the number of processes in the system. For each process p, the information in $primary\text{-}graph_p$ and $secondary_p$ requires at most $O(N^4 logN)$ bits to write down.*

Proof We know that $|\mathcal{L}|$ is $O(N^3)$. So each label in $\mathcal{L}$ can be written down using $O(\log N)$ bits. Similarly, each process name can be written down using $O(\log N)$ bits. So each tuple in the sets $latest_p$, $unack_p$, $received_p$ and $secondary_p$ requires only $O(\log N)$ bits to write down.

There are $O(N^2)$ entries in $primary_p$. So, the edge relation E_p of $primary\text{-}graph_p$ can be represented in terms of an adjacency matrix, using $O(N^4)$ bits. So, all of $primary\text{-}graph_p$ can be written down in $O(N^4)$ bits.

The real bottleneck turns out to be $secondary_p$. There are $O(N^2)$ elements in $primary_p$. For each of these elements $\bar{e}$, we have to maintain $primary(\bar{e}\downarrow)$, which requires $O(N^2 \log N)$ bits. So, overall $secondary_p$ requires $O(N^4 \log N)$ bits to write down.

□

Putting together all the results we have proved so far, we can state the following theorem

Theorem 9 *The protocol we have described solves the gossip problem for fifo, B-bounded computations with only a bounded amount of additional information being added to each message of the underlying computation.*

6 Non-fifo computations

We briefly describe how to extend our protocol to deal with B-bounded computations which are *not* fifo. We do not go into all the formal details here due to a lack of space. Details are available in the full paper [8].

The essential difference is in the procedure for collecting acknowledgements. In a non-fifo computation, messages may be received out of order. So, we cannot infer from the fact that $e = received_{q \leftarrow p}(I)$ that all messages sent by p to q

before e have actually reached q. In general, we have to explicitly acknowledge each message label to the sending process. This can be done by recording for each pair of processes p, q, the set $\textit{received-set}_{q \leftarrow p}(I)$ consisting of all events e of the form $s_{p \rightarrow q}$ such that $\phi^{-1}(e)$ belongs to I.

So, we extend the information maintained by each process p to include the sets $\textit{received-set}_{r \leftarrow q}(I_p)$ for each pair of processes q, r. These sets can be updated using an extension of Proposition 3. The problem is, of course, that these sets grow without bound as the computation proceeds.

However, it can be shown that once e is added to $\textit{received-set}_{q \leftarrow p}(I_q)$ by q, it will become known to p (i.e., it will appear in $\textit{received-set}_{q \leftarrow p}(I'_p)$ for $I' \supseteq I$) within a bounded number of communications in the system. This follows from the assumption that the underlying computation is B-bounded. Once e appears in $\textit{received-set}_{q \leftarrow p}(I'_p)$, it has been acknowledged and can be removed from q's list of received messages.

As a result, it turns out that each process p needs to retain only the B-most recent events in $\textit{received-set}_{r \leftarrow q}(I_p)$, where "most recent" refers to the order in which the messages were received by r and not the order in which they were sent by q. So, acknowledgments can still be propogated using a bounded amount of information, even when messages arrive out of order and our protocol can be extended to solve the gossip problem for arbitrary B-bounded computations.

References

[1] R. Cori, Y. Metivier: Approximations of a trace, asynchronous automata and the ordering of events in a distributed system, *Proc. ICALP '88, LNCS* **317** (1988) 147–161.

[2] R. Cori, Y. Metivier, W. Zielonka: Asynchronous mappings and asynchronous cellular automata, *Inform. and Comput.*, **106** (1993) 159–202.

[3] D. Dolev, N. Shavit: Bounded concurrent time-stamps are constructible, *Proc. ACM STOC* (1989) 454–466.

[4] A. Israeli, M. Li: Bounded time-stamps, *Proc. 28th IEEE FOCS* (1987) 371–382.

[5] R. Krishnan, S. Venkatesh: Optimizing the gossip automaton, *Report TCS-94-3*, School of Mathematics, SPIC Science Foundation, Madras, India (1994).

[6] L. Lamport: Time, clocks and the ordering of events in a distributed system, *Comm. ACM* **17**(8) (1978) 558–565.

[7] L. Lamport, N. Lynch: Distributed Computing: Models and Methods, in: J. van Leeuwen (ed.), *Handbook of Theoretical Computer Science: Volume B*, North-Holland, Amsterdam (1990) 1157–1200.

[8] M. Mukund, K. Narayan Kumar, M. Sohoni: Keeping track of the latest gossip in message-passing systems, *Report TCS-95-3*, School of Mathematics, SPIC Science Foundation, Madras, India (1995).

[9] M. Mukund, M. Sohoni: Keeping track of the latest gossip: Bounded time-stamps suffice, *Proc. FST&TCS '93, LNCS* **761** (1993) 388–399.

A Local Presentation of Synchronizing Systems

R. Ramanujam
The Institute of Mathematical Sciences
Madras - 600 113, India.
e-mail: jam@imsc.ernet.in

Abstract

We study systems of sequential agents which communicate by synchronization, whose behaviours are given by a subclass of event structures. Transition system models for such structures typically require global state information which cannot be obtained by taking products of local transition systems. We offer a presentation whereby the notion of local state is modified, and an appropriate product operation precisely captures this class of behaviours. This is shown using a back-and-forth construction.

1 Introduction

Consider a system of several *agents* which act autonomously and periodically exchange information in rendezvous. This is a typical model studied in concurrency theory. One way of specifying "handshake" synchronization is by having letters of a common alphabet of actions between concurrent programs. For example, in a TCSP-style syntax, we can find programs like $(a + aa)\ c \ \ || \ \ (b + bb)\ c$ where we see two agents which synchronize after doing either one action or two actions locally. Note that this allows behaviours where synchronization occurs when one agent has done one local action whereas the other has done two local actions.

The distributed system is specified here as a parallel composition of some sequential processes. We can ask the following question : given a description of the behaviours of a distributed system in some fashion, when (and how) can we decompose the given system into a parallel composition of sequential processes, preserving behaviour ? This becomes a nontrivial question, when the given specification of the distributed system admits complicated types of synchronization. For example, in Mazurkiewicz trace theory [Maz], consider a concurrency alphabet $\{a, b, c\}, \{(a, b), (b, a)\}$, and the trace language $\{[abc], [aabbc]\}$ over it. According to it, when the synchronization occurs, both the agents have done one local action, or both have done two local actions, disallowing the possibility of one local action by one agent and two by the other.

In this paper, we address the question above, and present a subclass of event structures which can be obtained as products of transition systems. The attempt here is *not* to propose a new model of concurrency, but rather to highlight possible ways of distributing global information.

Transition system presentation of non-sequential behaviour has been studied extensively; we refer the reader to [BC, Dro, Muk, NRT, WN] for a variety of interesting results. Typically, these presentations are at a global level. That

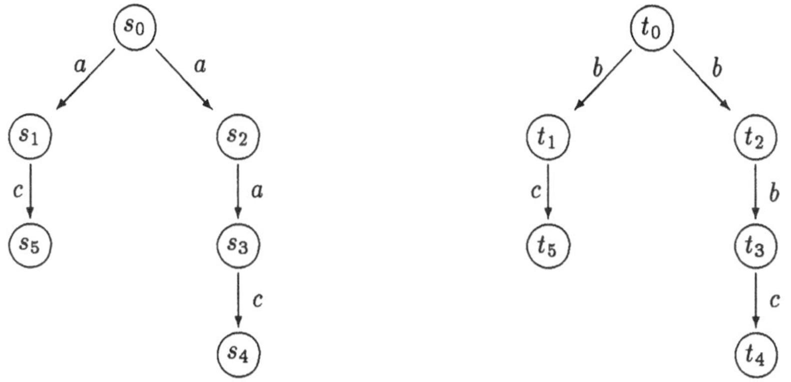

Figure 1: Two transition systems

is, the transition system consists of the set of all possible global states of the system and the allowed transitions between these states. There are models where the global states are obtained by taking products of local states, but even these require a global transition relation.

In a sense, this need for global information is inescapable. Suppose we consider a pair of transition systems as in Figure 1 to represent the situation above. When we consider the product transition system, a part of which is depicted in Figure 2, we find that c is enabled at the global state (s_1, t_3), when agent 1 has performed an a and agent 2 has done two b's. Indeed, the language $[abc] \cup [aabbc]$ cannot be obtained as a product of two transition systems, one over the alphabet $\{a, b\}$ and the other over $\{b, c\}$.

In this paper, we propose systems where local states can be *common* between agents. The idea is that a local state includes not only local information, but also information about the rest of the system available "as of date". Thus, an agent in a synchronizing system, has access not only to current values of local variables, but also values of other agents' variables as obtained by communication. This gives us a **local** presentation of synchronizing systems, where the understanding of the product operation is changed so that whenever a synchronization occurs, all agents participating in it reach an identical local state.

Formally, we consider the model of synchronizing systems presented as a subclass of event structures and show that these can be obtained as unfolded behaviours of transition systems with common local states. In a technical sense, this representation is faithful, as evidenced by a back-and-forth construction.

The paper is organized as follows: in the next section, we define the model of *Synchronization Structures* with finitely many agents and transition systems associated with such systems. In Section 3, we formally introduce the class of *Synchronizing Transition Systems*, and associate a synchronization structure with each such system. After this, we show that these associations preserve behaviour both ways. We end the paper with a discussion.

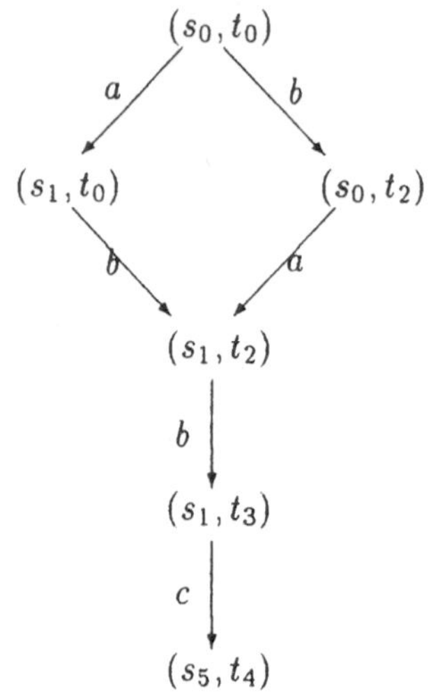

Figure 2: A part of the product transition system

2 Synchronization Structures

In [LRT], a model of distributed systems called Communicating Sequential Agents (CSAs) has been defined in the following manner: assume a collection of n agents, each of which is sequential, interacting by message passing. Each agent is 'tree-like', in the sense that its behaviour is given as a 'backwards-linear' poset of event occurrences. We present a subclass of CSAs below where interaction between agents is only by means of synchronization, modelled by shared event occurrences. Unlike [LRT], we study labelled systems here, for relating with transition systems in the next section.

Fix a finite set of *locations* $Loc = \{1, \ldots, n\}, n > 0$. The first notion we need is that of a **distributed alphabet**, a tuple $\widetilde{\Sigma} = (\Sigma_1, \ldots, \Sigma_n)$, where each Σ_i is a finite nonempty set of *actions of agent i*. When an action a is in $\Sigma_i \cap \Sigma_j, i \neq j$, we think of it as a synchronization action between i and j. Given a distributed alphabet $\widetilde{\Sigma}$, we often speak of the set $\Sigma \stackrel{\text{def}}{=} \Sigma_1 \cup \ldots \cup \Sigma_n$ as the alphabet of the system. We also make implicit use of the associated function $loc : \Sigma \rightarrow 2^{\{1,\ldots,n\}}$ defined by $loc(a) \stackrel{\text{def}}{=} \{i \mid a \in \Sigma_i\}$.

The formal definition of a synchronization structure is as follows:

Definition 2.1 *A Synchronization Structure with n agents (an n-SS) over a distributed alphabet* $\widetilde{\Sigma}$ *is a triple* $\mathcal{E} = (E, \leq, \lambda)$ *where* $n \in \mathcal{N}, n > 0$, *and*

(i) E *is a set of event occurrences,*

(ii) $\leq \subseteq E \times E$ *is a partial order called the causality relation,*

(iii) $\lambda : E \rightarrow \Sigma$ *is a labelling function such that*
$\forall e \forall i \in \{1, \ldots, n\}$, $\{e' | e' \leq e\} \cap E_i$ *is totally ordered by* $\leq$,
where $E_i \stackrel{\text{def}}{=} \{e' \in E \mid \lambda(e') \in \Sigma_i\}$, *and*

(iv) $\leq = (\bigcup_i \leq_i)^*$, *where*

$$\leq_i \stackrel{\text{def}}{=} \leq \bigcap (E_i \times E_i).$$

We will use the notation $\downarrow x$ for the initial segment of the partially ordered set $(X, \leq)$ up to the element x, i.e. for $x \in X$, $\downarrow x = \{y | y \leq x\}$; similarly for $X' \subseteq X$, $\downarrow X' = \{y | \exists x \in X', \ y \leq x\}$. We will also speak of SSs, leaving n implicit.

An agent is simply the set of event occurrences having the same name. We will often speak of the set E_i as agent i. We will also find it useful to define a *naming function* $\eta : E \to 2^{Loc}$ give by $\eta(e) \stackrel{\text{def}}{=} \{i \mid e \in E_i\}$. (Often, what we have called event occurrences are referred to as action occurrences; we do not need the distinction here.)

Note that condition (iii) imposes backward-linearity, making each agent tree-like. It is in this sense that the agents are sequential. When an event occurrence belongs to more than one agent, we interpret that as the occurrence of a synchronization.

$\leq$ is a causality relation in the sense that when $e_1 \leq e_2$, every observation of event occurrence e_2 necessarily implies an earlier observation of e_1. Note that due to condition (iv), when $\eta(e_1) \cap \eta(e_2) = \emptyset$, and $e_1 \leq e_2$, there must be a sequence of synchronization occurrences between e_1 and e_2 such that an agent in $\eta(e_1)$ participates in the first and an agent in $\eta(e_2)$ participates in the last.

λ and $\leq$ lead naturally to notions of conflict and concurrency in SSs. When we consider event occurrences which are incomparable under the causality ordering, conflicting ones are those which cannot both occur in the same computation, and concurrent ones are those which can. These are formally given below.

We say that event occurrences e_1 and e_2 are in *local conflict* when neither $e_1 \leq e_2$, nor $e_2 \leq e_1$, and $\eta(e_1) \cap \eta(e_2) \neq \emptyset$. Since agents are sequential, we do not interpret causal independence within agents as potential concurrency but as denoting choice made in computation. e_1 and e_2 are in *conflict*, if and only if there exist $e'_1 \leq e_1, e'_2 \leq e_2$ such that e'_1 and e'_2 are in local conflict.

When are two event occurrences concurrent ? We can say that e_1 and e_2 are *concurrent* if $\eta(e_1) \cap \eta(e_2) = \emptyset$, e_1 and e_2 are incomparable under the causal ordering, and $\downarrow e_1 \cup \downarrow e_2$ is conflict-free. Note that for any $e_1, e_2 \in E$, we have $e_1 \leq e_2$ or $e_2 \leq e_1$ or e_1 is in conflict with e_2 or e_1 and e_2 are concurrent.

We can now define a notion of a *global state* in an SS: $c \subseteq E$ is a global state iff $\downarrow c \subseteq c$ and c is conflict-free. Thus a state being downward-closed and conflict-free, can be thought of as a partial run. We will often refer to global states as *configurations*. Note that the empty set is always a configuration. More importantly, for any $e \in E$, the set $\downarrow e$ is a configuration, a fact which we will use crucially later on.

Consider the 2-SS in Figure 3. Events e_2 and e_6 are in local conflict, whereas e_2 and e_3 are concurrent. For this example, $\{e_2, e_3, e_5\}$ is a configuration, whereas $\{e_3, e_5\}$ and $\{e_2, e_3, e_5, e_6\}$ are not.

We say that an SS is *finitary* if and only if $\downarrow e$ is finite, for every $e \in E$. In the context of computation, it is natural to restrict attention to finitary SSs, and we will do precisely that. In fact, we will also be interested only in the *finite*

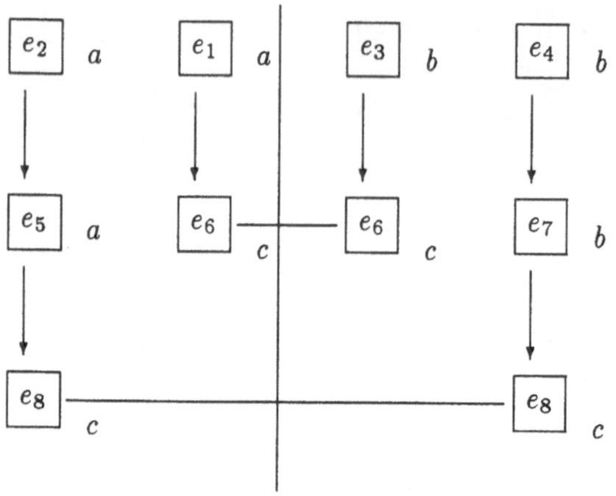

Figure 3: A synchronization structure over $\{a, c\}, \{b, c\}$

configurations of finitary SSs. Note that finitariness implies discreteness of the partial order and hence we will freely make use of $\lessdot$, the covering relation of $\leq$.

A configuration in an SS is a state in the sense that event occurrences cause state transitions. This is seen precisely in the transition system $T\mathcal{E}$ associated with the finitary SS $\mathcal{E}$. Let C denote the set of finite configurations of $\mathcal{E}$. (Here, and in what follows, we use the notation $c \oplus \{e\}$ to mean the set $c \cup \{e\}$ alongwith the assertion that $e \notin c$).

- $T\mathcal{E} \stackrel{\text{def}}{=} (C, \rightarrow, \emptyset)$, where
- $c \xrightarrow{a} c'$ iff $\exists e \in E, c' = c \oplus \{e\}$, and $\lambda(e) = a$.

The transition system describes a connected acyclic graph, with an initial state, namely the empty configuration (it has no in-coming edges, and every other node is reachable from it). Figure 4 shows a part of the transition system associated with the SS of Figure 3.

On the other hand, we can also define a notion of *local state* in an SS. Clearly, if $e \in E_i$, the set $\downarrow e$ can be thought of as a "local history" of agent i and this becomes a good candidate for a local state of that agent. Let $\mathcal{E}$ be a finitary SS. Define the transition system $T_i\mathcal{E}$ associated with $\mathcal{E}$ to be the tuple $(LC_i, \rightarrow_i)$, where the set of *local configurations* of agent i is given by :

- $LC_i \stackrel{\text{def}}{=} \{\emptyset\} \cup \{\downarrow e \mid e \in E_i\}$, and
- $x \xrightarrow{a}_i x'$ iff $\exists e \in E_i, x' \cap E_i = (x \cap E_i) \oplus \{e\}$, and $\lambda(e) = a, a \in \Sigma_i$.

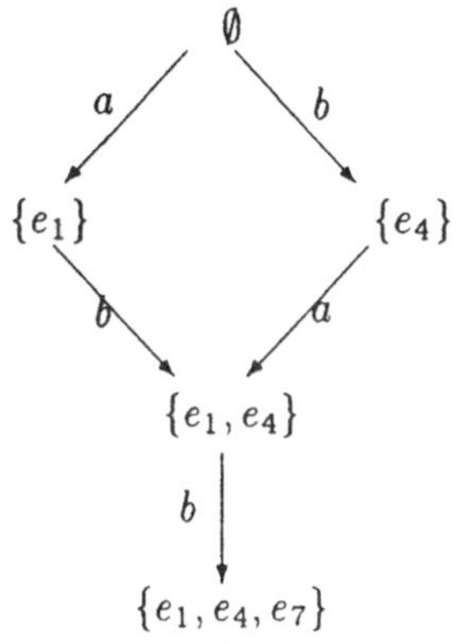

Figure 4: A part of transition system associated with the SS of Fig 3

On the other hand, we cannot simply take products of $T_i\mathcal{E}$ for all $i \in \{1, ..., n\}$ to obtain $T\mathcal{E}$. In the next section, we address these issues in the general context of transition systems.

In Figure 5, we see the local transition systems of the two agents in the SS of Figure 3. It looks almost the same as the pair of transition systems in Figure 1, but note that both s_5 and t_5 have been mapped to $\downarrow e_6$, and similarly for s_4 and t_4. At the global level, we find that Figure 4 is in correspondence with Figure 2, except for the missing transition. Thus, the synchronization structure in Figure 3 specifes the same behaviour as the trace language we mentioned in Section 1.

3 Synchronizing Transition Systems

A *labelled transition system* over a nonempty finite alphabet Σ is a tuple $TS = (Q, \rightarrow, q_0)$, where Q is an at most countable set of states, $q_0 \in Q$, and $\rightarrow \subseteq Q \times \Sigma \times Q$. We usually write $q \xrightarrow{a} q'$ to mean $(q, a, q') \in \rightarrow$, and interpret it to mean that the system can perform an action a at a state q and the resulting state is q'. A *run* of TS is a sequence $q_0 \xrightarrow{a_1} q_1 \xrightarrow{a_2} \ldots$ — a possible "execution" of the system. Labelled transition systems provide a natural model for the study of system behaviour.

We are interested in interactions between systems. A natural way to represent interactions among n agents is by having n transition systems, each working on its own alphabet of actions, except the system undergoes "joint" transitions when common actions are encountered. The obvious idea is to take products of transition systems. However, this will in general include more transitions than those present in transition systems associated with SSs. (For instance, compare Figure 2 with Figure 4.) We somehow need the ability to synchronize on transitions rather than on action labels. The model defined below introduces that ability — whenever two agents perform a synchronized transition, we insist that the resulting local states be identical. For this, it is

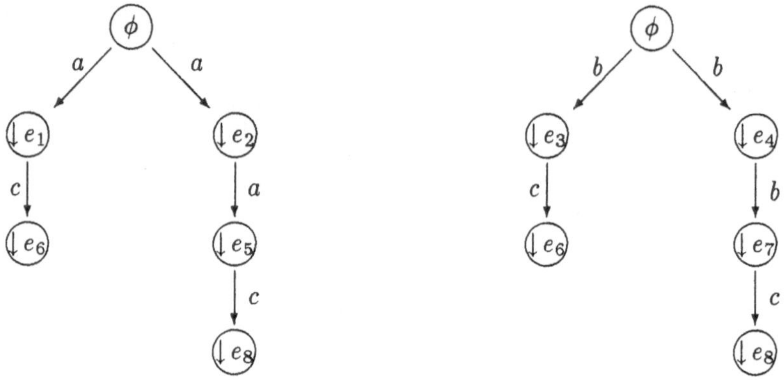

Figure 5: Local transition systems associated with structure of Fig 3

of course essential to consider agents which can "share" local states.

Definition 3.1 *Let* $\tilde{\Sigma} = (\Sigma_1, \ldots, \Sigma_n)$ *be a distributed alphabet. A* **Synchronizing System** *of* n *agents over* $\tilde{\Sigma}$ *is a tuple* $T = (TS_1, \ldots, TS_n)$*, where* $TS_i = (Q_i, \rightarrow_i, q_i^0)$ *is a labelled transition system over* Σ_i*, for* $i \in Loc$*, and* $q_1^0 = q_2^0 \ldots = q_n^0$.

Given a system as above, we will often refer to the set $Q = \bigcup_i Q_i$ as the set of states of the system. Let L_i denote the runs of TS_i, for $i \in Loc$. (We will avoid the awkward usage 'synchronizing system of transition systems' and abbreviate synchronizing systems as STSs.)

We have specified here that all systems begin at the identical initial state. We can read this as 'initial common knowledge' in the system. That is, the initial values of local variables of every agent is available to every other agent, and the situation is one of perfect information among agents. As the agents start making asynchronous moves, their information about others goes "out of sync", and synchronization is an occasion to update ("resync") information about the system.

The remarks above are formalized in the definition of global runs of synchronizing systems. Before we proceed to the definition, we need some notation. Let $Q = \bigcup_i Q_i$ and let $\sigma \in Q^*$ such that $\sigma = q_0\sigma'$. We say that q in σ is i-maximal for $i \in Loc$, when $\sigma = q_0 q_1 \ldots q_m$ and $q = q_k$, where k is the largest index such that $q_k \in Q_i$. Note that for all $i \in Loc$, such an i-maximal element exists, since $q_0 \in \bigcap_i Q_i$. We can refer to this state as $max_i(\sigma)$, and it can be thought of as the 'current' local state of agent i.

Definition 3.2 *Let* T *be a synchronizing system of* n *agents over* $\tilde{\Sigma}$. *A* **global run** *of* T *is a sequence* $q_0, a_0, q_1, a_1, q_2 \ldots$*, where*

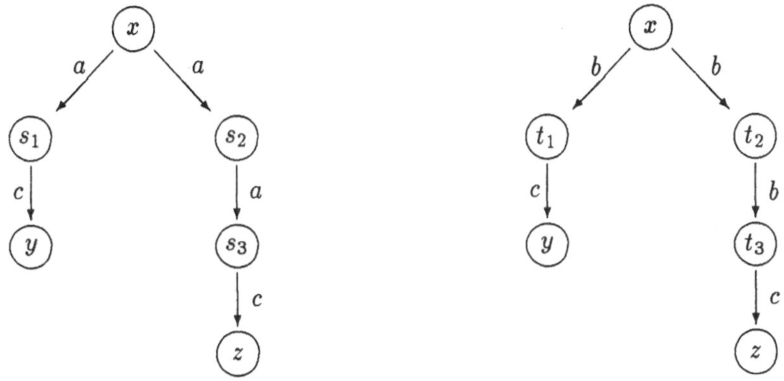

Figure 6: A 2-STS for Example 1

- $\{q_0, q_1, \ldots\} \subseteq Q = \bigcup_i Q_i$,
- $\{a_0, a_1, \ldots\} \subseteq \Sigma = \bigcup_i \Sigma_i$,
- q_0 *is the common initial state and*
- *for all* $k \geq 0$, *for* $j \in loc(a_k)$, $max_j(q_0 q_1 \ldots q_k) \xrightarrow{a_k}_j q_{k+1}$.

This definition is the main definition of the paper, in some sense. Global runs represent arbitrary interleavings of actions of different agents. The run is extended by an action a if and only if for every agent i participating in that action, a is enabled at the current local state of i and the resulting local state is identical for all such agents. Common local states have been crucially used here. Note that given a system $\mathcal{T}$, the singleton sequence q_0 may very well be the only global run of the system.

In Figure 6, we present an STS with 2 agents which represents the behaviour specified by the trace language of Section 1. Note that this is isomorphic to the pair of local transition systems of Figure 5.

Another way of presenting global runs is by defining products of synchronizing systems as given below. This product construction is different from the standard notion. However, the set of runs of the product system is in bijection with the set of global runs of the system. In that sense, the definition above carries the same content as the one below.

Definition 3.3 *Let* $\mathcal{T}$ *be a synchronizing system over* $\widetilde{\Sigma}$. *The* **product system** *induced by* $\mathcal{T}$ *is the* Σ*-labelled transition system* $TS = (Q, \Rightarrow, q_0)$, *where*

- $\widetilde{Q} \stackrel{\text{def}}{=} Q_1 \times \ldots \times Q_n$,
- $q_0 \stackrel{\text{def}}{=} q_1^0 = \ldots = q_n^0$, *and*

- $\Rightarrow \subseteq \tilde{Q} \times \Sigma \times \tilde{Q}$ *is the* **global** *transition function defined as follows:* $(q_1, \ldots, q_n) \stackrel{a}{\Rightarrow} (q'_1, \ldots, q'_n)$ *iff*
  - $\forall i \in loc(a), q_i \stackrel{a}{\rightarrow}_i q'_i,$
  - $\forall j \notin loc(a), q_j = q'_j,$ *and*
  - $\forall i, j \in loc(a), q'_i = q'_j.$

When we form the induced product system for the STS of Figure 6, we find (as one may expect) that it is isomorphic to the global configuration system of Figure 4. The following proposition, not surprisingly, asserts that this is no accident.

Proposition 3.4 *Let $\mathcal{E}$ be an SS over $\tilde{\Sigma}$. Then $T_\mathcal{E} \stackrel{\text{def}}{=} (T_1\mathcal{E}, \ldots, T_n\mathcal{E})$ is a synchronizing system over $\tilde{\Sigma}$, whose induced product is isomorphic to $T\mathcal{E}$.*

Having associated a synchronizing transition system with a given synchronization structure, we now proceed to construct the event structure given the transition system. This forms the main technical content of the paper, and the construction highlights the information transfer between agents discussed above. We will be working throughout with the global runs of Definition 3.2, rather than the product system of Definition 3.3. We will write global runs of the form $q_0, a_0, q_1, a_1, \ldots$ as $q_0 \stackrel{a_0}{\Longrightarrow} q_1 \stackrel{a_1}{\Longrightarrow} \ldots$.

For the rest of the section, fix T, a synchronizing system of n agents over a distributed alphabet $\tilde{\Sigma} = (\Sigma_1, \ldots, \Sigma_n)$. Let R be the set of *finite non-null* global runs of this system. We use δ, δ' etc. to denote elements of R. Note that non-null runs have at least one transition. Define $\tau : R \rightarrow \Sigma$ by: $\tau(\delta) \stackrel{\text{def}}{=} a_k$ where $\delta = \$q_0 \stackrel{a_0}{\Longrightarrow} q_1 \stackrel{a_1}{\Longrightarrow} \ldots q_k \stackrel{a_k}{\Longrightarrow} q_{k+1}$. We will also use $\eta : R \rightarrow 2^{Loc}$ defined by $\eta(\delta) \stackrel{\text{def}}{=} loc(\tau(\delta))$. We use the notation $\preceq$ to denote the prefix ordering on sequences.

Consider $\delta \in R$ such that $\delta = q_0 \stackrel{a_1}{\Longrightarrow} q_1 \stackrel{a_2}{\Longrightarrow} \ldots \stackrel{a_k}{\Longrightarrow} q_k$. For every $i \in Loc$, let m_i be the largest index such that $q_{m_i} \in Q_i \cap Q_j$ for some $j \in loc(a_k)$. Clearly for $i \in loc(a_k), m_i = k$. In general, call the prefix $q_0 \stackrel{a_1}{\Longrightarrow} q_1 \stackrel{a_2}{\Longrightarrow} \ldots q_{m_i}$, denoted $\delta(i)$, the *i-snapshot at δ*. We can think of this as all the information available to agents in $loc(a_k)$ about agent i pooled together. Even though different agents in $loc(a_k)$ may see earlier or later states of i before the synchronization a_k, at the state q_k, they all see the latest.

For example, consider a situation like this : agents 1 and 2 synchronize independently of a synchronization between agents 3 and 4. At this juncture, 2 has a recent view of 1's local state, but in 3's view, 1 is still in the initial state. On the other hand, 3 has the latest information about 4, but 2 does not. Now, agents 2 and 3 synchronize, and in the joint view of agents 2 and 3, the current states of 1 and 4 become available to both. In this sense, global runs of STSs depict the "gossip" scenarios of [MS].

Define the *projection maps* $\phi_i : R \rightarrow L_i$, for $i \in Loc$ as follows: $\phi_i(\delta)$ is the sequence obtained by erasing all states not in Q_i and actions not in Σ_i. It can be easily seen that ϕ_i is well-defined, though it is clearly not onto.

Let the map $\epsilon : R \rightarrow L_1 \times \ldots \times L_n$ be defined as follows:
$\epsilon(\delta) = (\phi_1(\delta_1), \ldots, \phi_n(\delta_n)),$

where δ_i is the i-snapshot at δ. Clearly the map ϵ is well-defined. Define $\approx \subseteq R \times R$ by: $\delta_1 \approx \delta_2$ iff $\epsilon(\delta_1) = \epsilon(\delta_2)$. Let $[\delta]$ denote the equivalence class of δ under $\approx$. We use $\approx$ to associate an SS with the synchronizing transition system.

Below, assume that $\preceq$ is extended pointwise: that is, for tuples, $(\rho_1, \ldots, \rho_n) \preceq (\rho'_1, \ldots, \rho'_n)$ iff $\forall i \in Loc, \rho_i \preceq \rho'_i$.

The following propositions are reasonably easy to show:

Proposition 3.5 *Suppose $\delta_1 \preceq \delta_2$ and $\eta(\delta_1) \cap \eta(\delta_2) \neq \emptyset$. Then $\epsilon(\delta_1) \preceq \epsilon(\delta_2)$.*

Proposition 3.6 *Suppose $\delta_1 \approx \delta_2$. Then $\tau(\delta_1) = \tau(\delta_2)$.*

Proposition 3.7 *Suppose $\delta_1 \preceq \delta_2$ and $\delta_2 \approx \delta'_2$. Then there exists $\delta'_1 \preceq \delta'_2$ such that $\delta_1 \approx \delta'_1$.*

Proposition 3.8 *Let $\mathcal{T} = (TS_1, \ldots, TS_n)$ be a synchronizing system over the distributed alphabet $\widetilde{\Sigma} = (\Sigma_1, \ldots, \Sigma_n)$. Define the structure $\mathcal{E} = (E, \leq, \lambda)$ by:*

$$E \stackrel{\text{def}}{=} \{[\delta] | \delta \in R_{\mathcal{T}}\},$$

$$\leq \stackrel{\text{def}}{=} \{([\delta_1], [\delta_2]) | \text{ there exists } \delta \preceq \delta_2, [\delta] = [\delta_1]\} \text{ and}$$

$$\lambda([\delta]) \stackrel{\text{def}}{=} \tau(\delta).$$

Then $\mathcal{E}$ is a finitary synchronization structure over $\widetilde{\Sigma}$.

Proof: Reflexivity of $\leq$ is trivial. Antisymmetry follows from Proposition 3.5 above. Transitivity and backward linearity (within agents) can be proved using Proposition 3.7. Proposition 3.6 ensures that λ is well-defined. For finitariness, we only need to observe that every element of R is a finite sequence, and hence has only finitely many prefixes.

We need to show that $\leq$ is generated from $\leq_i$. It suffices to prove that whenever $[\delta_1] \lessdot [\delta_2]$, we have that $\eta(\delta_1) \cap \eta(\delta_2) \neq \emptyset$. Now assume that $[\delta_1] \lessdot [\delta_2]$, but that $\eta(\delta_1) \cap \eta(\delta_2) = \emptyset$. Let $\delta \preceq \delta_2$ such that $[\delta] = [\delta_1]$ and $\tau(\delta) = \tau(\delta_1) = a$, say. Let $\delta = \delta' \stackrel{a}{\Longrightarrow} q_k$ and $\delta_2 = \delta \stackrel{a_1}{\Longrightarrow} q_{k+1} \ldots \stackrel{a_m}{\Longrightarrow} q_{k+m}$. Now suppose there exists j between 1 and m such that $loc(a) \cap loc(a_j) \neq \emptyset$ and $loc(a_j) \cap loc(a_m) \neq \emptyset$. Then, by proposition above, $\epsilon(\delta) \leq \epsilon(\delta_j) \leq \epsilon(\delta_2)$ contradicting the assumption that $[\delta] \lessdot [\delta_2]$. Thus, we have that $loc(a) \cap loc(a_j) = \emptyset$, for every j such that $loc(a_j) \cap loc(a_m) \neq \emptyset$. But then, for $i \in loc(a)$, the i-snapshot at δ_2 is a strictly smaller prefix of the i-snapshot at δ, which contradicts the assumption that $\delta \leq \delta_2$. Thus, we must have that $\eta(\delta_1)(= \eta(\delta)) \cap \eta(\delta_2) \neq \emptyset$, as required. □

We refer to the structure $\mathcal{E}$ as the SS associated with $\mathcal{T}$ and denote it $\mathcal{E}_{\mathcal{T}}$.

4 Back and Forth

Given an SS $\mathcal{E}$, we can associate an STS $\mathcal{T}_{\mathcal{E}}$ with it, and with this STS we can associate an SS $\mathcal{E}_{\mathcal{T}_{\mathcal{E}}}$. What is the relationship between $\mathcal{E}$ and $\mathcal{E}_{\mathcal{T}_{\mathcal{E}}}$?

Theorem 4.1 *Let $\mathcal{E} = (E, \leq, \lambda)$ be a finitary synchronization structure over a distributed alphabet $\tilde{\Sigma} = (\Sigma_1, \ldots, \Sigma_n)$, and let $\mathcal{E}_{\mathcal{T}_\mathcal{E}} = (E', \leq', \lambda')$. Then there is a bijection $f : E \rightarrow E'$ such that :*

$\forall e, e' \in E, e \leq e'$ *iff* $f(e) \leq' f(e')$ *and*

$\forall e \in E, \lambda(e) = \lambda'(f(e))$.

In other words, $\mathcal{E}_{\mathcal{T}_\mathcal{E}}$ is isomorphic to $\mathcal{E}$.

Proof: Let $\mathcal{E} = (E, \leq, \lambda)$ be the given SS, and fix $\mathcal{T}_\mathcal{E}$ to be the STS $\mathcal{T} = (TS_1, \ldots, TS_n)$. Note that TS_i is simply the transition system $T_i\mathcal{E}$ defined in Section 2. It is given by the tuple $(LC_i, \rightarrow_i)$, where the set of local configurations of agent i is given by:

- $LC_i \stackrel{\text{def}}{=} \{\emptyset\} \cup \{\downarrow e \mid e \in E_i\}$, and
- $x \stackrel{a}{\rightarrow} x'$ iff $\exists e \in E_i, x' \cap E_i = (x \cap E_i) \oplus \{e\}$, and $\lambda(e) = a, a \in \Sigma_i$.

Let C denote the set of finite configurations of $\mathcal{E}$. A *schedule* of a configuration $c \in C$ is a sequence $e_1 ... e_k$ such that $l \neq m$ implies $e_l \neq e_m, c = \{e_1, ..., e_k\}$, and if $e_l \leq e_m$, then $l \leq m$. Note that every schedule (of any configuration) corresponds to a global run in T. Conversely for every global run from $\emptyset$ to a local state $\downarrow e$ in T of the form $\emptyset \stackrel{\lambda(e_1)}{\Longrightarrow} \downarrow e_1 \ldots \stackrel{\lambda(e_k)}{\Longrightarrow} \downarrow e_k$, the set $\{e_1, \ldots, e_k\}$ is indeed a configuration in C, and $e_1 \ldots e_k$ is a schedule of that configuration.

Given $e \in E, i \in \eta(e)$, consider the set $\{\downarrow e' | e' \leq e, i \in \eta(e')\}$; since $\mathcal{E}$ is finitary and backwards linear within agents, this set can be written as a finite sequence of elements $\downarrow e_1, ..., \downarrow e_k$, where for $j \in \{1, ..., k-1\}, e_j < e_{j+1}$ and $e_k = e$; call this sequence, prefixed by $\{\emptyset\}$, the local history of i at e, denoted $lh_i(e)$.

Consider a schedule of a configuration c ending at event occurrence e, and let $\lambda(e) = a$. Now consider the tuple $< \rho_1, \ldots, \rho_n >$, where for $j \in \{1, ..., n\}, \rho_j$ is the null sequence if $\downarrow e \cap E_j = \emptyset$, and otherwise it is $lh_j(e')$, where e' is the maximal j-event in $\downarrow e$. Call this tuple the $\eta(e)$-view at e. (Note that when $\{i, j\} \subseteq \eta(e)$, the i-view is the same as the j-view.)

We claim that the $\eta(e)$-view at e is indeed an event occurrence in $\mathcal{T}$ and hence a member of E'. To see this, observe firstly that for any $e' \in E, j \in \eta(e')$, $lh_j(e')$ is a local j-run in TS_j when $|lh(e')| > 1$. We can easily check that every such $\eta(e)$-view is generated as $\epsilon(\delta)$ for some run δ in $\mathcal{T}$.

We can thus meaningfully define the map $F : E \rightarrow E'$: given by $F(e) =$ the $\eta(e)$-view at e.

To prove that F is injective, suppose $e_1 \neq e_2$. If $\eta(e_1) \cap \eta(e_2) \neq \emptyset$. Then for every $i \in \eta(e_1) \cap \eta(e_2)$, clearly $lh_i(e_1) \neq lh_i(e_2)$ and hence $\eta(e_1)$-view at e_1 is distinct from the $\eta(e_2)$-view at e_2. Otherwise, suppose that $\eta(e_1) \cap \eta(e_2) = \emptyset, i \in \eta(e_1), j \in \eta(e_2)$, but that the $\eta(e_1)$-view at e_1 is identical to the $\eta(e_2)$-view at e_2. This means that e_1 is the maximal i-event in $\downarrow e_2$, and that e_2 is the maximal j-event in $\downarrow e_1$. Then in particular, we get $e_1 \leq e_2, e_2 \leq e_1$ and $e_1 \neq e_2$, contradicting the antisymmetry of $\leq$. Thus F is injective.

Now consider $e' \in E', i \in \eta'(e')$. Then e' is an i-event occurrence in $\mathcal{T}_\mathcal{E}$ say, $< \sigma_1, ..., \sigma_n >$. Since $|\sigma_i| > 1$, it is of the form $\sigma x x'$, where for some

$e \in E_i, (x' \cap E_i) = (x \cap E_i) \oplus \{e\}$. It can then be easily checked that e' is indeed the $\eta(e)$-view at e, that is $F(e) = e'$, proving surjectivity of F.

The fact that $\lambda(e) = \lambda'(F(e))$ is trivial. Now consider $e_1, e_2 \in E$ such that $e_1 \leq e_2$. Then every schedule δ_2 for $\downarrow e_2$ includes as a prefix a schedule δ_1 for $\downarrow e_1$. Since $\delta_1 \preceq \delta_2$, $\epsilon(\delta_1) \leq' \epsilon(\delta_2)$, that is, $F(e) \leq' F(e')$, as required.

On the other hand, suppose $e_1, e_2 \in E$ such that $F(e_1) \leq' F(e_2)$. Again let $F(e_1)$ be an i-event occurrence in $\mathcal{T}_{\mathcal{E}}$ and $F(e_2)$ a j-event occurrence. Then if $F(e_2) =< \rho_1, ..., \rho_n >$, then ρ_j is of the form $\rho x x'$ where $x' \cap E_j = (x \cap E_j) \oplus \{e_2\}$, and ρ_i is of the form $\rho' x_1 x_2 \rho'' x_3 x_4$, where $x_2 \cap E_i = (x_1 \cap E_i) \oplus \{e_1\}, x_2 \subseteq x_3, x_4 \cap E_i = (x_3 \cap E_i) \oplus \{e_3\}$, and e_3 is the i-maximal event in $\downarrow e_2$. Thus we get $e_1 \leq e_3 \leq e_2$, and thus $e_1 \leq e_2$. Thus F is order-preserving too, and we have the result. □

We now consider the tour in the other direction : given an STS $\mathcal{T}$, we can associate an SS $\mathcal{E}_{\mathcal{T}}$ with it, and with this SS we can associate a STS $\mathcal{T}_{\mathcal{E}_{\mathcal{T}}}$. What is the relationship between $\mathcal{T}$, and $\mathcal{T}_{\mathcal{E}_{\mathcal{T}}}$? Obviously, we cannot again hope for an isomorphism, but we can expect an unfolding, and we define this first.

Definition 4.2 *Let $TS = (S, \rightarrow, s_0)$ and $TS' = (S', \rightarrow', s'_0)$ be labelled transition systems over the same alphabet Σ. We say that TS' is a* simulation *of TS iff there exists a map $f : S' \rightarrow S$: such that*

(i) $f(s'_0) = s_0$,

(ii) *for $s'_1, s'_2 \in S'$, if $s'_1 \xrightarrow{a}{}' s'_2$ then $f(s'_1) \xrightarrow{a} f(s'_2)$,*

(iii) *for $s_1, s_2 \in S$ such that $s_1 \xrightarrow{a} s_2$, if there exists s'_1 such that $f(s'_1) = s_1$, then there exists s'_2 such that $f(s'_2) = s_2$ and $s'_1 \xrightarrow{a}{}' s'_2$.*

A simulation is said to be strong *if it is also surjective. A map which only satisfies conditions (i) and (ii) above is called a* weak *simulation.*

Note that a weak simulation only says that the behaviour in TS' is generated by the structure of TS. The latter might in addition have many states which have no representatives at all in S', and even when f happens to be onto, we might well have $s_1 \xrightarrow{a} s_2$ in TS, and $f(s'_1) = s_1$, but no s'_2 such that $f(s'_2) = s_2$.

Definition 4.3 *Let $\mathcal{T} = (TS_1, \ldots, TS_n)$ and $\mathcal{T}' = (TS'_1, \ldots, TS'_n)$ be synchronizing transition systems over the same distributed alphabet $\widetilde{\Sigma} = (\Sigma_1, \ldots, \Sigma_n)$. We say that $\mathcal{T}'$ is a* local (weak / strong) simulation *of $\mathcal{T}$ iff for every $i \in Loc$, TS' is a (weak / strong) simulation of TS.*

For the rest of this section, fix an STS (over $\widetilde{\Sigma}$), $\mathcal{T} = (TS_1, \ldots, TS_n)$, where each TS_i is a tuple $TS_i = (Q_i, \rightarrow_i, q_i^0)$. Let $\mathcal{E}_{\mathcal{T}} = (E, \leq, \lambda)$, and $\mathcal{T}_{\mathcal{E}_{\mathcal{T}}} = (TS'_1, \ldots, TS'_n)$, denoted $\mathcal{T}'$.

We will show that the latter STS weakly simulates the former. The reason why we cannot expect any stronger simulation is this: in general, in a given STS, there can be many transitions which are "useless" in the sense that they are never part of any global run. For instance, consider a system with 2 agents. One has a single transition labelled a from the initial state, the other has no transitions, and the alphabet is the same for both, namely the single letter a. Clearly, the local transition can never be part of any global run. In fact, for this system, the unfolding consists of the trivial run with a single element – the initial state.

For each $i \in Loc$, we need to define the weak simulation f_i from TS'_i to TS_i. Now, TS'_i is nothing but $T_i\mathcal{E}$. Thus, we want a map from LC_i to Q_i. Obviously, we can set $f_i(\emptyset) = q_i^0$, where the latter is the initial state of TS_i. What about $f_i(\downarrow e), e \in E_i$?

Clearly, $e = [\delta]$, for some δ such that $\tau(\delta) = a \in \Sigma_i$. Notice that δ is of the form $\delta' q' \overset{a}{\Longrightarrow} q$ where $q \in Q_i$. Furthermore, for every other δ_1 such that $[\delta] = [\delta_1]$, δ_1 is of the form $\delta'_1 q'' \overset{a}{\Longrightarrow} q$. Thus the function f_i defined by: $f_i(\downarrow e) = q$, where $e = [\delta]$ and δ ends with q, is well-defined.

Now suppose that $\downarrow e_1 \overset{a}{\rightarrow} \downarrow e_2$ in $T_i\mathcal{E}$. Then, there exist δ_1, δ_2 such that $e_1 = [\delta_1], e_2 = [\delta_2]$ and $\delta'_2 \preceq \delta_2$ such that $[\delta_1] = [\delta'_2]$. Now suppose that $f_i(\downarrow e_1) = q, f_i(\downarrow e_2) = q'$. By the remarks above, we find that δ'_2 ends with q and δ_2 ends with q'. It is easy to see that δ'_2 is the i-maximal prefix of δ_2 (if not, there would be a prefix δ_3 of δ_2 suffix of δ'_2, and hence we would get $[\delta_3]$ between e_1 and e_2). Now, by definition of global run, for δ_2 to end with $q'' \overset{a}{\longrightarrow} q'$, we must have $q \overset{a}{\rightarrow}_i q'$, as required.

These remarks prove the following lemma, and the theorem below follows.

Lemma 4.4 *The global runs of T are in bijection with those of $T_{\mathcal{E}_T}$,*

Theorem 4.5 *Let T be an STS. Then $T_{\mathcal{E}_T}$ is a local weak simulation of T.*

In a sense, the image of f_i above consists of the actual local states reached by computations of the system, and every i-transition which figures in some global run is actually mimicked in the unfolding. Call a transition $q \overset{a}{\rightarrow}_i q'$ in TS_i in an STS T **useful** when there exists a global run $\delta \overset{a}{\Longrightarrow} q'$ in T, where q is i-maximal in δ. Call TS_i **complete** when every transition in TS_i is useful, and every state in Q_i reachable from the initial state.

Theorem 4.6 *If T is an STS made up of complete subsystems, then $T_{\mathcal{E}_T}$ is a local strong simulation of T.*

Unfortunately, this fact about useful transitions is not very useful. Deciding whether a transition in a finite STS is useful seems quite hard.

5 Discussion

The preceding section shows that SSs can be captured as products of STSs, where the product construction includes the condition that after any synchronization, the resulting local state for each of the participants in the synchronization must be identical. We can read this result as a way of decomposing synchronization structures.

Considered as subclasses of prime event structures, the structures we have defined, namely SSs, have an important restriction: the conflict relation between event occurrences is generated by local conflict. This rules out specification of a priori global conflict iin the system. Consider the prime event structure with 2 events e_1 and e_2, labelled a and b respectively, over the distributed alphabet $(\{e_1\}, \{e_2\})$, where e_1 and e_2 are in conflict. This cannot be obtained as an SS.

As it turns out, the construction given in the paper can be carried out for such structures as well, provided we allow STSs to have **multiple initial states**. Formally, we can define a *trace event structure* over a distributed alphabet $\widetilde{\Sigma}$ to be a λ-labelled finitary prime event structure $(E, \leq, \#)$ such that whenever for concurrent events e_1 and e_2, $loc(\lambda(e_1)) \cap loc(\lambda(e_2)) = \emptyset$, and whenever $e_1 \lessdot e_2, loc(\lambda(e_1)) \cap loc(\lambda(e_2)) \neq \emptyset$. On the other hand, we can define a *trace STS* to be a tuple $(TS_1, \ldots, TS_n)$, where $TS_i = (Q_i, \rightarrow_i, I_i)$ is a Σ_i-labelled transition system with $I_i \subseteq Q_i$, and $I_1 = \ldots = I_n$. We can then associate a trace-STS with a trace event structure and vice versa such that results similar to the ones in the last section hold. The details of this construction will be given in the full paper.

The more interesting question is regarding **finite state** STSs. Obviously, we can consider tuples of transition systems above, where each is given as $(Q_i, \rightarrow_i, I_i, F_i)$, and $F_i \subseteq Q_i$ are *final states.* This leads to a notion of accepting languages $L \subseteq \Sigma^*$, where a string is accepted if there is a run in the induced product system from one of the common initial states using that string to a global state where each of the local states is final. It can be easily seen that languages accepted by finite-state STSs in this sense, are regular and I-consistent (where I is the obvious independence relation associated with the distributed alphabet: aIb iff $loc(a) \cap loc(b) = \emptyset$). We believe that the converse is also true, though we do not as yet have a proof. (That would yield a local presentation for Zielonka automata [Zie].)

On a very different track, global runs of STSs can be seen as very natural candidates for temporal logics like TrPTL [Thi], where atomic propositions are interpreted over local ones, and global formulas are built up from local ones. TrPTL crucially makes use of the fact that the local views of agents are identical after synchronization, and this property is reflected in global runs of STSs.

We can also see the states of STSs as states of knowledge of agents and synchronization as perfect knowledge transfer. Can this intuition be made precise in some sense ? In the theory of knowledge in distributed systems, it is customary to define an agent i's knowledge by an equivalence relation $\approx_i$ on the global states of the system. The idea is that at any state s, the agent knows exactly those properties which hold for all states $s' \in [s]_i$. The relation $\approx_i$ can be seen as an indistiguishability relation for agent i.

In the context of distributed systems, knowledge is usually defined by equivalence relations on histories; two histories look the same to an agent if the projections to events which the agent participates in are the same ([HM]). Given such a notion of knowledge, we can go on to study the notion of *implicit group knowledge* — intuitively, a property is implicitly known to a group of agents, if some agent in the group knows it. Communication is necessary for making this knowledge explicit. Synchronous communication not only makes implicit knowledge explicit, but also makes it common knowledge to the members of the group who synchronize. Conversely, achieving this kind of knowledge requires synchronization.

What has all this got to do with STSs ? When we define the notion of knowledge (as above) on runs of STSs, it turns out that $\epsilon(\delta)$ in fact represents the group common knowledge at δ for the group $loc(\tau(\delta))$. This means that the local states of STSs are, in effect, states of implicit group knowledge. We believe that in a precise sense, the *Knowledge Transition Systems* of [KR] (suitably modified to rule out asynchronous communication) correspond to product systems induced by STSs.

Finally, this work needs to be done in a categorical set-up, so that the product construction given here can be seen as a categorical product operation.

References

[BC] Boudol, G. and Castellani, I., "Flow models of distributed computations: event structures and nets", *Rapports de Recherche 1482*, INRIA, July 1991.

[Dro] Droste, M., "Event structures and domains", *Theoretical Computer Science*, vol. 68, 1989, 37-47.

[HM] Halpern, J., and Moses, Y., "Knowledge and Common Knowledge in a Distributed Environment", *JACM*, vol. 37, 1991, 549-578.

[KR] Krasucki, P., and Ramanujam, R., "Knowledge and the ordering of events in distributed systems", *TARK V*, Theoretical Aspects of Reasoning about Knowledge, 1994, 267-283.

[LPRT] Lodaya, K., Parikh, R., Ramanujam, R. and Thiagarajan, P.S., "A logical study of distributed transition systems", *Report* **IMSc/92/07** 1992, to appear in *Information and Computation.*

[LRT] Lodaya, K., and Ramanujam, R., and Thiagarajan, P.S., "Temporal Logics for Communicating Sequential Agents: I", *Int. J. Found. Comp. Sci.*, vol. 3, 2, 1992, 117-159.

[Maz] Mazurkiewicz, A., "Basic notions of trace theory", *LNCS 354*, 1989, 285-363.

[Muk] Mukund, M., "Petri nets and step transition systems", *Int. J. Found. Comp. Sci.*, vol. 3, 4, 1992, 443-478.

[MS] Mukund, M. and Sohoni, M., "Keeping track of the latest gossip: bounded time-stamps suffice", *LNCS 761*, 1993, 388-399.

[NRT] Nielsen, M., Rozenberg, G., and Thiagarajan, P.S., "Behavioural notions for elementary net systems", *Distr. Comput.*, vol. 4, 1990, 45-57.

[Thi] Thiagarajan, P.S., "A trace based extension of propositional linear time temporal logic", *Proc IEEE LICS*, 1994, 438-447.

[WN] Winskel, G., and Nielsen, M., "Models for concurrency", in *Handbook of Logic in Computer Science*, Eds: Abramsky, Gabbay and Maibaum, Oxford Science Publications.

[Zie] Zielonka, W., "Notes on finite asynchronous automata", RAIRO-Inf. Theor. et Appli., vol. 21, 1987, 99-135.

On Well-formedness Analysis: The Case of Deterministic Systems of Sequential Processes *

Laura Recalde, Enrique Teruel, and Manuel Silva
C.P.S. Ingenieros, Universidad de Zaragoza
Zaragoza, Spain

Abstract

New results on structural analysis of well-formedness, that is, structural boundedness and structural liveness, of Place/Transition net systems, based on the rank of the incidence matrix, are introduced and related to previously known ones.

These results are a general sufficient condition, that follows from the Rank Theorem for Equal Conflict systems, and a polynomial-time characterisation of well-formedness for an enlarged version of the structured class of Deterministic Systems of Sequential Processes.

1 Introduction

A fruitful approach to overcome the inherent complexity of analysing concurrent systems is trying to obtain useful information directly from the structure of the model. A major concern for us is enhancing structural techniques for the analysis of boundedness and liveness.

When general net systems are considered, the obtained structural conditions happen to be only necessary *or* sufficient. In this case, the goal is strengthening the conditions as much as we can. In this paper we recall the best to our knowledge necessary condition for *well-formedness* (structural boundedness and liveness) [1], and introduce a sufficient condition, which is a generalisation of that in [2] to the weighted case. In this case, as it sometimes happens, the consideration of weights does not spoil clarity; in fact, it greatly simplifies the proof. Both conditions are formulated in terms of conservativeness and consistency plus some relation between the value of the *rank of the incidence matrix* and the number of certain net constructs related to conflicts. Unfortunately, there are many nets fulfilling the necessary but not the sufficient condition, so these are not enough to decide on well-formedness.

More satisfactory results are obtained when the scope is limited to restricted classes of systems. For instance, the aforementioned general necessary condition becomes also sufficient in the case of *Equal Conflict systems*, the weighted generalisation of Free Choice ones [1]. In this paper we extend this result to a somehow related but not comparable structured subclass: *Deterministic Systems of Sequential Processes* are net systems formed by a collection of strongly

*This work has been partially supported by the projects CICYT TIC-94-0242, Esprit BRA Project 7269 (QMIPS), and Esprit W.G. 6067 (CALIBAN).

connected State Machines — ordinary nets where every transition has exactly one input and one output place — marked with one token each, the *Sequential Processes*, interconnected by a set of places, the *buffers*, in a restricted way. This class was introduced in [3, 4] and further enlarged (introducing weights in arcs adjacent to buffers) and studied in [5, 6]. Among other properties and results, these works characterise structural liveness in terms of the absence of some substructures called *small rings* or *circles.* On one hand, it is interesting having a purely structural characterisation, but, on the other, it is still complex to determine absence of such rings: one is forced to consider every ring, and there could be an exponential number of them. Here we consider an enlarged version of the class (see [7]) where buffers receiving tokens from various Sequential Process are allowed, and derive a characterisation of well-formedness with *polynomial time complexity* wrt. the size of the net. This is a principal result within the structure theory of Deterministic Systems of Sequential Processes. There are several others, but they are out of the scope of this work. (The reader is referred to [7] for some indications.)

The rest of the paper is organised as follows: Section 2 recalls basic concepts and notation, while the specialised definitions related to structural conflicts are contained in Sect. 3. A necessary and a sufficient condition for well-formedness of general Place/Transition net systems form Sect. 4. The case of Deterministic Systems of Sequential Processes is studied in Sect. 5, where a characterisation of well-formedness is derived.

2 Preliminaries

The reader is assumed to be familiar with Petri net theory (see [8] for a survey). Nevertheless, in this section we recall the basic concepts and introduce the notation to be used. For the sake of readability, whenever a net or system is defined it "inherits" the definition of all the characteristic sets, functions, parameters... with names conveniently marked to identify whose is which.

Place/Transition net and related concepts. A P/T *net* is a triple $\mathcal{N} = (P, T, W)$ where P and T are disjoint finite sets of *places* and *transitions*, and $W : (P \times T) \cup (T \times P) \to \mathbb{N}$ defines the *weighted flow relation*: if $W(u,v) > 0$, then we say that there is an *arc* from u to v, with *weight* or *multiplicity* $W(u,v)$. *Ordinary* nets are those where $W : (P \times T) \cup (T \times P) \to \{0,1\}$. Since a P/T net can be seen, and drawn, as a bipartite weighted directed graph, several graph concepts, like circuits (directed and elementary, unless otherwise stated), connectedness (without loss of generality, nets are assumed to be connected), strong connectedness, etc. can be extended to nets. In particular, let $v \in P \cup T$; its *preset* and *postset* are given by: ${}^\bullet v = \{u \mid W(u,v) > 0\}$, and $v^\bullet = \{u \mid W(v,u) > 0\}$. The preset (postset) of a set of nodes is the union of presets (postsets) of its elements. The weighted flow relation can be alternatively defined by: $Pre(p,t) = W(p,t)$, $Post(p,t) = W(t,p)$. These functions can be represented by matrices[1]. The *incidence matrix* is defined as $C = Post - Pre$.

[1] Places and transitions are supposed to be arbitrarily, but fixedly, ordered. Therefore rows and columns can be indexed by the sets P and T. The submatrix of m corresponding to rows in $\pi \subseteq P$ and columns in $\tau \subseteq T$ is denoted by $m[\pi, \tau]$. Similarly for vectors. (Braces are omitted in singletons in this context.) The usual multiplication of scalars, vectors and/or

A net $\mathcal{N}'$ is *subnet* of $\mathcal{N}$ ($\mathcal{N}' \subseteq \mathcal{N}$) iff $P' \subseteq P$, $T' \subseteq T$ and W' is the restriction of W to P' and T'. Subnets are *generated by* subsets of nodes of both kinds. A subnet generated by a subset V of nodes of a single kind is assumed to be that generated by $V \cup {}^\bullet V \cup V^\bullet$. Subnets generated by a subset of places (transitions) are called *P-(T-)subnets.*

Place/Transition system and related concepts. A function $M: P \rightarrow I\!N$ is called *marking*, and can be represented by a vector. A P/T *system* is a pair $(\mathcal{N}, M_0)$ where $\mathcal{N}$ is a P/T net and M_0 is the *initial* marking. A transition t is *enabled* at M iff $M \geq Pre[P,t]$. Being enabled, t may *occur* (or *fire*) yielding a new marking $M' = M + C[P,t]$, and this is denoted by $M \xrightarrow{t} M'$. An *occurrence sequence* from M is a sequence $\sigma = t_1 \cdots t_k \cdots \in T^\omega$ such that $M \xrightarrow{t_1} M_1 \cdots M_{k-1} \xrightarrow{t_k} \cdots$. If the firing of sequence σ yields the marking M', this is denoted by $M \xrightarrow{\sigma} M'$; denoting by $\vec{\sigma}$ the *firing count vector* of σ, clearly $M' = M + C \cdot \vec{\sigma}$. The set of all the occurrence sequences from M_0, the *language*, is denoted by $\mathrm{L}(\mathcal{N}, M_0)$, and the set of all the markings reachable from M_0, the *reachability set*, is denoted by $\mathrm{R}(\mathcal{N}, M_0)$.

The state equation and the semiflows. Let $(\mathcal{N}, M_0)$ be a P/T system with incidence matrix C. The *state equation* is a linear algebraic equation that gives a *necessary* condition for a marking to be reachable: a vector $M \in I\!N^{|P|}$ such that $\exists \vec{\sigma} \in I\!N^{|T|} : M = M_0 + C \cdot \vec{\sigma}$ is said to be *potentially reachable*, denoted by $M \in \mathrm{PR}(\mathcal{N}, M_0) \supseteq \mathrm{R}(\mathcal{N}, M_0)$, a inclusion that may well be proper.

Flows (*semiflows*) are integer (natural) annullers of C. Right and left annullers are called T- and P-(semi)flows, respectively. We call a semiflow *minimal* when its support[2] is not a proper superset of the support of any other, and the greatest common divisor of its elements is one. Flows are important because they induce certain invariant relations which are useful for reasoning on the behaviour. Actually, several structural (str.) properties are defined in terms of the existence of certain annullers, or similar vectors:

$$\mathcal{N} \text{ is consistent (str. repetitive)} \Leftrightarrow \exists X > 0 \text{ such that } C \cdot X = (\geq)\, 0$$
$$\mathcal{N} \text{ is conservative (str. bounded)} \Leftrightarrow \exists Y > 0 \text{ such that } Y \cdot C = (\leq)\, 0 \;.$$

3 Conflicts and Net Subclasses

We consider a conflict as the situation where two transitions are simultaneously enabled, but the firing of one of them may modify the enabling of the other:

Definition 1 *Let $(\mathcal{N}, M_0)$ be a P/T system. Two transitions $t, t' \in T$ are in* Conflict Relation *at marking M iff there exist $k, k' \in I\!N$ such that $M \geq k \cdot Pre[P,t]$ and $M \geq k' \cdot Pre[P,t']$, but $M \not\geq k \cdot Pre[P,t] + k' \cdot Pre[P,t']$. This notion is extended to sets "pairwisely". (It is silly but correct saying that an enabled transition is in conflict with itself.)*

matrices a and b is denoted by $a \cdot b$. The (componentwise) comparisons of a and b are denoted by $a \geq b$ and $a > b$, while $a \gneq b$ denotes $a \geq b$ but $a \neq b$, not to be confused with $a \not\geq b$, meaning that $a \geq b$ is false.

[2]The set $\|v\|$ of the non-zero components of vector v.

We shall define some subclasses of net systems according to the generality of their possible conflicts. Also we want our definitions to be syntactical, that is, that the membership problem is reduced to examining the net without requiring any exploration of the behaviour. This is why we define some relations between transitions of the net, which are related to conflicts:

Definition 2 *Let $\mathcal{N}$ be a* P/T *net.*

1. *Two transitions, $t, t' \in T$, are in* Choice (or Structural Conflict) Relation *iff $t = t'$ or ${}^\bullet t \cap {}^\bullet t' \neq \emptyset$. This relation is not transitive.*

2. *Two transitions, $t, t' \in T$, are in* Coupled Conflict Relation *iff there exist $t_0, \ldots, t_k \in T$ such that $t = t_0$, $t' = t_k$ and, for $1 \leq i \leq k$, t_{i-1} and t_i are in Choice Relation. It is an equivalence relation on the set of transitions, and each equivalence class is a* Coupled Conflict Set. *We denote by $\mathcal{C}$ the set of Coupled Conflict Sets, and by $\overline{t}$ the equivalence class of t. This notation is extended to sets: $\overline{T} = \bigcup_{t \in T} \overline{t}$.*

3. *Two transitions, $t, t' \in T$, are in* Equal Conflict Relation *iff $t = t'$ or $Pre[P, t] = Pre[P, t'] \neq 0$. This is also an equivalence relation on the set of transitions, and each equivalence class is an* Equal Conflict Set. *We denote by $\mathcal{E}$ the set of Equal Conflict Sets.*

Choices (places with more than one output transition, these transitions being consequently in Choice Relation) are the "topological construct" making possible the existence of conflicts, because (it is obvious that) two transitions must be in Choice Relation if they are in Conflict Relation at some marking, although the converse is not true. The Coupled Conflict Relation is the transitive closure of the Choice Relation. An Equal Conflict Set is such that whenever any transition belonging to it is enabled, then all of them are. Note that whenever two transitions are in Equal Conflict Relation they are also in Choice Relation, and whenever two transitions are in Choice Relation they are also in Coupled Conflict Relation, but not the other way round.

A *T-allocation* is a function, α, that selects transitions not in Choice Relation out of the transitions in each Coupled Conflict Set, c (see [1]). In the case of an Equal Conflict Set, whenever one of its transitions is enabled all of them are, so $\alpha(c)$ is a single transition and the allocation can be interpreted as a "control function" that statically selects a transition to be fired when a conflict set is enabled:

Definition 3 *Let $\mathcal{E}$ be the set of Equal Conflict Sets of a* P/T *net, $\mathcal{N}$. A mapping $\alpha\colon \mathcal{E} \to T$ that assigns to each $e \in \mathcal{E}$ one of its transitions $t \in e$ is an* EC-allocation *over $\mathcal{N}$. The notation is extended to sets: $\alpha(\gamma)$ denotes $\bigcup_{e \in \gamma} \alpha(e)$.*

If no matter which control function is applied the net is capable of infinite activity, we call it *allocatable*:

Definition 4 *Let $\mathcal{N}$ be a* P/T *net. The net $\mathcal{N}$ is* EC-allocatable *iff for every* EC-*allocation over $\mathcal{N}$ the T-subnet generated by the allocated nodes has at least a T-semiflow.*

For the purpose of this work, let us recall here the definition of a couple of syntactical subclasses of P/T nets, namely nets with no (structural) conflicts and nets where all the conflicts are equal. These subclasses, which are studied in some detail in [9, 1], generalise the ordinary Marked Graphs and Extended Free Choice, respectively:

Definition 5 *Let $\mathcal{N}$ be a* P/T *net.*

1. *$\mathcal{N}$ is* Choice-free *(*CF*) iff* $\forall p \in P : |p^\bullet| \leq 1$.
2. *$\mathcal{N}$ is* Equal Conflict *(*EC*) iff* $^\bullet t \cap {}^\bullet t' \neq \emptyset \Rightarrow Pre[P,t] = Pre[P,t']$.

4 Well-formedness: General Necessary or Sufficient Conditions

A P/T system is *bounded* when every place is bounded, i.e., its token content is less than some bound at every reachable marking. It is *live* when every transition is live, i.e., it can ultimately occur from every reachable marking. Boundedness is necessary whenever the system is to be implemented, while liveness is often required, specially in reactive systems. A net $\mathcal{N}$ is *str. bounded* when $(\mathcal{N}, M_0)$ is bounded for *every* M_0, and it is *str. live* when *there exists* an M_0 such that $(\mathcal{N}, M_0)$ is live. Consequently, if a net $\mathcal{N}$ is str. bounded and str. live there exists some marking M_0 such that $(\mathcal{N}, M_0)$ is bounded and live (B&L). In such case, non B&L is exclusively imputable to the marking, and we say that the net is *well-formed.* Notice that, in general, well-formedness is *not* necessary for B&L, although it happens to be in some selected subclasses, as EC systems [1].

A well-known polynomial time necessary condition for well-formedness is str. boundedness and str. repetitiveness, or conservativeness and consistency (Cv&Ct). This condition has been improved by adding an upper bound for the rank of the incidence matrix:

Theorem 6 (General Nec. Cond. for Well-formedness [1])
Let $\mathcal{N}$ be a P/T *net.*
If $\mathcal{N}$ is well-formed, then it is Cv&Ct *and* $\text{rank}(C) < |\mathcal{E}|$.

In the EC case, where $\mathcal{E} = \mathcal{C}$, this is not only necessary but also sufficient:

Theorem 7 (EC Rank Theorem [1])
Let $\mathcal{N}$ be an EC *net.*
Then, $\mathcal{N}$ is well-formed iff it is Cv&Ct *and* $\text{rank}(C) = |\mathcal{C}| - 1$.

Based on the result for EC nets, next we prove a general sufficient condition for well-formedness, with a similar statement to the general necessary condition. It is closely related to a result for ordinary nets [2]. Thanks to allowing weights, our result has broader applicability and has a simple proof.

Theorem 8 (General Suff. Cond. for Well-formedness)
Let $\mathcal{N}$ be a P/T *net.*
If $\mathcal{N}$ is Cv&Ct *and* $\text{rank}(C) = |\mathcal{C}| - 1$, *then it is well-formed.*

Proof. For every $\tau \in \mathcal{C}$ and $p \in {}^\bullet\tau$, define $\mu(p,\tau) = \max\{W(p,t) \mid t \in \tau\}$. We build a new net, $\mathcal{N}' = (P,T,W')$, "equalling" the conflicts as follows: for every $t \in T$ and $p \in P$, $W'(p,t) = \mu(p,\bar{t})$ and $W'(t,p) = \mu(p,\bar{t}) - W(p,t) + W(t,p) \geq 0$.

Clearly $\mathcal{N}$ and $\mathcal{N}'$ have the same incidence matrix (i.e., $C' = C$) and the same set of Coupled Conflict Sets (i.e., $\mathcal{C}' = \mathcal{C}$), so $\mathcal{N}'$ is Cv&Ct and $\text{rank}(C') = |\mathcal{C}'| - 1$. Since $\mathcal{N}'$ is EC, it is well-formed (Th. 7). Therefore, there exists M_0 such that $(\mathcal{N}', M_0)$ is live. We prove next that $(\mathcal{N}, M_0)$ is also live. Take $M \in \text{R}(\mathcal{N}, M_0)$ and $t \in T$. We shall show that there exists $\tilde{M} \in \text{R}(\mathcal{N}, M)$ such that t is enabled at $\tilde{M}$. Being reachable, $M \in \text{PR}(\mathcal{N}, M_0)$, what is equivalent to $M \in \text{PR}(\mathcal{N}', M_0)$ because the incidence matrices are the same. Since $\mathcal{N}'$ is EC, $(\mathcal{N}', M)$ is live [9], so there exists $\tilde{M} \in \text{R}(\mathcal{N}', M)$ enabling t. Let $\sigma \in \text{L}(\mathcal{N}', M)$ be the sequence leading to $\tilde{M}$. Obviously, since preconditions are less restrictive in $\mathcal{N}$ than in $\mathcal{N}'$, $\sigma \in \text{L}(\mathcal{N}, M)$ too, and it leads to $\tilde{M}$, that enables t. ◇

Therefore, Cv&Ct nets with $\text{rank}(C)$ greater than $|\mathcal{E}| - 1$ are not well-formed, while nets with rank equal to $|\mathcal{C}| - 1$ are well-formed. There are no Cv&Ct nets with $\text{rank}(C) < |\mathcal{C}| - 1$ (because there are no Cv&Ct EC nets with $\text{rank}(C) < |\mathcal{E}| - 1$ [1]). But, unfortunately, there are many nets whose rank lies between $|\mathcal{C}|$ and $|\mathcal{E}| - 1$ for which we still cannot decide on well-formedness.

5 Well-formedness of Deterministic Systems of Sequential Processes

Definition 9 *A* P/T *system* $((P_1 \cup \ldots \cup P_q \cup B, T_1 \cup \ldots \cup T_q, W), M_0)$ *is a* Deterministic System of Sequential Processes *(*DSSP*) if the following conditions hold:*

1. $P_i \cap B = \emptyset$, *and for every* $i \neq j$: $P_i \cap P_j = \emptyset$ *and* $T_i \cap T_j = \emptyset$.
2. *If* W_i *and* M_{0i} *are the restrictions of* W *and* M_0 *to* P_i *and* T_i, *then* $(\mathcal{N}_i, M_{0i})$ *is a strongly connected State Machine marked with one token.*
3. $\forall b \in B$:

   (a) $b^\bullet \cap T_i \neq \emptyset \Rightarrow \forall j \neq i : b^\bullet \cap T_j = \emptyset$. *If* $b^\bullet \neq \emptyset$, $I(b) = i : b^\bullet \cap T_i \neq \emptyset$.
   (b) $\forall p \in P_1 \cup \ldots \cup P_q : t, t' \in p^\bullet \Rightarrow W(b,t) = W(b,t')$.

The two first items state that a DSSP is formed by a set of *Sequential Processes* (SP), $(\mathcal{N}_i, M_{0i})$, and a set of buffers, B. (In the DSSP of Fig. 1, each SP is identified by the superscript $i \in \{1,2,3\}$ at every node. The buffers have been shaded.) The last item in the definition restricts the buffers to be output-private, i.e., a buffer cannot have output transitions in two distinct SP, and prevents that a buffer affects the resolution of conflicts in an SP.

The class of DSSP is not comparable to that of EC systems, i.e., there are DSSP that are not EC and viceversa, but it somehow enjoys the "EC property": all the conflicts that may occur are equal, although there are structural conflicts that are not, namely those formed by output transitions of the same buffer which are output of distinct places of their SP. Consider, for instance, transitions t_1^2 and t_2^2 in the net of Fig. 1; although they share the input place b_1

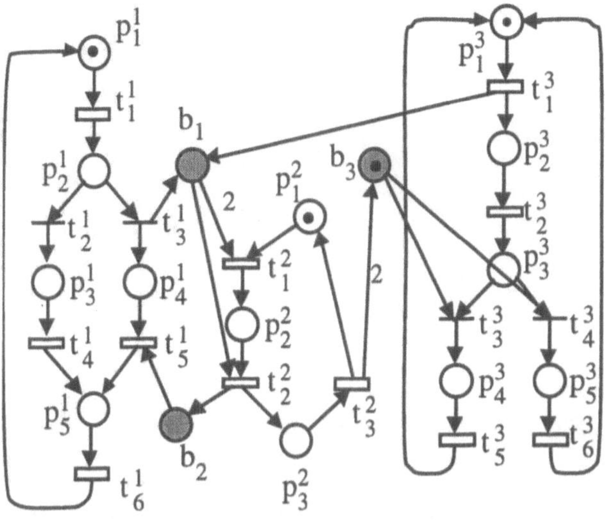

Figure 1: A B&L Deterministic System of Sequential Processes.

(a buffer) they can never be in conflict, because they belong to the same SP, so at most one of them can be enabled at a time. Otherwise stated, the only effect of buffers on the SP are possible *delays* due to a lack of tokens, but they can never affect private conflict resolutions, which are always free. This property supports the intuition that several results that are valid for EC systems may have a DSSP counterpart.

5.1 The Coarse Net of a DSSP

In order to concentrate on the *interconnection level* of the net, we want to obviate the details regarding the inner structure of the SP, while keeping relevant information concerning the effect on the buffers.

A first approach could be substituting a (coarse) transition for each SP, keeping the interconnection with buffers. For example, consider the DSSP depicted in Fig. 2 left. The SP $(\mathcal{N}_1, M_{01})$ is connected to $(\mathcal{N}_2, M_{02})$ through b_2, while $(\mathcal{N}_2, M_{02})$ is connected to $(\mathcal{N}_1, M_{01})$ through b_1. Therefore, the net of Fig. 2 right-top could be a coarse model of the system, in the sense that it reflects these interconnections. This is the *coarse structure* proposed in [3]. We claim, though, that it is not enough for analytical purposes. It is clear that the original DSSP is not well-formed: for every initial marking, if we keep on firing t_2^1 instead of t_1^1 when the conflict occurs, a deadlock is ultimately reached; or, if we fire always t_1^1, the marking of buffers can grow arbitrarily. Nevertheless, the coarse structure does not reflect this problem at all; apparently, tokens circulate through the buffers smoothly, as it would happen if we repeatedly fired in $(\mathcal{N}_1, M_{01})$ the sequence $t_1^1\, t_3^1\, t_2^1\, t_4^1$. So to say, we have obviated too much information about $(\mathcal{N}_1, M_{01})$, namely that it has a conflict depending on whose solution a different "run" of the SP is performed.

To keep this information, we give symbolic relative rates to conflicting tran-

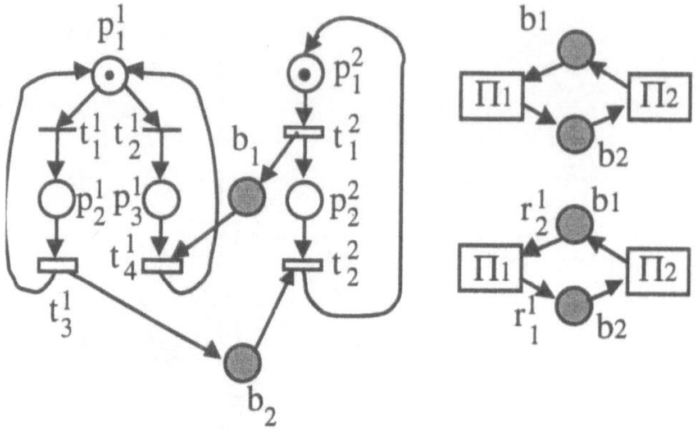

Figure 2: An ill-structured DSSP and its coarse net.

sitions (e.g., r_1^1, r_2^1 in the example net), which are assumed to be *normalised* (e.g., $r_1^1 + r_2^1 = 1$), and take them into account when computing the coarse model. The idea is to replace each SP by a transition whose effect on the buffers is the same as that of the original SP in the long term when respecting the given rates. Regarding the name for this net, in [7] it was called the *parametrically weighted control net*, in the sense that the buffer interconnection somehow "controls" the Sequential Processes. Now we would rather keep the "coarseness" notion, highlighting that the average token flow on the buffers is somehow preserved, and not only the interconnection structure. We call our coarse model *coarse net*, instead of *coarse structure*. The coarse model is not a net in a strict sense, because arc weights may not be natural numbers. Nevertheless, abusing notation, we still call it *net*, and even we speak of strong connectedness, conservativeness, or consistency: we would say, for instance, that the "net" in Fig. 3, which is the coarse net of the DSSP of Fig. 1, is strongly connected, conservative ($Y \cdot C = 0$ for $Y = [1\ 1\ 1]$), and consistent ($C \cdot X = 0$ for $X = [1\ r_3^1\ 2r_3^1]$).

The outline of the algorithm to compute such coarse net is as follows:

1. Combine conflicting transitions: for every (private) conflict of an SP, replace the conflicting transitions by their linear combination, using the rates as coefficients. The combined transition produces the same effect on the adjacent places as the original ones in the long term.

2. Perform the appropriate positive linear combinations of columns to annihilate the entries corresponding to places of each SP. For instance, use the entries in the minimal T-semiflow of each SP as coefficients (since after the previous step, there are no choices in the SP, there is just one minimal T-semiflow per SP). When completed, the resulting submatrix corresponding to buffers is the incidence matrix of the coarse net.

Let us illustrate this by means of the example in Fig. 2, whose incidence matrix is written below. First, we replace the columns corresponding to transitions t_1^1 and t_2^1 by the sum of them multiplied by their rates (see the second

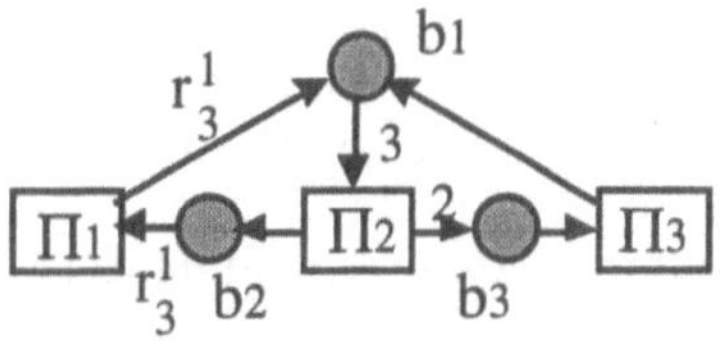

Figure 3: The coarse net of the DSSP of Fig. 1.

matrix). Then we perform the positive linear combinations required to annihilate the entries corresponding to places of the SP: the addition of column t^1_{12}, r^1_1 times t^1_3, and r^1_2 times t^1_4 leads to Π^1, and the addition of t^2_1 and t^2_2 leads to Π^2 (see the third matrix). The submatrix corresponding to the buffer rows is the incidence matrix of the coarse net depicted right-bottom in Fig. 2.

$$
\begin{array}{c}
\\ p^1_1 \\ p^1_2 \\ p^1_3 \\ p^2_1 \\ p^2_2 \\ b_1 \\ b_2
\end{array}
\begin{array}{c}
\begin{array}{cccccc} t^1_1 & t^1_2 & t^1_3 & t^1_4 & t^2_1 & t^2_2 \end{array} \\
\left(\begin{array}{cccc|cc}
-1 & -1 & 1 & 1 & 0 & 0 \\
1 & 0 & -1 & 0 & 0 & 0 \\
0 & 1 & 0 & -1 & 0 & 0 \\ \hline
0 & 0 & 0 & 0 & -1 & 1 \\
0 & 0 & 0 & 0 & 1 & -1 \\ \hline
0 & 0 & 0 & -1 & 1 & 0 \\
0 & 0 & 1 & 0 & 0 & -1
\end{array}\right)
\end{array}
\rightsquigarrow
\begin{array}{c}
\begin{array}{ccccc} t^1_{12} & t^1_3 & t^1_4 & t^2_1 & t^2_2 \end{array} \\
\left(\begin{array}{ccc|cc}
-1 & 1 & 1 & 0 & 0 \\
r^1_1 & -1 & 0 & 0 & 0 \\
r^1_2 & 0 & -1 & 0 & 0 \\ \hline
0 & 0 & 0 & -1 & 1 \\
0 & 0 & 0 & 1 & -1 \\ \hline
0 & 0 & -1 & 1 & 0 \\
0 & 1 & 0 & 0 & -1
\end{array}\right)
\end{array}
\rightsquigarrow
\begin{array}{c}
\begin{array}{cc} \Pi^1 & \Pi^2 \end{array} \\
\left(\begin{array}{cc}
0 & 0 \\
0 & 0 \\
0 & 0 \\
0 & 0 \\
0 & 0 \\ \hline
-r^1_2 & 1 \\
r^1_1 & -1
\end{array}\right)
\end{array}
\qquad (1)
$$

Observe that the resulting coarse net, which is CF, is not well-formed *for all* rates, what signals the interconnection problem we have in the original DSSP: if $r^1_1 < r^1_2$ it is not str. repetitive, while if $r^1_1 > r^1_2$ it is not str. bounded. Similarly, the reader may check that the coarse net of the DSSP of Fig. 1, depicted in Fig. 3, is a well-formed CF net *for all* (positive) rates.

A last comment about the computation of the coarse net: When there are buffers being both input and output of the same SP, the corresponding weights are subtracted (this is the typical problem of representing a non-pure net by its incidence matrix). In order to overcome this inconvenience, we just need to keep track separately of both the *Pre* and *Post* arcs between a buffer and a coarse transition instead of just subtracting their weights. (The example of Fig. 4 illustrates this matter: Applying the original procedure we would compute the coarse model in the middle; If we prevent cancellation of *Pre* and *Post* weights we obtain the right coarse net, which is the appropriate one for most purposes.)

5.2 Necessary Condition for Well-formedness

We start with a property about the coarse net of Cv&Ct DSSP:

Proposition 10 ([7]) *Let $\mathcal{N}$ be the net of a* DSSP. *If $\mathcal{N}$ is* Cv&Ct, *then:*

1. *The coarse net of $\mathcal{N}$ is strongly connected, conservative and* CF.
2. $\text{rank}(C) \geq |\mathcal{E}| - 1$.

Proof. Part 1 follows rather immediately from the way the coarse net was defined. Strong connectedness comes from strong connectedness of $\mathcal{N}$ (it is Cv&Ct). For conservativeness, consider the last matrix we obtained when computing the coarse net, as in (1). Its submatrix corresponding to places of the SP is zero, while its submatrix corresponding to the buffers is precisely the incidence matrix of the coarse net, C_C on the sequel. Since the columns of this matrix were obtained by linear combinations of the columns of C:

$$Y \cdot C = 0 \Rightarrow Y \cdot \left(\frac{0}{C_C}\right) = 0 \Rightarrow Y[B] \cdot C_C = 0 \ .$$

For Part 2, write the incidence matrix of a DSSP as follows:

$$\begin{array}{c} \quad T_1 \; T_2 \cdots T_q \\ C = \left(\begin{array}{c|c|c|c} C_1 & 0 & \cdots & 0 \\ \hline 0 & C_2 & \cdots & 0 \\ \hline \vdots & \vdots & \ddots & \vdots \\ \hline 0 & 0 & \cdots & C_q \\ \hline B_1 & B_2 & \cdots & B_q \end{array}\right) \begin{array}{c} P_1 \\ P_2 \\ \vdots \\ P_q \\ B \end{array} = \left(\frac{\{C_i\}}{C_B}\right) \end{array} \tag{2}$$

where C_i represents the incidence matrix of $\mathcal{N}_i$, so $\text{rank}(C_i) = |P_i| - 1 = |\mathcal{E}_i| - 1$ because it is a strongly connected State Machine. It is clear that:

$$\text{rank}(C) = \text{rank}\left(\begin{array}{c|c} \{C_i\} & 0 \\ \hline C_B & C_C \end{array}\right) \geq \text{rank}(\{C_i\}) + \text{rank}(C_C) \ . \tag{3}$$

Due to the restrictions to buffers, the Equal Conflict Sets of the whole net are exactly those of the SP, so:

$$\text{rank}(\{C_i\}) = \sum_{i=1}^{q} \text{rank}(C_i) = \sum_{i=1}^{q} (|\mathcal{E}_i| - 1) = |\mathcal{E}| - q \ . \tag{4}$$

Since C_C is the incidence matrix of a strongly connected and conservative CF net [9]:

$$\text{rank}(C_C) \geq q - 1 \ , \tag{5}$$

because q is the number of transitions of the coarse net.

Using (4) and (5) in (3), we get the announced result. ◊

The DSSP of Fig. 4 illustrates that the coarse net (the right one) of a Cv&Ct DSSP need not be Cv&Ct: it is a strongly connected and conservative CF net, but it is not consistent. In this case, $\text{rank}(C) = |\mathcal{E}|$. In fact, with this information we can decide ill-formedness of that net:

Corollary 11 (DSSP Nec. Cond. for Well-formedness [7])
Let $\mathcal{N}$ be the net of a DSSP. *If $\mathcal{N}$ is well-formed, then:*

1. *$\mathcal{N}$ is* Cv&Ct *and* $\text{rank}(C) = |\mathcal{E}| - 1$.
2. *The coarse net is a* Cv&Ct CF *net.*

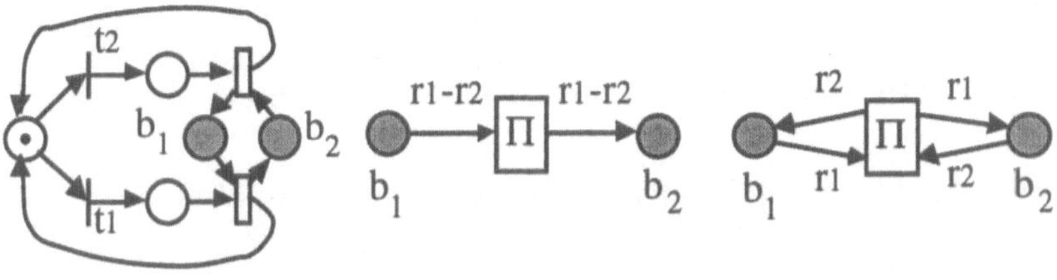

Figure 4: An ill-formed DSSP and its coarse net.

Proof. Part 1 is the particularisation of Th. 6 taking Prop. 10.2 into account.

For Part 2, $\text{rank}(C) = |\mathcal{E}| - 1$ implies $\text{rank}(C_C) = q - 1$. For strongly connected and conservative CF nets, the equality of the rank to the number of transitions minus one is equivalent to consistency [9]. $\diamond$

5.3 Sufficient Condition for Well-formedness

Before proving the characterisation, we point out that the general sufficient condition given by Th. 8 is not suitable for DSSP: In general, for a DSSP $\mathcal{N}$, unless it is EC, we have $|\mathcal{C}| < |\mathcal{E}|$. According to Prop. 10.2, $\text{rank}(C) \geq |\mathcal{E}| - 1$, so it is impossible that $\text{rank}(C) = |\mathcal{C}| - 1$, as required by Th. 8.

For our proof, we require some lemmata. The first two of them concern State Machines and their T-semiflows. They state that we can select an arbitrary minimal T-semiflow by an appropriate EC-allocation, and that the only way to fire a transition twice in a monomarked State Machine is by completing first the firing of a minimal T-semiflow.

Lemma 12 *Let $\mathcal{N}$ be a strongly connected State Machine.*

1. *Let $\mathcal{N}' = (P', T', W')$ be the T-subnet generated by the image of an* EC-*allocation over $\mathcal{N}$. If $\mathcal{N}'$ is connected, then it has a unique minimal T-semiflow, X, and every T-flow of $\mathcal{N}'$ is proportional to it.*

2. *If X is a minimal T-semiflow of $\mathcal{N}$, there exists an* EC-*allocation over $\mathcal{N}$, α, such that $\|X\| \subseteq \alpha(\mathcal{C})$ and whose image generates a connected T-subnet.*

Proof. For Part 1, $\mathcal{N}'$ is a (CF) State Machine, with $P' = P$ and $|T'| = |\mathcal{E}|$ by the definition of allocations. Being a connected State Machine, $\text{rank}(C') = |P'| - 1 = |P| - 1 = |\mathcal{E}| - 1 = |T'| - 1$. Then, there exists $X \neq 0$ such that $C' \cdot X = 0$. Without loss of generality, the flow X can be decomposed as $X = X' - X''$, where $X' \gneq 0$, $X'' \geq 0$, and $\|X'\| \cap \|X''\| = \emptyset$. We claim that X' is indeed a T-semiflow of $\mathcal{N}'$. Since $C' \cdot X = 0$, it follows that $C' \cdot X' = C' \cdot X''$. We prove first that $C' \cdot X' \geq 0$. Assume the contrary, i.e., there exists $p \in P$ such that $C'[p, T'] \cdot X' = C'[p, T'] \cdot X'' < 0$. Since $\mathcal{N}'$ is CF, there exists a unique $t \in p^\bullet$, so there is just one negative entry in row $C'[p, T']$, so it must be $t \in \|X'\| \cap \|X''\|$, contradiction. As every State Machine is conservative, we cannot have $C' \cdot X' \gneq 0$, so $C' \cdot X' = 0$. Finally, since $\text{rank}(C') = |T'| - 1$ the space of right annullers is one-dimensional, so every T-flow is proportional to X.

For Part 2, the sought allocation can be obtained applying the algorithm in Theorem 18 of [1] with $T'_0 = \|X\|$, which does not contain conflicting transitions because it is the support of a minimal T-semiflow of a State Machine. $\diamond$

Lemma 13 *Let $(\mathcal{N}, M_0)$ be a State Machine marked with one token, and let $\{X_1, \ldots, X_r\}$ be the set of minimal T-semiflows. If $\sigma \in \mathrm{L}(\mathcal{N}, M_0)$ and for every $1 \leq i \leq r : \vec{\sigma} \ngeq X_i$, then $\vec{\sigma} \leq \mathbb{1}$.*

Proof. Assume the contrary: there is $t \in T$ such that $\sigma = \sigma_1 t \sigma_2 t$ is a prefix of σ and $\vec{\sigma}_1[t] = \vec{\sigma}_2[t] = 0$. Then, $M_0 \xrightarrow{\sigma_1} M_1 \xrightarrow{t\sigma_2} M_2$, and $M_1 = M_2$ because both enable t. Therefore $\overrightarrow{t\sigma_2}$ is a T-semiflow, contradiction. $\diamond$

The following lemma states that we can build a T-semiflow of the whole DSSP out of one minimal T-semiflow from each SP:

Lemma 14 *Let $\mathcal{N}$ be the net of a* Cv&Ct DSSP *with* $\mathrm{rank}(C) < |\mathcal{E}|$.

1. *Let α be an* EC*-allocation over $\mathcal{N}$ and $\mathcal{N}'$ the T-subnet generated by the image of α, $\alpha(\mathcal{C})$. If for every* SP *$\mathcal{N}_i$, $\alpha(\mathcal{C}) \downarrow_{T_i}$ generates a connected T-subnet, then $\mathcal{N}'$ has at least a T-semiflow.*

2. *For every* SP *$\mathcal{N}_i$, $1 \leq i \leq q$, let $X_{i,J(i)}$ be one of its minimal T-semiflows. If we define $\widehat{X_{i,J(i)}}[T_i] = X_{i,J(i)}$ and $\widehat{X_{i,J(i)}}[T - T_i] = 0$, then $X = \sum_{i=1}^{q} a_i \cdot \widehat{X_{i,J(i)}}$ is a T-semiflow of $\mathcal{N}$ for some $a_i \in \mathbb{N}$.*

Proof. For Part 1, as $\mathrm{rank}(C') \leq \mathrm{rank}(C) < |\mathcal{E}| = |T'|$, there exists a T-flow of $\mathcal{N}'$, $X \neq 0$. Without loss of generality, the flow X can be decomposed as $X = X' - X''$, where $X' \gneq 0$, $X'' \geq 0$, and $\|X'\| \cap \|X''\| = \emptyset$. We claim that X' is indeed a T-semiflow of $\mathcal{N}'$. Since $C' \cdot X = 0$, it follows that $C' \cdot X' = C' \cdot X''$. We prove first that $C' \cdot X' \geq 0$. Asume contrary: exists $p \in P' = P = (\bigcup_{i=1}^{q} P_i) \cup B$ such that $C'[p, T'] \cdot X' = C'[p, T'] \cdot X'' < 0$.

- If $p \in P_i$, since $\mathcal{N}'_i$ is CF there is only one negative entry in row $C'[p, T']$, so there exists $t \in \|X'\| \cap \|X''\|$, contradiction;

- If $p \in B$, then let $\mathcal{N}_i$, whith $i = I(p)$, be the SP that takes tokens from p. As $C' \cdot X = 0$, then $C'_i \cdot X[T_i] = 0$. Applying Lemma 12.1, $\mathcal{N}'_i$ has a unique minimal T-semiflow and every T-flow is proportional to it, in particular $X[T_i]$. Then, either $X[T_i] \geq 0$ or $X[T_i] \leq 0$, so $\|X[T_i]\| \subseteq \|X'\|$ or $\|X[T_i]\| \subseteq \|X''\|$, hence it is impossible $C'[p, T'] \cdot X' = C'[p, T'] \cdot X'' < 0$.

Then, $C' \cdot X' \geq 0$ and, since $\mathcal{N}'$ is conservative (because $\mathcal{N}$ is), $C' \cdot X' = 0$.

For Part 2, by Lemma 12, for every SP $\mathcal{N}_i$, given a minimal T-semiflow $X_{i,J(i)}$, there is an EC-allocation such that the T-subnet it generates, $\mathcal{N}'_i$, is connected and $X_{i,J(i)}$ is the unique minimal T-semiflow. Merging all these EC-allocations, we get an EC-allocation over $\mathcal{N}$. This allocation fulfils the conditions of Part 1, so there exists X', T-semiflow of $\mathcal{N}'$ (the T-subnet generated by the image of the allocation). As $C' \cdot X' = 0$, then $C'_i \cdot X'[T_i] = 0$. Since $X_{i,J(i)}$ is the only minimal T-semiflow of $\mathcal{N}'_i$, then $X'[T_i] = a_i \cdot X_{i,J(i)}$ for some $a_i \in \mathbb{N}$. The sought X is $X[T'] = X'$ and $X[T - T'] = 0$. $\diamond$

We are ready now for the sufficient condition for well-formedness of a DSSP. In fact, it is guaranteed that there are markings *respecting the* DSSP *definition* (i.e., one token on each SP) for which the system is B&L:

Proposition 15 (DSSP Suff. Cond. for Well-formedness)
Let $\mathcal{N}$ *be the net of a* DSSP. *If* $\mathcal{N}$ *is* Cv&Ct *and* $\mathrm{rank}(C) = |\mathcal{E}| - 1$, *then there exists* M_0 *such that* $(\mathcal{N}, M_0)$ *is a* B&L DSSP.

Proof. We define an M_0 and later we prove that $(\mathcal{N}, M_0)$ is a live DSSP.

Let $\{X_1, \ldots, X_r\}$ be the set of minimal T-semiflows of $\mathcal{N}$. For every $1 \leq l \leq r$, $X_l[T_i]$ is a T-semiflow of $\mathcal{N}_i$, so $X_l[T_i] = \sum_{j=1}^{r_i} \lambda_{i,j}^l \cdot X_{i,j}$, where $\{X_{i,1}, \ldots X_{i,r_i}\}$ is the set of minimal T-semiflows of $\mathcal{N}_i$ and $\lambda_{i,j}^l \in \mathbb{N}$. Let $\mu_{i,j} = \max_{1 \leq l \leq r} \{\lambda_{i,j}^l\}$. Define the number of tokens taken from b when $X_{I(b),j}$ is fired:

$$k(b,j) = Pre[b, T_{I(b)}] \cdot X_{I(b),j} \geq -C[b, T_{I(b)}] \cdot X_{I(b),j} \ . \tag{6}$$

For every $b \in B$, let $M_0[b] = \sum_{j=1}^{r_{I(b)}} (\mu_{I(b),j} + 1) \cdot k(b,j)$, that is, enough tokens to enable at once any minimal T-semiflow involving transitions of $I(b)$, and let M_0 put one token on each SP.

Assume $(\mathcal{N}, M_0)$ is not live. Then it can reach a deadlock marking M_d firing some sequence σ (Th. 5 in [7]). We can write $\vec{\sigma} = \sum_{j=1}^{r} \lambda_j \cdot X_j + X_R$, for some $\lambda_j \in \mathbb{N}$ and $X_R \ngeq X_j$ for every $1 \leq j \leq r$. Then, for every $b \in B$:

$$M_d[b] - M_0[b] = C[b,T] \cdot \vec{\sigma} = C[b,T] \cdot X_R \geq C[b, T_{I(b)}] \cdot X_R[T_{I(b)}] \ , \tag{7}$$

because if $t \notin T_{I(b)}$ then $W(b,t) = 0$, so $C[b,t] \geq 0$. Fix an SP, $(\mathcal{N}_i, M_{0i})$. Clearly:

$$X_R[T_i] = \sum_{j=1}^{r_i} \lambda_{i,j} \cdot X_{i,j} + X_{R_i} \ , \tag{8}$$

for some $\lambda_{i,j} \in \mathbb{N}$ and $X_{R_i} \ngeq X_{i,j}$ for every $1 \leq j \leq r_i$. Since $\mathcal{N}_i$ is consistent, $\sum_{j=1}^{r_i} X_{i,j} \geq \mathbb{1}$. On the other hand, by the way X_{R_i} has been defined, since $(\mathcal{N}_i, M_{0i})$ is an SP, it is guaranteed that there exists a sequence in $\mathrm{L}(\mathcal{N}_i, M_{0i})$ having X_{R_i} as firing count vector. Then, by Lemma 13, $X_{R_i} \leq \mathbb{1}$. Therefore:

$$X_{R_i} \leq \sum_{j=1}^{r_i} X_{i,j} \ . \tag{9}$$

Let $b \in B \cap {}^\bullet T_i$.

$$\begin{aligned}
M_d[b] &\overset{(7)}{\geq} M_0[b] + C[b,T_i] \cdot X_R[T_i] \\
&\overset{(8)}{=} M_0[b] + \sum_{j=1}^{r_i} \lambda_{i,j} \cdot C[b,T_i] \cdot X_{i,j} + C[b,T_i] \cdot X_{R_i} \\
&\overset{(6)}{\geq} M_0[b] - \sum_{j=1}^{r_i} \lambda_{i,j} \cdot k(b,j) - Pre[b,T_i] \cdot X_{R_i} \\
&\overset{(9)}{\geq} M_0[b] - \sum_{j=1}^{r_i} \lambda_{i,j} \cdot k(b,j) - \sum_{j=1}^{r_i} Pre[b,T_i] \cdot X_{i,j} \\
&\overset{(6)}{=} M_0[b] - \sum_{j=1}^{r_i} \lambda_{i,j} \cdot k(b,j) - \sum_{j=1}^{r_i} k(b,j) \\
&= \sum_{j=1}^{r_i} (\mu_{i,j} - \lambda_{i,j}) \cdot k(b,j),
\end{aligned}$$

by our definition of M_0.

If $\mu_{i,j} > \lambda_{i,j}$ for every $1 \leq j \leq r_i$, then $M_d[b] \geq \sum_{j=1}^{r_i} k(b,j)$ for every $b \in B \cap {}^\bullet T_i$, so the buffers would have enough tokens to fire every T-semiflow in $\mathcal{N}_i$. In such case M_d could not be a deadlock, against the hypothesis. Therefore, there exists $1 \leq J(i) \leq r_i$, with $\mu_{i,J(i)} \leq \lambda_{i,J(i)}$. We can repeat this for every SP. By Lemma 14.2, there exists $X = \sum_{i=1}^{q} a_i \cdot \widehat{X_{i,J(i)}}$ T-semiflow of $\mathcal{N}$, which can be assumed to be minimal. By the way $\mu_{i,j}$ was defined, $a_i \leq \mu_{i,J(i)}$, so $\lambda_{i,J(i)} \geq a_i$. Let $X_{R_i}^{T_i}[T_i] = X_R[T_i]$ and $X_{R_i}^{T_i}[T - T_i] = 0$. Since $X_{R_i}^{T_i} \geq \lambda_{i,J(i)} \cdot X_{i,J(i)}$, then:

$$X_R = \sum_{i=1}^{q} X_{R_i}^{T_i} \geq \sum_{i=1}^{q} \lambda_{i,J(i)} \cdot \widehat{X_{i,J(i)}} \geq \sum_{i=1}^{q} a_i \cdot \widehat{X_{i,J(i)}} = X \ ,$$

contradiction. ◊

Linking together Prop. 10 and 15 we get the announced polynomial-time characterisation of well-formedness for the class:

Theorem 16 (DSSP Rank Theorem) *Let $\mathcal{N}$ be the net of a* DSSP.
Then, $\mathcal{N}$ is well-formed iff it is Cv&Ct *and* $\text{rank}(C) = |\mathcal{E}| - 1$.

Corollary 17 *Well-formedness of the net of a* DSSP *can be decided in polynomial time on the size of the net.*

6 Conclusions

Well-formedness is an important structural property of Place/Transition net systems for which no efficient characterisation is available for general nets, although there are both a necessary (Th. 6, see also [1]) and a sufficient condition

(Th. 8) with polynomial-time complexity. It is possible, in general, that a net fulfils the necessary but not the sufficient condition. Nevertheless, for some subclasses we have a complete characterisation. One example are Equal Conflict nets, where both conditions coincide (Th. 7, see also [1]). Another are Deterministic Systems of Sequential Processes, for which the general sufficient condition does not apply in general, but the necessary condition happens to be also sufficient. Therefore we have for this class a polynomial-time characterisation of well-formedness (Th. 16). It improves, for structurally bounded nets, the previously known characterisations — valid for strict subclasses of the one studied here — which had exponential complexity [3, 4, 5].

References

[1] E. Teruel and M. Silva. Well-formedness of Equal Conflict systems. In R. Valette, editor, *Application and Theory of Petri Nets 1994*, volume 815 of *Lecture Notes in Computer Science*, pages 491–510. Springer Verlag, 1994.

[2] J. Desel. Regular marked Petri nets. In J. Leeuwen, editor, *WG' 93: 19th Int. Workshop on Graph-Theoretic Concepts in Computer Science*, volume 790 of *Lecture Notes in Computer Science*, pages 264–275. Springer Verlag, 1993.

[3] W. Reisig. Deterministic buffer synchronization of sequential processes. *Acta Informatica*, 18:117–134, 1982.

[4] G. Berthelot, G. Memmi, and W. Reisig. A control structure for sequential processes synchronized by buffers. In *Proc. 4th European Workshop on Application and Theory of Petri Nets*, pages 43–58, 1983.

[5] Y. Souissi and N. Beldiceanu. Deterministic systems of sequential processes: Theory and tools. In *Concurrency 88*, volume 335 of *Lecture Notes in Computer Science*, pages 380–400. Springer Verlag, 1988.

[6] Y. Souissi. Deterministic systems of sequential processes: A class of structured Petri nets. In G. Rozenberg, editor, *Advances in Petri Nets 1993*, volume 674 of *Lecture Notes in Computer Science*, pages 406–426. Springer Verlag, 1993.

[7] J. Campos, J. M. Colom, M. Silva, and E. Teruel. Functional and performance analysis of cooperating sequential processes. In O. J. Boxma and G. M. Koole, editors, *Performance Evaluation of Parallel and Distributed Systems. Solution Methods*, volume 106 of *CWI Tracts*, pages 233–251. CWI, 1994.

[8] T. Murata. Petri nets: Properties, analysis and applications. *Proceedings of the IEEE*, 77(4):541–580, 1989.

[9] E. Teruel. *Structure Theory of Weighted Place/Transition Net Systems: The Equal Conflict Hiatus*. PhD thesis, DIEI. Univ. Zaragoza, June 1994.

Acknowledgement

We thank the five anonymous referees for their comments and suggestions.

An Event-Based SOS for a Language with Refinement*

Arend Rensink
Institut für Informatik, University of Hildesheim
Hildesheim, Germany

Abstract

The notion of *action refinement* has been studied intensively in the past few years. It is usually introduced in the form of an operator in a process algebraic language, for which a denotational semantics in a suitable model is then given. In this paper we complement this approach by defining a corresponding *operational* semantics for refinement, in the form of derivation rules for a transition relation. Because of the (well-known) fact that ordinary transition systems are not expressive enough to capture the effects of refinement, we use an *event-based* transition system model described elsewhere in the literature. The operational semantics of refinement thus defined is equivalent (in fact event isomorphic) to the usual denotational semantics.

1 Introduction

Process algebras form a well-known specification paradigm for concurrent systems. Typical operators describe such constructions as parallel composition, and such implementation mechanisms as sequential composition. One operator that has been studied in depth in the past six years is *action refinement*, which basically has the effect of substituting actions in a given behavioural specification with more complex behaviour that in some sense implements those actions. This operator can be seen on the one hand as allowing a *top-down design strategy* in which activities can first be specified on a very abstract level as single actions and then turned into more concrete, detailed behaviour; and on the other as corresponding to the implementation mechanism of *procedure call* in declarative languages.

Action refinement has been studied mostly on the basis of constructions on denotational models. Except for a small subclass of refinements (see Czaja et al. [9]), it turns out that the standard model of *labelled transition systems* does not allow a satisfactory definition: several intuitive properties of refinement such as distribution over parallel composition (without synchronisation) cannot be satisfied. Successful denotational definitions have however been given on several types of partial order models: cf. Van Glabbeek and Goltz [14], Vogler [26], Jategaonkar and Meyer [17], Best et al. [4], Darondeau and Degano [10].

---

*Based on [22, Chapter 3]. This work was initiated while the author was staying at the University of Twente, and is partially supported by the Esprit BRA Project 3096 SPEC (Formal Methods and Tools for the Development of Distributed and Real Time Systems) and the Esprit Working Group 6067 CALIBAN (Causal Calculi Based on Nets)

It should be remarked at this point that there is no general agreement on the question whether refinement should distribute over *synchronisation* as well (in addition to independent parallel composition). If such distribution is allowed, refinement is interpreted wholly syntactically; this is arguably more faithful to the notion of top-down design, but we know of no models that are compositional with respect to such an operation. The denotational constructions mentioned above, on the other hand, interpret refinement as *semantic substitution* (in an appropriate model); it does not in general distribute over synchronisation. Goltz, Gorrieri and Rensink have investigated in [15] when these two approaches coincide. In this paper, we adhere to the latter.

The failure of transition systems to model action refinement immediately implies that the standard use of *structural operational semantics*, where only actions are used as labels, will not work for action refinement. However, several transition system extensions with corresponding operational semantics are known whose expressivity equals that of partial order models, for instance Degano et al. [11], Degano and Gorrieri [12], Boudol and Castellani [6]. On the basis of such extensions it should be possible to give an operational semantics of refinement. One such definition is indeed given in [12].

In this paper we, too, define an operational semantics for refinement, this time on the basis of an approach developed by Langerak [18]: transition labels are extended with *event names*, derived from *annotations* that are added to terms of the language before evaluation. This makes for a very smooth extension of the standard semantics, at the cost of the auxiliary machinery for annotation. Refinement can be captured by three operational rules, respectively for the case that the refined action is not yet terminated after a given transition, that it is terminated, and for termination of the term under refinement as a whole. We use the auxiliary concepts of *independent transitions* and *busy refinements*, the latter being exactly those action occurrences whose refinement has started but not yet terminated. We claim that the ensuing definition is intuitive and easy to use; we derive some axioms for refinement. As proof of its correctness we compare the semantics with a construction on an event-based model called *families of posets*, developed by us in [21].

We proceed as follows: Section 2 presents the language under consideration, its standard semantics and its extended semantics using Langerak's event annotations for the fragment without refinement. Section 3 discusses and defines the semantics of the refinement operator. Section 4 presents the corresponding denotational construction and sketches the correctness proof. In Section 5 we discuss related work and draw some conclusions.

2 Language and semantics

We consider a language $\mathbf{L}$ generated by the following grammar:

$$B ::= \delta \mid \varepsilon \mid a \mid B + B \mid B;B \mid B \parallel_A B \mid B[r] \mid X \ .$$

Here $a \in \mathbf{A}$ is an arbitrary action. We use a special pseudo-action $\checkmark \notin \mathbf{A}$ to denote termination; $\mathbf{A}_\checkmark := \mathbf{A} \cup \{\checkmark\}$ is ranged over by μ. $B_1 + B_2$ denotes choice between B_1 and B_2, with neutral element δ denoting deadlock. $B_1; B_2$ is sequential composition, with neutral element ε denoting termination; it is assumed to bind stronger than $+$. We also use $\mathbf{L}^{-\varepsilon}$ to denote the fragment

	$\vdash$	$a \xrightarrow{a} \varepsilon$
	$\vdash$	$\varepsilon \xrightarrow{\checkmark} \delta$
$B_1 \xrightarrow{\mu} B'$	$\vdash$	$B_1 + B_2 \xrightarrow{\mu} B'$
$B_2 \xrightarrow{\mu} B'$	$\vdash$	$B_1 + B_2 \xrightarrow{\mu} B'$
$B_1 \xrightarrow{a} B_1'$	$\vdash$	$B_1; B_2 \xrightarrow{a} B_1'; B_2$
$B_1 \xrightarrow{\checkmark} B_1' \quad B_2 \xrightarrow{\mu} B_2'$	$\vdash$	$B_1; B_2 \xrightarrow{\mu} B_2'$
$B_1 \xrightarrow{a} B_1' \quad a \notin A$	$\vdash$	$B_1 \parallel_A B_2 \xrightarrow{a} B_1' \parallel_A B_2$
$B_2 \xrightarrow{a} B_2' \quad a \notin A$	$\vdash$	$B_1 \parallel_A B_2 \xrightarrow{a} B_1 \parallel_A B_2'$
$B_1 \xrightarrow{\mu} B_1' \quad B_2 \xrightarrow{\mu} B_2' \quad \mu \in A \cup \{\checkmark\}$	$\vdash$	$B_1 \parallel_A B_2 \xrightarrow{\mu} B_1' \parallel_A B_2'$
$B_1 \xrightarrow{\mu} B_1' \quad \phi(\mu) \xrightarrow{\lambda} B_2'$	$\vdash$	$B_1[\phi] \xrightarrow{\lambda} B_1'[\phi]$
$\theta(X) \xrightarrow{\mu} B$	$\vdash$	$X \xrightarrow{\mu} B$

Table 1: Standard operational semantics for the flat fragment of $\mathbf{L}$

of $\mathbf{L}$ without ε. $B_1 \parallel_A B_2$ is synchronisation of B_1 and B_2 over the actions in A, where $B_1 ||| B_2$ denotes the special case that $A = \varnothing$. $B[r]$ is refinement of B according to the refinement function $r: \mathbf{A} \rightarrow \mathbf{L}^{-\varepsilon}$ (hence ε may not be used in refinement terms; the reason for this restriction will be discussed later). Refinement functions are implicitly extended to $r: \checkmark \mapsto \varepsilon$ (this therefore being the only r-image in which ε occurs). Finally, $X \in \mathbf{X}$ is a process name. The meaning of process names is determined by a *process environment* $\theta: \mathbf{X} \rightarrow \mathbf{L}$ which generates a recursive system of equations.

Except for refinement, all the operators in $\mathbf{L}$ stem from well-known process algebras. Refinement is also known in the restricted case where all the images of r are simple actions: then it is alternatively called a *renaming* operator and denoted ϕ. We call a term *flat* if all instances of refinements are renamings. For the flat fragment of $\mathbf{L}$, the standard operational semantics is given in Table 1.[1] (Note that the rule for renaming is formulated in a nonstandard way; this is in order to make the generalisation to refinement more direct.)

To obtain a partial order operational semantics, the transition labels of the standard semantics can be enriched with additional information, essentially to encode the *causal dependencies* among the transitions. The question is in what form this additional information should be provided. Here we apply an approach developed by Langerak [18]. The remainder of this section basically describes the approach, extended only to model sequential composition. Our own contribution, described in the next sections, is the definition of operational rules for refinement, and the choice of denotational model.

First, we replace the action labels by *pairs of events and actions*. In other words, a transition in our extended semantics will be of the form $B \xrightarrow{e,\mu} B'$, where $e \in \mathbf{E}$ is some arbitrary *event*. The universe of events $\mathbf{E}$ is assumed

[1]Not included are operators for hiding and restriction of actions. The former has been omitted because we are working in a setting where all actions are visible. The latter is implicitly present in the synchronisation operator.

	$\vdash$	${}_ea \xrightarrow{e,a} {}_{(*,e)}\varepsilon$
	$\vdash$	${}_e\varepsilon \xrightarrow{e,\surd} \delta$
$C_1 \xrightarrow{e,\mu} C'$	$\vdash$	$C_1 + C_2 \xrightarrow{e,\mu} C'$
$C_2 \xrightarrow{e,\mu} C'$	$\vdash$	$C_1 + C_2 \xrightarrow{e,\mu} C'$
$C_1 \xrightarrow{e_1,a} C_1'$	$\vdash$	$C_1; C_2 \xrightarrow{e_1,a} C_1'; C_2$
$C_1 \xrightarrow{e_1,\surd} C_1' \quad C_2 \xrightarrow{e_2,\mu} C_2'$	$\vdash$	$C_1; C_2 \xrightarrow{e_2,\mu} C_2'$
$C_1 \xrightarrow{e_1,a} C_1' \quad a \notin A$	$\vdash$	$C_1 \parallel_A C_2 \xrightarrow{(e_1,*),a} C_1' \parallel_A C_2$
$C_2 \xrightarrow{e_2,a} C_2' \quad a \notin A$	$\vdash$	$C_1 \parallel_A C_2 \xrightarrow{(*,e_2),a} C_1 \parallel_A C_2'$
$C_1 \xrightarrow{e_1,\mu} C_1' \quad C_2 \xrightarrow{e_2,\mu} C_2' \quad \mu \in A \cup \{\surd\}$	$\vdash$	$C_1 \parallel_A C_2 \xrightarrow{(e_1,e_2),\mu} C_1' \parallel_A C_2'$
$C_1 \xrightarrow{e,\mu} C_1' \quad \phi(\mu) \xrightarrow{d,\lambda} C_2'$	$\vdash$	$C[\phi] \xrightarrow{(e,d),\lambda} C'[\phi]$
$\theta(X) \xrightarrow{e,\mu} C$	$\vdash$	${}_dX \xrightarrow{(d,e),\mu} k_d(C)$
$C \xrightarrow{e,\mu} C'$	$\vdash$	$k_d(C) \xrightarrow{(d,e),\mu} k_d(C')$

Table 2: Event-based operational semantics for the flat part of **L**

to be closed under pairing, in order to allow the construction of new events: $(\mathbf{E} \times \mathbf{E}) \cup (\mathbf{E} \times \{*\}) \cup (\{*\} \times \mathbf{E}) \subset \mathbf{E}$, where $* \notin \mathbf{E}$ is a special symbol. These events are generated by *annotating* the terms of **L** according to some scheme, such that all the actions and process names are augmented with a distinguished event. In other words, we do not evaluate **L** directly but rather the annotated language **L(E)** with the following grammar:

$$C ::= \delta \mid {}_e\varepsilon \mid {}_ea \mid C + C \mid C;C \mid C \parallel_A C \mid C[r] \mid {}_eX \mid k_e(C) \ .$$

Refinement functions r as well as the process environment θ now range over **L(E)** rather than **L**. For the operational characterisation of recursive behaviour we need auxiliary operators $k_e(C)$ where $k_e = \lambda d.\,(e,d)$ is a function $\mathbf{E} \rightarrow \mathbf{E}$ for all $e \in \mathbf{E}$, which glues e to all the event transitions in the execution of C, making the events distinct even in infinite behaviour. To ensure that event names do not occur more than once in annotated terms, we restrict ourselves to those terms where the annotation is *sound*, in the sense that different elements of the term are annotated differently.

It is relatively straightforward to write down an intuitively reasonable operational semantics in the Langerak format for the flat part of **L(E)**: see Table 2. Note that in this setup, the images of renaming functions, being a special case of refinement functions, have to be annotated as well.

A sound annotation of a given term $B \in \mathbf{L}$ is easy to construct. For instance, for any "seed" event $e \in \mathbf{E}$, the function $ann_e\colon \mathbf{L} \rightarrow \mathbf{L(E)}$ defined in Table 3 will correctly annotate B. The function $strip\colon \mathbf{L(E)} \rightarrow \mathbf{L}$ removes annotations; it should be clear that $strip(ann_e(B)) = B$ for all $e \in \mathbf{E}$. Both *ann* and *strip* are pointwise extended to refinement functions.

In practice, rather than apply ann_e we will simply enumerate all actions, εs and process names in a given term, hence for instance obtaining $({}_0a + {}_1\varepsilon); {}_2b$ from $(a + \varepsilon); b$, rather than the much more complex $({}_{((e,*),*)}a + {}_{(*,(e,*))}\varepsilon); {}_{(*,e)}b$

$$\begin{array}{rcl}
ann_e(\delta) & := & \delta \\
ann_e(B) & := & {}_eB \;\; (B \in \{\varepsilon\} \cup \mathbf{A} \cup \mathbf{X}) \\
ann_e(B_1 \diamond B_2) & := & ann_{(e,*)}(B_1) \diamond ann_{(*,e)}(B_2) \;\; (\diamond \in \{+\,,;\,,\|_A\}) \\
ann_e(B[r]) & := & ann_e(B)[ann_e(r)] \\
\hline
strip(\delta) & := & \delta \\
strip({}_eB) & := & B \;\; (B \in \{\varepsilon\} \cup \mathbf{A} \cup \mathbf{X}) \\
strip(C_1 \diamond C_2) & := & strip(C_1) \diamond strip(C_2) \;\; (\diamond \in \{+\,,;\,,\|_A\}) \\
strip(C[r]) & := & strip(C)[strip(r)] \\
strip(k_d(C)) & := & strip(C)
\end{array}$$

Table 3: An example annotation function

returned by ann_e.[2]

The question the becomes how we intend to interpret the operational semantics, and in particular the event labels. It is obvious that different annotations of a given term will result in different transition systems; this difference does not have anything to do with the actual behaviour. To obtain a sensible level of abstraction, we therefore interpret the semantics up to a bijective event renaming.

1 Definition. Two annotated terms $C_1, C_2 \in \mathbf{L}(\mathbf{E})$ are called *event isomorphic*, denoted $C_1 \cong_o C_2$, if there exist a bijection $f: E_1 \to E_2$ and a relation $\rho \subseteq \mathbf{L}(\mathbf{E}) \times \mathbf{L}(\mathbf{E})$ such that $C_1 \,\rho\, C_2$ and for all $C_1' \,\rho\, C_2'$

- If $C_1' \xrightarrow{e,a} C_1''$ then $e \in E_1$ and $\exists C_2''.\, C_2' \xrightarrow{f(e),a} C_2'' \,\rho^{-1}\, C_1''$;
- If $C_2' \xrightarrow{e,a} C_2''$ then $e \in E_2$ and $\exists C_1''.\, C_1' \xrightarrow{f^{-1}(e),a} C_1'' \,\rho\, C_2''$.

We call the above bisimulation-like relation an *isomorphism* because for any annotated term C, the outgoing event transitions are deterministic:

$$C \xrightarrow{e,a} C' \wedge C \xrightarrow{e,a} C'' \Longrightarrow C' = C'' \ .$$

This follows from the distinctness of the events in annotated terms. It follows that the *event traces* of C, defined as those strings $\sigma \in (\mathbf{E} \times \mathbf{A})^*$ such that $C = C_0 \xrightarrow{\sigma_0}\xrightarrow{\sigma_1} \cdots$, completely characterise the operational behaviour, and if we denote the set of these traces by $ET(C)$ then

$$f: C_1 \cong_o C_2 \iff ET(C_2) = \{\, f^*(\sigma) \mid \sigma \in ET(C_1) \,\}$$

where f^* denotes the pointwise extension of f to event traces. More importantly however, we have

$$strip(C_1) = strip(C_2) \implies C_1 \cong_o C_2 \ . \qquad (1)$$

[2] The need to annotate terms before interpreting them makes the semantics non-compositional. A compositional variant is obtained if annotations are generated dynamically by the operational rules, like in the proved transitions of Boudol and Castellani [5]. This does not affect the tenets of this paper; we prefer Langerak's presentation for simplicity.

This implies that we have indeed abstracted from the particular annotation mechanism. Since also for all $B \in \mathbf{L}$ there is a $C \in \mathbf{L}(\mathbf{E})$ such that $strip(C) = B$ (for instance, $C = ann_e(B)$), it follows that the following extension of $\cong_o$ to $\mathbf{L}$ is well-defined:

$$strip(C_1) \cong_o strip(C_2) :\Leftrightarrow C_1 \cong_o C_2 \ .$$

Event isomorphism is a congruence over $\mathbf{L}(\mathbf{E})$ and through this definition also over $\mathbf{L}$; moreover it subsumes commutativity and associativity of choice, associativity of sequential composition and commutativity of synchronisation; also ε is a neutral element with respect to sequential composition and δ with respect to choice. On the other hand, for instance $a;c+b;c \not\cong_o (a+b);c$ and $a;b+b;a \not\cong_o a \,|||\, b$. Especially the former shows that $\cong_o$ is still a very strong notion: it negates the right-distributivity of sequential composition over choice common to almost all equivalences known from the literature. For our purpose this is in fact beneficial, since we will be using $\cong_o$ to show correspondence of the above operational semantics to an event-based denotational semantics; the correspondence will remain valid under any more abstract interpretation than $\cong_o$, hence the stronger this relation is the better.

A very important question is whether the semantics is in some sense "correct". One immediate observation is that by stripping the terms in Table 2 and removing the event labels from the transitions, we get back Table 1 exactly. In other words, for all annotated terms C_1, C_2 we have

$$strip(C_1) \xrightarrow{a} B_2 \Leftrightarrow (\exists e \in \mathbf{E}, C_2 \in \mathbf{L}(\mathbf{E}).\, C_1 \xrightarrow{e,a} C_2 \wedge strip(C_2) = B_2 \quad (2)$$

This shows that we have directly extended the standard semantics. In fact it is not difficult to prove that $\cong_o$ implies isomorphism of the standard semantics derived according to Table 1. A second, more important criterion is the existence of a *denotational* semantics to go with Table 2. We return to this issue in Section 4.

3 Refinement

Let us discuss the problems involved in extending our operational semantics to refinement. As mentioned in the introduction, we take the traditional view, put forward initially by Aceto and Hennessy [2] and Van Glabbeek and Goltz [13], in which refinement equates to *substitution* of abstract actions by the concrete behaviour to which they are mapped. For instance, we expect to obtain

$$B_1[r] = (a;b+(a \,|||\, c))[a \leadsto a_1;a_2] \cong_o a_1;a_2;b+(a_1;a_2 \,|||\, c) \quad (3)$$

where $a \leadsto a_1;a_2$ denotes the refinement function mapping a to $a_1;a_2$ and all other actions to themselves. In the presence of synchronisation however, it turns out that straightforward syntactic substitution sometimes gives unexpected results; here we follow Van Glabbeek and Goltz in moving to *semantic* substitution in some sufficiently expressive denotational model (see Section 4). It is this kind of substitution, then, that we wish to capture in operational rules. We aim at rules of the following approximate form:

$$B_1 \xrightarrow{a} B_1' \quad r(a) \xrightarrow{b} B_2' \quad [\cdots] \quad \vdash \quad B_1[r] \xrightarrow{b} B_1''[r'] \quad (4)$$

(for the moment disregarding annotations). The important part is choosing representations for B_1'' and r'.

3.1 Basic intuitions

We take $B_1[r]$ in (3) as an example. $B_1[r]$ may do a_1, thereby resolving the choice between the two a's and moving to either $a_2; b$ or $a_2 \,|||\, c$. To denote these behaviours as terms of the form $B_1''[r']$, we need to encode which occurrence of a is involved, and the intermediate state of its refinement, which is now at a_2. Fortunately, a pointer to the individual action occurrences is already available in the form of the *annotation*. We use these to *extend the refinement functions* with "busy" refinements. Hence r' in (4) will extend r by mapping the relevant occurrence of a to a_2 rather than $a_1; a_2$.

The need for a compositional semantics forces us to make a further choice: either the action occurrence involved in the refinement should be removed from the term B_1 under refinement or it should be left standing. (Note that in particular, we cannot (syntactically) replace the actiona occurrence within B_1 by the remainder B_2', since this would not be compositional.) More precisely, B_1'' in (4) equals either B_1' or B_1. In (3), if we were to set $B_1'' = B_1'$ this would yield $B_1[a \rightsquigarrow a_1; a_2] \xrightarrow{a_1} b[r']$, after which it is no longer visible that action b depends on the action a_2 still outstanding. Instead we will use $B_1'' = B_1$, and hence $B_1[a \rightsquigarrow a_1; a_2] \xrightarrow{a_1} (a; b + (a \,|||\, c))[r']$. But this causes its own problems, since now it is not clear that the choice in the right hand term between $a; b$ and $a \,|||\, c$ has been resolved. Fortunately, here we can use the information in r' regarding the "busy refinements"; this tells us which occurrence of a is being refined, hence we can impose a side condition on the rule (4) to ensure that transitions may not be in conflict with busy refinements.

This in turn raises the question how to detect such conflicts. The answer once more lies in the annotations: if a given term may do two transitions with different annotations, those transitions are *independent* if they do not rule out each other, i.e. if each of them may still occur after the other; otherwise they are in *conflict*. For instance, if we take an annotated version of B_1 in (3) then

$$
\begin{array}{ll}
{}_0a; {}_1b + ({}_2a \,|||\, {}_3c) & \xrightarrow{(2,*),a} \xrightarrow{(*,3),c} \\
{}_0a; {}_1b + ({}_2a \,|||\, {}_3c) & \xrightarrow{(*,3),c} \xrightarrow{(2,*),a}
\end{array}
$$

whereas on the other hand ${}_0a; {}_1b + ({}_2a \,|||\, {}_3c) \xrightarrow{0,a} \not\xrightarrow{(*,3),c}$. In other words, events $(2, *)$ and $(*, 3)$ are independent whereas 0 and $(*, 3)$ are in conflict. This corresponds to the intuition that if the a_1-transition of $B_1[r]$ is due to ${}_0a$ then c may not occur anymore, whereas c is still allowed if a_1 is due to ${}_2a$.

3.2 Auxiliary concepts

It follows that in order to give operational rules for refinement we need two auxiliary concepts: "busy" refinements and independence (or conflict) between annotations. We will now formalise these concepts.

Intuitively, the outgoing transitions of a given term are independent if they arise out of different and non-synchronising parallel components of that term, and conflicting otherwise. One technique to decide this might be to investigate the internal structure of the events, imparted by the rules for synchronisation in Table 2. This however would be against the notion of abstraction we adhere to, according to which events may be renamed in arbitrary fashion, and hence their internal structure cannot hold information. Instead we use the local structure of the transition system to define independence: for all $C \in \mathbf{L}(\mathbf{E})$ and $E \subseteq \mathbf{E}$

$$C \xrightarrow{E} \;:\Leftrightarrow\; (\forall d, e \in E.\, d = e \vee \exists C', C''.\, C \xrightarrow{d} C' \xrightarrow{e} C'') \tag{5}$$

If $C \xrightarrow{E}$ we say that *the events of E are independent in C*. For instance, in our running example we have ${}_0a; {}_1b + ({}_2a \parallel\!\mid {}_3c) \xrightarrow{\{(2,*),(*,3)\}}$ but ${}_0a; {}_1b + ({}_2a \parallel\!\mid {}_3c) \not\xrightarrow{\{0,(*,3)\}}$. This definition uses the principle that in event-based models, independence corresponds to *confluent diamonds* in the state space. Note that in (5) we do not quite test for confluency, since we do not require the end states to coincide. Conceivably one could have $C \xrightarrow{d}\xrightarrow{e} C'$ and $C \xrightarrow{e}\xrightarrow{d} C''$ where $C' \neq C''$. In Section 4, however, we show that such a situation cannot occur in models of $\mathbf{L}$, and that $C \xrightarrow{E}$ indeed signifies independence of the events in E.

To account for intermediate, "busy" refinements, for the purpose of the operational semantics we extend the domain of refinement functions to subsets of $\mathbf{E}$. A refinement function r will henceforth be a function from $\mathbf{A} \cup E$ to $\mathbf{L}(\mathbf{E})$, where $E \subseteq_{\mathrm{fin}} \mathbf{E}$ is the finite set of *busy events* of r, which we will usually denote $busy(r)$. (Infinite sets of busy events cannot come into existence.) Furthermore we will use the following constructions on such extended refinement functions:

$$r \setminus d \;:=\; r \restriction (\mathbf{A} \cup (busy(r) \setminus \{d\})) \tag{6}$$

$$r_d^B \;:=\; (r \setminus d) \cup \{(d, B)\}\ . \tag{7}$$

In (6) an event is removed from the set of busy events, effectively throwing away the corresponding residual (presumably because it is terminated); in (7) the image of a busy event is changed, or possibly added if the event was not busy before. Note that $r \setminus d = r$ if $d \notin busy(r)$. It follows that $busy(r \setminus d) = busy(r) \setminus \{d\}$ and $busy(r_d^B) = busy(r) \cup \{d\}$. Now for convenience we also let such extended r range over annotated actions:

$$r = \lambda(e, a).\ \text{if } e \in busy(r) \text{ then } r(e) \text{ else } r(a).$$

In terms of this auxiliary notation, the constructions (6) and (7) can be characterised as follows:

$$r \setminus d \;:=\; \lambda(e, a).\ \text{if } d = e \text{ then } r(a) \text{ else } r(e, a)$$

$$r_d^B \;:=\; \lambda(e, a).\ \text{if } d = e \text{ then } B \text{ else } r(e, a).$$

To complete the running example of (3), we get $r' = r_d^{B_2'}$ where $d = 0$ or $d = (2, *)$ depending on the occurrence of a, and $B_2' = {}_{(*,4)}\varepsilon; {}_5a_2$.

3.3 Operational rules

With the necessary preparations out of the way, the actual statement of the operational rules for refinement becomes straightforward: see Table 4. Note that they reduce to the single rule in Table 2 if r is a renaming function.

The expression $C_1 \xrightarrow{\{d\} \cup busy(r)}$ tests the independence of the event about to be refined with respect to the existing busy events. Note that for all reachable terms, if $d \in busy(r)$ then $C_1 \xrightarrow{\{d\} \cup busy(r)}$ is fulfilled automatically (proved by

$$C_1 \xrightarrow{d,a} C_1' \quad r(d,a) \xrightarrow{e,b} C_2' \xrightarrow{\checkmark} \quad C_1 \xrightarrow{\{d\}\cup busy(r)} \;\vdash\; C_1[r] \xrightarrow{(d,e),b} C_1'[r \setminus d]$$

$$C_1 \xrightarrow{d,a} C_1' \quad r(d,a) \xrightarrow{e,b} C_2' \not\xrightarrow{\checkmark} \quad C_1 \xrightarrow{\{d\}\cup busy(r)} \;\vdash\; C_1[r] \xrightarrow{(d,e),b} C_1[r_d^{C_2'}]$$

$$C_1 \xrightarrow{d,\checkmark} C_1' \quad r(d,\checkmark) \xrightarrow{e,\checkmark} C_2' \quad C_1 \xrightarrow{\{d\}\cup busy(r)} \;\vdash\; C_1[r] \xrightarrow{(d,e),\checkmark} C_1'[r]$$

Table 4: Operational rules for refinement

induction on the length of the derivation); furthermore if $busy(r) = \varnothing$ then the condition is implied by $C_1 \xrightarrow{d,a}$.

Note that the second rule of Table 4 contains a negative premise concerning $\checkmark$-transitions. This could potentially lead to problems of well-definedness; cf. Groote [16]. However, in our system $\checkmark$-transitions can be derived independently from non-$\checkmark$-transitions, and so a stratification in the sense of Groote is immediate. Moreover, due to the fact that refinement functions are restricted to range over $\mathbf{L}^{-\varepsilon}$, if C is derived from an r-image then it cannot be the case that both $C \xrightarrow{\checkmark}$ and $C \xrightarrow{a}$ (where $a \neq \checkmark$).

We can now derive the behaviour for our example term (3). Let $r\colon a \mapsto {}_4a_1; {}_5a_2$ and $r\colon \mu \mapsto {}_6\mu$ for all $\mu \neq a$, and $B = {}_{(*,4)}\varepsilon; {}_5a_2$; then $r(a) \xrightarrow{4,a_1} B \not\xrightarrow{\checkmark}$ and $B \xrightarrow{5,a_2}\xrightarrow{\checkmark}$ and $r(\mu) \xrightarrow{6,\mu}\xrightarrow{\checkmark}$ for all $\mu \neq a$ and hence

$$\begin{array}{lll}
({}_0a; {}_1b + ({}_2a \mathbin{|||} {}_3c))[r] & \xrightarrow{(0,4),a_1} & ({}_0a; {}_1b + ({}_2a \mathbin{|||} {}_3c))[r_0^B] \\
& \xrightarrow{(0,5),a_2} & ({}_{(*,0)}\varepsilon; {}_1b)[r] \\
& \xrightarrow{(1,6),b} & {}_{(*,1)}\varepsilon[r] \\
& & \\
({}_0a; {}_1b + ({}_2a \mathbin{|||} {}_3c))[r] & \xrightarrow{((2,*),4),a_1} & ({}_0a; {}_1b + ({}_2a \mathbin{|||} {}_3c))[r_{(2,*)}^B] \\
& \xrightarrow{((2,*),5),a_2} & ({}_{(*,2)}\varepsilon \mathbin{|||} {}_3c)[r] \\
& \xrightarrow{((*,3),6),c} & ({}_{(*,2)}\varepsilon \mathbin{|||} {}_{(*,3)}\varepsilon)[r] \\
& & \\
({}_0a; {}_1b + ({}_2a \mathbin{|||} {}_3c))[r] & \xrightarrow{((2,*),4),a_1} & ({}_0a; {}_1b + ({}_2a \mathbin{|||} {}_3c))[r_{(2,*)}^B] \\
& \xrightarrow{((*,3),6),c} & ({}_2a \mathbin{|||} {}_{(*,3)}\varepsilon)[r_{(2,*)}^B] \\
& \xrightarrow{((2,*),5),a_2} & ({}_{(*,2)}\varepsilon \mathbin{|||} {}_{(*,3)}\varepsilon)[r] \\
& & \\
({}_0a; {}_1b + ({}_2a \mathbin{|||} {}_3c))[r] & \xrightarrow{((*,3),6),c} & ({}_2a \mathbin{|||} {}_{(*,3)}\varepsilon)[r]
\end{array}$$

It follows that the $\cong_o$-property in (3) indeed holds. In fact it is not hard to prove the axioms in Table 5, where $id\colon \mathbf{A} \to \mathbf{A}$ is the identity function over $\mathbf{A}$ and $r_2 \circ r_1$ denotes *concatenation* of refinement functions, defined by $(r_2 \circ r_1)\colon \mu \mapsto r_1(\mu)[r_2]$.

4 Denotational semantics

The main tool we have to establish the correctness of the above operational rules is to prove consistency with some denotational semantics.[3] Event-based

[3] In some sense this is the wrong order: one would rather expect a denotational semantics to be measured against an existing operational one. For the case of refinement, however,

$B[r] = B \quad (B \in \{\delta, \varepsilon\})$	
$a[r] = r(a)$	
$(B_1 \diamond B_2)[r] = B_1[r] \diamond B_2[r] \quad (\diamond \in \{+, ;, \vert\vert\vert\})$	
$X[r] = \theta(X)[r]$	
$B[id] = B$	
$B[r_1][r_2] = B[r_2 \circ r_1]$	

Table 5: Axioms of refinement

models that can deal appropriately with the flat part of **L** are *event automata* by Pinna and Poigné [20], *bundle event structures* by Langerak [18] and our own *families of lposets* [21, 22]. We will use the latter.

A *labelled poset* (*lposet*, for short) is a triple $p = \langle E_p, \leq_p, \ell_p \rangle$ where $E_p \subseteq \mathbf{E}$ is a finite set of events, $\leq_p \subseteq E_p \times E_p$ is a partial ordering relation over E_p, and $\ell_p : E_p \to \mathbf{A}_\checkmark$ is a labelling function such that $\ell_p(e) = \checkmark$ iff e is the unique maximal event of p. We denote $A_p = \ell_p^*(E_p)$. The class of labelled posets is denoted **LPO**. If $p, q \in \mathbf{LPO}$ then p and q are called *labelling consistent* if $\ell_p \restriction (E_p \cap E_q) = \ell_q \restriction (E_p \cap E_q)$, and p is a *prefix* of q if $E_p \subseteq E_q$ is left-closed according to $\leq_q$. The union of labelling consistent lposets is defined by $\bigcup P = \langle \bigcup E_p, \bigcup \leq_p, \bigcup \ell_p \rangle$. This yields an lposet only under certain conditions, which however will always be fulfilled in this paper.

A family of lposets (*flpo*, for short) is a nonempty, prefix closed set of labelling consistent lposets. The class of flpos is denoted **FPO**. A flpo $\mathcal{P}$ is called *confluent* if for all $p, q \in \mathcal{P}$, $E_p = E_q$ implies $p = q$, and *coherent* if for all $P \subseteq \mathcal{P}$, $\forall p, q \in P.\, p \cup q \in \mathcal{P}$ implies $\bigcup P \in \mathcal{P}$.

To interpret flpos we again use *event isomorphism*: $\mathcal{P}, \mathcal{Q} \in \mathbf{FPO}$ are event isomorphic, denoted $\mathcal{P} \cong_d \mathcal{Q}$, if there exists a bijection $f : E_\mathcal{P} \to E_\mathcal{Q}$ such that $\mathcal{Q} = \{ f^*(p) \mid p \in \mathcal{P} \}$ (where f^* is the pointwise extension of f to lposets). We then also denote $\mathcal{Q} = f^*(\mathcal{P})$. Note that this relation is defined over flpos while Definition 1 concerns event-labelled transition systems. However, a given flpo $\mathcal{P}$ gives rise to an event-labelled transition system $ETS(\mathcal{P}) = \langle \mathcal{P}, p_\varnothing, \to_\mathcal{P} \rangle$, where $\mathcal{P}$ is the set of states, $p_\varnothing \in \mathcal{P}$ is the initial state and $\to_\mathcal{P} \subseteq \mathcal{P} \times \mathbf{E} \times (\mathbf{A} \cup \{\checkmark\}) \times \mathcal{P}$ is a transition relation defined by

$$p \xrightarrow{e,\mu}_\mathcal{P} q \;:\Leftrightarrow\; (p \preceq q) \wedge (E_q \setminus E_p = \{e\}) \wedge (\ell_p(e) = \mu) \ .$$

In addition we use $p\checkmark_\mathcal{P}$ to denote $\exists e.\, p \xrightarrow{e,\checkmark}_\mathcal{P}$; intuitively $p\checkmark_\mathcal{P}$ means that p is a terminated state of $\mathcal{P}$. There is in general a mismatch between event isomorphism of flpos and of transition systems: although $\mathcal{P} \cong_d \mathcal{Q}$ implies $ETS(\mathcal{P}) \cong_o ETS(\mathcal{Q})$, the inverse does *not* hold in general. However, we do have the following:

2 Proposition. If $\mathcal{P}, \mathcal{Q} \in \mathbf{FPO}$ are confluent then $ETS(\mathcal{P}) \cong_o ETS(\mathcal{Q})$ iff $\mathcal{P} \cong_d \mathcal{Q}$.

denotational constructions have been in existence for some time, whereas the operational characterisation is the main contribution of this paper.

In Section 3 we introduced the notation $C \xrightarrow{E}$ to express that from state C, the events in E may be executed concurrently. The following proposition states that a certain subclass of **FPO**, including, as we will see, the denotational models of **L**(**E**), this notation indeed has the required meaning.

3 Proposition. If $\mathcal{P} \in \mathbf{FPO}$ is confluent and coherent then for all $p \in \mathcal{P}$ and $E \subseteq \mathbf{E}$, $p \xrightarrow{E}_{\mathcal{P}}$ iff there exists a $q \in \mathcal{P}$ such that $p \preceq q$, $E_q \setminus E_p = E$ and $E \subseteq \max_{\leq_q} E_q$. In other words, from the state p the events in E can be executed concurrently, resulting in the state q.

We define a denotational semantics $[\![\cdot]\!]: \mathbf{L}(\mathbf{E}) \to \mathbf{FPO}$ such that $strip(C_1) = strip(C_2)$ implies $[\![C_1]\!] \cong_d [\![C_2]\!]$; hence $[\![\cdot]\!]$ and $\cong_d$ can be extended to **L**. The operational and denotational semantics are then proved consistent:

4 Theorem. For all $B_1, B_2 \in \mathbf{L}$, $B_1 \cong_o B_2$ if and only if $[\![B_1]\!] \cong_d [\![B_2]\!]$.

We briefly define the denotational semantics and give a sketch of the proof. First we need a number of additional lposet concepts. If $E_p \cap E_q = \varnothing$ then the *sequential composition* of p and q is defined by $p;q$ where

$$\begin{aligned} E_{p;q} &= (E_p \setminus \ell_p^{-1}(\checkmark)) \cup \{ E_q \mid \checkmark \in A_p \} \\ \leq_{p;q} &= \leq_p \restriction (E_p \cap E_{p;q}) \cup (E_p \cap E_{p;q}) \times (E_q \cap E_{p;q}) \cup \leq_q \restriction (E_q \cap E_{p;q}) \\ \ell_{p;q} &= \ell_p \restriction (E_p \cap E_{p;q}) \cup \ell_q \restriction (E_q \cap E_{p;q}) \ . \end{aligned}$$

To capture *synchronisation* of terms, for lposets p and q and $E \subseteq (E_p \cup \{*\}) \times (E_q \cup \{*\})$ we define the (partial) *product* $p \times_E q$ by

$$\begin{aligned} E_{p \times_E q} &= E \\ \leq_{p \times_E q} &= \{ ((d,e),(d',e')) \mid d \leq_p d' \vee e \leq_p e' \}^* \\ \ell_{p \times_E q} &= \{ ((d,e),\mu) \mid \ell_p(d) = \mu \vee \ell_q(e) = \mu \} \ . \end{aligned}$$

This is only partially defined because the transitive closure of the ordering relation may fail to be antisymmetric. Moreover, we will only use this construct if E *A-synchronises* p and q for some $A \subseteq \mathbf{A}_\checkmark$, which is said to be the case when for all $(d,e) \in E$, either $\ell_p(d) \notin A \cup \{\checkmark\}$ and $e = *$, or $d = *$ and $\ell_q(e) \notin A \cup \{\checkmark\}$, or $\ell_p(d) = \ell_q(e) \in A \cup \{\checkmark\}$. Finally, we define the *refinement* of p by a function $w: E_p \to \mathbf{LPO}$ by $w(p)$ where

$$\begin{aligned} E_{w(p)} &= \{ (e,e') \in E_p \times \mathbf{E} \mid e' \in E_{w(e)} \} \\ \leq_{w(p)} &= \{ ((d,d'),(e,e')) \mid d <_p e \vee (d = e \wedge d' \leq_{w(d)} e') \} \\ \ell_{w(p)} &= \{ ((e,e'),\mu) \mid \mu = \ell_{w(e)}(e') \} \end{aligned}$$

This yields an lposet, provided that w behaves well. If $\mathcal{R}: \mathbf{A}_\checkmark \to \mathbf{FPO}$ is given then $w: E_p \to \mathbf{LPO}$ is a *p-witness* of $\mathcal{R}$ if for all $e \in E_p$, $w(e)\checkmark_{\mathcal{R}(\ell_p(e))}$ if e is non-maximal in p, $w(e) \in \mathcal{R}(\ell_p(e))$ if e is arbitrary, and $\checkmark \in A_{w(e)}$ implies $\ell_p(e) = \checkmark$. Refinement is then extended to flpos by

$$\mathcal{R}(\mathcal{P}) = \{ w(p) \mid p \in \mathcal{P}, w \text{ a } p\text{-witness of } \mathcal{R} \}$$

For the construction of infinite behaviour we use the fact that **FPO** is a complete partial order under $\subseteq$ in which least upper bounds correspond to

$[\![\delta]\!]$	:=	$\{\langle\varnothing,\varnothing,\varnothing\rangle\}$
$[\![_e\varepsilon]\!]$	:=	$\{\langle\{e\},\{(e,e)\},\{(e,\checkmark)\}\rangle\}$
$[\![_e a]\!]$	:=	$\{\langle\{e,d\},\{(e,e),(e,d),(d,d)\},\{(e,a),(d,\checkmark)\}\rangle\}$ $(d=(*,e))$
$[\![C_1+C_2]\!]$	:=	$[\![C_1]\!]\cup[\![C_2]\!]$
$[\![C_1;C_2]\!]$	:=	$\{\,p;q \mid p\in[\![C_1]\!], q\in[\![C_2]\!]\,\}$
$[\![C_1 \parallel_A C_2]\!]$	:=	$\{\,p\times_E q \mid p\in[\![C_1]\!], q\in[\![C_2]\!], E$ A-synchronises $p,q\,\}$
$[\![C[r]]\!]$	:=	$(\lambda a.\,[\![r(a)]\!])[\![C]\!]$
$[\![_e X]\!]$	:=	$k_e^*(\bigcup_{i\in\mathbb{N}}[\![X^i]\!])$
$[\![k_e(C)]\!]$	:=	$k_e^*([\![C]\!])$

Table 6: Denotational semantics of $\mathbf{L}(\mathbf{E})$

unions, and that the operators above are continuous. Hence the usual fixpoint construction, using standard approximants X^i, is applicable. The denotational semantics is now summed up in Table 6. (Note that $k_e^*(\mathcal{P})$ denotes the application of the event isomorphism $k_e = \lambda d.\,(d,e)$ to $\mathcal{P}$.) The following proposition states that this semantics yields confluent and coherent flpos, so that Propositions 2 and 3 are applicable.

5 Proposition. For all $C\in\mathbf{L}(\mathbf{E})$, $[\![C]\!]$ is a confluent and coherent flpo.

Hence we have that for all $C_1,C_2\in\mathbf{L}(\mathbf{E})$, $[\![C_1]\!]\cong_d[\![C_2]\!]$ holds if and only if $ETS[\![C_1]\!]\cong_o ETS[\![C_2]\!]$. To prove Theorem 4 we therefore only have to show $ETS[\![C]\!]\cong_o C$ for all $C\in\mathbf{L}(\mathbf{E})$. The proof is by induction on the term structure. The most interesting is the case of refinement refinement, which is stated below. The complete proof can be found in [22, Chapter 3].

6 Theorem. For all $C\in\mathbf{L}(\mathbf{E})$ and $r\colon\mathbf{A}\to\mathbf{L}(\mathbf{E})$, if $ETS[\![C]\!]\cong_o C$ and $ETS[\![r(a)]\!]\cong_o r(a)$ for all $a\in\mathbf{A}$ then $ETS[\![C[r]]\!]\cong_o C[r]$.

Proof sketch. We have to define a bijection $f\colon E_1\to E_2$ and a relation $\rho\subseteq[\![C[r]]\!]\times\mathbf{L}(\mathbf{E})$ such that $p_\varnothing\ \rho\ C[r]$ and the simulation conditions in Definition 1 are fulfilled. The proof uses the f_C and ρ_C proving $ETS[\![C]\!]\cong_o C$, and for all $a\in\mathbf{A}$ the f_a and ρ_a proving $ETS[\![r(a)]\!]\cong_o r(a)$ (see Definition 1). We denote $\mathcal{R}=\lambda a.\,[\![r(a)]\!]$, $\mathcal{P}=[\![C]\!]$ and $\mathcal{Q}=\mathcal{R}(\mathcal{P})$. It turns out that in general we can assume that f_C and the f_a equal the identity over $\mathbf{E}$; this allows us to define $f=id_{\mathbf{E}}$ as well. Moreover, it turns out that the relations $\rho_C\subseteq[\![C]\!]\times\mathbf{L}(\mathbf{E})$ and $\rho_a\subseteq\mathcal{R}(a)\times\mathbf{L}(\mathbf{E})$ for all $a\in\mathbf{A}$ are injective, which means that we can regard them as functions; we will also construct ρ as a function.

For arbitrary $w(p)\in[\![C[r]]\!]$ we construct the prefix of p where the witness w is already *complete*, i.e. on which all $w(e)$ are terminated. By construction this includes at least all the non-maximal events of p, but possibly some maximal events as well. Intuitively, the events in p that are *not* complete are still "busy". We define

$$busy(p,w) \;:=\; \{\,e\in E_p \mid \neg w(e)\checkmark_{\mathcal{R}(\ell_p(e))}\,\}$$
$$cmpl(p,w) \;:=\; p\restriction(E_p\setminus busy(p,w))\ .$$

It follows that $cmpl(p,w) \preceq p$, and, by Proposition 3, $p \xrightarrow{busy(p,w)}_{\mathcal{P}}$. We now construct an extended refinement function $r_{p,w}: (\mathbf{A} \cup E) \to \mathbf{L}(\mathbf{E})$ where $busy(r) = E = busy(p,w)$: let $r_{p,w} \restriction \mathbf{A} = r$ and for all $e \in busy(p,w)$

$$r_{p,w}: e \mapsto \rho_{\ell_p(e)}(w(e)) \ .$$

We are now ready to define ρ: for all $w(p) \in \mathcal{Q}$,

$$\rho: w(p) \mapsto \rho_C(cmpl(p,w))[r_{p,w}] \ .$$

This is well-defined because, as mentioned above, w and p are uniquely determined for all $w(p) \in \mathcal{Q}$. If $p = p_\varnothing$ then $w = \varnothing$ we get $cmpl(p,w) = p_\varnothing = p$ and $r_{p,w} = r$. It follows that $\rho(p_\varnothing) = \rho_C(p_\varnothing)[r] = C[r]$ as required. Due to the functional nature of ρ, the simulation properties of Definition 1 collapse to

$$w(p) \xrightarrow{(d,e),\mu}_{\mathcal{Q}} w'(p') \ \Rightarrow \ \rho(w(p)) \xrightarrow{(d,e),\mu} \rho(w'(p')) \tag{8}$$

$$\rho(w(p)) \xrightarrow{(d,e),\mu} C'[r'] \ \Rightarrow \ \exists w'(p') \in \rho^{-1}(C'[r']).\, w(p) \xrightarrow{(d,e),\mu}_{\mathcal{Q}} w'(p') \tag{9}$$

The proof of this naturally divides into three cases, one for each of the operational rules for refinement in Table 4: (1) $\mu \in \mathbf{A}$ and $w'(e)\surd_{\mathcal{R}(\mu)}$ resp. $e \notin busy(r')$; (2) $\mu \in \mathbf{A}$ and $\neg w'(e)\surd_{\mathcal{R}(\mu)}$ resp. $e \in busy(r')$; or (3) $\mu = \surd$. We sketch the proof of (8) for the first case. From the assumptions it follows that $cmpl(p,w) \xrightarrow{d,a} cmpl(p',w')$ where $a = \ell_{p'}(d)$, and hence $\rho_C(cmpl(p,w)) \xrightarrow{d,a} \rho_C(cmpl(p',w'))$; moreover $busy(r_{p,w}) \cup \{d\} = E_{cmpl(p',w')} \setminus E_{cmpl(p,w)}$ and hence $\rho_C(cmpl(p,w)) \xrightarrow{\{d\}\cup busy(r_{p,w})}$; finally, $q \xrightarrow{e,\mu}_{\mathcal{R}(a)} w'(d)\surd_{\mathcal{R}(a)}$ where $q = w(d)$ if $d \in busy(p,w)$ and $q = p_\varnothing$ otherwise; hence $r_{p,w}({}_d a) = \rho_a(q) \xrightarrow{e,\mu}\xrightarrow{\surd}$. It follows that according to Table 4, $\rho(w(p)) = \rho_C(cmpl(p,w))[r_{w,p}] \xrightarrow{(d,e),\mu} \rho_C(cmpl(p',w'))[r_{p,w} \setminus d] = \rho_C(cmpl(p',w'))[r_{p',w'}] = \rho(w'(p'))$. This concludes the proof. □

5 Conclusions

On the basis of an approach developed by Langerak in [18], we have defined an event-based operational semantics for a rich process algebraic language **L**, and extended it with an operator for action refinement. The definition has been proved correct modulo *event isomorphism* by comparing it with a denotational semantics based on families of posets. The rules for refinement are based on two auxiliary concepts: *independence of transitions* and *intermediate* or *busy refinements*.

5.1 Related work

Similar studies can be found elsewhere in the literature. Degano et al. [11] and Boudol and Castellani [6] compare event-based operational and denotational semantics for CCS, which differs from **L** in that it has a different form of synchronisation, action prefixing rather than sequential composition, and no refinement. Extending these approaches to sequential composition, especially including the neutral element ε, will cause grave difficulties: for instance, it is unclear how to model ε denotationally in the flow event structures underlying [6]. The extension to refinement may however be less problematic. We conjecture that the concepts of event independence and busy refinements can be translated with relative ease to the operational setting in [6], and our refinement rules may remain essentially unchanged.

Degano and Gorrieri [12] also study operational and denotational semantics for a language with refinement, but with action prefixing and without recursion. The denotational model is again flow event structures. An important difference of our paper with [12] is that correctness there is modulo *history-preserving bisimulation*, which is weaker than event isomorphism and may in some circumstances be more suitable. Busi, Van Glabbeek and Gorrieri [7] follow the same programme with respect to *ST-bisimulation*, which is weaker yet. Below we discuss the possibility of characterising weaker equivalences on the basis of our operational semantics.

Aceto and Hennessy deal with refinement syntactically rather than semantically (see the introduction, where we have briefly discussed the syntactic approach, or [15] for an exhaustive comparison of semantic and syntactic refinement). In Section 3 we show that for the synchronisation-free fragment of **L** we can likewise interpret refinement syntactically, with results comparable to [2]; in the presence of synchronisation [1] the approaches diverge.

In the introduction we have already listed a number of denotational constructions for refinement; apart from the ones mentioned above however, no corresponding operational semantics has been developed.

5.2 Evaluation and future work

An advantage of our approach is the relative ease of proving equivalences between terms on the basis of our operational semantics. For instance, the axioms in Table 5 are straightforward to prove. An interesting test case is

$$(B_1 \parallel_A B_2)[r] \equiv (B_1[r]) \parallel_A (B_2[r]) \ .$$

In [15], necessary and sufficient conditions are given under which this is sound for event isomorphism, but the proof, based on denotational constructions, is rather involved. An alternative proof on the basis of the operational semantics in this paper should show a decisive improvement.

The equivalence relation in our correctness criterion, event isomorphism, is rather strong. For instance, it does not satisfy the right distributivity of sequential composition over choice: $(x + y); z \neq x; z + y; z$ for event isomorphism. For our purpose this is not at all a disadvantage since we use the equivalence exclusively to compare operational and denotational semantics: the stronger the equivalence relation, the stronger this correspondence result. In other circumstances however, a weaker equivalence could be preferable. This is a matter of defining such on the basis of the event-labelled transition system we have used here. For instance, (2) shows that by ignoring the event information in the labels we have access to the entire world of interleaving equivalence. On the other hand, we have given an impossibility result in [23] which shows that weaker event-based equivalences do not preserve global properties like event independence. However, it should be possible to construct a *transition system with independence* in the sense of Nielsen et al. [19] from a given event-labelled transition system, on which the characterisation of equivalences looks more promising.

The event-labelled transition systems we have taken from [18] can be found in many variations in the literature, for instance *asynchronous automata* [3, 24] and *trace automata* [25]. In Section 2 we have already commented on the possibility of employing a more compositional formalism such as the *proved transitions* of [5].

References

[1] L. Aceto and M. Hennessy. Adding action refinement to a finite process algebra. In J. Leach Albert, B. Monien, and M. R. Artalejo, editors, *Automata, Languages and Programming*, volume 510 of *Lecture Notes in Computer Science*, pages 506–519. Springer-Verlag, 1991. To apear in Information and Computation.

[2] L. Aceto and M. Hennessy. Towards action-refinement in process algebras. *Information and Computation*, 103:204–269, 1993.

[3] M. A. Bednarczyk. *Categories of Asynchronous Systems.* PhD thesis, University of Sussex, Oct. 1987.

[4] E. Best, R. Devillers, and J. Esparza. General refinement and recursion operators for the Petri box calculus. In P. Enjalbert, A. Finkel, and K. W. Wagner, editors, *STACS 93*, volume 665 of *Lecture Notes in Computer Science*, pages 130–140. Springer-Verlag, 1993.

[5] G. Boudol and I. Castellani. Permutations of transitions: An event structure semantics for CCS and SCCS. In J. W. de Bakker, W.-P. de Roever, and G. Rozenberg, editors, *Linear Time, Branching Time and Partial Order in Logics and Models for Concurrency*, volume 354 of *Lecture Notes in Computer Science*, pages 411–427. Springer-Verlag, 1989.

[6] G. Boudol and I. Castellani. Flow models of distributed computations: Three equivalent semantics for CCS. Rapports de Recherche 1484, INRIA, July 1991. To appear in *Information and Computation.*

[7] N. Busi, R. van Glabbeek, and R. Gorrieri. Axiomatising ST bisimulation equivalence. In E.-R. Olderog, editor, *Programming Concepts, Methods and Calculi*, volume A–56 of *IFIP Transactions*, pages 169–188. IFIP, 1994.

[8] W. R. Cleaveland, editor. *Concur '92*, volume 630 of *Lecture Notes in Computer Science.* Springer-Verlag, 1992.

[9] I. Czaja, R. J. van Glabbeek, and U. Goltz. Interleaving semantics and action refinement with atomic choice. In G. Rozenberg, editor, *Advances in Petri Nets 1992*, volume 609 of *Lecture Notes in Computer Science*, pages 89–109. Springer-Verlag, 1992.

[10] P. Darondeau and P. Degano. Refinement of actions in event structures and causal trees. *Theoretical Comput. Sci.*, 118:21–48, 1993.

[11] P. Degano, R. De Nicola, and U. Montanari. A partial ordering semantics for CCS. *Theoretical Comput. Sci.*, 75:223–262, 1991.

[12] P. Degano and R. Gorrieri. An operational definition of action refinement. Technical Report TR–28/92, Università di Pisa, 1992. To appear in Information and Computation.

[13] R. van Glabbeek and U. Goltz. Equivalence notions for concurrent systems and refinement of actions. In A. Kreczmar and G. Mirkowska, editors, *Mathematical Foundations of Computer Science 1989*, volume 379 of *Lecture Notes in Computer Science*, pages 237–248. Springer-Verlag, 1989.

[14] R. van Glabbeek and U. Goltz. Refinement of actions in causality based models. In J. W. de Bakker, W.-P. de Roever, and G. Rozenberg, editors, *Stepwise Refinement of Distributed Systems — Models, Formalisms, Correctness*, volume 430 of *Lecture Notes in Computer Science*, pages 267–300. Springer-Verlag, 1990.

[15] U. Goltz, R. Gorrieri, and A. Rensink. On syntactic and semantic action refinement. In M. Hagiya and J. C. Mitchell, editors, *Theoretical Aspects of Computer Software*, volume 789 of *Lecture Notes in Computer Science*, pages 385–404. Springer-Verlag, Apr. 1994.

[16] J. F. Groote. Transition system specifications with negative premises. *Theoretical Comput. Sci.*, 118:263–299, 1993.

[17] L. Jategaonkar and A. Meyer. Testing equivalences for Petri nets with action refinement. In Cleaveland [8], pages 17–31.

[18] R. Langerak. *Transformations and Semantics for LOTOS*. PhD thesis, University of Twente, Nov. 1992.

[19] M. Nielsen, V. Sassone, and G. Winskel. Relationships between models for concurrency. In J. W. de Bakker, W.-P. de Roever, and G. Rozenberg, editors, *A Decade of Concurrency*, volume 803 of *Lecture Notes in Computer Science*, pages 425–476. Springer-Verlag, 1994.

[20] G. M. Pinna and A. Poigné. On the nature of events. In *Mathematical Foundations of Computer Science 1992*, Lecture Notes in Computer Science. Springer-Verlag, 1992.

[21] A. Rensink. Posets for configurations! In Cleaveland [8], pages 269–285.

[22] A. Rensink. *Models and Methods for Action Refinement*. PhD thesis, University of Twente, Enschede, Netherlands, Aug. 1993.

[23] A. Rensink. Order isomorphism does not preserve independence. *Bull. Eur. Ass. Theoret. Comput. Sci.*, 51:228–235, Oct. 1993.

[24] M. W. Shields. Concurrent machines. *The Computer Journal*, 28(5):449–465, 1985.

[25] E. W. Stark. Connections between a concrete and an abstract model of concurrent systems. In M. Main, A. Melton, M. Mislove, and D. Schmidt, editors, *Mathematical Foundations of Programming Semantics*, volume 442 of *Lecture Notes in Computer Science*, pages 53–79. Springer-Verlag, 1990.

[26] W. Vogler. Failures semantics based on interval semiwords is a congruence for refinement. In C. Choffrut and T. Lengauer, editors, *STACS 90*, volume 415 of *Lecture Notes in Computer Science*, pages 285–297. Springer-Verlag, 1990.

On the Computation of Place Invariants for Algebraic Petri Nets

Karsten Schmidt
Humboldt–Universität zu Berlin, Institut für Informatik
Unter den Linden 6, 10099 Berlin, Germany
e-mail: kschmidt@informatik.hu-berlin.de

Abstract

The paper is concerned with the computation of a generator set for the space of all place invariants for a given algebraic net. We will show that the problem can be divided into two major steps. First we trace back the problem to a set of equations between terms. Then we combine the solutions of these equations to obtain the solution of the original problem. For both steps we present a solution for a restricted class of algebraic nets, where the algebraic specification contains no equations and at most unary operation symbols.

1 Introduction

Computer aided analysis of concurrent systems is a difficult problem. For Petri nets with individual token most of the usual analysis problems become undecidable as soon as there are infinitely many possible inscriptions for a token. Such nets appear for instance as models of concurrent algorithms. Nevertheless there are methods for the analysis of such models, among them the invariant method, which has been established for almost every high–level net calculus. With the help of this calculus several system properties can be expressed and automatically verified. This verification yields a set of "user intended" place invariants. Additionally it is possible to use place invariants to prove the non–reachability of many markings, since the application of a place invariant yields the same value for all *reachable* markings. This non–reachability test can be done automatically. The more independent place invariants are involved in this test, the more states can be classified to be non–reachable. Therefore it would be useful to have the possibility to compute a generator set of *all* place invariants of a high–level Petri net, including possibly some which are not user intended. In the sequel we will discuss an approach how the problem of computing a generator set of all place invariants can be solved for algebraic Petri nets. The idea is to trace back the problem to a set of equations between terms which appear in the incidence matrix of the net. We present a solution of these equations for a restricted class of Petri nets. We assume an equation free specification and the absence of operation symbols with an arity greater than one. This restriction is very strong but covers a lot of interesting nets, among them the famous philosophers nets and all kinds of counter automata (including their composition via different kinds of synchronization). After having solved the above mentioned system of equations its solutions must be combined to obtain invariants. Again we present a solution for the mentioned net class. Finally we discuss the problems which appear when the restrictions are weakened.

2 Basic Definitions

The definitions are based on [EM85] and [Re91].

Definition 2.1 (Specifications) *A* **signature** $\Sigma = [S, \Omega]$ *consists of a set* S *of* **sorts** *and a family* $\Omega = \{\Omega_{w,s}\}_{w \in S^*, s \in S}$ *of* **operation symbols**. *For* e *being the empty word,* $\Omega_{e,s}$ *is the set of* **constant symbols** *of sort* s. *A set of* Σ*–***variables** *is a family* $X = \{X_s\}_{s \in S}$ *of* **variables**. *The set* $T_{\Omega,s}(X)$ *of* (Ω, X)*–***terms of sort** s *is inductively defined by 1.* $X_s \cup \Omega_{e,s} \subseteq T_{\Omega,s}(X)$ *and 2. for* $\omega \in \Omega_{s_1 \cdots s_n, s}$ *and* $T_i \in T_{\Omega,s_i}(X)$, $\omega(T_1, \cdots, T_n) \in T_{\Omega,s}(X)$. *The set* $T_{\Omega,s} := T_{\Omega,s}(\emptyset)$ *contains the* **ground terms of sort** s, $T_\Omega(X) := \bigcup_{s \in S} T_{\Omega,s}(X)$ *is the set of* Σ*–***terms over** X, *and* $T_\Omega := T_\Omega(\emptyset)$ *is the set of* Σ*–***ground terms**. *A* Σ*–***equation** *of sort* s *over* X *is a pair* $[L, R]$ *of terms* $L, R \in T_{\Omega,s}(X)$. *A* **specification** $D = [\Sigma, E]$ *consists of a signature* Σ *and a set* E *of* Σ*–equations.*

Definition 2.2 (Algebras) *A* Σ*–***algebra** $A = [S_A, \Omega_A]$ *consists of a family* $S_A = \{s_A\}_{s \in S}$ *of* **domains** *and a set* $\Omega_A = \{\omega_A \mid \omega \in \Omega\}$ *of* **operations**, *where* $\omega_A : s_{1A} \times \cdots \times s_{nA} \to s_A$ *for* $\omega \in \Omega_{s_1 \cdots s_n, s}$. *The elements* ω_A *for* $\omega \in \Omega_{e,s}$ *can be identified with elements of* s_A. *An* **assignment** *is a family* $\alpha = \{\alpha_s\}_{s \in S}$ *of mappings* $\alpha_s : X_s \to s_A$. *An* **evaluation** *according to an assignment* α *is a family of mappings* $\{\alpha_s^\#\}_{s \in S}$ *with* $\alpha_s^\# : T_{\Omega,s}(X) \to s_A$ *which is defined inductively by 1.* $\alpha_s^\#(x) := \alpha_s(x)$ *for* $x \in X_s$, *and 2.* $\alpha_s^\#(\omega(T_1, \cdots, T_n)) := \omega_A(\alpha_{s_1}^\#(T_1), \cdots, \alpha_{s_n}^\#(T_n))$ *for* $\omega \in \Omega_{s_1 \cdots s_n, s}$. *For ground terms* $T \in T_{\Omega,s}$ *we define the* **value** *of* T *in* A $\#_A(T) := \alpha_s^\#(T)$ *for an arbitrary assignment* α *(the value is actually not dependent on* α, *since ground terms do not contain variables). A* Σ*–equation* $[L, R]$ *is* **valid** *in a* Σ*–algebra* A *iff for all assignments* α, $\alpha^\#(L) = \alpha^\#(R)$. *For a specification* $D = [\Sigma, E]$ *the* Σ*–algebra* A *is a* D*–***algebra** *(or a* **model** *of* D*) iff all the equations in* E *are valid in* A.

Definition 2.3 (Substitutions) *Let* X *and* Y *be two sets of* Σ*–variables. A* **substitution** σ *over* X *is an assignment* $\sigma : X \to T_\Omega(Y)$, $(X_s \to T_{\Omega,s}(Y))$. *A* **ground substitution** *is a substitution* $\sigma : X \to T_\Omega$. *An injective substitution* $\sigma : X \to Y$ *is called* **renaming**. *For a term* T *and a substitution* σ *the term* $T\sigma$ *results from simultaneously replacing the variables in* T *by their corresponding* σ*–values. The application of a substitution* σ *with* $\sigma(x) = T$ *and* $\sigma(y) = y$ *on all variables except* x *to a term* T' *is written* $T'_{<x \leftarrow T>}$.

Definition 2.4 (Term Equivalence) *Two terms* T_1 *and* T_2 *are* **equivalent** *according to a specification* $D = [\Sigma, E]$ $(T_1 \equiv_E T_2)$ *iff for all* D*–algebras* A *and all assignments* α *in* A, $\alpha_A^\#(T_1) = \alpha_A^\#(T_2)$.

$\equiv_E$ is an equivalence relation on $T_\Omega(X)$. It is actually a congruence relation, i.e. $T_1 \equiv_E T_2$ implies $T_1\sigma \equiv_E T_2\sigma$ for arbitrary substitutions σ. With $[T]_E$ we denote the equivalence class of the term T according to the relation $\equiv_E$.

Definition 2.5 (Initial Algebra) *Let* $D = [\Sigma, E]$ *be a specification. The* **initial algebra** I *of* D *consists of the domains* $s_I := \{[T]_E \mid T \in T_{\Omega,s}\}$ *and the operations* ω_I *with* $\omega_I([T_1]_E, \cdots, [T_n]_E) := [\omega(T_1, \cdots, T_n)]_E$.

Due to the properties of the relation $\equiv_E$ the initial algebra is a model of D. Furthermore it satisfies the "no junk" property (every element is represented by a ground term) and the "no confusion" property (there are no valid equations except those implied by E). Though there are several models for a specification and it is challenging to obtain results which are valid for several models, we will consider exclusively initial algebras in the sequel.

Definition 2.6 (Multisets) *For a set M, a* **multiset** *over M is a mapping from M into the integer numbers. A multiset is* **semipositive** *iff all the values are greater or equal 0. A multiset is* **finite** *iff it has finite support. The* **empty** *multiset over M, denoted by ϑ_M, assigns 0 to every element of M. For an element $m \in M$, the multiset $\underline{m}$ assigns 1 to m and 0 to every other $m' \in M$. The multisets $\mu_1+\mu_2$ and $\mu_1-\mu_2$ are defined by $(\mu_1+\mu_2)(m) := \mu_1(m)+\mu_2(m)$ and $(\mu_1 - \mu_2)(m) := \mu_1(m) - \mu_2(m)$. This way every finite multiset can be represented as a* **formal sum** *of the $\underline{m}(m \in M)$. In such formal sums we usually write m instead of $\underline{m}$. A multiset μ_1 is* **less or equal** *to μ_2 iff for all $m \in M$, $\mu_1(m) \leq \mu_2(m)$.* We write $2 \cdot a + 3 \cdot b$ for $a + a + b + b + b$.

Definition 2.7 (Algebraic Petri Nets) *$AN = [D; \mathbf{P}, \mathbf{T}, \mathbf{F}; \psi, \xi, \lambda; m_0]$ is an* **algebraic Petri net** *iff (1) $D = [\Sigma, E]$ is a specification with $\Sigma = [S, \Omega]$; (2) $[\mathbf{P}, \mathbf{T}, \mathbf{F}]$ is a* **net**, *i.e.* **P** *and* **T** *are finite and disjoint sets called* **places** *and* **transitions**, *respectively, and* **F** *is a relation $\mathbf{F} \subseteq (\mathbf{P} \times \mathbf{T}) \cup (\mathbf{T} \times \mathbf{P})$, the elements of which are called* **arcs**; *(3) ψ is a* **sort assignment** *$\psi : \mathbf{P} \to S$; (4) ξ assigns a finite set of Σ-variables $\xi(t)$ to each transition $t \in \mathbf{T}$; (5) λ is the* **arc inscription** *such that for $f = [p,t]$ or $f = [t,p]$ in* **F**, *$\lambda(f)$ is a finite multiterm over $T_{\Omega,\psi(p)}(\xi(t))$; (6) m_0 is a* **marking**, *i.e. it assigns a finite multiterm over $T_{\Omega,\psi(p)}$ to every $p \in \mathbf{P}$. m_0 is called the* **initial marking**.

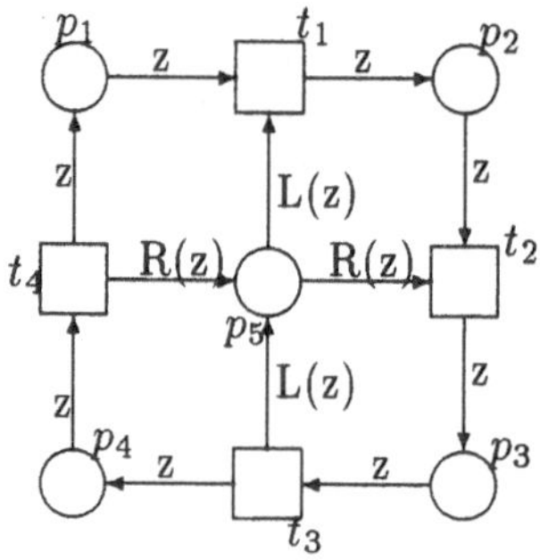

Figure 1:

Example. Figure 1 shows an algebraic Petri net. It is based on a specification where $S = \{phil, fork\}$, $\Omega_{\epsilon,phil} = \{a,b,c,d,e\}$, $\Omega_{phil,phil} = \{succ\}$, $\Omega_{phil,fork} = \{L, R\}$, $X_{phil} = \{x\}$ and $E = \{succ(a) = b, succ(b) = c, succ(c) = d, succ(d) = e, succ^5(x) = x, R(x) = L(succ(x))\}$. Furthermore assume, that $\psi(p_1) = \cdots = \psi(p_4) = phil$, $\psi(p_5) = fork$, $\xi(t_1) = \cdots = \xi(t_4) = \{z\}$ and λ is defined as depicted in the figure (for instance: $\lambda([p_5, t_2]) = R(z); \lambda([t_4, p_1]) = z$). The initial marking shall be $a + \cdots + e$ on p_1 and $L(a) + \cdots + L(e)$ on p_5.

For $f \notin \mathbf{F}$ we define $\lambda(f) := \vartheta$. With t^- and t^+ we denote the **P**-indexed vectors defined by $t^-(p) := \lambda([p,t])$ and $t^+(p) := \lambda([t,p])$, respectively. We write Δt for $t^+ - t^-$. The $(\mathbf{P} \times \mathbf{T})$–matrix where the column belonging to t is Δt is called the **incidence matrix** of the algebraic net in question. It is possible to interpret an algebraic Petri net according to an arbitrary model of the specification D. The result is a colored net. This way all the behavioral aspects of an algebraic net can be traced back to colored nets. We define the transition rule of an algebraic net for its initial algebra model only.

Definition 2.8 (Transition Rule) *Any ground substitution β of $\xi(t)$ is an* **occurrence mode** *of transition $t \in \mathbf{T}$. A transition $t \in \mathbf{T}$ is* **enabled** *in an occurrence mode β at a marking m iff for all $p \in \mathbf{P}$ with $[p,t] \in \mathbf{F}$, $[\lambda([p,t])\beta]_E \leq [m(p)]_E$. If t is enabled in β at m, then t may* **fire** *yielding the marking m', where for all $p \in \mathbf{P}$, $m'(p) = m(p) - \lambda([p,t])\beta + \lambda([t,p])\beta$. We write $m \xrightarrow{t,\beta} m'$ in this case. The set $R_{AN}(m^*)$ of markings* **reachable** *from a target marking m^* is the smallest set of markings, which contains m^* and if $m \in R_{AN}(m^*)$ and $m \xrightarrow{t,\beta} m'$ for some occurrence mode β, then there is a marking in $R_{AN}(m^*)$ which is componentwise equivalent to m' w. r. t. $\equiv_E$.*

3 Place Invariants

Place invariants provide weight functions for every place in such a way, that the weighted sum of all tokens remains invariant under transition occurrences. For algebraic nets (as well as for coloured nets, conf. [Je81b]) the weights are multisets of a color domain. Syntactically the weight functions are expressed as multiterms with one distinguished variable which we will always call x or "the argument variable" in the sequel. Other variables may occur in the multiterms. These variables will be written y and called "parameter variables". For the place invariant approach it is of major importance to distinguish clearly between these two completely different kinds of variables. The value of the weight function F for a given argument T is the multiterm which results from F by replacing every occurrence of x in F by T (this is the special task of the variable x which makes it necessary to distinguish it from the remaining variables). The formalism has to pay attention to the different sorts.

Definition 3.1 *Let $\Sigma = [S, \Omega]$ be a signature with $s_1, s_2 \in S$, Y a family of Σ–variables and x_s $(s \in S)$ a variable of sort s with $x_s \notin Y$. The set of* **weight terms from** *s_1* **to** *s_2 is the set $T_{\Omega,s_2}(Y \cup \{x_{s_1}\})$ of terms of sort s_2 with the* **argument variable** *x_{s_1} and* **parameter variables** *from Y. The set of* **weight functions from** *s_1* **to** *s_2 $W_{s_1,s_2}(Y)$ is the set of all finite multiterms over $T_{\Omega,s_2}(Y \cup \{x_{s_1}\})$, i.e. all finite multisets of weight terms from s_1 to s_2.*

Now we define the application of a weight term W to a term T (written $W \cdot T$).

Definition 3.2 *For $W \in W_{s_1,s_2}(Y)$ and $T \in T_{\Omega,s_1}(Y)$, $W \cdot T := W_{<x_{s_1} \leftarrow T>}$.*

Consequently, a weight term represents a function which assigns an element of sort s_2 (namely $W_{<x_{s_1} \leftarrow T>}$) to every element of sort s_1 (represented by T). The parameter variables are free parameters of the invariant. The definition allows weight terms where x_{s_1} does not appear. In this case the value of the weight terms is independent from the argument, that is a constant function. The application of a weight function F to a multiterm M is defined as the linear extension of the above product. Furthermore the product can be linear extended to a product between vectors of weight functions and vectors of multiterms. Usually the sort of the argument variable x_{s_1} is clear from the context and therefore we skip the sort index in the remaining part of the paper.

Definition 3.3 ([Re91]) *Let $AN = [S, \Omega, E; \mathbf{P}, \mathbf{T}, \mathbf{F}; \psi, \xi, \lambda, m_0]$ be an algebraic net and Y a family of Σ-variables being disjoint to all $\xi(t), t \in \mathbf{T}$. A* **place invariant** *of sort s is a $\mathbf{P}$-indexed vector I where $I(p)$ is a weight function from $\psi(p)$ to s such that for every marking m' reachable from a marking m it holds $I \cdot m \equiv_E I \cdot m'$.*

Example. The vector $\left(\begin{array}{c|l} p_2 & L(x) \\ p_3 & L(x)+R(x) \\ p_4 & R(x) \\ p_5 & x \end{array}\right)^{*}$ is a place invariant for the dining philosopher's model. The weight of the initial marking is $\vartheta_{<x \leftarrow (a+\cdots+e)>} + x_{<x \leftarrow (L(a)+\cdots+L(e))>} = \vartheta + L(a) + \cdots + L(e) = L(a) + \cdots + L(e)$ The weight of a marking with some token x and $succ(x)$ on p_3 would contain $2 \cdot L(succ(x))$. Hence the weight of such a marking differs from the weight of the initial marking where all elements appear only once and therefore it cannot be reachable.

Of course the arguments of the weight function for a place p are token values on p, that is terms of sort $\psi(p)$. The result of the application of a place invariant I to a marking m is a multiterm of sort s — the weight of m with respect to I. A vector I of weight functions is a place invariant iff transition occurrences do not change the weights of markings. The following theorem characterizes place invariants (C^T is the transposed of the incidence matrix C).

Theorem 1 ([Re91]) *I is a place invariant of an algebraic net having the incidence matrix C iff $C^T \cdot I \equiv_E \underline{\vartheta}$.* □

The weight of markings remains invariant under transition occurrences iff for every transition $t \in \mathbf{T}$ the *change* of the weight caused by t is the empty multiterm. The changes of weights by single transition occurrences are expressed by a product $C^T \cdot I$ (the linear extension of the above defined product between multiterms and weight functions).

Example. For the above invariant it holds $C^T \cdot I =$

$$= \left(\begin{array}{c|llll} & t_1 & t_2 & t_3 & t_4 \\ \hline p_1 & -z & & & z \\ p_2 & z & -z & & \\ p_3 & & z & -z & \\ p_4 & & & z & -z \\ p_5 & -L(z) & -R(z) & L(z) & R(z) \end{array}\right)^{\mathrm{T}} \cdot \left(\begin{array}{c|l} p_1 & \\ p_2 & L(x) \\ p_3 & L(x)+R(x) \\ p_4 & R(x) \\ p_5 & x \end{array}\right)$$

$$= \left(\begin{array}{c|l} t_1 & L(x) \cdot z - x \cdot L(z) \\ t_2 & -L(x) \cdot z + (L(x) \\ & +R(x)) \cdot z - x \cdot R(z) \\ t_3 & -(L(x)+R(x)) \cdot z \\ & +R(x \cdot z + x \cdot L(z) \\ t_4 & -R(x) \cdot z + x \cdot R(z) \end{array}\right) = \left(\begin{array}{c|l} t_1 & L(z) - L(z) \\ t_2 & -L(z) + L(z) + \\ & R(z) - R(z) \\ t_3 & -L(z) - R(z) + \\ & R(z) + L(z) \\ t_4 & -R(z) + R(z) \end{array}\right) = \underline{\vartheta}$$

---

*Unmentioned components (here: p_1) are assumed to be empty (or ϑ, resp.)

Consequently the problem we have to consider in the sequel is to find a generator set for all solutions of the system of equations $C^T \cdot I \equiv_E \underline{\vartheta}$, where I is the vector of unknowns. For the computation of generator sets it is important to have in mind the operations which preserve invariance. Among these operations are obviously linear combinations.

Theorem 2 ([Re91]) *With I_1 and I_2, also $I_1 + I_2$ and $I_1 - I_2$ are place invariants.* □

Additionally we exploit the congruence properties of $\equiv_E$.

Lemma 3 *Let W_1 and W_2 be two weight terms from a sort s_1 to a sort s_2 with a set Y of parameter variables. If for some terms T_1 and T_2 taken from $T_\Omega(X)$ (X and Y are assumed to be disjoint) it holds $W_1 \cdot T_1 \equiv_E W_2 \cdot T_2$, then it holds also (1.) $(W_1\sigma) \cdot T_1 \equiv_E (W_2\sigma) \cdot T_2$ for arbitrary substitutions σ over Y (without substituting the argument variable x!) and (2.) $(W \cdot W_1) \cdot T_1 \equiv_E (W \cdot W_2) \cdot T_2$ for arbitrary weight terms W from s_2 to some sort s_3.*

Proof. Follows immediately from the congruence properties of $\equiv_E$ and the definition of the product between terms and weight functions. □

This lemma immediately leads to the observation that instantiation of parameter variables and concatenation of weight functions preserve invariance.

Theorem 4 *Let Y be a family of Σ-variables, s a sort and I an invariant where $I(p)$ is a weight function from $\psi(p)$ to s. Let $\sigma : Y \to T_\Omega(Y)$ be a substitution of the parameter variables in I. Then $I\sigma$ defined by $I\sigma(p) := I(p)\sigma$ is a place invariant.* (Straightforward from Lemma 3) □

Theorem 5 *Let I be a place invariant of sort s_1 and W a weight function from sort s_1 to sort s_2. Then $W \cdot I$ defined by $(W \cdot I)(p) := W \cdot I(p)$ is a place invariant of sort s_2.* (Straightforward from Lemma 3.) □

4 Incidence Matrix and Term Equations

The only input for the computation of place invariants is the incidence matrix. We have to exploit this matrix to obtain conditions which distinguish invariants from place vectors of multiterms which are no invariants. For this purpose we regard the process of verifying a given invariant. Let I be an appropriate **P**–indexed vector of weight functions, that is a **P**–indexed vector which satisfies all syntactical conditions for a place invariant. According to Theorem 1 we build $C^T \cdot I$. The result is a **T**–indexed vector of multiterms. Thereby *every* term appearing in a component of this vector results from applying a weight term from the **P**–indexed vector to a term of the incidence matrix. Its multiplicity is the product of multiplicities of the weight term and the matrix term. Then we have to check whether the resulting vector is equivalent to the vector of empty multiterms. Therefore we have to build the equivalence classes of terms in every component of the **T**–indexed vector and to check, whether the sum of the multiplicities in every resulting equivalence class is 0.

Example. Consider the philosopher's net and the invariant considered in the previous section. The **T**–indexed vector resulting from the product of C^T

and I is $\left(\begin{array}{c|l} t_1 & L(z) - L(z) \\ t_2 & -L(z) + L(z) + R(z) - R(z) \\ t_3 & -L(z) - R(z) + R(z) + L(z) \\ t_4 & -R(z) + R(z) \end{array}\right)$. The equivalence classes of terms in the second component of this vector are $[L(z)]$ and $[R(z)]$ while in the first component there is only the equivalence class $[L(z)]$. Obviously the sum of all multiplicities of terms belonging to one and the same equivalence class is always zero.

In this example the detection of equivalence classes is very simple. Usually there are equivalent terms which are syntactically different (due to the specified set of equations). From this observation we learn, that the application of a weight term to a matrix term results in a term which is equivalent to the results of the application of other weight terms to matrix terms. In this sense we can consider the weight terms contained in the invariant to be solutions of sets of equations like $X_1 \cdot T_1 \equiv_E \cdots \equiv_E X_n \cdot T_n$ where $X_1, \cdots, X_n$ are the unknown weight terms, "$\cdot$" is the product between weight terms and matrix terms which has been defined in the previous section and $T_1, \cdots, T_n$ are terms appearing in one and the same column of the incidence matrix (otherwise the results of applying weight terms to them would not appear in the same component of $C^T \cdot I$).

Example. The weight terms $X_1 := L(x)$ and $X_2 := x$, occurring in the components p_2 and p_5, resp. in the running example, are solutions of the equation $X_1 \cdot z \equiv_E X_2 \cdot L(z)$. This equation reflects the equivalence class $[L(z)]$ in the component t_1 of $C^T \cdot I$ in the above example where one of the two "$L(z)$" origins from $L(x) \cdot z$ and the other "$L(z)$" from $x \cdot L(z)$.

The first major idea for computing a generator set of all place invariants for a given net is to find "most general solutions" for the above kind of equations for all sets of terms appearing in the same column of the incidence matrix and to construct invariants based on those most general solutions.

In the sequel we discuss how to find the most general solutions of equations for weight terms in the case, that the specification consists only of constant symbols and unary operation symbols and the set of equations is empty. For reasons of simplicity we consider only single equations (i.e. $X_1 \cdot T_1 \equiv_E X_2 \cdot T_2$) but the results can be straightforward generalized to n unknowns. We introduce two notations for operating on terms which can be built in the restricted kind of specifications. Let $T_1 = \omega_m(\omega_{m-1}(\cdots(\omega_1(r))\cdots))$ and $T_2 = \omega'_n(\omega'_{n-1}(\cdots(\omega'_1(r'))\cdots))$ be terms where r and r' are variables or constant symbols. We say $T_1 \sqsubseteq T_2$ iff $m \leq n$, $r = r'$ and for all i $(1 \leq i \leq m)$ it holds $\omega_i = \omega'_i$. If $T_1 \sqsubseteq T_2$ the term $T_2 \setminus T_1$ is defined as $\omega'_n(\cdots(\omega'_{m+1}(x))\cdots)$. Thereby $T_2 \setminus T_1$ is always considered to be a weight term and its variable x is the argument variable of the sort of T_1. Thus for terms T_1 and T_2 with $T_1 \sqsubseteq T_2$ we have always the relation $(T_2 \setminus T_1) \cdot T_1 = T_2$ ($=$ is the syntactical identity of terms). These operations are sufficient to construct a set of most general solutions of the above kind of equations.

Definition 4.1 *The* **set of most general solutions** MGS *for the equation* $X_1 \cdot T_1 \equiv_\emptyset X_2 \cdot T_2$ *is defined as follows:*
1. $[X_1 = y, X_2 = y] \in MGS$ *(y is an arbitrary* parameter *variable);*
2. If T_1 *is a ground term, then* $[X_1 = x, X_2 = T_1] \in MGS$*;*
3. If T_2 *is a ground term, then* $[X_1 = T_2, X_2 = x] \in MGS$*;*
4. If $T_1 \sqsubseteq T_2$*, then* $[X_1 = T_2 \setminus T_1, X_2 = x] \in MGS$ *;*

5. If $T_2 \sqsubseteq T_1$, then $[X_1 = x, X_2 = T_1 \setminus T_2] \in MGS$;
6. MGS contains no elements except those according to the first five items.

Theorem 6 *All elements of MGS are solutions of the given equation.* □

Due to Lemma 3 instantiating the variable y and concatenating the solutions with other weight terms lead to solutions again. More interesting is the following justification for the name "most general solutions".

Theorem 7 *Every solution of the equation $X_1 \cdot T_1 \equiv_\emptyset X_2 \cdot T_2$ can be obtained from a solution in the corresponding MGS by instantiating the parameter variable y or by concatenating a solution with an arbitrary weight term.*

Proof. Let $[X_1 = W_1, X_2 = W_2]$ be a solution of the given equation, that is $W_1 \cdot T_1 \equiv_\emptyset W_2 \cdot T_2$. Assume first that both W_1 and W_2 are terms containing unary operation symbols, say $W_1 = \omega_1(W_1')$ and $W_2 = \omega_2(W_2')$. We obtain $\omega_1(W_1')_{<x \leftarrow T_1>} \equiv_\emptyset \omega_2(W_2')_{<x \leftarrow T_2>}$ leading to $\omega_1(W'_{1<x \leftarrow T_1>}) \equiv_\emptyset \omega_2(W'_{2<x \leftarrow T_2>})$. From this equation we may conclude (1.) $\omega_1 = \omega_2$ (since $\equiv_\emptyset$ is the syntactical identity on terms); (2.) $W'_{1<x \leftarrow T_1>} \equiv_\emptyset W'_{2<x \leftarrow T_2>}$, that is $[X_1 = W_1', X_2 = W_2']$ is a solution of the given equation. (3.) The solution $[X_1 = W_1, X_2 = W_2]$ can be obtained from the solution $[X_1 = W_1', X_2 = W_2']$ by concatenating it with $\omega_1(x)$. Hence every solution where both weight function contain unary operation symbols can be obtained by concatenation from a solution where at least one of the involved weight functions does not contain unary operation symbols. It remains to show that these solutions are generated by MGS. Since all the considerations for W_1 and W_2 are symmetric, we may assume, that W_1 is the weight function which does not contain unary operation symbols. Due to the restricted set of operation symbols there remain three cases for W_1.

Case 1: $W_1 = x$ (the argument variable). In this case it holds $W_1 \cdot T_1 = T_1$. On the other hand W_2 does or does not contain the variable x, too. If it does, the term T_2 appears at the bottom of $W_2 \cdot T_2$. Therefore it holds $T_2 \sqsubseteq T_1$ and due to the syntactical identity of T_1 and $W_2 \cdot T_2$ we obtain $W_2 = T_1 \setminus T_2$. This is covered by item 4 (or 5, resp.) of Def. 4.1. If W_2 does not contain x, then it holds $W_2 \cdot T_2 = W_2$ leading to $T_1 \equiv_\emptyset W_2$. Since the variable sets of weight functions are disjoint from the variable sets of terms from the incidence matrix, T_1 must be a ground term and it holds $W_2 = T_1$ (syntactically) due to the absence of equations. This is covered by item 2 (and 3, resp.) of Def. 4.1.

Case 2: $W_1 = y$ (a parameter variable). Then $W_1 \cdot T_1 = y$. Therefore, since y must appear in $W_2 \cdot T_2$, too but cannot appear in T_2, it must appear in W_2. Thus we have $W_2 \cdot T_2 = W_2$ and $y = W_1 = W_2 \cdot T_2 = W_2$, that is $W_2 = y$. This is covered by item 1 of Def. 4.1.

Case 3: $W_1 = \omega$ for some constant symbol ω. Then we have $W_1 \cdot T_1 = \omega \equiv_\emptyset W_2 \cdot T_2$. Since $W_2 \cdot T_2$ contains at least the operation symbols of W_2, W_2 cannot contain unary operation symbols (else there would be no way to obtain syntactical identity of ω and $W_2 \cdot T_2$). Therefore it remains to consider the case that W_2 consists of a constant symbol only, since all other cases can be covered by one of the first to cases through swapping W_1 and W_2. If W_2 is a constant symbol, then we have $W_2 \cdot T_2 = W_2$ leading to $W_2 = \omega$. This solution thus can be obtained from the solution $[X_1 = y, X_2 = y]$ by instantiating the variable y with ω. □

At this stage we will demonstrate the problems which appear when the above restrictions are weakened. The problem caused by the introduction of equations seems to be roughly the same as for the well-known technique of unification modulo a set of equations. That is, the existence of solutions is probably undecidable and the number of solutions (finitely many or infinitely many independent solutions) will probably depend on the design of the specification. We will concentrate on the second restriction. Assume there is a binary operation symbol f and there are at least two different ground terms T_1 and T_2. Then there are infinitely many solutions for the equation $X_1 \cdot T_1 \equiv_\emptyset X_2 \cdot T_2$. Among others there are the following solutions:
$[X_1 = f(x, T_2), X_2 = f(T_1, x)]$
$[X_1 = f(x, f(y_1, T_2)), X_2 = f(T_1, f(y_1, x))]$
$[X_1 = f(x, f(y_1, f(y_2, T_2))), X_2 = f(T_1, f(y_1, f(y_2, x)))] \cdots$

All these solutions are independent and cannot be covered by a finite set of solutions. Therefore there is no suitable way to weaken the restriction unless a) a complete different way of computing the invariants is found, b) the concept of invariants is generalized in a way, that all the solutions can be represented finitely or c) we have some additional information on how many of these solutions are sufficient to get enough invariants for a strong non-reachability test. Up to now we have not considered any of these possibilities and so we must leave their consideration to the future. In the next section we describe how to build invariants from the solutions considered here.

5 From Term Equations to Invariants

In the previous section we considered term equations in order to characterize the equivalence classes of terms in the components of the **T**-indexed vector $C^T \cdot I$ and to constrain the weight terms from which invariants can be composed.

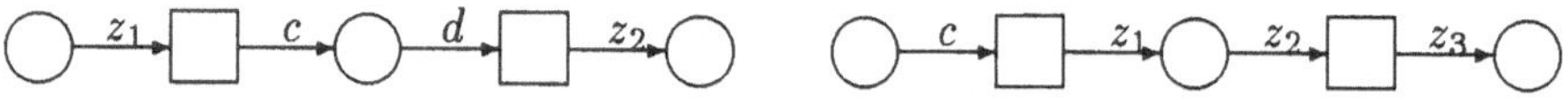

(a) (b)

Figure 2:

Example. Consider the net in Figure 2(a). The only interesting details concerning the specification are that c and d are constants while z_1 and z_2 are variables. The incidence matrix of this net is $\begin{pmatrix} -z_1 & \\ c & -d \\ & z_2 \end{pmatrix}$. Thereby places and transitions are assumed to be numbered consecutively from left to right in the figure. The only reasonable equations obtainable from the columns of this matrix are $X_1 \cdot c \equiv_\emptyset X_2 \cdot z_1$ for t_1 and $X_1 \cdot z_2 \equiv_\emptyset X_2 \cdot d$ for t_2. The MGS are $\{[X_1 = y, X_2 = y], [X_1 = x, X_2 = c]\}$ for the first and $\{[X_1 = y, X_2 = y], [X_1 = d, X_2 = x]\}$ for the second equation. But it is impossible to build invariants from c (for p_1 — z_1 belongs to an arc from p_1) and x (for p_2 — c belongs to an arc to p_2) only. In the component t_2 of $C^T \cdot I$ there would appear the term d $(= x \cdot d)$ with a non–zero multiplicity. Nevertheless there is an invariant for this net providing an equivalence class which is characterized by this equation and the considered solution, namely $(c, x, d)^T$. For this invariant it holds $C^T \cdot I = (c - c, -d + d)^T$. The equivalence class $[c]$ in the first component corresponds to the equation $X_1 \cdot z_1 \equiv_\emptyset X_2 \cdot c$ with the solution $[X_1 = c, X_2 = x]$, just the considered one. The crucial point is that in this invariant for the product of the weight term x in the second component and the matrix term d there is another equivalence class, namely $[d]$ in the second component of $C^T \cdot I$. That is, the weight term x is shared between the solutions of two different equations, namely as X_2 in the considered solution of $X_1 \cdot z_1 \equiv_E X_2 \cdot c$ and as X_1 in the solution $[X_1 = x, X_2 = d]$ of the equation $X_1 \cdot d \equiv_\emptyset X_2 \cdot z_2$.

A closer observation of how we multiply C^T and I shows that it is unavoidable to share weight terms in solutions of different term equations, since *every weight term must be applied to every term of the corresponding row of the incidence matrix and the results of this application usually appear in different equivalence classes.* Remember that the term equations correspond to the equivalence classes in $C^T \cdot I$. This leads us to a stronger relation between the weight terms which occur in a place invariant. We define a special concept for this relation.

Definition 5.1 *Let U be a set of ordered pairs $[Q, S]$ where Q is an equation of the kind $X_1 \cdot T_1 \equiv_E \cdots \equiv_E X_n \cdot T_n$ with $T_1, \cdots, T_n$ occurring in one and the same column of the incidence matrix C and S is a solution of Q. U is called* **complete system of solutions** *of the system $\{Q \mid \exists S : [Q, S] \in U\}$ iff for every weight term W appearing in a solution S of a $[Q, S] \in U$ corresponding to the part $X_i \equiv_E T_i$ of Q and for every term T which appears in the same row of C as T_i there is a $[Q', S'] \in U$ where for some j the equation $X_j \equiv_E T_j$ is part of Q' and $X_j = W$ is the corresponding solution in S'.*

Corollary 8 *Weight terms occurring in a place invariant always appear in a complete system of solutions of a set of equations where every equation contains only terms of one and the same column of C.*

From a complete system of solutions the component of I where a weight term W appears is determined by the rows for which the completeness condition has to be established.

Example. Consider the net in Figure 2(a). $U = \{[X_1 \cdot z_1 \equiv_\emptyset X_2 \cdot c, [X_1 = c, X_2 = x]], [X_1 \cdot d \equiv_\emptyset X_2 \cdot z_2, [X_1 = x, X_2 = d]]\}$ is a complete system of solutions since c appears corresponding to every term of the first row of C (only z_1), x appears corresponding to every term of the second row (for c in the first and d in the second pair of U) and d appears corresponding to the only term z_2 of the last row of c. In comparison, $U' = \{[X_1 \cdot z_1 \equiv_\emptyset X_2 \cdot c, [X_1 = c, X_2 = x]]\}$ is not complete. x appears there as solution for c but it does not appear as a solution for d, though both c and $-d$ occur in one and the same row of C.

Once having a complete system of solutions corresponding to a given net it is very easy to calculate invariants. We simply collect the weight terms in the components of the invariant and compute their still unknown multiplicities from a system of linear equations on the integer numbers. This system grows from the requirement, that the sum of multiplicities in every equivalence class of $C^T \cdot I$ must equal 0.

Example. Consider the above complete system of solutions U. The weight terms which occur in U are c corresponding to the matrix term z_1, that is to the first place, furthermore x corresponding to the matrix terms c and d, that is belonging to the second place and finally d belonging to the last place. The raw invariant is $(a_1 \cdot c, a_2 \cdot x, a_3 \cdot d)^T$, where the a_i are the multiplicities which are still unknown. The application of I to C^T results in $(-a_1 \cdot c + a_2 \cdot c, -a_2 \cdot d + a_3 \cdot d)^T$. The following equations can be derived: $a_1 = a_2$ and $a_2 = a_3$. A generator set for all integer solutions of this system of equations is $a_1 = a_2 = a_3 = 1$. Therefore every invariant of the above raw kind is generated by $(1 \cdot c, 1 \cdot x, 1 \cdot d)^T$. The remaining problem is to compute complete systems of solutions. A suitable algorithmic way for obtaining complete systems of MGS is to start with a single equation and a solution and then repeatedly to search for weight terms which violate the completeness conditions. These violations can be removed either by instantiating solutions which are already contained in the considered system of solutions or by adding new equations with their solutions to the system.

Example The incidence matrix of the net in Figure 2(b) is $\begin{pmatrix} -c & \\ z_1 & -z_2 \\ & z_3 \end{pmatrix}$.
We compute a complete system of solutions corresponding to this net. For this purpose we start with the equation $X_1 \cdot c \equiv_\emptyset X_2 \cdot z_1$ obtained from the first column of the incidence matrix, and an element of MGS, $[X_1 = x, X_2 = c]$. The completeness condition is violated for the weight term c, since there is no equation for z_2 which appears in the same row as z_1. Therefore we have to add an equation covering z_2. The only suitable equation for this purpose is $X_1 \cdot z_2 \equiv_\emptyset X_2 \cdot z_3$. This equation has only a singleton MGS, $[X_1 = y, X_2 = y]$. But with this solution the completeness condition is still violated. This violation can be removed by instantiating y to c. The resulting system of solutions consisting of $[X_1 = x, X_2 = c]$ for the first equation and $[X_1 = c, X_2 = c]$ for the second equation is obviously complete. This complete system of solutions is related to an invariant, namely $(x, c, c)^T$.

When we remove violations of the completeness condition by adding new equations it may happen that new violations of the condition appear. Therefore the termination of the algorithm cannot be guaranteed. Up to now we have no criterion when the computation of a complete system of solutions can be stopped, we also do not know whether for a given algebraic net there is always a finite generator set for the place invariants. But of course the algorithmic idea is sufficient to *enumerate* the complete systems of solutions. Therefore we will be able to enumerate generator sets of invariants which is often sufficient for the reachability test to be performed with place invariants. With the algorithm which we proposed we do not obtain *every* complete system of solutions.

Example Consider the net in Figure 2(b) and the following complete system of solutions U consisting of

$$\begin{aligned} X_1 \cdot c \equiv_E X_2 \cdot z_1 / [X_1 = x, X_2 = c] \\ X_1 \cdot z_2 \equiv_E X_2 \cdot z_3 / [X_1 = c, X_2 = c] \\ X_1 \cdot c \equiv_E X_2 \cdot z_1 / [X_1 = y, X_2 = y] \\ X_1 \cdot z_2 \equiv_E X_2 \cdot z_3 / [X_1 = y, X_2 = y] \end{aligned}$$

This system is related to the invariant $(1 \cdot x + 1 \cdot y, 1 \cdot c + 1 \cdot y, 1 \cdot c + 1 \cdot y)^T$.

Obviously the proposed algorithm started on the first element of U would stop after having added the second element since no completeness restrictions are violated (this is just the computation discussed in the previous example). Starting the algorithm with another element always leads to complete systems of solutions which either contain only the first two or the last two elements of U. The reason is that the weight terms x and c on the one hand and y on the other are completely independent in the sense that equivalence classes in $C^T \cdot I$ containing terms which are made from applications of x and c to matrix terms are disjoint to the classes of terms which result from the application of a weight term y. Thus the invariant can be represented as $(1 \cdot x, 1 \cdot c, 1 \cdot c)^T + (1 \cdot y, 1 \cdot y, 1 \cdot y)^T$ where both **P**–indexed vectors are invariants.

Definition 5.2 *A complete system of solutions is called* **basic** *iff none of its proper subsets is a complete system of solutions.*

There are two important properties holding for basic complete systems of solutions: (1) Every invariant can be represented as a linear combination of invariants constructible from basic complete systems of solutions; (2) With the above algorithmic approach it is possible to find every basic complete system of solutions. We do not prove these claims here since a formal proof would require a couple of additional notations. However there are dual theorems for the case of transition invariants which have been proved in [Sc94]. The arguments used there can be established for place invariants as well. The only problem we still have to overcome is to decide if and how we can instantiate or concatenate solutions in a most general way such that two distinguished weight terms appearing in a solution are equivalent (that is, a violation of the completeness condition can be removed). The only way to force this equivalence is to instantiate the solutions or to concatenate them with other weight functions. As well as for the term equation problem we have a solution for this problem only for the case that the set of equations occurring in the specification is empty and the specified operation symbols are constant symbol or unary operation symbols. Consider the two equations

$$X_1 \cdot T_1 = \cdots = X_m \cdot T_m \qquad (1)$$

$$X_1' \cdot T_1' = \cdots = X_n' \cdot T_n' \qquad (2)$$

Assume we have a generator set for a set of solutions of both equations. We aim in a generator set for *all* solutions which are generated by the given generator sets and which satisfy the additional requirement that $X_1 = X_1'$. Every solution of this problem must be obtainable as an instance of an element of the first generator set and an element of the second generator set. Let $[X_1 = W_1, \cdots X_m = W_m]$ and $[X_1' = W_1', \cdots, X_n' = W_n']$ be elements of the generator sets of solutions for the two equations. We consider 3 cases for W_1.
Case 1: W_1 is a parameter variable y. In this case it cannot be unified with W_1' if W_1' contains the argument variable x. In all other cases, that is W_1' contains no variable or a parameter variable, it is possible to instantiate the solution of equation (1) by $[y \rightarrow W_1']$. This solution and the original solution of equation (2) satisfy the considered requirements. Furthermore it is obvious that every pair of solutions which a) are instances of the two given solutions and b) satisfy the requirement $X_1 = X_1'$ are instances of the constructed solution.
Case 2: W_1 is ground term. Then there is no common solution if W_1' contains

the argument variable x. If W_1' is a parameter variable y then we can swap the two solutions and return to case 1. If W_1' is a ground term, too, W_1 and W_1' can only be unified when $W_1 \sqsubseteq W_1'$ or $W_1' \sqsubseteq W_1$, since the only way to modify the solutions is to concatenate the solutions with weight terms. If $W_1 \sqsubseteq W_1'$ then obviously we have to concatenate the solution of equation (1) with $W_1' \setminus W_1$, while in the case $W_1' \sqsubseteq W_1$ the two weight functions can be unified by concatenating the solution of equation (2) with $W_1 \setminus W_1'$.
Case 3: W_1 contains the argument variable x. Then it is impossible to build common instances of both solutions when W_1' does not contain the argument variable x. Therefore the requirement $X_1 = X_1'$ can be satisfied only if W_1' contains x, too. In this case the only way to unify W_1 and W_1' is to concatenate them with weight terms. Thus it holds $W_1 \sqsubseteq W_1'$ or $W_1' \sqsubseteq W_1$ and the requirement can be satisfied by concatenating the solution of equation (1) with $W_1' \setminus W_1$ or the solution of equation (2) with $W_1 \setminus W_1'$, resp.
Note, that we did not consider the case that W_1 is a term which consists of the parameter variable y and unary operation symbols. This is however not necessary, since neither the MGS of equations considered in the previous section nor the instantiation of solutions discussed here provide such terms. The case consideration leads immediately to the following

Theorem 9 *Let S_1 and S_2 be sets of solutions of the equations $X_1 \cdot T_1 = \cdots = X_m \cdot T_m$ and $X_1' \cdot T_1' = \cdots = X_n' \cdot T_n'$. Then every pair of solutions $[X_1 = W_1, \cdots, X_m = W_m]$ and $[X_1' = W_1', \cdots, X_n' = W_n']$ which can be generated from S_1 and S_2, respectively and for which it holds $W_1 = W_1'$ can be generated from pairs of solutions which can be constructed by solutions of S_1 and S_2 according to the above case consideration.*

With these considerations we have a complete calculus to enumerate place invariants for nets based on simple specifications. In the next section we will present an example where all the steps of calculation are demonstrated again for the philosopher's example.

6 Example

We consider the philosopher's example as described earlier, but without equations in the specification. The incidence matrix of the net in Figure 1 is

$$\left(\begin{array}{c|cccc} & t_1 & t_2 & t_3 & t_4 \\ \hline p_1 & -z & & & z \\ p_2 & z & -z & & \\ p_3 & & z & -z & \\ p_4 & & & z & -z \\ p_5 & -L(z) & -R(z) & L(z) & R(z) \end{array}\right).$$

In order to distinguish the different occurrences of syntactically identical terms in this matrix we superscribe the row and the column to every term. For instance we write z^{32} for the term z appearing in the row p_3 and the column t_2 of the matrix.

According to the algorithmical ideas discussed in the previous sections we have to compute complete systems of MGS for systems of term equations. For this purpose let us choose a starting equation. For t_1 we have the following starting equations: $X_1 \cdot z^{11} \equiv_\emptyset X_2 \cdot z^{21}$, $X_1 \cdot z^{11} \equiv_\emptyset X_2 \cdot L(z)^{51}$, $X_1 \cdot z^{21} \equiv_\emptyset X_2 \cdot L(z)^{51}$ and $X_1 \cdot z^{11} \equiv_\emptyset X_2 \cdot z^{21} \equiv_\emptyset X_3 \cdot L(z)^{51}$. Every of these four equations as well as the corresponding equations for the other columns of the matrix are possible starting equations. If we aim in a complete generator set we have to perform the following steps beginning with every of these starting equations. We choose the third equation. The next step is to compute the system of MGS for this equation. It consists of the two solutions $[X_1 = L(x), X_2 = x]$ and $[X_1 = y, X_2 = y]$. The first solution is more interesting (of course, for a complete generator set we have to consider both solutions). We get the first system of solutions U containing the equation $X_1 \cdot z^{21} \equiv_\emptyset X_2 \cdot L(z)^{51}$ with its solution $[X_1 = L(x), X_2 = x]$. The completeness condition is violated twice, since first $L(x)$ is a solution corresponding to z^{21} but there is no equation for z^{22} appearing in the same row and second x is a solution corresponding to $L(z)^{51}$ and there is no equation for $R(z)^{52}$, $L(z)^{53}$ and $R(z)^{54}$. Let us try to remove the first violation. Therefore we need an equation covering z^{22}. This equation cannot involve $R(z)^{52}$ since the MGS of such an equation would provide for z^{22} either the weight term y or the weight term $R(x)$ both not having common instances with the currently considered weight term $L(x)$. The only possible equation to solve the problem is $X_1 \cdot z^{22} \equiv_\emptyset X_2 \cdot z^{32}$. The MGS of this equation consists of $[X_1 = x, X_2 = x]$ and $[X_1 = y, X_2 = y]$ the first of which has common instances with the solution already contained in U. Thereby the only way to obtain a most general common instance between the two solutions is to concatenate the second solutions with $L(x)$. Therefore the system of MGS currently is

$$X_1 \cdot z^{21} \equiv_\emptyset X_2 \cdot L(z)^{51}/[X_1 = L(x), X_2 = x]$$
$$X_1 \cdot z^{22} \equiv_\emptyset X_2 \cdot z^{32}/[X_1 = L(x), X_2 = L(x)]$$

The considered violation has been removed but a new violation of the completeness condition appeared for z^{32}. In the same manner we remove all the violations step by step including those which are caused by the inserted elements.

Choosing one of the different possibilities at every stage of the computation we result in the following value for U:

$$X_1 \cdot z^{21} \equiv_\emptyset X_2 \cdot L(z)^{51}/[X_1 = L(x), X_2 = x] \text{ (starting eq.)}$$
$$X_1 \cdot z^{22} \equiv_\emptyset X_2 \cdot z^{32}/[X_1 = L(x), X_2 = L(x)] \text{ (viol. for } p_2)$$
$$X_1 \cdot z^{33} \equiv_\emptyset X_2 \cdot L(z)^{53}/[X_1 = L(x), X_2 = x] \text{ (viol. for } p_3)$$
$$X_1 \cdot z^{32} \equiv_\emptyset X_2 \cdot R(z)^{52}/[X_1 = R(x), X_2 = x] \text{ (viol. for } p_5)$$
$$X_1 \cdot z^{33} \equiv_\emptyset X_2 \cdot z^{43}/[X_1 = R(x), X_2 = R(x)] \text{ (viol. for } p_3)$$
$$X_1 \cdot z^{44} \equiv_\emptyset X_2 \cdot R(z)^{54}/[X_1 = R(x), X_2 = x] \text{ (viol. for } p_4 \text{ and } p_5)$$

This system is complete; all weight terms which appear in U appear for every term of the corresponding row of the incidence matrix. Therefore we may start to construct an invariant. For this purpose we collect the weight terms from U and arrange them with unknown $\left(\begin{array}{c|c} p_2 & a_1 \cdot L(x) \\ p_3 & a_2 \cdot L(x) + a_3 \cdot R(x) \\ p_4 & a_4 \cdot R(x) \\ p_5 & a_5 \cdot x \end{array}\right)$. In order to determine the multiplicities $a_1, \cdots, a_5$, we remember that every equation in U corresponds to an equivalence class of the vector $C^T \cdot I$ for the invariant which we want to construct. The sum of multiplicities in every equivalence class must be

0. Therefore every equation in U provides an equation for the a_i. For instance, the first equation in U provides $1 \cdot a_1 + (-1) \cdot a_5 = 0$, since the application of $a_1 \cdot L(x)$ to $1 \cdot z^{21}$ and the application of $a_5 \cdot x$ to $(-1) \cdot L(z)^{51}$ form the equivalence class $[L(z)]$ in the component t_1 of $C^T \cdot I$ and the sum of multiplicities is just $a_1 - a_5$. Additionally we obtain $a_2 - a_1 = 0$, $-a_2 + a_5 = 0$, $a_3 - a_5 = 0$, $-a_3 + a_4 = 0$ and $-a_4 + a_5 = 0$. Every solution of this system of equations is a linear combination of $a_1 = \cdots = a_5 = 1$. Therefore every invariant with of computed structure can be generated from $\left(\begin{array}{c|c} p_2 & 1 \cdot L(x) \\ p_3 & 1 \cdot L(x) + 1 \cdot R(x) \\ p_4 & 1 \cdot R(x) \\ p_5 & 1 \cdot x \end{array}\right)$.

With this invariant one branch of the computation has been finished. The consideration of the other branches leads to the invariants

$$\left(\begin{array}{c|c} p_1 & 1 \cdot x \\ p_2 & 1 \cdot x \\ p_3 & 1 \cdot x \\ p_4 & 1 \cdot x \end{array}\right), \left(\begin{array}{c|c} p_1 & 1 \cdot y \\ p_2 & 1 \cdot y \\ p_3 & 1 \cdot y \\ p_4 & 1 \cdot y \end{array}\right), \left(\begin{array}{c|c} p_2 & 1 \cdot y \\ p_3 & 2 \cdot y \\ p_4 & 1 \cdot y \\ p_5 & 1 \cdot y \end{array}\right)$$

and some linear combinations of these invariants such as

$$\left(\begin{array}{c|c} p_1 & -1 \cdot R(x) \\ p_2 & 1 \cdot L(x) - 1 \cdot R(x) \\ p_3 & 1 \cdot L(x) \\ p_5 & 1 \cdot x \end{array}\right) = \left(\begin{array}{c|c} p_2 & L(x) \\ p_3 & L(x) + R(x) \\ p_4 & R(x) \\ p_5 & x \end{array}\right) - R(x) \circ \left(\begin{array}{c|c} p_1 & x \\ p_2 & x \\ p_3 & x \\ p_4 & x \end{array}\right).$$

Consequently our generator set is not minimal. Especially our algorithm cannot detect whether an invariant can be combined from other ones when the equivalence classes caused by the combined invariants are merged. For instance the application of both invariants above leads to the equivalence class $R(z)$ in the component t_3 of the resulting $\mathbf{T}$–indexed vectors. But the weight terms which are responsible for this equivalence class cancel out each other in the combined invariant. Nevertheless the algorithm computes a small and fortunately a finite generator set of all invariants for the philosopher's net.

7 Conclusions

The problem of computing place invariants for algebraic nets can be traced back to equations between terms. We can solve these term equations in the case of equation free specifications with at most unary operation symbols. Though these restrictions are very hard several important problems can be specified this way. The generator sets we obtain with the proposed method are small in the following sense. A) We do not compute invariants which can be generated from other ones by simple concatenation or instantiation, we restrict ourselves to most general solutions and most general common instances. B) We avoid the computation of linear combinations, when the equivalence classes of the involved invariants do not influence each other. C) We only compute generator sets for the multiplicities of the weight terms occurring in a constructed invariant. Therefore, though the computed generator sets are not minimal, the computed set is suitable for the non–reachability test we aim in. Nevertheless a lot of problems remain. We only enumerate complete sets of solutions. We do not have a termination criterion for the generation of such sets (for T–invariants we know that there is no such criterion). It requires a lot of additional work to be able to cover specifications with equations or arbitrary operation symbols. Perhaps it will be necessary to modify the concept of place invariants for this purpose. All in all the algorithm may be a suitable element of an analysis tool for algebraic nets and a useful supply for their human driven analysis.

References

[Co90] J.M. Couvreur. The general computations of flows for coloured nets. *Proc. of the 11th Int. Workshop on Appl. and Theory of Petri Nets*, 1990.

[CM91] J.M. Couvreur, J. Martinez. Linear invariants in commutative high level nets. *High-level Petri nets, Theory and Application*, 1991.

[EM85] H. Ehrig, B. Mahr. *Fundamentals of Algebraic Specifications 1. EATCS Monographs on Theor. Comp. Science 6.* Springer 1985.

[GL83] H. Genrich, K. Lautenbach. S–invariance in Pr/T–nets. *Informatik-Fachberichte*, (66), 1983.

[HC88] S. Haddad, J.M. Couvreur. Towards a general and powerful computation of flows for parameterized coloured nets. *Proc. of the 9th European Workshop on Appl. and Theory of Petri Nets*, 1988.

[Je81a] K. Jensen. Coloured Petri nets and the invariant–method. *Theor. Comp. Science*, 14:317–336, 1981.

[Je81b] K. Jensen. How to find invariants for coloured Petri nets. *LNCS*, 118:327–338, 1981.

[LP85] K. Lautenbach, A. Pagnoni. Invariance and duality in predicate/transition nets and in coloured nets. *Arbeitspapiere der GMD*, 132, 1985.

[MV87] G. Memmi, J. Vautherin. Analysing nets by the invariant method. In: Rozenberg, Brauer, Reisig (eds.), *Advances in Petri nets 1986*, pages 300–336, 1987.

[Re91] W. Reisig. Petri nets and algebraic specifications. *Theor. Comp. Science*, 80 (1991):1–34.

[Sc94] K. Schmidt. T–invariants of algebraic petri nets. *Informatik–Bericht der HU Berlin*, 31, 1994.

[S+91] M. Silva, J. Martinez, P. Ladet, H. Alla. Generalized inverses and the calculations of symbolic invariants for coloured nets. *High-level Petri nets, Theory and Application*, 1991.

[Va86] J. Vautherin. Parallel systems specification with colored Petri nets and abstract data types. *Proc. of the 7th European Workshop on ATPN* Oxford 1986, 5–23.

[Va87] J. Vautherin. Calculation of semi–flows for Pr/T–systems. In T. Murata, editor, *Int. Workshop on Petri Nets and Performance Models.* IEEE Computer Society Press, 1987.

Failure-based Equivalences Are Faster Than Many Believe

Antti Valmari
Tampere University of Technology, Software Systems Laboratory
PO Box 553, FIN-33101 Tampere, FINLAND
ava@cs.tut.fi

Abstract

It has been known at least since 1983 that the checking of failure equivalence of two labelled transition systems is a PSPACE-complete problem, while the same problem for observation equivalence can be solved in roughly cubic time. Correspondingly, a widespread belief has arisen that failure-based equivalences are slower for the verification of concurrent systems than observation equivalence. In this article it is shown that this belief is based on a misinterpretation of the theoretical complexity result. Both theoretical and experimental evidence is given to the claim that failure-based equivalences are often faster in practice than observation equivalence. It is argued that the weakest congruence preserving given properties is in a certain sense optimal for the verification of those properties. Results giving weakest congruences for several verification tasks are surveyed.

1 Introduction

Consider two labelled transition systems (abbreviated *LTS*) with a total of n states and m transitions. In 1983 Kanellakis and Smolka published an article [1, 2] containing proofs that (1) whether the LTSs are *observation-equivalent* [3] can be checked in time $O(mn^3)$, and (2) the checking if they are *failure-equivalent* [4] is a PSPACE-complete problem. The PSPACE-complexity of failure equivalence checking was proven also in [5]. The upper limit $O(mn^3)$ for observation equivalence is not even tight. The observation equivalence checking algorithm can be used also for constructing a state-minimal observation-equivalent LTS from a given LTS (e.g. [6]). Furthermore, the result can be made transition-minimal and unique up to the naming of states [7]. On the other hand, failure equivalence does not guarantee the existence of unique minimal LTSs, and finding an equivalent LTS with the smallest possible number of states is PSPACE-hard.

Perhaps because of this reason, observation equivalence has become more popular in contemporary process algebra verification research of concurrent systems than failure-based equivalences. For instance, the majority of the tools listed on p. 175 in [8] have a feature for checking observation equivalence, but only some of them support failure equivalence or its variants. When discussing with other researchers of the field, the present author has several times encountered the attitude that failure-based equivalences are less promising for practical verification because of the above-mentioned complexity results.

The goal of the present article is to show that failure-based equivalences are not as bad as the general opinion seems to suggest. In Section 2 we investigate the structure of typical concurrent system verification problems, define the basic concepts we will need later, and discuss the state explosion problem and some process algebraic techniques with which it can be alleviated. Section 3 is devoted to a discussion of the use of equivalences as tools for the verification of other properties, such as mutual exclusion or the absence of deadlocks. We demonstrate that the above complexity results affect only one step in the verification process. Furthermore, this step does not determine the complexity of the overall verification problem. We argue that if there are several equivalences available which preserve the interesting properties, then in a certain practical sense a weaker one (i.e. one which makes less distinctions between LTSs) cannot have a disadvantage but may have an advantage over a stronger one. We give examples supporting this claim.

The results in [1, 2] do not directly apply to concurrent systems because, as the size of a system, they use the size of its (global) LTS. In Section 4 we first cite some results concerning the complexity of equivalence checking when the size of a concurrent system is interpreted as the sum of the sizes of its component processes. Although the results are not decisive, they suggest that the theoretical complexity advantage of observation equivalence remains valid in this new setting. Then we point out that despite the theoretical result, in a typical practical verification situation, it is not at all clear that observation equivalence is faster than failure-based equivalences. We are not claiming, however, that the opposite would be true; we are just claiming that there is no decisive a-priori evidence that observation equivalence would be faster in the general case (although there is strong evidence that it is faster in the worst case).

The results in Section 3 make it clear that the weakest congruence preserving a certain class of properties is, in a certain sense, optimal for the verification of general properties within that class. Therefore, in Section 5 we survey some results giving weakest congruences for various classes of problems. Section 6 concludes the article.

We would like to stress at this point that we are not claiming that observation equivalence is bad or that it should not be used. We are only trying to say that failure-based equivalences are not bad and they should be used more often than they have been used so far. There are good reasons for using observation equivalence: the notion of bisimulation it is based on is intuitively appealing; it has nice mathematical properties; and fast, useful algorithms have been developed for it. However, the arguments in favour of observation equivalence are well known, so we concentrate in this article on the less well known negative arguments.

2 Concurrent System Verification Problems

The theoretical complexity results cited in the introduction apply to situations where two explicitly given LTSs have to be compared or an explicitly given LTS has to be minimised. In practice, when verifying concurrent systems, we seldom start with this kind of a problem. Instead, we usually have some system whose behaviour we want to analyse or verify. We will call it the *implementation* in this article. We may also have a *specification* which is another system or a collection of logical formulas; or we may just have a set of queries such as "Are there any deadlocks in the implementation?" or "Does mutual exclusion hold in it?". We assume that the implementation has been given as a parallel composition of more than one processes. This is most of the time true in practice, and when it is not, we can hardly talk about the verification of *concurrent* systems. A system representing the specification is allowed to consist of only one process.

To represent verification problems formally, we define labelled transition systems and the *parallel composition* "||" and *hiding* "**hide** A **in** ..." operators for combining them. According to the definition of "||", a visible action a is executed simultaneously by all LTSs connected together with it. That is, all LTSs with a in their sets of visible actions synchronise on a. The LTSs do not synchronise on the invisible action τ. The hiding operator transforms into the invisible action all actions which belong to the set A specified in the operator. It can be used for abstracting irrelevant details away, and it is a central operator in many process algebraic theories. (In the case of CCS [3], there is no explicit hiding operator, but the parallel composition operator also hides actions.)

Definition 2.1

- τ is the *invisible* action.
- A *labelled transition system* (or *LTS*, for short) is the quadruple $(S, \Sigma_\tau, \Delta, s_0)$, where S is a set of *states*, Σ is the set of *visible* actions chosen such that $\tau \notin \Sigma$, $\Sigma_\tau = \Sigma \cup \{\tau\}$, the *next state relation* $\Delta \subseteq S \times \Sigma_\tau \times S$, and the *initial state* $s_0 \in S$. The elements of Δ are called *transitions*.
- The *size* of an LTS $L = (S, \Sigma_\tau, \Delta, s_0)$ is $size(L) = |S| + |\Sigma_\tau| + |\Delta|$.
- Let $L = (S, \Sigma_\tau, \Delta, s_0)$ be an LTS, and $a \in \Sigma_\tau$. L may execute a and transform itself into $L' = (S, \Sigma_\tau, \Delta, s)$, if and only if $(s_0,a,s) \in \Delta$. We denote this by $L \text{-}a\rightarrow L'$. When S, Σ_τ, and Δ are obvious from the context and s and $s' \in S$, we may also write $s \text{-}a\rightarrow s'$ to denote that $(s,a,s') \in \Delta$.
- If L is an LTS, then *reach*(L) is the LTS obtained from L by removing all states and transitions not reachable from the initial state.
- If $L_1 = (S_1, \Sigma_{1\tau}, \Delta_1, s_{01})$, ..., $L_n = (S_n, \Sigma_{n\tau}, \Delta_n, s_{0n})$ are LTSs, then $L_1 \parallel \ldots \parallel L_n$ is the LTS *reach*($(S', \Sigma'_\tau, \Delta', s'_0)$), where $S' = S_1 \times \ldots \times S_n$, $\Sigma' = \Sigma_1 \cup \ldots \cup \Sigma_n$, $s'_0 = (s_{01}, \ldots, s_{0n})$, and $((s_1, \ldots, s_n), a, (s'_1, \ldots, s'_n)) \in \Delta'$ if and only if either $a = \tau$ and $(s_i,\tau,s'_i) \in \Delta_i$ for some $1 \le i \le n$ and $s_j = s'_j$ when $j \ne i$; or $a \ne \tau$ and $(s_i,a,s'_i) \in \Delta_i$ for every i such that $a \in \Sigma_i$ and $s_i = s'_i$ for every i such that $a \notin \Sigma_i$.
- If $L = (S, \Sigma_\tau, \Delta, s_0)$ is an LTS and $A \subseteq \Sigma$, then **hide** A **in** L is the LTS $(S, \Sigma'_\tau, \Delta', s_0)$, where $\Sigma' = \Sigma - A$, and $(s,a,s') \in \Delta'$ if and only if $a = \tau \wedge \exists\, b \in A$: $(s,b,s') \in \Delta$, or $a \notin A \wedge (s,a,s') \in \Delta$.

□

The following notation will be used for talking about the sequences of actions a process may execute.

Definition 2.2 Let P, Q, P_0, P_1, ..., P_n be LTSs with Σ as their set of visible actions, and let $a_1, a_2, \ldots, a_n \in \Sigma_\tau$.

- Σ^* is the set of finite and Σ^ω is the set of infinite strings of symbols from Σ. In particular, Σ^* contains the *empty string* ε.
- $P -a_1a_2\ldots a_n\rightarrow Q$ iff $\exists\, P_0, P_1, \ldots, P_n$: $P_0 = P \wedge P_n = Q \wedge P_0 -a_1\rightarrow P_1 -a_2\rightarrow \ldots -a_n\rightarrow P_n$.
- $P =a_1a_2\ldots a_n\Rightarrow Q$ iff $P -\tau^*a_1\tau^*a_2\tau^*\ldots\tau^*a_n\tau^*\rightarrow Q$, where τ^* denotes any sequence consisting of zero or more τ-symbols.
- $P -a_1a_2\ldots a_n\rightarrow$ iff $\exists\, Q$: $P -a_1a_2\ldots a_n\rightarrow Q$, and similarly for $P =a_1a_2\ldots a_n\Rightarrow$.
- $P \not-a_1a_2\ldots a_n\not\rightarrow Q$ iff not $P -a_1a_2\ldots a_n\rightarrow Q$, and similarly for $P \not-a_1a_2\ldots a_n\not\rightarrow$, $P \neq a_1a_2\ldots a_n\not\Rightarrow Q$, and $P \neq a_1a_2\ldots a_n\not\Rightarrow$.

□

The following concepts are needed in the definitions of many of the equivalences we will talk about in this article.

Definition 2.3 Let P be a LTS with Σ as its set of visible actions.

- P is *stable*, denoted by *stable*(P), iff $P \not-\tau\not\rightarrow$.
- The set of *traces* of P is $tr(P) = \{\, \sigma \in \Sigma^* \mid P =\sigma\Rightarrow \,\}$.
- The set of *infinite traces* of P is $inftr(P) =$
  $\{\, a_1a_2a_3\ldots \in \Sigma^\omega \mid$
  $\exists\, P_0, P_1, P_2, \ldots$: $P = P_0 \wedge P_0 =a_1\Rightarrow P_1 =a_2\Rightarrow P_2 =a_3\Rightarrow \ldots \,\}$.
- The set of *divergence traces* of P is $divtr(P) =$
  $\{\, \sigma \in \Sigma^* \mid \exists\, Q_0, Q_1, Q_2, \ldots$: $P =\sigma\Rightarrow Q_0 \wedge Q_0 -\tau\rightarrow Q_1 -\tau\rightarrow Q_2 -\tau\rightarrow \ldots \,\}$.
- The set of *failures* of P is
  $fail(P) = \{\, (\sigma, A) \in \Sigma^* \times 2^\Sigma \mid \exists\, Q$: $P =\sigma\Rightarrow Q \wedge \forall\, a \in A$: $Q \neq a\not\Rightarrow \,\}$.
- The set of *stable failures* of P is
  $sfail(P) = \{\, (\sigma, A) \in \Sigma^* \times 2^\Sigma \mid \exists\, Q$: $P =\sigma\Rightarrow Q \wedge \forall\, a \in A \cup \{\tau\}$: $Q \not-a\not\rightarrow \,\}$.
- The set of *nondivergent failures* of P is
  $ndfail(P) = \{\, (\sigma, A) \in sfail(P) \mid \sigma \notin divtr(P) \,\}$.

□

Trace, failure, and observation equivalence may be defined as follows.

Definition 2.4 Let $P = (S_P, \Sigma_\tau, \Delta_P, s_{0P})$ and $Q = (S_Q, \Sigma_\tau, \Delta_Q, s_{0Q})$ be two LTSs with a common set of visible actions.

- P and Q are *trace-equivalent* iff $tr(P) = tr(Q)$.
- P and Q are *failure-equivalent* iff $fail(P) = fail(Q)$.
- P and Q are *observation-equivalent*, iff there is a relation "$\approx$" $\subseteq S_P \times S_Q$ such that
  - $(s_{0P}, s_{0Q}) \in$ "$\approx$";
  - if $(s_P, s_Q) \in$ "$\approx$", $a \in \Sigma \cup \{\varepsilon\}$, and $s_P =a\Rightarrow s'_P$, then there is s'_Q such that $s_Q =a\Rightarrow s'_Q$ and $(s'_P, s'_Q) \in$ "$\approx$"; and
  - if $(s_P, s_Q) \in$ "$\approx$", $a \in \Sigma \cup \{\varepsilon\}$, and $s_Q =a\Rightarrow s'_Q$, then there is s'_P such that $s_P =a\Rightarrow s'_P$ and $(s'_P, s'_Q) \in$ "$\approx$".

□

The analysis and verification of systems consisting of several parallel processes suffer from the well-known *state explosion* problem. That is, the number of states tends to grow exponentially in the number of processes, with the consequence that typical systems have far more states than can be processed in a realistic computer. Several techniques have been developed for attacking the state explosion problem. Three examples of techniques applicable in the process algebra world are the direct generation of bisimulation-minimised LTSs using symbolic representation techniques [9]; the generation of a failure- and divergence-preserving reduced LTS using stubborn sets [10]; and on-the-fly checking of preorder relations with tester processes such as those in [11] and algorithms like those in [12, 13]. Yet another technique, *compositional LTS construction* (see e.g. [14, 15, 16]), deserves to be discussed in some detail here.

In compositional LTS construction the system is divided into subsystems, and LTSs for the subsystems are constructed. Then, for each subsystem LTS, an equivalent but smaller LTS is constructed (we call this *LTS reduction*). Finally the reduced subsystem LTSs are composed together to yield a reduced LTS of the system as a whole. This way the construction of the huge complete LTS of the system as a whole is avoided. For the results to be correct, it is required that the equivalence used is a congruence with respect to the operators used in building up the system from its components. Typically these operators include some forms of the parallel composition and hiding operators. If the number of processes is big enough, then even more savings may be obtained by constructing the subsystem LTSs compositionally.

When the specification is given in the form of another system, then the verification problem is typically the problem of checking whether the implementation is equivalent to or conforms to the specification. In this case, checking of equivalence or preorder is a goal in itself. We will investigate this situation in Section 4. When the specification is a collection of logical formulas or we have a collection of queries instead of a specification, then equivalences are used only as tools in the verification. For instance, instead of constructing the huge complete global LTS of a system, we may use some equivalence to compositionally construct a much smaller LTS, and run our model checker tool on it or search for deadlocks in it. This approach requires that the equivalence preserves the properties expressed by the formulas or queries.

To give a concrete example, in [15] an advanced version of compositional LTS construction was used for verifying that the token rotates correctly in a certain kind of a token ring system with n customers. The full LTS of the system contained an exponential number of states ($9n2^{n-2}$, to be exact), but only n LTSs with a linear number of states (apparently $4n+4$) in the biggest of them had to be constructed with the method in the article. The equivalence used was observation equivalence.

In the next section we discuss the use of different equivalences as tools in the verification of other kinds of properties.

3 Equivalences as a Verification Aid

The following results can be found more or less directly in [17, 18, 19]. We prove them again here, because the proof is easy, and the reader can easily adapt it to show that many other verification problems of concurrent systems are PSPACE-complete.

Theorem 3.1 Let P be a system of the form $P = L_1 \parallel \ldots \parallel L_n$, where $L_1, \ldots, L_n$ are LTSs. If the size of P is considered to be $size(L_1) + \ldots + size(L_n)$, then the following problems are PSPACE-complete:

(a) Can P deadlock, i.e. reach a state s such that $s \not\!-a\!\not\rightarrow$ for every $a \in \Sigma_\tau$?

(b) Can P ever execute action a, i.e. is there any $\sigma \in \Sigma_\tau^*$ such that $P -\sigma a\rightarrow$?

(c) Is P guaranteed to eventually execute a at least once?

(d) Is P guaranteed to execute a infinitely many times?

Proof An execution leading to a deadlock; ending with the execution of a; deadlocking or looping before executing a; or deadlocking or looping without a occurring in the loop can be simulated without using much extra space, so these problems are in nondeterministic polynomial space (NPSPACE). But NPSPACE = PSPACE, so the problems are in PSPACE, too.

A *linear bounded automaton* (*LBA*) is a nondeterministic Turing machine with a fixed-length tape. That is, the tape contains two endmarkers ">" and "<", which the machine does not overtake or write over. The problem whether an LBA accepts the string originally on its tape is PSPACE-complete [20], p. 347. We shall prove that (a) and (b) are PSPACE-hard by transforming the LBA acceptance problem to them.

An LBA with n states, tape length m (including the endmarkers), and k elements in its alphabet (including the endmarkers) can be simulated by $m+2$ LTSs connected in parallel. One of the LTSs "L_S" has n states, and it represents the state of the LBA. Another LTS "L_H" has m states, and it is used to store the location of the tape head of the LBA. The remaining m LTSs "L_1", ..., "L_m" represent the m tape locations, and each of them has k states. Each transition of the LBA is represented by m actions, one for each possible tape head location. Each action is synchronised to by L_S, L_H, and exactly one of the L_i. The corresponding local transitions model the testing and change of state of the finite control of the LBA, the tape head location, and the symbol on the tape. The names of these actions should be other than "a".

Acceptance can be transformed to the reachability of deadlocks or the executability of a by adding to L_S an extra state s_d with no output transitions, an a-transition to s_d from each state of L_S corresponding to an accepting state of the LBA, and a τ-transition from each state of L_S to itself. So (a) and (b) are PSPACE-hard.

The proof of (c) and (d) consists of a more complicated LBA simulation construction containing a binary counter, which keeps track of the number of simulated steps, and resets the system into its initial state each time the number exceeds nmk^m. This reset action is given the name "a". The system is reset to the initial state also each time it reaches an accepting state, but the name of this reset action is different from a. The system is prevented from deadlocking by adding, where needed, to the transition relation of the LBA idle transitions not changing the LBA state, the symbol on the tape, or the tape head location. Because the LBA has at most nmk^m configurations, the system may avoid ever executing a if and only if the LBA accepts the string originally on the tape. In the opposite case the system is bound to execute a infinitely many times. The counter needs only a polynomial amount of space, because it consists of several LTSs, each storing only one bit of the count. A detailed proof can be found in [19]. □

So we see that when the system consists of several processes connected in parallel, then typical verification problems are PSPACE-hard. It would thus be irrational to avoid a failure-based equivalence as a tool for verifying these properties only on the basis of the argument that with it, equivalence checking is PSPACE-complete.

Of course, the high complexity of the verification problems manifests itself in the state explosion problem. Do we, then, have both the state explosion problem and over-expensive LTS comparison and/or minimisation when failure-based equivalences are used, while with observation equivalence we can avoid the latter? To answer this question it is necessary to take a more detailed look at the use of an equivalence as a tool in the verification of other properties.

For concreteness, consider the verification of the absence of deadlocks using compositional LTS construction. So we construct an equivalent reduced LTS instead of the full LTS, and check it for deadlocks. For the results to be correct, it is essential that the equivalence preserves deadlocks. That is, if two LTSs are equivalent, then either both or none of them contains a deadlock.

Definition 3.2 Equivalence "$\approx_1$" is *weaker* than equivalence "$\approx_2$", if and only if for all LTSs P and Q, $P \approx_2 Q \Rightarrow P \approx_1 Q$, but not necessarily vice versa. □

We claim that if both "$\approx_1$" and "$\approx_2$" preserve deadlocks and "$\approx_1$" is weaker than "$\approx_2$", then "$\approx_1$" is better for the verification of the absence of deadlocks. This is because the equivalence classes induced by "$\approx_2$" are subsets of those induced by "$\approx_1$". Therefore, with "$\approx_1$" it is always possible to find at least as small and often possible to find a smaller reduced LTS than with "$\approx_2$". Thus "$\approx_1$" has potential for more savings of states during compositional LTS construction than "$\approx_2$".

But, if we assume that the weaker equivalence "$\approx_1$" is failure equivalence and the stronger equivalence "$\approx_2$" is observation equivalence, doesn't the high complexity of finding the smallest failure-equivalent LTS rule out the extra savings of states? No, because the reduced LTS need not be the smallest equivalent one. Minimisation is needed only for LTS comparison; for compositional LTS construction, reduction is sufficient. Therefore, we can use any heuristic reduction algorithm which is cheap and powerful enough. For instance, we can beat observation equivalence with its own weapons by first minimising an LTS with respect to it, then performing some heuristic trick preserving failure equivalence but not observation equivalence (for instance, the τ-collapse tricks in Section 3.4 of [16]), and finally minimising again with respect to observation equivalence. This algorithm is less than three times as slow as minimisation with respect to observation equivalence. Because the last two steps cannot make the number of states grow, the result contains at most as many states as the observation-equivalence-minimised LTS. However, because the middle step does not preserve observation equivalence, the result may be much smaller than the observation-equivalence-minimised LTS.

In general, if we can use several equivalences in the verification of some property φ, then weaker equivalences offer more tools for conducting the verification efficiently than strong ones. Any verification of φ conducted using an equivalence "$\approx_2$" can be thought of as a legal verification conducted using a strictly weaker equivalence "$\approx_1$", but "$\approx_1$" offers additional possibilities, because it allows more kinds of manipulation of the LTSs. Although some of the extra possibilities may be prohibitively expensive (like the minimisation of states with respect to failure equivalence), there may be other possibilities which are more powerful but no more expensive than the best ways provided by the stronger equivalence (like the LTS reduction algorithm for failure equivalence discussed above).

Let us consider again the token-ring verification problem in [15]. The article [10] presents a fully automatic reduced LTS construction method which is valid for failure equivalence and the CFFD- and NDFD-equivalences presented later in this article, but is not valid for observation equivalence. From the token-ring system this method constructs only one LTS with only $5n$ states and $6n$ transitions. So the result is even better than with the observation-equivalence-preserving method described in [15].

We have to remember also that the theoretical complexity results in [1, 2] are worst-case results. An algorithm may work well in practice although its worst-case complexity is bad. A CFFD-equivalence-preserving LTS normalisation algorithm based on determinisation and minimisation of the input LTS was described in [21]. Although the determinisation step may make the LTS grow, most of the time in practice the output LTS is smaller than the input LTS, making the algorithm also suitable for LTS reduction. In [22] a series of experiments was reported, where this algorithm was applied to 66 LTSs which had already been minimised with respect to a divergence-preserving variant of observation equivalence. In 76 % of the experiments the algorithm reduced the LTS further still, and the average reduction (over the cases where there was extra reduction) was 44 %.

So we see that the weaker an equivalence is, the more possibilities it offers for efficient verification. Of course, the equivalence has to preserve the properties of interest. Furthermore, many practical verification techniques require that the equivalence is a congruence. Therefore, given a class of properties, it is interesting to know the weakest congruence preserving it. In Section 5 we survey results giving the weakest congruences for several classes of properties.

4 Verification of Equivalence

In this section we discuss the complexity of checking the equivalence of two systems. The situation is different from the previous section in that if we change from one equivalence to another, then we do not only change the verification algorithm, but also the verification problem. Even so, comparing the complexities is interesting.

The complexity results in [1, 2] make it almost certain that the checking of observation equivalence of two LTSs is, in the worst case, much faster than the checking of their failure equivalence. Not much has been published about the complexity of comparing two systems consisting of several LTSs connected in parallel. The following theorem is from [23]. The article contains a proof sketch for the trace-part, and refers to an unpublished result by Stockmeyer for the observation-equivalence-part.

Theorem 4.1 Let $P =$ **hide** A_P **in** ($L_{P1} \| \ldots \| L_{Pn}$) and $Q =$ **hide** A_Q **in** ($L_{Q1} \| \ldots \| L_{Qm}$), where $L_{P1}, \ldots, L_{Pn}, L_{Q1}, \ldots, L_{Qm}$ are LTSs, A_P is some set of visible actions of $L_{P1}, \ldots, L_{Pn}$, and A_Q is some set of visible actions of $L_{Q1}, \ldots, L_{Qm}$. The checking of whether P and Q are trace-equivalent is EXPSPACE-complete, and the same problem for observation equivalence is EXPTIME-complete. (The size of a problem instance is considered to be $size(L_{P1}) + \ldots + size(L_{Pn}) + |A_P| + size(L_{Q1}) \ldots + size(L_{Qm}) + |A_Q|$.) □

This theorem leaves open the complexity of checking failure equivalence of P and Q, but it is likely that it is the same as for trace equivalence. Open remains also the special case where m has been fixed to 1, that is, Q consists of one explicitly given LTS. This special case is interesting in practice because often the specification of a system has been given as an explicit LTS, but the system itself is a parallel composition of several LTSs. However, we knew already at the start that this special case is at least PSPACE-hard for failure equivalence. It is not difficult to see that it is at least PSPACE-hard also for observation equivalence.

Theorem 4.2 Let L be an LTS, and let P be a system of the form P = **hide** A **in** ($L_1 \parallel \ldots \parallel L_n$), where $L_1, \ldots, L_n$ are LTSs, and A is some set of visible actions of $L_1, \ldots, L_n$. The checking of whether P and L are observation-equivalent is PSPACE-hard.

Proof Let Σ be the set of all visible actions of $L_1, \ldots, L_n$ and let a be some visible action. We can check whether $L_1 \parallel \ldots \parallel L_n$ can ever execute a by checking if **hide** $\Sigma - \{a\}$ **in** ($L_1 \parallel \ldots \parallel L_n$) is not observation-equivalent with the LTS "**stop**" consisting of one state and no transitions. By Theorem 3.1 (b) the former problem is PSPACE-complete, so the latter is PSPACE-hard. □

Of course, it is likely that the problem is much harder both for failure equivalence and for observation equivalence.

What is the practical significance of these results? When we are comparing two systems using failure equivalence, then the comparison algorithm is potentially much more expensive than with observation equivalence. On the other hand, we saw in the previous section that failure equivalence allows more LTS reduction than observation equivalence. Therefore, the more expensive algorithm is applied to smaller LTSs. The above complexity results suggest that in the worst case, failure equivalence does not provide enough reduction to compensate for the greater comparison costs. They do not, however, say anything about the practical cases. The experiences reported in [24] p. 29, [25], and [22] suggest that the checking of failure and CFFD-equivalences is often, but not always, feasible in practice.

5 Weakest Congruences

We showed in Section 3 that the weaker an equivalence is, the more tools it offers for the efficient verification of a property, as long as it preserves that property. We also mentioned that many verification techniques (most notably compositional LTS construction) require that the equivalence is a congruence. The weakest congruence preserving a property is thus optimal for the verification of that property in the sense that among all applicable congruences, the set of tools it offers is the largest. It would thus be nice to know the weakest congruences for various classes of verification problems.

Unfortunately, the notion of "congruence" is relative to the set of operators allowed for building a system up from its components.

Definition 5.1 Let f be a function from n systems to systems. An equivalence "$\approx$" is a *congruence* with respect to f, iff $P_1 \approx Q_1 \wedge \ldots \wedge P_n \approx Q_n$ implies $f(P_1, \ldots, P_n) \approx f(Q_1, \ldots, Q_n)$. □

Fortunately, if an equivalence is a congruence with respect to some set of system composition operators, then it is a congruence with respect to all functions composable of those operators. Even so, when claiming that some equivalence is a congruence, we have to fix the set of system composition operators. The congruence properties of various failure-based equivalences with respect to all operators of the specification language Basic Lotos [26] were analysed in [27]. The congruence claims given below are taken from [27], and apply to all operators of Basic Lotos, unless explicitly said otherwise.

Our first weakest congruence result is not surprising. We give and prove it in order to demonstrate a technique which may be used in proofs that a congruence is the *weakest* preserving some property.

Theorem 5.2 The weakest congruence (with respect to any set of Basic Lotos operators containing parallel composition) preserving the property "P may ever execute action a" is trace equivalence.

Proof It is well known that trace equivalence is a congruence with respect to almost all system composition operators found in process algebraic languages. In particular, it is a congruence with respect to all operators in Basic Lotos. Furthermore, it is obvious that it preserves the property. We now show that any congruence preserving the property must preserve all traces of the system. Let Σ be the alphabet of P, and $a_1a_2\ldots a_n \in \Sigma^*$. Let x be some symbol not in Σ, and let $test(a_1a_2\ldots a_n)$ be the LTS with alphabet $\Sigma \cup \{x\}$ and the states and transitions as shown in Figure 1. The system $P \parallel test(a_1a_2\ldots a_n)$ can ever execute x if and only if $a_1a_2\ldots a_n \in tr(P)$. If "$\approx$" is a congruence preserving the property in the theorem and $P \approx Q$, then $P \parallel test(a_1a_2\ldots a_n) \approx Q \parallel test(a_1a_2\ldots a_n)$ by the congruence requirement. Thus either none or both of $P \parallel test(a_1a_2\ldots a_n)$ and $Q \parallel test(a_1a_2\ldots a_n)$ can ever execute x, and we conclude that either none or both of P and Q have $a_1a_2\ldots a_n$ as their trace. Because the same reasoning applies to all elements of Σ^*, we conclude that $tr(P) = tr(Q)$. □

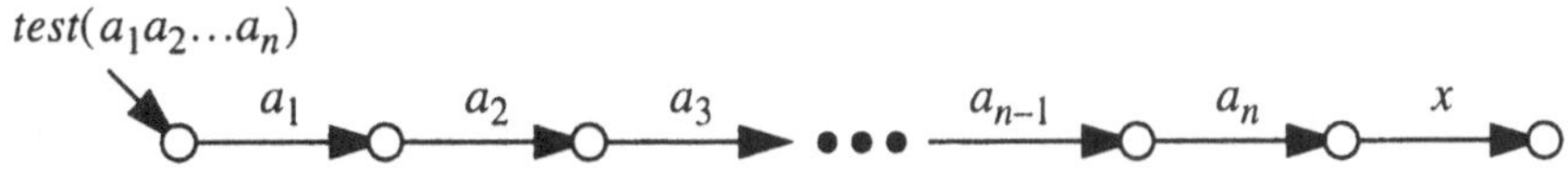

Figure 1

The next result concerns the preservation of deadlocks. Notice that the problem is different from the well-known problem of finding the weakest congruence refining the completed-trace-equivalence, although the congruence happens to be the same.

Definition 5.3 System P is a *deadlock*, iff $\forall\, a \in \Sigma_\tau$: $P \not\xrightarrow{a}$. P is *deadlocking*, iff there are $\sigma \in \Sigma^*$ and a deadlock D such that $P = \sigma \Rightarrow D$. We define $dlg(P) =$ **true**, if P is deadlocking, and $dlg(P) =$ **false**, otherwise. An equivalence "$\approx$" is *deadlock-preserving*, if $P \approx Q$ implies $dlg(P) = dlg(Q)$. □

The weakest congruence which preserves deadlocks depends heavily on the set of system composition operators, as can be seen from Theorem 5.4. The theorem has been taken from [28]. The *action prefix* operator "$a;$" is the same as "$a.$" in CCS. The *renaming* operator "$[h_1,\ldots,h_k / g_1,\ldots,g_k]$" transforms all occurrences of action g_i to occurrences of the corresponding h_i and leaves the LTS otherwise intact. It is the same as the CCS operator "$[f]$". The *choice* operator "[]" is the same as the CCS "+". Finally, "[>" is the *disabling* operator. Its essential feature is that initially P [> Q behaves like P and may behave like P forever or until termination. However, at any instant of time it may start behaving like Q, and from then on it may behave only like Q. The intuition is that the execution of P may be interrupted and aborted at any time by starting the execution of Q. The disabling operator resembles the CSP interrupt "$\wedge$" [29] p. 180, [3] p. 192. For a formal definition of these operators see [26], for instance.

Theorem 5.4

(a) Let "$\approx_{sf}$", "stable failure equivalence", be the equivalence defined by $P \approx_{sf} Q$ iff $sfail(P) = sfail(Q)$. "$\approx_{sf}$" is the weakest deadlock-preserving congruence w.r.t. any subset of {"$a;$", "||", "**hide**", "$[h_1,\ldots,h_k / g_1,\ldots,g_k]$"} containing "||".

(b) The "stability and stable failure equivalence" "$\approx_{ss}$" defined by
$$P \approx_{ss} Q \Leftrightarrow sfail(P) = sfail(Q) \wedge stable(P) = stable(Q)$$
is the weakest deadlock-preserving congruence w.r.t. any subset of {"$a;$", "||", "**hide**", "$[h_1,\ldots,h_k / g_1,\ldots,g_k]$", "[]"} containing "||" and "[]".

(c) The "traces, stability, and stable failure equivalence" "$\approx_{tss}$" defined by
$$P \approx_{tss} Q \Leftrightarrow tr(P) = tr(Q) \wedge sfail(P) = sfail(Q) \wedge stable(P) = stable(Q)$$
is the weakest deadlock-preserving congruence w.r.t. any subset of Basic Lotos operators containing "||" and "[>".

□

The equivalence "$\approx_{tss}$" is essentially the same as the equivalence "$\equiv$*" in [30].

Finally, assume that we want to use the classic state-based linear time temporal logic for specifying properties of a system [31]. We remove the *next state* operator "O" from the logic, because it is not very meaningful in the context of concurrent systems, and call the resulting logic *LTL′*. In order to make process algebraic techniques applicable, we assume that our specification does not refer to the local states of all LTSs of the system. (For instance, the system may be a model of a communication protocol with customers, and the specification may describe the service expected to be provided by the protocol. Then the specification refers to the local states of the customers but not to the local states of the LTSs implementing the protocol.) Therefore, let the system be of the form P || Q, where P is a subsystem containing at least the LTSs whose local states are talked about by the specification, and Q contains the remaining LTSs. It is now meaningful to use process algebraic techniques such as compositional LTS construction for replacing Q by another, smaller LTS, and ask whether the validity of any LTL′ formula changes.

The question is thus: what is the weakest congruence preserving the validity of all LTL′ formulas talking about the local states of P in the system P || Q, when Q is replaced by an equivalent LTS? The answer was given in [32]:

Theorem 5.5 Let $P \parallel Q$ be a system, and let "≈" be an equivalence. We say that "≈" *preserves* LTL′, if and only if for every Q' such that $Q' \approx Q$, the truth-value of every LTL′-formula talking about the local states of only the LTSs in P is the same in $P \parallel Q'$ as in $P \parallel Q$. The following holds:

(a) The weakest congruence preserving LTL′ is the *NDFD-equivalence* (*Nondivergent Failures Divergences equivalence*) defined as follows:

$$P \approx_{NDFD} Q \Leftrightarrow stable(P) = stable(Q) \wedge ndfail(P) = ndfail(Q) \wedge divtr(P) = divtr(Q) \wedge inftr(P) = inftr(Q).$$

(b) The weakest congruence preserving LTL′ and the ability of deadlocking is the *CFFD-equivalence* (*Chaos-Free Failures Divergences equivalence*) defined as follows:

$$P \approx_{CFFD} Q \Leftrightarrow stable(P) = stable(Q) \wedge sfail(P) = sfail(Q) \wedge divtr(P) = divtr(Q) \wedge inftr(P) = inftr(Q).$$

□

It may seem surprising that adding deadlock-preservation to LTL′ yields a different weakest congruence, because it has the counter-intuitive consequence that ability to deadlock cannot be specified in LTL′. The explanation is that in the above setting, the logic is not allowed to look at the local states of Q. Therefore, if Q does not ever synchronise with P, the logic cannot see whether it is because Q is in a deadlock or because it is executing an infinite sequence of τ-actions.

6 Conclusions

We demonstrated that despite widespread belief, failure-based equivalences are not necessarily any slower in practical verification of concurrent systems than observation equivalence. Instead, sometimes they may be faster. We presented both theoretical (analysis of the complexity of some verification problems) and practical (measurements, verification techniques applicable to failure-based equivalences but not to observation equivalence) evidence to our claim. Our results are not in contradiction with earlier complexity results because we assumed that the system under analysis is not given as an explicit LTS, but as a parallel composition of several LTSs. This is actually the case most of the time when equivalences are used in verification, and when it is not, we can hardly talk about the verification of *concurrent* systems.

Outside computational efficiency, failure-based equivalences have two practical advantages. First, when an attempt to verify some property reveals that the property does not hold, then it would be helpful to get information about what is wrong. For instance, if the system deadlocks, it would be helpful to get a sample execution leading to a deadlock. When two systems are not failure-equivalent, then there is always an execution with which this can be demonstrated. The same does not hold for observation equivalence because of its "branching" nature. Second, if LTSs are investigated manually like in [33], then it is important to have as small LTSs as possible. As we demonstrated in Sections 3 and 5, weaker equivalences yield smaller LTSs, and failure-based equivalences are typically weaker than observation equivalence.

Finally, we stress once more that we are not claiming that observation equivalence is bad. What we are claiming is that failure-based equivalences are not as bad as generally thought. The arguments in favour of observation equivalence are well-known, so we did not repeat them here, but concentrated on the arguments in favour of failure-based equivalences.

Acknowledgements

This work was partly funded by the Technology Development Centre of Finland (TEKES) and the Technical Research Centre of Finland (VTT), in connection with the European Community ESPRIT BRA Project REACT (6021). This article has evolved from an unpublished talk I gave in the 5th Nordic Workshop on Program Correctness, Turku, Finland, October 1993. I thank Hans Hüttel, Kim Larsen, Frits Vaandrager, Pierre Wolper, and the anonymous referees for fruitful discussions and helpful remarks.

References

1. Kanellakis, P. C. & Smolka, S. A.: *CCS Expressions, Finite State Processes, and Three Problems of Equivalence.* Proceedings of the 2nd Annual ACM Symposium on Principles of Distributed Computing, Montreal, Canada, August 1983, pp. 228–240.
2. Kanellakis, P. C. & Smolka, S. A.: *CCS Expressions, Finite State Processes, and Three Problems of Equivalence.* Information and Computation **86** (1990) pp. 43–68.
3. Milner, R.: *Communication and Concurrency.* Prentice-Hall 1989, 260 p.
4. Brookes, S. D., Hoare, C. A. R. & Roscoe, A. W.: *A Theory of Communicating Sequential Processes.* Journal of the ACM, 31 (3) 1984, pp. 560–599.
5. Brookes, S. D. & Rounds, W. C.: *Behavioural Equivalence Relationships Induced by Programming Logics.* Proceedings of 10th International Conference on Automata, Languages, and Programming, Barcelona, Spain, Lecture Notes in Computer Science 154, Springer-Verlag 1983, pp. 97–108.
6. Fernandez, J.-C.: *An Implementation of an Efficient Algorithm for Bisimulation Equivalence.* Science of Computer Programming 13 (1989/90) 219–236.
7. Eloranta, J.: *Minimizing the Number of Transitions with Respect to Observation Equivalence.* BIT 31 (1991), 576–590.
8. Inverardi, P. & Priami, C.: *Evaluation of Tools for the Analysis of Communicating Systems.* EATCS Bulletin 45, October 1991, pp. 158–185.
9. Bouajjani, A., Fernandez, J.-C. & Halbwachs, N.: *Minimal Model Generation.* Computer-Aided Verification '90 (Proceedings of a Workshop), AMS-ACM DIMACS Series in Discrete Mathematics and Theoretical Computer Science, Vol. 3, American Mathematical Society 1991, pp. 85–91.

10. Valmari, A.: *Alleviating State Explosion during Verification of Behavioural Equivalence.* Department of Computer Science, University of Helsinki, Report A-1992-4, Helsinki, Finland 1992, 57 p.

11. Brinksma, E.: *A Theory for the Derivation of Tests.* Protocol Specification, Testing and Verification VIII (Proceedings of International IFIP WG 6.1 Symposium, 1988), North-Holland 1988, pp. 63–74.

12. Courcoubetis, C., Vardi, M., Wolper, P. & Yannakakis, M.: *Memory Efficient Algorithms for the Verification of Temporal Properties.* Formal Methods in System Design, 1: 275–288 (1992).

13. Valmari, A.: *On-the-fly Verification with Stubborn Sets.* Proceedings of CAV'93, 5th International Conference on Computer-Aided Verification, Elounda, Greece, Lecture Notes in Computer Science 697, Springer-Verlag 1993, pp. 397-408.

14. Madelaine, E. & Vergamini, D.: *AUTO: A Verification Tool for Distributed Systems Using Reduction of Finite Automata Networks.* Formal Description Techniques II (Proceedings of FORTE '89), North-Holland 1990, pp. 61–66.

15. Graf, S. & Steffen, B.: *Compositional Minimization of Finite State Processes.* Computer-Aided Verification '90 (Proceedings of a workshop), AMS-ACM DIMACS Series in Discrete Mathematics and Theoretical Computer Science, Vol. 3, American Mathematical Society 1991, pp. 57–73.

16. Valmari, A.: *Compositional State Space Generation.* Advances in Petri Nets 1993, Lecture Notes in Computer Science 674, Springer-Verlag 1993, pp. 427–457.

17. Räuchle, T. & Toueg, S.: *Exposure to Deadlock for Communicating Processes is Hard to Detect.* Information Processing Letters 21 (1985) pp. 63–68.

18. Jones, N. D., Landweber, L. H. & Lien, Y. E.: *Complexity of Some Problems in Petri Nets.* Theoretical Computer Science 4 (1977) pp. 277–299.

19. Valmari, A.: *Some Polynomial Space Complete Concurrency Problems.* Tampere University of Technology, Software Systems Laboratory, Report 4, 1988, 34 p.

20. Hopcroft, J. E. & Ullman, J. D.: *Introduction to Automata Theory, Languages and Computation.* Addison-Wesley 1979, 418 p.

21. Valmari, A. & Tienari, M.: *An Improved Failures Equivalence for Finite-State Systems with a Reduction Algorithm.* Protocol Specification, Testing and Verification XI, North-Holland 1991, pp. 3–18.

22. Eloranta, J.: *Minimal Transition Systems with Respect to Divergence Preserving Behavioural Equivalences.* Doctoral thesis, University of Helsinki, Department of Computer Science, Report A-1994-1, Helsinki, Finland 1994, 162 p.

23. Rabinovich, A.: *Checking Equivalences Between Concurrent Systems of Finite Agents (Extended Abstract).* Proceedings of ICALP 92, Lecture Notes in Computer Science 623, Springer-Verlag 1992, pp. 696–707.

24. Cleaveland, R., Parrow, J. & Steffen, B.: *The Concurrency Workbench.* Proceedings of the Workshop on Automatic Verification Methods for Finite-State Systems, Lecture Notes in Computer Science 407, Springer-Verlag 1990, pp. 24–37.

25. Valmari, A., Kemppainen, J., Clegg, M. & Levanto, M.: *Putting Advanced Reachability Analysis Techniques Together: the "ARA" Tool.* Proceedings of Formal Methods Europe '93, Lecture Notes in Computer Science 670, Springer-Verlag 1993, pp. 597–616.

26. Bolognesi, T. & Brinksma, E.: *Introduction to the ISO Specification Language LOTOS.* Computer Networks and ISDN Systems 14, 1987, pp. 25–59. Also in: The Formal Description Technique LOTOS, North-Holland 1989, pp. 23–73.

27. Valmari, A. & Tienari, M.: *Compositional Failure-Based Semantic Models for Basic LOTOS.* To appear in Formal Aspects of Computing, 29 p.

28. Valmari, A.: *The Weakest Deadlock-Preserving Congruence.* To appear in Information Processing Letters.

29. Hoare, C. A. R.: *Communicating Sequential Processes.* Prentice-Hall 1985, 256 p.

30. Bergstra, J. A., Klop, J. W. & Olderog, E.-R.: *Failures without Chaos: A New Process Semantics for Fair Abstraction.* Formal Description of Programming Concepts III, North-Holland, 1987, pp. 77–103.

31. Manna, Z. & Pnueli, A.: *The Temporal Logic of Reactive and Concurrent Systems: Specification.* Springer-Verlag 1992, 427 p.

32. Kaivola, R. & Valmari, A.: *The Weakest Compositional Semantic Equivalence Preserving Nexttime-less Linear Temporal Logic.* Proceedings of CONCUR '92, Lecture Notes in Computer Science 630, Springer-Verlag 1992, pp. 207–221.

33. Setälä, M. & Valmari, A.: *Validation and Verification with Weak Process Semantics.* Proceedings of Nordic Seminar on Dependable Computing Systems 1994, Lyngby, Denmark, August 1994, pp. 15–26.

Partial Order Semantics and Weak Fairness

Walter Vogler *
Institut für Mathematik, Universität Augsburg
D-86135 Augsburg, Germany

Abstract

Causality-based partial order semantics allows an easy formulation of weak fairness. It is demonstrated that this is true also for other partial order semantics, namely for partial words and interval semiwords.

1 Introduction

The interleaving approach describes a run of a concurrent system as a sequence of actions. In contrast to this, we can also model a run as a partial order of actions; this way we additionally give information on the independence of actions. Most common are partial orders that describe the causality in system runs, e.g processes in the model of safe Petri nets; but there are also others like partial words [Gra81, Sta81] or interval semiwords. Partial order semantics is intuitively appealing, but it has to be asked whether the effort to handle partial orders instead of sequences is really worthwhile; and when comparing the respective merits of partial order and interleaving semantics, one should also take into account partial orders that are not so much causality-based. For example, it has been argued that processes are needed to deal with action refinement [CDMP87]; and indeed, processes give a congruence for action refinement, but so do partial words and interval semiwords – the latter being fully abstract and thus really adequate; see [Vog92], also for further references.

Often, system runs are only of interest if they satisfy (weak) fairness, which requires that no component of a concurrent system unnecessarily stops, i.e. that an action is eventually taken if all needed resources are continuously available. Processes allow a particularly simple characterization of fairness: let π be a process and w one of its linearizations; then π is maximal (w.r.t. the prefix-relation) if and only if w is fair [Rei84]. The purpose of this note is to show that this result depends on the prefix-relation for partial orders and not on the concept of causality; it holds just as well for partial words and interval semiwords. We also consider fairness for ST-traces [Gla90], which are sequential representations of interval semiwords [Vog92], and establish an analogous characterization of fair ST-traces with interval semiwords.

*This work was partially supported by the DFG (Project 'Halbordnungstesten' and the ESPRIT Basic Research Working Group 6067 CALIBAN (CAusal calcuLI BAsed on Nets). A journal version of this paper will appear in Information Processing Letter.

In Section 2, we define some Petri net notions and ST-traces. In Section 3, we introduce partial words and interval semiwords and relate the latter to ST-traces. These relations are known for the finite case; see [Vog92] for a presentation and more detailed explanations of all the behaviour notions. Since we are concerned with fairness, we have to lift the finite to the infinite case. Fairness is considered in Section 4, where the announced results are presented.

2 Petri nets and ST-sequences

In this paper, a *(Petri) net* $N = (S, T, W, M_N)$ consists of finite disjoint sets S of *places* and T of *transitions*, the *weight function* $W : S \times T \cup T \times S \rightarrow \{0, 1\}$, and the *initial marking* $M_N : S \rightarrow I\!N_0$. We assume that the reader knows some Petri net notions like firing sequence, step, reachable marking, presets ${}^\bullet t$ etc. A net is *safe* if for all places s and reachable markings M we have $M(s) \leq 1$. In a safe net, markings can be regarded as sets of places and a transition t is *enabled* under a marking M, $M[t\rangle$, if ${}^\bullet t \subseteq M$.

General assumption We assume that, for the rest of this paper, some fixed net N is given which is safe and has no transitions with empty preset.

We will need a refined version of firing sequences, where we separate the *start* t^+ and the *end* t^- of a transition t; t^+ and t^- are *transition parts*. Correspondingly, a state is not described by a marking alone, but additionally by the set of currently firing transitions: an *ST-marking* is a pair (M, C) consisting of a marking M and a subset C of T.

A finite or infinite sequence w of transition parts is called an *ST-sequence*, if for every $t \in T$ the start and end of t occur alternatingly beginning with t^+. If every t^+ is followed by a t^-, then w is called *closed*.

A transition start t^+ is *enabled* under an ST-marking (M, C), if $M[t\rangle$; if in this case $M'(s) = M(s) - W(s, t)$ for all $s \in S$, then we denote this by $(M, C)[t^+\rangle(M', C \cup \{t\})$ and say that t^+ can *fire* under (M, C) *reaching* $(M', C \cup \{t\})$. Similarly, t^- is *enabled* under (M, C), if $t \in C$; if in this case $M'(s) = M(s) + W(t, s)$ for all $s \in S$, then we denote this by $(M, C)[t^-\rangle(M', C - \{t\})$ and say that t^- can *fire* under (M, C) *reaching* $(M', C - \{t\})$. This definition of enabling and firing can be extended to finite and infinite sequences as usual. A sequence of transition parts enabled under $(M_N, \emptyset)$ is called an *ST-trace*. The following proposition is fairly obvious from the definitions.

Proposition 2.1 *Each ST-trace is an ST-sequence. A finite or infinite sequence $t_1 t_2 \ldots$ of transitions is a firing sequence if and only if $t_1^+ t_1^- t_2^+ t_2^- \ldots$ is an ST-trace.*

3 Partial order semantics and interval orders

We are interested in modelling system runs by partial orders. Hence, for us a (finitely preceded, labelled) *partial order* $p = (E, <, l)$ consists of a set E of *events*, a labelling $l : E \to T$, and an irreflexive, transitive relation $<$ on E such that for each $e' \in E$ there are only finitely many e with $e < e'$; we do not distinguish isomorphic partial orders. If $e < e'$, we call e a *predecessor* of e' and e' a *successor* of e. If $e < e'$ or $e = e'$, we write $e \leq e'$. Two events e and e' are *concurrent*, e *co* e', if neither $e < e'$ nor $e' < e$. (We regard each event as concurrent to itself; this way a partial word as defined below corresponds to a step sequence if and only if *co* is an equivalence relation.)

$E' \subseteq E$ is *left-closed*, if for all $e' < e \in E'$ we have $e' \in E'$; in this case, $(E', <', l')$ is a *prefix* of $(E, <, l)$, where $<'$ and l' are the appropriate restrictions of $<$ and l. $(E, <, l)$ is an *augmentation* of $(E, <', l)$, if $<'$ is a subset of $<$. If $<$ is *total*, i.e. for all events e and e' we have $e < e'$, $e = e'$ or $e' < e$, then $(E, <, l)$ is a *linearization* of $(E, <', l)$. In this case, we can view $(E, <, l)$ as a sequence of events; this is also true for infinite E, since all our partial orders are finitely preceded by definition.

Petri net theory has a long tradition of studying 'true concurrency' using partial orders: concurrent events correspond to independent transition firings. Most often a partial order semantics is given by so-called processes; in this approach the partial order models causality. Another view is that concurrency is more than arbitrary interleaving but includes it. This idea is formalized in the partial words of [Gra81], which coincide with the semiwords of [Sta81] for the nets we consider here: in such a partial order, any set of pairwise concurrent elements represents a step that can be fired provided the precedences prescribed by the partial order are observed.

- We call a partial order $(E, <, l)$ a *partial word* of N if for all finite disjoint subsets B and C of E we have: if all elements of C are pairwise concurrent and B and $B \cup C$ are left-closed, then the transitions in $l(C)$ are *concurrently enabled* under $M_N + \sum_{e \in B} W(l(e), .) - W(., l(e))$, i.e.

$$\sum_{e \in C} W(., l(e)) \leq M_N + \sum_{e \in B} W(l(e), .) - W(., l(e)).$$

From the finite case it is known that the marking mentioned is a reachable marking, so by the general assumption for N equally labelled events cannot be concurrent. Thus, a set C as above and, hence, any set of pairwise concurrent events in a partial word has at most $|T|$ elements. Therefore, a partial word is at most countable and it has a linearization. (The latter is not necessarily true for uncountable partial orders, since linearizations have to be finitely preceded.)

It is not difficult to see that the set of partial words is closed under augmentation; i.e., if p is an augmentation of a partial word q, then p is a partial word, too; see [Vog92] for the finite case. The latter shows that concurrency in partial words is just seen as a possibility; possibly concurrent transitions can also be performed sequentially.

The above is an extension of the original definition to infinite partial orders. The extension is sensible as the following result shows. (This result also holds for unsafe or infinite N.)

Proposition 3.1 *A partial order is a partial word if and only if all finite prefixes are partial words.*

$(E, <, l)$ is an *interval order*, if for all $a, b, c, d \in E$ we have: if $a < b$ and $c < d$, then $a \leq d$ or $c \leq b$; a basic reference for interval orders is [Fis85, Chapter 2]. A partial word is an *interval semiword* if it is an interval order. An interval semiword can intuitively be seen as the observation of a system run where each firing takes some time, as the next result shows (which also explains the name 'interval order').

Theorem 3.2 *[Fis85] A countable partial order $(E, <, l)$ is an interval order if and only if there are closed intervals $[r_1(e), r_2(e)] \subseteq \mathbb{R}$, $e \in E$, such that for all events e and e' we have $e < e'$ if and only if $r_2(e) < r_1(e')$.* □

We can think of event e as starting at time $r_1(e)$ and lasting until $r_2(e)$. Event e is smaller than event e' if the interval of e lies completely before the interval of e'. An ST-sequence also assigns intervals to occurrences of transitions: each t^+ is the start of an interval that extends to the next t^- (or to infinity if no t^- follows) and represents a firing of t. Therefore we define: an ST-sequence $w = \rho_1\rho_2\ldots$ (or $w = \rho_1\rho_2\ldots\rho_n$) *represents* an interval order $(E, <, l)$, if there are injective partial functions $+, - : E \to \mathbb{N}$ (or $+, - : E \to \{1, \ldots, n\}$) with the following properties:

- $+$ is total; $+(E) \cup -(E) = \mathbb{N}$ (or $= \{1, \ldots, n\}$), i.e. each transition part in w marks the start or the end of an event in E.
- For all $e \in E$, if $l(e) = t$, then $\rho_{+(e)} = t^+$ and $-(e)$ is the index of the next t^- in w, if it exists.
- For all $e, e' \in E$, $e < e'$ if and only if $-(e) < +(e')$.

We can think of a representation as a sequential observation where we see transitions start and end; thus, we observe the firings as intervals in time and from this we can conclude that some partial order occurred, namely the represented interval order. If, in the last requirement above, we replace 'if and only if' by 'only if', then w is called *compatible* with $(E, <, l)$.

The next two results relate the firing rules for interval semiwords and ST-traces.

Theorem 3.3 *i) Let some ST-sequence w represent the interval order p. Then, w is an ST-trace if and only if p is an interval semiword.*

ii) Let some ST-sequence w be compatible with some interval order p. Then, w is an ST-trace if p is an interval semiword.

Proof: i) For the finite case, this has been shown in [Vog92, Theorem 5.5.3]. For the infinite case, let first p be an interval semiword. Each finite prefix of w represents a finite prefix of p, which is an interval semiword. From the finite case we conclude that each finite prefix of w and, thus, w itself is an ST-trace.

Secondly, let w be an ST-trace. To check that p is a partial word, choose B and C according to the definition. Let n be the maximal value of $+$ on $B \cup C$; then the prefix v of w of length n represents a prefix q of p that contains $B \cup C$. Since v is an ST-trace, q is an interval semiword by the finite case, and B and C satisfy the requirement.

ii) Follows from i), since w represents an augmentation of p, which is an interval semiword, too. □

The relationship between processes and firing sequences is particularly pleasing since processes induce a partitioning of firing sequences via linearization; we show an analogous result for interval semiwords and ST-traces.

Theorem 3.4 *Let each interval semiword correspond to the set of ST-traces that represent it; this way, interval semiwords induce a partitioning of the set of ST-traces; in particular, each interval semiword has a representation.*

Proof: Since by definition and by Theorem 3.3 an ST-trace represents a unique interval semiword, we only have to show that each infinite interval semiword $(E, <, l)$ has a representation. (The finite case is dealt with in [Vog92, 5.5.2].) Let us call an event e' a *co-predecessor* of e if there is some $e'' < e$ with $e' \, co \, e''$.

We first show that each $e \in E$ has only finitely many *co*-predecessors. Assume to the contrary. Since e cannot be a predecessor of one of its *co*-predecessors and since e has only finitely many predecessors, there are infinitely many *co*-predecessors of e which are concurrent to e; among them, there are infinitely many which are concurrent to the same e'' of the finitely many $e'' < e$. By definition of an interval order, these must be pairwise concurrent. This is a contradiction, since there are at most $|T|$ pairwise concurrent events.

Generalizing [JK93] it is shown in [Vog93, remark after 2.4], that the above finiteness property implies the existence of a sequence $w = \eta_1\eta_2 \ldots$ which satisfies:

- Each η_i is an e^+ or an e^- with $e \in E$.
- For all $e \in E$, e^+ occurs exactly once, while e^- occurs at most once, and in this case after e^+.
- For all $e, e' \in E$, $e < e'$ if and only if e^- occurs before e'^+ in w.

To construct v from w, we replace each e^+ by $l(e)^+$ and each e^- by $l(e)^-$. Since equally labelled events are not concurrent in $(E, <, l)$, the start and end of $t \in T$ occur in v alternatingly beginning with t^+. Hence, v is an ST-sequence, and by the above properties of w it represents $(E, <, l)$. □

Naturally, fairness requires that a transition that has started must end at some time; hence, a fair ST-trace should be closed. We call an interval order *closed*, if for each event the number of concurrent events is finite.

Theorem 3.5 *An interval order p has a closed representation if and only if it is closed; if in this case p is infinite, then all representations are closed.*

Proof: If p has a closed representation, then for each event e there are only finitely many e' with $+(e') < -(e)$, i.e. we have $e < e'$ for all but finitely many e'; thus, p is closed.

Let p be closed. If p is finite, then it has a closed representation [Vog92]; so let p be infinite. Theorem 3.4 shows that p has a representation. If in some representation some t^+ is not succeeded by a t^-, then, for the corresponding event e, e is concurrent to the infinitely many e' with $+(e) < +(e')$, a contradiction. □

4 Fairness

A system run is (weakly) fair, if each action occurs eventually provided the necessary resources are continuously available. Most often, a fair firing sequence is defined as follows:

Definition 4.1 A firing sequence $t_1t_2\ldots$ is *fair** if either it is finite and no transition is enabled under the marking reached at the end or it is infinite and for $M_N[t_1\rangle M_1[t_2\rangle M_2\ldots$ we have: if for some $t \in T$ and $i \in I\!N$ we have $M_j[t\rangle$ for all $j \geq i$, then $t = t_j$ for some $j > i$. □

This definition is not fully adequate in the presence of loops. Take a net with a single, marked place and two transitions t and t', where t is on a loop with the place, while t' is just in its postset. Intuitively, an infinite sequence of t's should be fair, since the only resource is repeatedly in use; but it is in fact not fair*. The reason is, intuitively, that Definition 4.1 checks all markings after M_i to see whether t is continuously enabled, but it disregards the points in time at which some transition is in the middle of firing. Therefore, we use the following refined (and more complicated) version of fairness.

Definition 4.2 A firing sequence $t_1t_2\ldots$ is *fair* if either it is finite and no transition is enabled under the marking reached at the end or it is infinite and for $M_N[t_1\rangle M_1[t_2\rangle M_2\ldots$ we have: if for some $t \in T$ and $i \in I\!N$ we have that t and t_{j+1} are concurrently enabled under M_j for all $j \geq i$, then some t_j with $j > i$ equals t. □

With this definition, maximality of a process corresponds to fairness of an arbitrary linearization [Rei84]. Thus, fairness for processes can be defined much more easily than for firing sequences. Our aim is to show analogous results for partial words and interval semiwords. We start with a result which demonstrates the close relation between fairness and the prefix-relation on partial orders.

Theorem 4.3 *A firing sequence w is fair if and only if* – regarded as a partial word – *it is maximal w.r.t. the prefix-relation.*

Proof: For finite w this is obvious. If an infinite w is not maximal, we can extend it with one event e. The predecessors of e form a prefix v of w, and by definition of a partial word $l(e)$ is concurrently enabled with each transition after v under the appropriate marking; hence, w is not fair.

If w is not fair due to some $t \in T$ and $i \in \mathbb{N}$, we add a t-labelled event e succeeding the first i events of w. We have to check the requirement for all appropriate B and C. If $e \in B$, it is sufficient to check $B - \{e\}$ and $C \cup \{e\}$ since e is a maximal event. Hence, in the only interesting case C consists of e and a concurrent event, and $l(C)$ is enabled by the assumption on t and i. □

The next theorem is immediately implied by this result and the following lemma.

Lemma 4.4 *Let the partial order $p = (E, <, l)$ be an augmentation of a partial word $q = (E, \prec, l)$. Then q is a maximal partial word if and only if p is.*

Proof: If q is a proper prefix of another partial word $q' = (E', \prec', l')$, then $(E', (\prec' \cup <)^+, l')$ is – as augmentation of q' – a partial word and has p as proper prefix. (By $(.)^+$ we denote the transitive closure of a relation.)

If p is a proper prefix of a partial word $p' = (E', <', l')$, then we may assume that p' has just one additional event e'. Then q is a proper prefix of $q' = (E', \prec \cup \{(e, e') \mid e \in E, e <' e'\}, l')$. The ordering relation of q' is transitive, since e' is a maximal event and $e_0 \prec e <' e'$ implies $e_0 < e$, $e_0 <' e$ and $e_0 <' e'$. To check that q' is a partial word, choose B and C accordingly; as in the last proof, we only have to consider the case $e' \in C$. Let $B' \subseteq B$ consist of the predecessors of e' in p'. Since p' is a partial word, $l(e')$ is enabled after the occurrence of B' and $^\bullet l(e')$ has no place in common with any $^\bullet l(e)$, $e \in E - B'$. Thus, $l(e')$ is enabled after B without conflict with $l(C - \{e'\})$, which is concurrently enabled after B since q is a partial word. □

Theorem 4.5 *Let w be a linearization of a partial word p. Then p is maximal w.r.t. the prefix-relation if and only if w is fair.*

This is also an independent proof of the result in [Rei84], since processes can be identified with partial words which are not proper augmentations of others. This relation is shown in [Kie88] for the finite case constructing for each partial word p a process that (more precisely, whose event structure) has p as an augmentation; see also [Vog92]. Since the construction involved has a unique result for the net we consider, this carries easily over to the infinite case.

The following should be stressed: Theorem 4.5 does not mainly say that we can recover fair firing sequences from partial words; from the results of [Kie88] and [Rei84] it is clear that in the set of partial words we can find the processes and that from these the fair firing sequences can be read off. Instead, the main point is to demonstrate that fairness for partial words should be defined as maximality; i.e. fairness for partial words is defined as easily as for processes and much more easily than for firing sequences.

Of course, Theorem 4.5 applies also to interval semiwords, but instead of maximality in the set of partial words we want a result for maximality in the set of interval semiwords. We first define fairness for ST-traces; we have already argued that a fair ST-trace should be closed.

Definition 4.6 An ST-trace $\rho_1\rho_2\ldots$ is *fair* if it is closed and either it is finite and no transition start is enabled under the ST-marking reached or it is infinite and for $(M_N,\emptyset)[\rho_1\rangle(M_1,C_1)\ [\rho_2\rangle(M_2,C_2)\ldots$ we have: if for some $t\in T$ and $i\in I\!N$ we have $M_j[t\rangle$ for all $j\geq i$, then some ρ_j with $j>i$ equals t^+. □

Equivalently, we could omit closedness from this definition and require instead that each *transition part* that is continuously enabled after a prefix has to occur after this prefix. Note that this definition takes the simpler form of Definition 4.1; nevertheless, it corresponds to the more complicated Definition 4.2, since a transition is enabled after the firing of some t^+ if and only if it is concurrently enabled to t before t^+:

Proposition 4.7 *A firing sequence $t_1t_2\ldots$ is fair if and only if the ST-trace $t_1^+t_1^-t_2^+t_2^-\ldots$ is fair.*

We now relate fair firing sequences and ST-traces to interval semiwords. The first result shows again that the simple characterization of fairness depends on the tool 'partial orders' and their prefix-relation. It is particularly interesting since interval semiwords and ST-traces are equivalent in the following sense: from an interval semiword we can determine its representations, but – different from the case of processes and their linearizations – we can also determine an interval semiword from a single representation.

Theorem 4.8 *Let w be an ST-trace that represents some interval semiword $p=(E,<,l)$, and let w be closed if it is finite. Then, p is maximal w.r.t. the prefix-relation (in the set of interval semiwords) and closed if and only if w is fair.*

Proof: The finite case is clear, since the markings reached after p and w coincide. Hence, let w and p be infinite.

First, let w be fair. By Theorem 3.5 p is closed. Assume that p is a proper prefix of $p'=(E',<',l')$, which has just one additional event e'. Since w is closed, $-(e)$ is defined for all predecessors e of e'; let i be the maximum of these values. Since p' is a partial word, ${}^\bullet l'(e')$ is marked after the occurrences of the predecessors, and no other event removes a token from ${}^\bullet l'(e')$. Thus, $l'(e')^+$ is enabled under the ST-marking reached after the prefix of w of length i and all following ST-markings. This contradiction to the fairness of w implies that p is maximal even as a partial word.

Secondly, let p be maximal and closed. By Theorem 3.5 w is closed. Assume that some t and i violate the fairness of w. Then we can insert t^+ into w after the prefix of length i and obtain an ST-trace. This ST-trace represents an interval semiword, which has p as a prefix and one additional t-labelled event, a contradiction. □

For the last two results we need the following lemma.

Lemma 4.9 *Let p be a closed interval semiword. Then, p is maximal as an interval semiword if and only if it is maximal as a partial word.*

Proof: The 'if'-part is obvious. If p is a maximal interval semiword, it has a fair representation by Theorem 4.8. Now the first part of the proof of Theorem 4.8 shows that p is also maximal as a partial word. □

Theorem 4.10 *Let w be a closed ST-trace that is compatible with some closed interval semiword p. Then, p is a maximal interval semiword w.r.t. the prefix-relation if and only if w is fair.*

Proof: w represents an augmentation q of p, which is also closed. Now, p is a maximal interval semiword if and only if p is a maximal partial word by Lemma 4.9 if and only if q is a maximal partial word by Lemma 4.4 if and only if q is a maximal interval semiword by Lemma 4.9 if and only if w is fair by Theorem 4.8. □

Theorem 4.11 *Let w be a linearization of some closed interval semiword p. Then, p is a maximal interval semiword w.r.t. the prefix-relation if and only if w is fair.*

Proof: p is a maximal interval semiword if and only if p is a maximal partial word by Lemma 4.9 if and only if w is fair by Theorem 4.5. □

Remark: The closedness assumption in Lemma 4.9 and the two theorems cannot be omitted as the following counterexample shows. Consider an interval semiword p consisting of an infinite sequence and an additional independent event e. Assume that $l(e)$ enables another transition; hence, we can construct partial words from p by adding a new event that has e and finitely many events from the sequence as predecessors. None of these partial words is an interval order. Thus, p is maximal as an interval semiword, but not as a partial word. □

References

[CDMP87] L. Castellano, G. De Michelis, and L. Pomello. Concurrency vs. interleaving: An instructive example. *Bull. EATCS*, 31:12–15, 1987.

[Fis85] P.C. Fishburn. *Interval Orders and Interval Graphs*. J. Wiley, 1985.

[Gla90] R.J. v. Glabbeek. The refinement theorem for ST-bisimulation semantics. In M. Broy and C.B. Jones, editors, *Programming Concepts and Methods, Proc. IFIP Working Conference*, 27–52. Elsevier Science Publisher(North-Holland), 1990.

[Gra81] J. Grabowski. On partial languages. *Fundamenta Informaticae*, IV.2:428–498, 1981.

[JK93] R. Janicki and M. Koutny. Representations of discrete interval orders and semi-orders. Technical Report 93-02, Dept. Comp. Sci. Sys., McMaster University, Hamilton, Ontario, 1993.

[Kie88] A. Kiehn. On the interrelationship between synchronized and non-synchronized behaviour of Petri nets. *J. Inf. Process. Cybern. EIK*, 24:3–18, 1988.

[Rei84] W. Reisig. Partial order semantics versus interleaving semantics for CSP-like languages and its impact on fairness. In J. Paredaens, editor, *Automata, Languages and Programming*, Lect. Notes Comp. Sci. 172, 403–413. Springer, 1984.

[Sta81] P.H. Starke. Processes in Petri nets. *J. Inf. Process. Cybern. EIK*, 17:389–416, 1981.

[Vog92] W. Vogler. *Modular Construction and Partial Order Semantics of Petri Nets*. Lect. Notes Comp. Sci. 625. Springer, 1992.

[Vog93] W. Vogler. The limit of $split_n$-language equivalence. Technical Report Nr. 288, Inst. f. Mathematik, Univ. Augsburg, 1993.

Author Index

Published in 1990–93

Security and Persistence, Proceedings of the International Workshop on Computer Architectures to Support Security and Persistence of Information, Bremen, West Germany, 8–11 May 1990
John Rosenberg and J. Leslie Keedy (Eds)

Women into Computing: Selected Papers 1988–1990
Gillian Lovegrove and Barbara Segal (Eds)

3rd Refinement Workshop (organised by BCS-FACS, and sponsored by IBM UK Laboratories, Hursley Park and the Programming Research Group, University of Oxford), Hursley Park, 9–11 January 1990
Carroll Morgan and J. C. P. Woodcock (Eds)

Designing Correct Circuits, Workshop jointly organised by the Universities of Oxford and Glasgow, Oxford, 26–28 September 1990
Geraint Jones and Mary Sheeran (Eds)

Functional Programming, Glasgow 1990
Proceedings of the 1990 Glasgow Workshop on Functional Programming, Ullapool, Scotland, 13–15 August 1990
Simon L. Peyton Jones, Graham Hutton and Carsten Kehler Holst (Eds)

4th Refinement Workshop, Proceedings of the 4th Refinement Workshop, organised by BCS-FACS, Cambridge, 9–11 January 1991
Joseph M. Morris and Roger C. Shaw (Eds)

AI and Cognitive Science '90, University of Ulster at Jordanstown, 20–21 September 1990
Michael F. McTear and Norman Creaney (Eds)

Software Re-use, Utrecht 1989, Proceedings of the Software Re-use Workshop, Utrecht, The Netherlands, 23–24 November 1989
Liesbeth Dusink and Patrick Hall (Eds)

Z User Workshop, 1990, Proceedings of the Fifth Annual Z User Meeting, Oxford, 17–18 December 1990
J.E. Nicholls (Ed.)

IV Higher Order Workshop, Banff 1990
Proceedings of the IV Higher Order Workshop, Banff, Alberta, Canada, 10–14 September 1990
Graham Birtwistle (Ed.)

ALPUK91, Proceedings of the 3rd UK Annual Conference on Logic Programming, Edinburgh, 10–12 April 1991
Geraint A.Wiggins, Chris Mellish and Tim Duncan (Eds)

Specifications of Database Systems
International Workshop on Specifications of Database Systems, Glasgow, 3–5 July 1991
David J. Harper and Moira C. Norrie (Eds)

7th UK Computer and Telecommunications Performance Engineering Workshop
Edinburgh, 22–23 July 1991
J. Hillston, P.J.B. King and R.J. Pooley (Eds)

Logic Program Synthesis and Transformation
Proceedings of LOPSTR 91, International Workshop on Logic Program Synthesis and Transformation, University of Manchester, 4–5 July 1991
T.P. Clement and K.-K. Lau (Eds)

Declarative Programming, Sasbachwalden 1991
PHOENIX Seminar and Workshop on Declarative Programming, Sasbachwalden, Black Forest, Germany, 18–22 November 1991
John Darlington and Roland Dietrich (Eds)

Building Interactive Systems: Architectures and Tools
Philip Gray and Roger Took (Eds)

Functional Programming, Glasgow 1991
Proceedings of the 1991 Glasgow Workshop on Functional Programming, Portree, Isle of Skye, 12–14 August 1991
Rogardt Heldal, Carsten Kehler Holst and Philip Wadler (Eds)

Object Orientation in Z
Susan Stepney, Rosalind Barden and David Cooper (Eds)

Code Generation – Concepts, Tools, Techniques
Proceedings of the International Workshop on Code Generation, Dagstuhl, Germany, 20–24 May 1991
Robert Giegerich and Susan L. Graham (Eds)

Z User Workshop, York 1991, Proceedings of the Sixth Annual Z User Meeting, York, 16–17 December 1991
J.E. Nicholls (Ed.)

Formal Aspects of Measurement
Proceedings of the BCS-FACS Workshop on Formal Aspects of Measurement, South Bank University, London, 5 May 1991
Tim Denvir, Ros Herman and R.W. Whitty (Eds)

AI and Cognitive Science '91 University College, Cork, 19–20 September 1991
Humphrey Sorensen (Ed.)

5th Refinement Workshop, Proceedings of the 5th Refinement Workshop, organised by BCS-FACS, London, 8–10 January 1992
Cliff B. Jones, Roger C. Shaw and Tim Denvir (Eds)

Algebraic Methodology and Software Technology (AMAST'91)
Proceedings of the Second International Conference on Algebraic Methodology and Software Technology, Iowa City, USA, 22–25 May 1991
M. Nivat, C. Rattray, T. Rus and G. Scollo (Eds)

ALPUK92, Proceedings of the 4th UK Conference on Logic Programming, London, 30 March – 1 April 1992
Krysia Broda (Ed.)

Logic Program Synthesis and Transformation
Proceedings of LOPSTR 92, International Workshop on Logic Program Synthesis and Transformation, University of Manchester, 2–3 July 1992
Kung-Kiu Lau and Tim Clement (Eds)

NAPAW 92, Proceedings of the First North American Process Algebra Workshop, Stony Brook, New York, USA, 28 August 1992
S. Purushothaman and Amy Zwarico (Eds)

First International Workshop on Larch
Proceedings of the First International Workshop on Larch, Dedham, Massachusetts, USA, 13–15 July 1992
Ursula Martin and Jeannette M. Wing (Eds)

Persistent Object Systems
Proceedings of the Fifth International Workshop on Persistent Object Systems, San Miniato (Pisa), Italy, 1–4 September 1992
Antonio Albano and Ron Morrison (Eds)

Formal Methods in Databases and Software Engineering, Proceedings of the Workshop on Formal Methods in Databases and Software Engineering, Montreal, Canada, 15–16 May 1992
V.S. Alagar, Laks V.S. Lakshmanan and F. Sadri (Eds)

Modelling Database Dynamics
Selected Papers from the Fourth International Workshop on Foundations of Models and Languages for Data and Objects, Volkse, Germany, 19–22 October 1992
Udo W. Lipeck and Bernhard Thalheim (Eds)

14th Information Retrieval Colloquium
Proceedings of the BCS 14th Information Retrieval Colloquium, University of Lancaster, 13–14 April 1992
Tony McEnery and Chris Paice (Eds)

Functional Programming, Glasgow 1992
Proceedings of the 1992 Glasgow Workshop on Functional Programming, Ayr, Scotland, 6–8 July 1992
John Launchbury and Patrick Sansom (Eds)

Z User Workshop, London 1992
Proceedings of the Seventh Annual Z User Meeting, London, 14–15 December 1992
J.P. Bowen and J.E. Nicholls (Eds)

Interfaces to Database Systems (IDS92)
Proceedings of the First International Workshop on Interfaces to Database Systems, Glasgow, 1–3 July 1992
Richard Cooper (Ed.)

AI and Cognitive Science '92
University of Limerick, 10–11 September 1992
Kevin Ryan and Richard F.E. Sutcliffe (Eds)

Theory and Formal Methods 1993
Proceedings of the First Imperial College Department of Computing Workshop on Theory and Formal Methods, Isle of Thorns Conference Centre, Chelwood Gate, Sussex, UK, 29–31 March 1993
Geoffrey Burn, Simon Gay and Mark Ryan (Eds)

Algebraic Methodology and Software Technology (AMAST'93)
Proceedings of the Third International Conference on Algebraic Methodology and Software Technology, University of Twente, Enschede, The Netherlands, 21–25 June 1993
M. Nivat, C. Rattray, T. Rus and G. Scollo (Eds)

Logic Program Synthesis and Transformation
Proceedings of LOPSTR 93, International Workshop on Logic Program Synthesis and Transformation, Louvain-la-Neuve, Belgium, 7–9 July 1993
Yves Deville (Ed.)

Database Programming Languages (DBPL-4)
Proceedings of the Fourth International Workshop on Database Programming Languages – Object Models and Languages, Manhattan, New York City, USA, 30 August–1 September 1993
Catriel Beeri, Atsushi Ohori and Dennis E. Shasha (Eds)

Music Education: An Artificial Intelligence Approach, Proceedings of a Workshop held as part of AI-ED 93, World Conference on Artificial Intelligence in Education, Edinburgh, Scotland, 25 August 1993
Matt Smith, Alan Smaill and Geraint A. Wiggins (Eds)

Rules in Database Systems
Proceedings of the 1st International Workshop on Rules in Database Systems, Edinburgh, Scotland, 30 August–1 September 1993
Norman W. Paton and M. Howard Williams (Eds)